SOLID STATE PHYSICS

VOLUME 25

Contributors to This Volume

S. G. Davison

J. D. Levine

R. L. Parker

Duane C. Wallace

SOLID STATE PHYSICS

Advances in
Research and Applications

Editors

HENRY EHRENREICH

*Division of Engineering and Applied Physics
Harvard University, Cambridge, Massachusetts*

FREDERICK SEITZ

The Rockefeller University, New York, New York

DAVID TURNBULL

*Division of Engineering and Applied Physics
Harvard University, Cambridge, Massachusetts*

VOLUME 25

1970

ACADEMIC PRESS • NEW YORK AND LONDON

Copyright © 1970, by Academic Press, Inc.
ALL RIGHTS RESERVED
NO PART OF THIS BOOK MAY BE REPRODUCED IN ANY FORM,
BY PHOTOSTAT, MICROFILM, RETRIEVAL SYSTEM, OR ANY
OTHER MEANS, WITHOUT WRITTEN PERMISSION FROM
THE PUBLISHERS.

ACADEMIC PRESS, INC.
111 Fifth Avenue, New York, New York 10003

United Kingdom Edition published by
ACADEMIC PRESS, INC. (LONDON) LTD.
Berkeley Square House, London W1X 6BA

LIBRARY OF CONGRESS CATALOG CARD NUMBER: 55-12200

PRINTED IN THE UNITED STATES OF AMERICA

Contents

CONTRIBUTORS TO VOLUME 25	vii
PREFACE	ix
CONTENTS OF PREVIOUS VOLUMES	xi
SUPPLEMENTARY MONOGRAPHS	xvi
ARTICLES TO APPEAR SHORTLY	xvii

Surface States

S. G. DAVISON AND J. D. LEVINE

I.	Introduction	2
II.	Brief Historical Review	3
III.	Crystal Orbital Methods	5
IV.	Crystal Potential Methods	31
V.	Three-Dimensional Calculations	75
VI.	Semiclassical Methods	80
VII.	Comparison between Theory and Experiment	88
VIII.	Concluding Remarks	148
	List of Abbreviations	149

Crystal Growth Mechanisms: Energetics, Kinetics, and Transport

R. L. PARKER

	Introduction	152
I.	Thermodynamics of Phase Changes: Phenomenology of the Problem	155
II.	Statistical Mechanics of Crystal Growth: Theoretical Physics of the Problem	161
III.	Stefan Problem: Mathematical Physics of Crystal Growth Transport Processes	170
IV.	Nucleation	197
V.	Interface Structures and Interface Kinetics	208
VI.	Morphological Stability	253
VII.	Effects of Impurities on Crystal Growth Mechanisms	272
VIII.	Fluid Flow Effects	284

Thermoelastic Theory of Stressed Crystals and Higher-Order Elastic Constants

Duane C. Wallace

   I. Introduction... 302
  II. Thermoelasticity of Stressed Materials................................ 305
 III. Expansions for Small Initial Stress................................... 336
  IV. Crystal Symmetry Properties... 347
   V. Measurement of Third-Order Elastic Constants.......................... 363
  VI. Fourth-Order Elastic Constants of Cubic Crystals...................... 380
 VII. Interpretation of Measurements in Terms of Atomic Theories............ 391

Author Index... 405

Subject Index.. 415

Contributors to Volume 25

S. G. DAVISON,* *Quantum Theory Group, Departments of Applied Mathematics and Physics, University of Waterloo, Ontario, Canada*

J. D. LEVINE, *RCA Laboratories, David Sarnoff Research Center, Princeton, New Jersey*

R. L. PARKER, *National Bureau of Standards, Washington, D. C.*

DUANE C. WALLACE, *Sandia Laboratories, Albuquerque, New Mexico*

---

* *Present address*: Department of Physics, Institute of Colloid and Surface Science, Clarkson College of Technology, Potsdam, New York.

Preface

The contents of this volume are devoted to three quite diverse and important subjects: the electronic structure of surfaces, crystal growth, and the higher-order elastic constants.

In contrast with the theory for the infinite crystal the quantum-mechanical theory of solid surfaces is still in the early stages of development. However, the developments have been significant and they are reviewed by Davison and Levine in the first article of this volume. This article also includes a review of experimental studies of the distribution of surface states.

Important aspects of the subject of crystal growth have been surveyed in earlier articles in this series, especially in the articles of Pfann (Volume 4) and Laudise and Neilson (Volume 12). A somewhat more comprehensive review of crystal growth phenomena, with emphasis on some recent developments, such as those on the theory of morphological stability, is presented by Parker in the second article of this volume.

In the final article Wallace reviews the theory of thermoelasticity, with special emphasis on the higher-order elastic constants. The theory for these constants is particularly important to the understanding of anharmonic effects in solids. Wallace's treatment complements the earlier articles by Huntington (Volume 7) and by Leibfried and Ludwig (Volume 12).

Contents of Previous Volumes

Volume 1, 1955

Methods of the One-Electron Theory of Solids
JOHN R. REITZ

Qualitative Analysis of the Cohesion in Metals
EUGENE P. WIGNER AND
FREDERICK SEITZ

The Quantum Defect Method
FRANK S. HAM

The Theory of Order-Disorder Transitions in Alloys
TOSHINOSUKE MUTO AND
YUTAKA TAKAGI

Valence Semiconductors, Germanium, and Silicon
H. Y. FAN

Electron Interaction in Metals
DAVID PINES

Volume 2, 1956

Nuclear Magnetic Resonance
G. E. PAKE

Electron Paramagnetism and Nuclear Magnetic Resonance in Metals
W. D. KNIGHT

Applications of Neutron Diffraction to Solid State Problems
C. G. SHULL AND E. O. WOLLAN

The Theory of Specific Heats and Lattice Vibrations
JULES DE LAUNAY

Displacement of Atoms during Irradiation
FREDERICK SEITZ AND J. S. KOEHLER

Volume 3, 1956

Group III—Group V Compounds
H. WELKER AND H. WEISS

The Continuum Theory of Lattice Defects
J. D. ESHELBY

Order-Disorder Phenomena in Metals
LESTER GUTTMAN

Phase Changes
DAVID TURNBULL

Relations between the Concentrations of Imperfections in Crystalline Solids
F. A. KRÖGER AND H. J. VINK

Ferromagnetic Domain Theory
C. KITTEL AND J. K. GALT

Volume 4, 1957

Ferroelectrics and Antiferroelectrics
WERNER KÄNZIG

Theory of Mobility of Electrons in Solids
FRANK J. BLATT

The Orthogonalized Plane-Wave Method
TRUMAN O. WOODRUFF

Bibliography of Atomic Wave Functions
ROBERT S. KNOX

Techniques of Zone Melting and Crystal Growing
W. G. PFANN

Volume 5, 1957

Galvanomagnetic and Thermomagnetic Effects in Metals
J.-P. JAN

Luminescence in Solids
CLIFFORD C. KLICK AND
JAMES H. SCHULMAN

Space Groups and Their Representations
G. K. KOSTER

Shallow Impurity States in Silicon and Germanium
W. KOHN

Quadrupole Effects in Nuclear Magnetic Resonance Studies in Solids
M. H. COHEN AND F. REIF

Volume 6, 1958

Compression of Solids by Strong Shock Waves
M. H. Rice, R. G. McQueen, and J. M. Walsh

Changes of State of Simple Solid and Liquid Metals
G. Borelius

Electroluminescence
W. W. Piper and F. E. Williams

Macroscopic Symmetry and Properties of Crystals
Charles S. Smith

Secondary Electron Emission
A. J. Dekker

Optical Properties of Metals
M. Parker Givens

Theory of the Optical Properties of Imperfections in Nonmetals
D. L. Dexter

Volume 7, 1958

Thermal Conductivity and Lattice Vibrational Modes
P. G. Klemens

Electron Energy Bands in Solids
Joseph Callaway

The Elastic Constants of Crystals
H. B. Huntington

Wave Packets and Transport of Electrons in Metals
H. W. Lewis

Study of Surfaces by Using New Tools
J. A. Becker

The Structures of Crystals
A. F. Wells

Volume 8, 1959

Electronic Spectra of Molecules and Ions in Crystals
Part I. Molecular Crystals
Donald S. McClure

Photoconductivity in Germanium
R. Newman and W. W. Tyler

Interaction of Thermal Neutrons with Solids
L. S. Kothari and K. S. Singwi

Electronic Processes in Zinc Oxide
G. Heiland, E. Mollwo, and F. Stöckmann

The Structure and Properties of Grain Boundaries
S. Amelinckx and W. Dekeyser

Volume 9, 1959

The Electronic Spectra of Aromatic Molecular Crystals
H. C. Wolf

Polar Semiconductors
W. W. Scanlon

Static Electrification of Solids
D. J. Montgomery

The Interdependence of Solid State Physics and Angular Distribution of Nuclear Radiations
Ernst Heer and Theodore B. Novey

Oscillatory Behavior of Magnetic Susceptibility and Electronic Conductivity
A. H. Kahn and H. P. R. Frederikse

Heterogeneities in Solid Solutions
Andre Guinier

Electronic Spectra of Molecules and Ions in Crystals
Part II. Spectra of Ions in Crystals
Donald S. McClure

Volume 10, 1960

Positron Annihilation in Solids and Liquids
Philip R. Wallace

Diffusion in Metals
David Lazarus

Wave Functions for Electron-Excess Color Centers in Alkali Halide Crystals
Barry S. Gourary and Frank J. Adrian

The Continuum Theory of Stationary Dislocations
ROLAND DE WIT

Theoretical Aspects of Superconductivity
M. R. SCHAFROTH

Volume 11, 1960

Semiconducting Properties of Gray Tin
G. A. BUSCH AND R. KERN

Physics at High Pressure
C. A. SWENSON

The Effects of Elastic Deformation on the Electrical Conductivity of Semiconductors
ROBERT W. KEYES

Imperfection Ionization Energies in CdS-Type Materials by Photoelectronic Techniques
RICHARD H. BUBE

Cyclotron Resonance
BENJAMIN LAX AND JOHN G. MAVROIDES

Volume 12, 1961

Group Theory and Crystal Field Theory
CHARLES M. HERZFELD AND
PAUL H. E. MEIJER

Electrical Conductivity of Organic Semiconductors
HIROO INOKUCHI AND HIDEO AKAMATU

Hydrothermal Crystal Growth
R. A. LAUDISE AND J. W. NIELSEN

The Thermal Conductivity of Metals at Low Temperatures
K. MENDELSSOHN AND H. M. ROSENBERG

Theory of Anharmonic Effects in Crystals
G. LEIBFRIED AND W. LUDWIG

Volume 13, 1962

Vibration Spectra of Solids
SHASHANKA S. MITRA

Behavior of Metals at High Temperatures and Pressures
F. P. BUNDY AND H. M. STRONG

Dislocations in Lithium Fluoride Crystals
J. J. GILMAN AND W. G. JOHNSTON

Electron Spin Resonance in Semiconductors
G. W. LUDWIG AND H. H. WOODBURY

Formalisms of Band Theory
E. I. BLOUNT

Chemical Bonding Inferred from Visible and Ultraviolet Absorption Spectra
CHR. KLIXBÜLL JØRGENSEN

Volume 14, 1963

g Factors and Spin-Lattice Relaxation of Conduction Electrons
Y. YAFET

Theory of Magnetic Exchange Interactions: Exchange in Insulators and Semiconductors
PHILIP W. ANDERSON

Electron Spin Resonance Spectroscopy in Molecular Solids
H. S. JARRETT

Molecular Motion in Solid State Polymers
N. SAITO, K. OKANO, S. IWAYANAGI, AND T. HIDESHIMA

Volume 15, 1963

The Changes in Energy Content, Volume, and Resistivity with Temperature in Simple Solids and Liquids
G. BORELIUS

The Dynamical Theory of X-Ray Diffraction
R. W. JAMES

The Electron-Phonon Interaction
L. J. SHAM AND J. M. ZIMAN

Elementary Theory of the Optical Properties of Solids
FRANK STERN

Spin Temperature and Nuclear Relaxation in Solids
L. C. HEBEL, JR.

Volume 16, 1964

Cohesion of Ionic Solids in the Born Model
MARIO P. TOSI

F-Aggregate Centers in Alkali Halide Crystals
W. DALE COMPTON AND HERBERT RABIN

Point-Charge Calculations of Energy Levels of Magnetic Ions in Crystalline Electric Fields
M. T. HUTCHINGS

Physical Properties and Interrelationships of Metallic and Semimetallic Elements
KARL A. GSCHNEIDNER, JR.

Volume 17, 1965

The Effects of High Pressure on the Electronic Structure of Solids
H. G. DRICKAMER

Electron Spin Resonance of Magnetic Ions in Complex Oxides. Review of ESR Results in Rutile, Perovskites, Spinel, and Garnet Structures
W. LOW AND E. L. OFFENBACHER

Ultrasonic Effects in Semiconductors
NORMAN G. EINSPRUCH

Quantum Theory of Galvanomagnetic Effect at Extremely Strong Magnetic Fields
RYOGO KUBO, SATORU J. MIYAKE, AND NATSUKI HASHITSUME

Volume 18, 1966

Energy Loss and Range of Energetic Neutral Atoms in Solids
D. K. NICHOLS AND V. A. J. VAN LINT

The Fundamental Optical Spectra of Solids
J. C. PHILLIPS

Crystal Symmetry, Group Theory, and Band Structure Calculations
ALLEN NUSSBAUM

Theoretical and Experimental Aspects of the Effects of Point Defects and Disorder on the Vibrations of Crystals—1
A. A. MARADUDIN

Volume 19, 1966

Theoretical and Experimental Aspects of the Effects of Point Defects and Disorder on the Vibrations of Crystals—2
A. A. MARADUDIN

X-Ray Diffraction Studies of the Lattice Parameters of Solids under Very High Pressure
H. G. DRICKAMER, R. W. LYNCH, R. L. CLENDENEN, AND E. A. PEREZ-ALBUERNE

Shock Effects in Solids
DONALD G. DORAN AND RONALD K. LINDE

Interaction of Acoustic Waves and Conduction Electrons
HAROLD N. SPECTOR

Volume 20, 1967

Determination of Electron Distribution in Crystals by Means of X Rays
R. BRILL

Electronic Effects in the Elastic Properties of Semiconductors
ROBERT W. KEYES

The Jahn-Teller Effect in Solids
M. D. STURGE

Green's Function Method in Lattice Dynamics
PHILIP C. K. KWOK

Helical Spin Ordering—1 Theory of Helical Spin Configurations
TAKEO NAGAMIYA

Volume 21, 1968

Insulating and Metallic States in Transition Metal Oxides
DAVID ADLER

The Excitonic State at the Semiconductor-Semimetal Transition
B. I. HALPERIN AND T. M. RICE

Polarons
J. APPEL

Magnetic Properties of Rare Earth Metals
BERNARD R. COOPER

Volume 22, 1968

Indirect Exchange Interactions in Metals
C. KITTEL

Dislocations and Plastic Flow in the Diamond Structure
H. ALEXANDER AND P. HAASEN

Lattice Dynamics of Metals
S. K. JOSHI AND A. K. RAJAGOPAL

Aspects of Group Theory in Electron Dynamics
E. BROWN

Diffusion in Metals
N. L. PETERSON

Volume 23, 1969

Effects of Electron-Electron and Electron-Phonon Interactions on One-Electron States of Solids
LARS HEDIN AND STIG LUNDQVIST

Theory of Dilute Magnetic Alloys
J. KONDO

Localized Moments and Nonmoments in Metals: The Kondo Effect
A. J. HEEGER

The Physics of Quantum Crystals
R. A. GUYER

Volume 24, 1970

The Pseudopotential Concept
VOLKER HEINE

The Fitting of Pseudopotentials to Experimental Data and Their Subsequent Application
MARVIN L. COHEN AND VOLKER HEINE

Pseudopotential Theory of Cohesion and Structure
VOLKER HEINE AND D. WEAIRE

Supplementary Monographs

Supplement 1: T. P. Das and E. L. Hahn
Nuclear Quadrupole Resonance Spectroscopy, 1958

Supplement 2: William Low
Paramagnetic Resonance in Solids, 1960

Supplement 3: A. A. Maradudin, E. W. Montroll, and G. H. Weiss
Theory of Lattice Dynamics in the Harmonic Approximation, 1963

Supplement 4: Albert C. Beer
Galvanomagnetic Effects in Semiconductors, 1963

Supplement 5: R. S. Knox
Theory of Excitons, 1963

Supplement 6: S. Amelinckx
The Direct Observation of Dislocations, 1964

Supplement 7: J. W. Corbett
Electron Radiation Damage in Semiconductors and Metals, 1966

Supplement 8: Jordan J. Markham
F-Centers in Alkali Halides, 1966

Supplement 9: Esther M. Conwell
High Field Transport in Semiconductors, 1967

Supplement 10: C. B. Duke
Tunneling in Solids, 1969

Supplement 11: Manuel Cardona
Optical Modulation Spectroscopy of Solids, 1969

Articles to Appear Shortly

A. A. Abrikosov	Introduction to the Theory of Metals (Supplement 12)
William A. Barker	Production and Detection of Nuclear Orientation
P. G. Bray—S. G. Bishop—T. Oja	Nuclear Magnetic Resonance Studies of the Structure of Glasses
J. I. Budnick	Hyperfine Interactions and Spin Densities in Magnetically Ordered Metals
H. Callen—R. Tahir-Kheli	Collective Properties of Pure and Impure Magnetic Insulators
A. Corciovei—G. Costache—D. Vamanu	Ferromagnetic Thin Films
J. O. Dimmock	Calculations of Electronic Energy Bands by the APW Method
J. E. Enderby	Electronic Properties of Liquid Metals
M. Glicksman	Plasmas in Solids
J. R. Hardy—A. M. Karo	The Lattice Dynamics and Statics of Alkali Halide Crystals
I. P. Ipatova—A. A. Maradudin—E. W. Montroll—G. H. Weiss	Theory of Lattice Dynamics in the Harmonic Approximation (Second Edition of Supplement 3)
G. D. Mahan	Many Body Theory of X-Ray Emission in Solids
S. C. Moss	Structure of Amorphous Solids
T. Nagamiya	Helical Spin Ordering—2
R. O. Simmons	Noble Gas Crystals
G. Slack—C. T. Walker	Thermal Conductivity in Nonmetals
L. Slifkin	The Photographic Process

R. Smoluchowski	Off-Center Substitutional Ions
J. Tauc	Electronic Properties of Amorphous Semiconductors
M. B. Webb—M. Lagally	Interactions of Electrons with Surfaces
M. Weger	Properties of the β-Tungstens
I. S. Zheludev	Ferroelectricity and Symmetry
J. Ziman	The Band Structure Problem

SOLID STATE PHYSICS

VOLUME 25

Surface States

S. G. DAVISON*†

*Quantum Theory Group,‡ Departments of Applied Mathematics and Physics
University of Waterloo, Ontario, Canada*

AND

J. D. LEVINE

RCA Laboratories, David Sarnoff Research Center, Princeton, New Jersey

I. Introduction	2
II. Brief Historical Review	3
III. Crystal Orbital Methods	5
1. Direct Approach	5
2. Resolvent Technique	18
IV. Crystal Potential Methods	31
3. Kronig–Penney Model	32
4. Mathieu Problem	54
V. Three-Dimensional Calculations	75
5. Direct Extension of One-Dimensional Picture	76
6. Analytical Continuation	77
7. Complications at the Boundary Region	78
8. Density of States	79
VI. Semiclassical Methods	80
9. Dangling Bond Approach	81
10. Madelung Potential Method	84
VII. Comparison between Theory and Experiment	88
11. Basic Phenomena	90
12. Surface State Distributions	94
13. Experimental Methods	102
14. Surface Preparation	108
15. Characterization by LEED	110
16. Line Defects	112
17. Data and Interpretation	113
VIII. Concluding Remarks	148
List of Abbreviations	149

* Work supported by the National Research Council of Canada, the Ontario Department of University Affairs, and the University of Waterloo Research Committee.

† *Present address*: Department of Physics, Institute of Colloid and Surface Science, Clarkson College of Technology, Potsdam, New York.

‡ QTG article S-154.

I. Introduction

Until now, surveys[1-5] dealing with the quantum-mechanical aspects of surface phenomena have been restricted essentially to a discussion of a few particular topics or methods. In the present review, no such restriction is made, and most of the topics and methods of interest to the surface scientist are included. This is achieved by selecting a variety of problems and analyzing them by different approaches. The article is concerned with both the theoretical description and experimental characterization of crystal surfaces, with particular reference to intrinsic surface states. Thus, it is hoped that a fairly complete picture is presented of the current understanding of the behavior of electrons at clean solid surfaces. For auxiliary reading, theoreticians are referred to Davison and Stęślicka[6] and experimentalists to Many et al.[7] and Frankl.[8] Vol'kenshtein[9] is also in this category, and his work is recommended to those interested in catalysis, which is largely beyond the scope of the present review. This survey article should also be of interest to the scientists working in other areas, such as spin waves, phonons, excitons, etc, in which surface problems are important.

Since a crystal is a many-particle system, containing nuclei and electrons in continual motion, the states of these particles are described by the solutions of the *complete* many-particle Schroedinger equation. By utilizing the Born–Oppenheimer approximation,[10] the nuclear and electronic parts of the complete Schroedinger equation can be separated.[11] The *electronic* Schroedinger equation, which is to be studied here, is a many-electron wave equation describing the motions of the electrons within a fixed nuclear framework. By adopting the independent-particle model of the Hartree–Fock scheme,[11,12] where each of the electrons is assumed to move under the influence of the static nuclear potential and the *average* field of all the other

---

[1] T. B. Grimley, *Advan. Catalysis* **12**, 1 (1960).
[2] J. Koutecký, *Advan. Chem. Phys.* **9**, 85 (1965).
[3] P. Mark, *Catalysis Rev.* **1**, 165 (1967).
[4] S. G. Davison, *Intern. J. Quantum Chem.* **2S**, 291 (1968).
[5] J. D. Levine and P. Mark, *Phys. Rev.* **182**, 926 (1969).
[6] S. G. Davison and M. Stęślicka, "Quantum Surface Physics." Pergamon Press, Oxford, 1971 (to be published).
[7] A. Many, Y. Goldstein, and N. B. Grover, "Semiconductor Surfaces." North-Holland Publ., Amsterdam, 1965.
[8] D. R. Frankl, "Electrical Properties of Semiconductor Surfaces." Pergamon Press, Oxford, 1967.
[9] F. F. Vol'kenshtein, "The Electronic Theory of Catalysis on Semiconductors." Pergamon Press, Oxford, 1963.
[10] M. Born and J. R. Oppenheimer, *Ann. Physik* **84**, 457 (1927).
[11] S. G. Davison, *in* "Physical Chemistry" (H. Eyring, D. J. Henderson, and W. Jost, eds.), Vol. 3, Chapter 2. Academic Press, New York, 1969.
[12] P. O. Löwdin, *Advan. Chem. Phys.* **2**, 207 (1959).

electrons, the many-electron Schroedinger equation can be reduced to a one-electron form. In investigating the electronic properties of crystal surfaces, the one-electron Schroedinger equation is solved, together with the boundary conditions describing the system, by means of the various methods discussed in the following sections.

II. Brief Historical Review

Among the earliest applications of quantum theory to periodic crystals was the Kronig–Penney (KP) model[13] (a linear array of δ function repulsive potentials) introduced in 1931. One year later, Tamm[14] considered this potential to be terminated at a free surface. The termination was taken as a step discontinuity in the potential, and the wavefunction and its derivative were matched across the discontinuity. Tamm was thus the first to show that, under certain conditions, surface states appear. Maue[15] considered the nearly free electron model of a crystal and expanded the crystal potential in a Fourier series. Again the surface was represented as a step discontinuity at the potential maximum or minimum and the conditions for surface states found. In 1939, Goodwin[16] published a series of three papers. The first was a generalization of Maue's work, but the other two developed approaches which were new at the time; they were based on the linear-combination-of-atomic orbitals (LCAO) method. Essentially, the discontinuity was represented in a different way; the surface atoms were simply presumed to have different Coulomb and resonance integrals from the bulk. Goodwin considered crystals with only one orbital per atom as well as crystals with s–p hybridization. Also in 1939, Shockley[17] examined the properties of a general one-dimensional periodic potential, terminated at its potential maximum by a step. His method was to match the wave functions across the discontinuity. He showed that if the bulk bands were "crossed," then surface states appeared when the surface "perturbation" was sufficiently small. This was in contrast to previous studies, in which the surface states appeared when the surface "perturbation" was sufficiently large. Later these surface states came to be called "Shockley states" and "Tamm states," respectively. Shockley speculated that "crossed bands" and negligible surface "perturbation" corresponded to the situation found in Si and Ge crystals.

In the 1940's, little significant surface state theory was developed. By contrast, the experimental evidence for surface states greatly increased with

---

[13] R. de L. Kronig and W. G. Penney, *Proc. Roy. Soc.* **A130** 499 (1931).
[14] I. Tamm, *Z. Physik* **76**, 849 (1932); *Phys. Z. Sowjet.* **1**, 733 (1932).
[15] A. W. Maue, *Z. Physik* **94**, 717 (1935).
[16] E. T. Goodwin, *Proc. Cambridge Phil. Soc.* **35**, 205, 221, 232 (1939).
[17] W. Shockley, *Phys. Rev.* **56**, 317 (1939).

the growth of rectifier and transistor technology in the late 1940's. Bardeen[18] in 1947 explained Meyerhof's "anomalous" results[19] on metal-silicon contacts in terms of surface states. In 1948, Shockley and Pearson[20] tested Bardeen's hypothesis. Using a field plate, they modulated the potential of a semiconductor free surface—the so-called "field-effect" experiment. They showed that part of the surface space charge was mobile, but most was immobile when a secondary field was applied parallel to the surface. The immobile charge was ascribed to charges trapped in surface states. Later, it was found that slow and fast traps were present; these are related to the oxide formation on air-exposed Si and Ge surfaces.

In the 1950's, these experimental results spurred new interest in surface state theory, particularly those related to Si and Ge surfaces. Statz,[21] Artmann,[22] Hoffmann,[23] and Lippmann,[24] among others, elaborated on the various aspects of Tamm and Shockley states. A step toward a sophisticated general theory of localized surface states, applicable to real systems, was taken in 1957 by Koutecký,[25] and in 1960 by Koutecký and Tomášek,[26] who succeeded in treating all kinds of localized states on ideal crystal surfaces. The method was called the one-electron resolvent method and most of the results were obtained in the LCAO representation. At the end of this period, Antončik[27] modified Maue's work and applied the pseudo-potential method to treat one type of electronic surface states, namely, Shockley states.

In the early 1960's, however, low-energy-electron-diffraction (LEED) experiments showed that some free surfaces (e.g., Si {111} cleaved in ultrahigh vacuum) are reconstructed. The reconstruction (extra spots on the screen) is also unstable and temperature dependent. This shows that silicon surfaces have *severe* surface perturbations—not *negligible* surface perturbations as inferred by Shockley and others. By contrast, LEED showed that the ionic faces (i.e., electrostatically neutral faces) of III–V, II–VI, and I–VII compounds are not reconstructed (Section 15). Such faces are found on NaCl (100), zinc blende (110), and wurtzite ($10\bar{1}0$). Theoretical surface state studies on these mixed or partially ionic crystals have appeared only in the last few years. The techniques that have been used, which are discussed in the subsequent sections, include dangling

[18] J. Bardeen, *Phys. Rev.* **71**, 717 (1947).
[19] W. E. Meyerhof, *Phys. Rev.* **71**, 727 (1947).
[20] W. Shockley and G. Pearson, *Phys. Rev.* **74**, 232 (1948).
[21] H. Statz, *Z. Naturforsch.* **5a**, 534 (1950).
[22] K. Artmann, *Z. Physik* **131**, 244 (1952).
[23] T. A. Hoffmann, *Acta Phys. Acad. Sci. Hung.* **2**, 195 (1952).
[24] B. A. Lippmann, *Ann. Phys.* **2**, 16 (1957).
[25] J. Koutecký, *Phys. Rev.* **108**, 13 (1957).
[26] J. Koutecký and M. Tomášek, *Phys. Rev.* **120**, 1212 (1960).
[27] E. Antončik, *J. Phys. Chem. Solids* **21**, 137 (1961).

bonds,[28,29] Madelung constant,[30] LCAO methods,[31–33] and Mathieu potential.[34]

III. Crystal Orbital Methods

One of the earliest and most fruitful methods used for solving surface problems is the crystal orbital (CO) approach. By "digitizing" the crystal potential into a network of Coulomb and resonance integrals, the CO approach enables the one-electron Schroedinger differential equation to be converted into an equivalent difference equation. The surface is then distinguished from the bulk by modifying the Coulomb integrals in its vicinity and eliminating certain coupling terms (resonance integrals). With the method being extremely flexible, it is applicable to a wide range of surface state problems. In addition to the electronic spectra, the vibrational or phonon spectra of solids can also be investigated by the difference equation technique.[35]

Initially, the straightforward approach of solving the difference equations in terms of Bloch-type functions[36–39] is presented. The general three-dimensional (3-D) case cannot be solved directly,[1] so it is necessary to resort to the more powerful resolvent technique. Since this method has recently been discussed at length by Koutecký,[2] only an illustrative example is treated here, after the basic theory has been formulated.

1. DIRECT APPROACH

a. Basic Theory

In the Dirac notation,[40] the one-electron Schroedinger equation is

$$(H - E) \mid \psi \rangle = 0, \tag{1.1}$$

[28] H. C. Gatos and M. C. Lavine, *J. Electrochem. Soc.* **107**, 427, 433 (1960).
[29] H. R. Huff, S. Kawaji, and H. C. Gatos, *Surface Sci.* **10**, 232 (1968).
[30] J. D. Levine and P. Mark, *Phys. Rev.* **144**, 751 (1966).
[31] S. G. Davison and J. Koutecký, *Proc. Phys. Soc.* **89**, 237 (1966).
[32] J. Koutecký and S. G. Davison, *Intern. J. Quantum Chem.* **2**, 73 (1968).
[33] J. D. Levine and S. G. Davison, *Phys. Rev.* **174**, 911 (1968).
[34] J. D. Levine, *Phys. Rev.* **171**, 701 (1968).
[35] S. G. Davison and Y. C. Cheng, *Phys. Letters* **27A**, 592 (1968).
[36] F. Bloch, *Z. Physik* **52**, 555 (1928).
[37] J. M. Ziman, "Principles of the Theory of Solids." Cambridge Univ. Press, London and New York, 1965.
[38] M. Sachs, "Solid State Theory." McGraw-Hill, New York, 1963.
[39] R. A. Smith, "Wave Mechanics of Crystalline Solids." Chapman and Hall, London, 1967.
[40] P. A. M. Dirac, "Quantum Mechanics." Oxford Univ. Press, London and New York, 1959.

H being the effective Hamiltonian, $|\psi\rangle$ the wave function, and E the corresponding energy. If $|\psi\rangle$ is expanded as a linear combination of atomic orbitals (LCAO) $|\phi_n\rangle$, i.e.,

$$|\psi\rangle = \sum_n c_n |\phi_n\rangle \qquad (1.2)$$

where $|\phi_n\rangle$ describes the motion of an electron in the field of the nth atom only and c_n are the coefficients of the LCAO, then inserting (1.2) in (1.1) and multiplying through on the left by $\langle \phi_m |$ leads to a system of linear homogeneous equations for c_n, namely,

$$\sum_n (H_{mn} - ES_{mn}) c_n = 0 \qquad (1.3)$$

in which

$$H_{mn} = \langle \phi_m | H | \phi_n \rangle \qquad (1.4)$$

$$S_{mn} = \langle \phi_m | \phi_n \rangle. \qquad (1.5)$$

Before attempting to solve (1.3) for a monatomic system, two simplifying assumptions are made. First, the AO $\{\phi_n\}$ are assumed to form an orthonormal set, so that the *overlap matrix* has the form

$$\begin{aligned} S_{mn} = \delta_{mn} &= 1, \quad m = n, \\ &= 0, \quad m \neq n. \end{aligned} \qquad (1.6)$$

Second, when $m = n$ the Hamiltonian matrix elements in (1.4) are taken to be constant, while for $m \neq n$ only those elements representing the interaction between nearest-neighbor (NN) atoms are assumed to be nonzero and constant. Mathematically, these assumptions, which form the basis of the well-known *tight-binding* (TB)[37–39] or *Hückel*[41,42] *approximation*, are written as

$$\begin{aligned} H_{mn} &= \alpha, \quad m = n, \\ &= \beta, \quad m = n \pm 1, \\ &= 0, \quad \text{otherwise} \end{aligned} \qquad (1.7)$$

where α and β denote the *Coulomb* and *resonance integrals*, respectively. With the aid of (1.6) and (1.7), Eq. (1.3) becomes

$$2Xc_n = c_{n+1} + c_{n-1}, \qquad (1.8)$$

$$2X = (E - \alpha)/\beta \qquad (1.9)$$

[41] A. Streitwieser, Jr., "Molecular Orbital Theory for Organic Chemists," Chapter 2. Wiley, New York, 1966.
[42] L. Salem, "The Molecular Orbital Theory of Conjugated Systems," Chapter 1. Benjamin, New York, 1966.

being the *reduced* energy. Since α represents the effective energy of an electron in an AO, and β the bond energy between NN atoms, it follows that X in (1.9) is a *dimensionless* quantity.

Having transformed the one-electron Schroedinger equation (1.1) into a homogeneous second-order linear difference equation with constant coefficients (1.8)[43,44] by neglecting overlap and making use of the LCAO–TB approximation, the remainder of this section is devoted to seeking solutions to equations of the type (1.8) when they are subjected to the boundary conditions describing the system under consideration. It should also be noted that, for normalized solutions

$$\langle \psi \mid \psi \rangle = 1, \tag{1.10}$$

so by (1.2) and (1.6) the coefficients must satisfy

$$\sum_n \mid c_n \mid^2 = 1. \tag{1.11}$$

b. *Volume and Surface States of* AB-*Type Crystals*

(*i*) *Formulation.* The motivation behind choosing an AB-type crystal model is that, in addition to representing a diatomic or ionic crystal, it is also possible to reduce the results obtained to those for a monatomic system. The simplest way to construct an AB-type crystal is to take a linear array of equally spaced lattice sites, numbered 1 to N (say), and on each of the odd and even sites place an A and B atom, respectively. If an s- (p) orbital is associated with each odd (even) atom, then the A and B atoms can be characterized by the Coulomb integrals $\alpha_{A,B}$ and the bonds between the NN atoms by the resonance integral β, whose sign alternates along the chain (Fig. 1). In this case, (1.8) is replaced by[33]

$$(X - z_1)c_n = c_{n-1} - c_{n+1}, \quad n \text{ odd,}$$
$$(X + z_1)c_n = c_{n+1} - c_{n-1}, \quad n \text{ even,} \tag{1.12}$$

where

$$X = (E - \bar{\alpha})/\beta, \quad \bar{\alpha} = (\alpha_A + \alpha_B)/2, \tag{1.13}$$

and

$$z_1 = (\alpha_A - \alpha_B)/2\beta \tag{1.14}$$

is the *composition parameter*.[45]

---

[43] L. M. Milne-Thomson, "The Calculus of Finite Differences," p. 387. Macmillan, New York, 1951.
[44] L. Brand, "Differential and Difference Equations," Chapter 9. Wiley, New York, 1966.
[45] For convenience, the A(B) ions are taken as being electropositive (electronegative), with $\alpha_A > \alpha_B$ so that $z_1 > 0$.

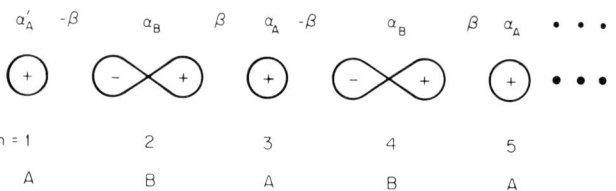

FIG. 1. Orbital model of a semi-infinite linear ionic crystal.

For a semi-infinite crystal ($N \to \infty$), Eqs. (1.12) are valid for $n > 1$ (2) for an A- (B)-atom termination. The presence of a surface at $n = 1$ (2) is assumed to perturb the electronic environment of the end atom *only*. Thus, the surface effect can be accounted for by merely changing the Coulomb integral of the atom at $n = 1$ (2) from α_A (α_B) to α_A' (α_B'), so that (1.12) reduces to the boundary conditions

$$(X - z_A)c_1 = -c_2 \tag{1.15}$$

$$(X + z_B)c_2 = c_3, \tag{1.16}$$

$$z_A = (\alpha_A' - \bar{\alpha})/\beta, \qquad z_B = (\bar{\alpha} - \alpha_B')/\beta \tag{1.17}$$

being the *perturbation parameters*. Furthermore, since the crystal is semi-infinite, the surface effect at $n = N$ can be neglected,[1] i.e.,

$$c_m = 0, \qquad m = N + 1. \tag{1.18}$$

(*ii*) *Solutions.* The solutions of (1.12) are found by putting[46]

$$c_n = u^n, \qquad n \text{ odd},$$

$$c_n = v^n, \qquad n \text{ even}. \tag{1.19}$$

Substituting (1.19) into (1.12) and eliminating v leads to the quartic

$$u^4 + \gamma u^2 + 1 = 0 \tag{1.20}$$

where

$$\gamma = X^2 - z_1^2 - 2. \tag{1.21}$$

Choosing

$$\gamma = -2\cos\theta \tag{1.22}$$

means that the general solution of (1.12) can be written as

$$c_n = P\cos n\theta/2 + Q\sin n\theta/2, \qquad n \text{ odd}, \tag{1.23}$$

which by (1.18) becomes

$$c_n = R\sin(m-n)\theta/2, \qquad n \text{ odd}, \tag{1.24}$$

$$R = P/\sin m\theta/2 \tag{1.25}$$

[46] A. T. Amos and S. G. Davison, *Physica* **30**, 905 (1964).

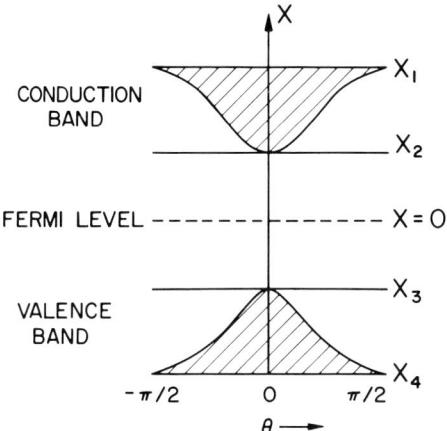

FIG. 2. Reduced energy band-structure diagram.

being the normalization factor. Inserting (1.24) into the second equation in (1.12) yields

$$c_n = KR \cos(m - n)\theta/2, \quad n \text{ even}, \tag{1.26}$$

where

$$K = -2 \sin(\theta/2)/(X + z_1). \tag{1.27}$$

From (1.21) and (1.22) the corresponding reduced energies are[33]

$$X = \pm (z_1^2 + 4 \sin^2 \theta/2)^{1/2}. \tag{1.28}$$

The unknown parameter θ, introduced in (1.22), is determined by the eigenvalue equations obtained from (1.15) and (1.16); for A- and B-atom terminations, respectively, these are

$$(X - z_A) \sin N\theta/2 = -K \cos(N - 1)\theta/2 \tag{1.29}$$

$$(X + z_B) K \cos(N - 1)\theta/2 = \sin(N - 2)\theta/2. \tag{1.30}$$

Analysis of these equations reveals that all their solutions[47] are real except two which, depending on the values of $z_{1,A,B}$, have the possibility of being real or complex.

(*iii*) *Volume States.* For θ real, the coefficients (1.24) and (1.26) are periodic throughout the crystal (i.e., delocalized), with different amplitudes on the A and B atoms. Hence, the probability of finding an electron at an A atom differs from that at a B atom. This is to be expected since, as indicated by $\alpha_{A,B}$, the crystal potential field is different at the A and B atoms.

---

[47] Only the solutions in the range $0 \leq \theta \leq \pi$ need be considered, because the remainder merely repeat these solutions.

The reduced energy levels (1.28) lie in two bands (Fig. 2) whose edges are

$$X_{1,4} = \pm(z_1^2 + 4)^{1/2}, \qquad X_{2,3} = \pm z_1. \tag{1.31}$$

The band and gap widths can easily be found from (1.31) and depend only on the composition parameter z_1 of the crystal. The two bands are centered on the line (Fermi level) $X = 0$, and are separated by a forbidden energy gap (FEG) of width $2z_1$.[48] In the case of monatomic crystals, $\alpha_A = \alpha_B$ (i.e., $z_1 = 0$), so the FEG vanishes and the two band energy spectrum reduces to the single band one.[1,3,14,16,23]

(iv) *Surface States.* The complex θ solutions of (1.29) and (1.30) are investigated by setting[31]

$$\theta = (\zeta + i\mu); \qquad \zeta, \mu \text{ real} > 0. \tag{1.32}$$

The μ is chosen positive, to ensure that the wave function decays exponentially into the crystal. Since the energy X in (1.28) must be real, it follows that Im $\sin^2 \theta/2 = 0$, so

$$\sin \zeta \sinh \mu = 0. \tag{1.33}$$

However, $\sinh \mu \neq 0$ for $\mu > 0$, so

$$\zeta = k\pi; \qquad k = 0, 1, \ldots, \tag{1.34}$$

where only the solutions for $k = 0$ and 1 need be considered, because all the other values of k merely repeat these solutions. Thus, the values of θ are

$$\theta = i\mu, \quad (\pi + i\mu) \tag{1.35}$$

so the energies (1.28) become

$$X = \pm(z_1^2 - 4\sinh^2 \mu/2)^{1/2} \tag{1.36}$$

$$X = \pm(z_1^2 + 4\cosh^2 \mu/2)^{1/2}. \tag{1.37}$$

With the levels (1.36) and (1.37) being located *inside* the FEG and *outside* the bands (Fig. 2), they are called *inner* and *outer* levels,[31,33] respectively. Moreover, when $X > 0$ ($X < 0$) the level is said to be $P(N)$ type, which should not be confused with p- and n-type semiconductor doping terminology.

(a) *Inner states.* For $\theta = i\mu$ and $N \to \infty$, (1.24) and (1.26) reduce to

$$c_n = PL^{n/2}, \qquad n \text{ odd},$$
$$c_n = KPL^{n/2}, \qquad n \text{ even}, \tag{1.38}$$

where

$$L = \exp(-\mu), \qquad 0 < L < 1, \tag{1.39}$$

[48] T. A. Hoffmann, *Acta Phys. Acad. Sci. Hung.* **1**, 175 (1951).

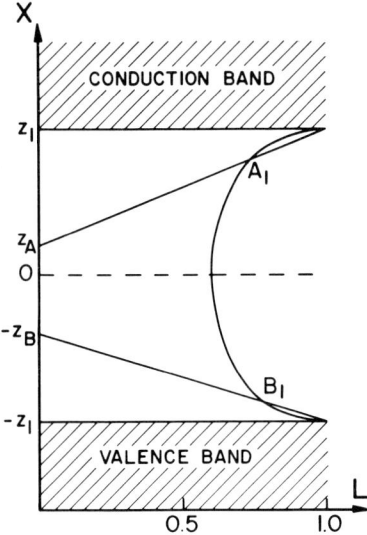

FIG. 3. Location of the A- and B-like energy levels for inner surface states.

can be compared with λ of Davison and Koutecký.[31] The negative exponential factor in (1.38) means that the coefficients are largest near $n = 1$ and decrease progressively as n increases. Hence, the electron is said to be *localized* at the crystal surface, and becomes more localized as μ increases from zero.

When $\theta = i\mu$, Eqs. (1.29) and (1.30) together with (1.21), (1.22), and (1.27) give

$$X = z_A + (z_1 - z_A)L \qquad (1.40)$$

$$X = -[z_B + (z_1 - z_B)L]. \qquad (1.41)$$

In terms of L, (1.36) becomes

$$X = \pm[z_1^2 - (1-L)^2 L^{-1}]^{1/2} \qquad (1.42)$$

so the positions of the inner levels in the X vs L plane (Fig. 3) are located by the intersections A_1 and B_1 of the lines (1.40) and (1.41) with the curve (1.42) in the range $0 < L < 1$. The intersections at $X = \pm z_1$, for which $L = 1$ or $\mu = 0$, are trivial, because they correspond to the conduction and valence band-edge states.

Eliminating X between (1.42) and (1.40) or (1.41) yields[31,33]

$$f_r L^2 + g_r L - 1 = 0, \qquad r = A \text{ or } B, \qquad (1.43)$$

where
$$f_r = (z_1 - z_r)^2 \qquad (1.44)$$
$$g_r = 1 + z_1^2 - z_r^2. \qquad (1.45)$$

The root of (1.43) for $L > 0$ is
$$L = [-g_r + (g_r^2 + 4f_r)^{1/2}]/2f_r \qquad (1.46)$$

which inserted in (1.40), (1.41) or (1.42) enables X to be expressed directly in terms of the composition and perturbation parameters. From (1.17), (1.46) and Fig. 3 it follows that, for $L > 0$,
$$z_1 > z_r, \qquad r = \text{A or B}. \qquad (1.47)$$

These inequalities are the *existence* or *threshold conditions* for the formation of inner states; i.e., when satisfied they ensure that the intersections A_1 and B_1 occur in Fig. 3. For monatomic crystals, where $z_1 = 0$ and no FEG appears in the main band, (1.47) is not valid and inner states do not exist, but become volume states.

When referred to the conduction and valence band edges, (1.40) and (1.41) show that
$$L = [(z_1 - z_r) - (z_1 - X)]/(z_1 - z_r). \qquad (1.48)$$

Physically, the r-like surface state has a "trap" energy of $(z_1 - X)$ and a "potential energy" (PE) of $(z_1 - z_r)$. Thus, the numerator in (1.48) is the "zero-point energy" (ZE) of the electron in the surface state trap, so $L = \text{ZE/PE}$. Also, since $0 < L < 1$, it follows that ZE $<$ PE. For fairly shallow surface states, f_r in (1.44) is small, so the square root in (1.46) can be expanded to give
$$L \sim g_r^{-1} \qquad (1.49)$$

which by (1.45) shows that the ratio of ZE/PE is smallest when $z_1 \gg z_r$. Such a situation is found in crystals with a wide FEG ($\sim 2z_1$) and a deep surface state PE ($\sim z_1 - z_r$).

(b) *Outer states.* Here $\theta = (\pi + i\mu)$ and the relevant equations are obtained from those for the inner states by merely replacing L by $-L^{31}$. Thus, (1.38) becomes
$$\begin{aligned} c_n &= P(-1)^{n/2}L^{n/2}, & n \text{ odd,} \\ c_n &= KP(-1)^{n/2}L^{n/2}, & n \text{ even,} \end{aligned} \qquad (1.50)$$

which, apart from the $(-1)^{n/2}$ factor, are the same as the exponentially damped coefficients for the inner states. The eigenvalue equations (1.40)

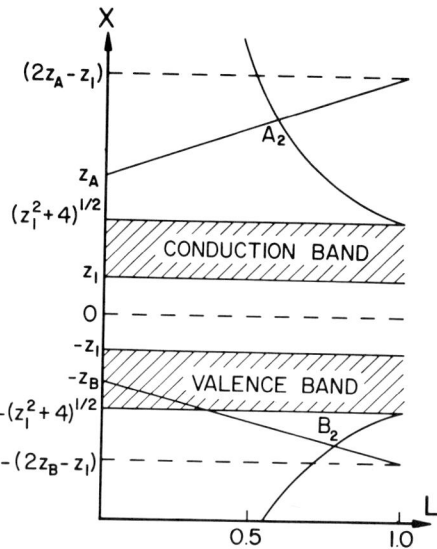

FIG. 4. Location of the A- and B-like energy levels for outer surface states.

and (1.41) now read

$$X = z_A - (z_1 - z_A)L \tag{1.51}$$

$$X = -[z_B - (z_1 - z_B)L] \tag{1.52}$$

while the reduced energy expression (1.42) becomes

$$X = \pm[z_1^2 + (1+L)^2 L^{-1}]^{1/2}. \tag{1.53}$$

The position of the outer levels can be found graphically by plotting X vs L for the lines (1.51) and (1.52) together with the curve (1.53) (Fig. 4). The actual value of L (and therefore μ) is given by the points of intersection A_2 and B_2 of the graphs or by the positive root of [cf. (1.43)]

$$f_r L^2 - g_r L - 1 = 0, \qquad r = A \text{ or } B, \tag{1.54}$$

namely [cf. (1.46)],

$$L = [g_r + (g_r^2 + 4f_r)^{1/2}]/2f_r. \tag{1.55}$$

Substituting (1.55) in (1.51), (1.52) or (1.53) gives X as a function of the z's.

For $L < 1$, (1.55) gives the existence condition[31]

$$z_r(z_r - z_1) > 1, \qquad r = A \text{ or } B, \tag{1.56}$$

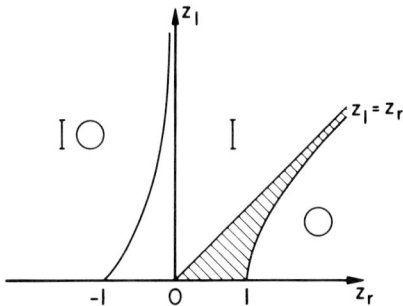

FIG. 5. Mapping of the existence regions for inner (I) and outer (O) states in the $z_1 z_r$ plane. Volume states occur in the shaded area.

which can also be obtained by noting that, at $L = 1$, the surface energy $|2z_r - z_1|$ in (1.51) or (1.52) is greater than the volume band-edge energy $|z_1^2 + 4|^{1/2}$ in (1.53). On putting $z_1 = 0$, (1.56) reduces to the well-known condition[1,3,16,23]

$$z_r^2 > 1 \qquad (1.57)$$

for the existence of surface states in 1-D monatomic crystals. Now $z_1 > 0$ so, when $z_r > 0$, the presence of the z_1 term in (1.56) means that outer surface states are more likely to arise in monatomic crystals than in AB-type ones; this is particularly true when z_1 is large, i.e., for wide FEG materials, where the two types of atoms are very different. Conversely, when $z_r < 0$, outer surface states are more likely to occur in AB-type crystals than monatomic ones. In this case, when z_1 is large, surface states can exist in AB-type crystals even when $z_r^2 < 1$.

The various regions of the $z_1 z_r$ plane, which fulfill or violate the inequalities (1.47) and (1.56), are mapped in Fig. 5.[31] Volume states *alone* exist, when the point (z_1, z_r) lies in the shaded area, where *both* (1.47) and (1.56) are violated. In the three remaining regions inner (I) and outer (O) states occur as indicated.

Experimentally, the inner states are more easily detected than the outer states. The reason being that the states in the valence-conduction FEG of semiconductors can appreciably shift the intrinsic Fermi level ($X = 0$). For theoretical calculations, however, it is useful to consider the existence of both types of surface states.

The surface states of AB-type alloys have also been investigated[49] using the scattering-matrix (SM) technique, which is described in Section 3,a.

[49] E. Aerts, *Physica* **26**, 1057 (1960).

c. Other Work on Binary Crystals

The discussion in the preceding section was purposely confined to a detailed analysis of the volume and surface states of AB-type binary or ionic crystals, so that a clear picture could be presented of the *application* of the direct crystal orbital (DCO) approach to a specific problem. In the present section, only an outline is given of the various other kinds of binary crystal problems that have been tackled by the DCO method.

(*i*) *Chemisorption States.* Besides treating volume and surface states, Davison and Koutecký[31] also investigated chemisorption states of AB-type crystals. These states occur when a foreign atom (adatom) interacts with a crystal surface in such a way that a chemical bond is formed between the adatom and the crystal. For an A-atom terminated crystal, the boundary conditions are (1.18), together with

$$(X - z_1 - z_2)c_1 = c_2 + \eta c_\nu \quad (1.58)$$

$$(X - z_1 - z^*)c_\nu = c_1 \quad (1.59)$$

where c_ν is the coefficient of the adatom ν and

$$z_2 = (\alpha_A' - \alpha_A)/\beta, \quad z^* = (\alpha^* - \alpha_A)/\beta, \quad \eta = \beta^*/\beta, \quad (1.60)$$

α^* being the adatom Coulomb integral and β^* the resonance integral of the bond between the adatom and the crystal. In addition to the volume state coefficients, the new boundary condition (1.59) yields the expression for c_ν, which has a different functional form from that of the bulk coefficients. Although the main features of the reduced energy spectrum are the same as before, i.e., the band edges are given by (1.31) and outer and inner states can occur, (1.58) leads to a new θ-eigenvalue equation, so the individual volume and chemisorption levels are located at different positions in the energy spectrum, and their existence conditions are more complicated than in the clean surface case. When z_1 vanishes, (1.58) and (1.59) describe the process of adsorbing an adatom onto a monatomic crystal surface.[1,2,8,23,50,51]

(*ii*) *Deformed Crystals.* As Rymer[52] and Zdhanov[53] pointed out, there is both experimental and theoretical evidence that the positions of the surface atoms are different from those in the interior. Careful LEED experiments on germanium[54] and diamond[55] also support this evidence.

[50] J. Koutecký, *Trans. Faraday Soc.* **54**, 1038 (1958).
[51] T. B. Grimley, *Proc. Phys. Soc.* **72**, 103 (1958).
[52] T. B. Rymer, *Nuovo Cimento Suppl.* **6**, Ser. X, **1**, 294 (1957).
[53] G. S. Zdhanov, "Crystal Physics," p. 202. Academic Press, New York, 1965.
[54] J. J. Lander, G. W. Gobeli, and J. Morrison, *J. Appl. Phys.* **34**, 2298 (1963).
[55] J. B. Marsh and H. E. Farnsworth, *Surface Sci.* **1**, 3 (1964).

Theoretically, the effect of lattice surface distortion on the surface states of AB-type crystals has been studied by the DCO approach,[56] while for monatomic crystals both crystal orbital[25,57] and crystal potential[58-60] methods have been used. The present status of Tamm states in surface reconstructed crystals has recently been reviewed by Davison and Steślicka.[60a]

The boundary conditions for a deformed AB-type crystal can easily be obtained from (1.58) and (1.59) by setting $z_2 = 0$ and reinterpreting the various parameters.[56] In this case, c_ν is the coefficient of the first atom in the crystal, which is connected to the remainder of the crystal by a bond (β^*), whose length differs from that of the bonds (β) between the atoms of the undeformed part of the crystal. The deformation suffered by the crystal, on the formation of the surface, manifests itself in the functional form of c_ν, which is different from that of the bulk atom coefficients. Perhaps one of the most interesting aspects of the work on deformed AB-type crystals is the results obtained when the effects of the surface deformation (η) and perturbation (z^*) are studied separately, namely, (1) the positions of the band edges (1.31) are unaffected by the replacing of a surface perturbation by a deformation, because they depend only on the crystal composition; (2) no inner states occur when only a surface deformation is present (i.e., only outer states exist); (3) inner states are produced only by a surface perturbation, while outer states are produced by a surface perturbation or deformation; and (4) if both a surface perturbation and deformation are present, then the positions of the inner state energy levels are insensitive to the deformation effect and are solely determined by the perturbation effect, whereas the locations of the outer levels depend on both the perturbation and deformation effects. Only displacements normal to the surface are considered here.

Koutecký,[25] using the resolvent crystal orbital (RCO) technique (Section 2), examined the surface states of a monatomic crystal, in which the perturbing potential due to the surface penetrated to a depth of $(Z + 1)$ cells below the crystal surface. The findings show that as many as $(Z + 1)$ surface states can exist, but no more. For $Z = 1$, Koutecký found that $|c_\nu/c_1| < 1$, so the coefficients are maximum *below*, and not at, the surface. Such *subsurface* states may have lower energy than the true surface states. Utilizing the Green function method (Section 3,b), in conjunction with a perturbed KP model,[13] Phariseau[58] has confirmed these findings of Koutecký.

[56] S. G. Davison, *Physica* **37**, 539 (1967); *Surface Sci.* **10**, 369 (1968).
[57] M. Steślicka, *Acta Physica. Polon.* **33**, 981 (1968); **34**, 875 (1968).
[58] P. Phariseau, *Physica* **26**, 1192 (1960). See also the preceding article.
[59] M. Steślicka and K. F. Wojciechowski, *Physica* **32**, 1274 (1966).
[60] M. Steślicka, *Acta Physiol. Polon.* **30**, 883 (1966).
[60a] S. G. Davison and M. Steślicka, *Int. J. Quantum Chem.* **4S**, (in press) (1970).

A rather extensive investigation of the surface states of monatomic crystals has also been carried out by Stęślicka, who made use of both the DCO [57] and SM [59,60] methods. Of special interest here are the 3-D DCO calculations on Li, which showed that, if the surface deformation caused the lattice constant at the surface to be reduced by an amount $\epsilon \geq 0.7$ Å (or $\epsilon \geq 0.30$ Å when overlap is included), then an additional band of surface states appears in the crystal energy spectrum. Furthermore, since the surface deformation has a marked effect on the existence of the surface states, Stęślicka's work gives some credence to Henzler's suggestion [61] regarding the origin of surface states, namely, that the special arrangement of the surface atoms, with respect to their nearest neighbors, could be mainly responsible for the observed states. The experimental aspects of surface deformation are discussed at greater length in Part VII.

(*iii*) A_rB_s-*Type Crystals.* Throughout his detailed treatment of the *volume* states of various binary systems, Hoffmann [48] employed a generating function technique, in the context of the DCO method, to solve the set of coupled linear difference equations, whose solutions described the electronic states of the system in question. Like the AB case, the number of equations in each set was the same as the number of atoms, $(r + s)$, in the unit cell of the crystal and the number of bands in the energy spectrum. The widths of the volume bands were found to depend on crystal composition. Hoffmann also derived expressions for the density of states, examined the effect of including the overlap integral in his calculations, and extended [62] the work on AB-type crystals to 2- and 3-D.

The *complete* energy spectrum (volume and surface levels) of Hoffmann's series of binary alloys has been investigated by Davison, Ling, and Ghosh (DLG) [63] via the DCO approach. They showed that a maximum of $2(r + s)$ surface levels could appear outside the $(r + s)$ bands and obtained the existence equations determining the positions and number of these levels. Furthermore, when $(r + s)$ is odd (even), they found that it is possible (impossible) for the surface levels to occur in all (some) of the FEG. However, it should be pointed out that, since DLG only considered A-atom terminations of the crystals, only $(r + s)$ surface levels should occur; the other $(r + s)$ levels correspond to B-atom terminations (Section 1,b,iv).

The above treatment of A_rB_s alloys led to the formulation of the RCO theory [32] for the electronic states of a general mixed crystal, where the unit cell contains any number of different atoms in any ordered sequence, with different bonds between each pair of atoms. This formulation is sufficiently general to encompass all the previous cases.

[61] M. Henzler, *Surface Sci.* **9**, 31 (1968).
[62] T. A. Hoffmann, *Acta Phys. Acad. Sci. Hung.* **2**, 107 (1952).
[63] S. G. Davison, T. Y. Ling, and U. S. Ghosh, *J. Phys. Chem. Solids* **28**, 1921 (1967).

2. Resolvent Technique

a. Basic Theory

Consider a situation in which a system of molecules or crystals is either noninteracting or interacting. Let H^0 and H be the effective one-electron Hamiltonians describing the system in these two cases, where

$$H = H^0 + V, \qquad (2.1)$$

V being the interaction potential. If the eigenvector and eigenvalue of H^0 are $|\psi_j^0\rangle$ and E_j^0 for the jth state, then for the noninteracting (unperturbed) system, (1.3) can be replaced by

$$\sum_l H_{kl}^0 c_{lj}^0 = E_j^0 \sum_l \delta_{kl} c_{lj}^0 \qquad (2.2)$$

where c_{lj}^0 are the coefficients in the expansion

$$|\psi_j^0\rangle = \sum_l c_{lj}^0 |\phi_l^0\rangle \qquad (2.3)$$

corresponding to (1.2). On taking $|\psi_j^0\rangle$ and $|\phi_l^0\rangle$ to be orthonormal, it turns out that

$$\sum_m c_{km}^{0\dagger} c_{ml}^0 = \delta_{kl}, \qquad (2.4)$$

i.e., c_{km}^0 are the elements of a *unitary* matrix.[64] With the aid of (2.4), Eq. (2.2) becomes

$$\sum_l H_{kl}^0 c_{lj}^0 = E_j^0 \sum_m c_{km}^0 \delta_{mj}$$

or, in matrix notation,

$$\mathbf{H}^0 \mathbf{c}^0 = \mathbf{c}^0 \mathbf{E}^0 \qquad (2.5)$$

where $\mathbf{c}^0$ is a matrix whose columns are eigenvectors of $\mathbf{H}^0$, and $\mathbf{E}^0$ is a diagonal matrix whose elements E_j^0 are the corresponding eigenvalues.

For the interacting (perturbed) system, (2.5) can be written as

$$(\mathbf{H}^0 + \mathbf{V})\mathbf{c} = E\mathbf{1}\mathbf{c} \qquad (2.6)$$

which, after some rearranging, yields

$$\mathbf{c} = (E\mathbf{1} - \mathbf{H}^0)^{-1} \mathbf{V} \mathbf{c}. \qquad (2.7)$$

From (2.5) and (2.4), it can be shown that

$$(E\mathbf{1} - \mathbf{H}^0)^{-1} = \mathbf{c}^0 (E\mathbf{1} - \mathbf{E}^0)^{-1} \mathbf{c}^{0\dagger} \qquad (2.8)$$

[64] G. Arfken, "Mathematical Methods for Physicists," 2nd ed. Academic Press, New York, 1970.

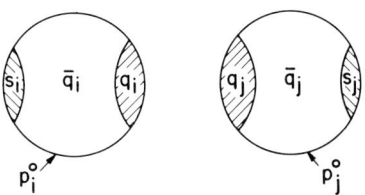

FIG. 6. Diagrammatic representation of the ith and jth subspaces of interacting and noninteracting systems.

which in (2.7) gives[65]

$$\mathbf{c} = \mathbf{GVc} \qquad (2.9)$$

where

$$\mathbf{G} = \mathbf{c}^0(E\mathbf{1} - \mathbf{E}^0)^{-1}\mathbf{c}^{0\dagger} \qquad (2.10)$$

is called the *resolvent* (or *Greenian*) *matrix*. Assuming that only the elements $V_1, \ldots, V_M$ of the diagonal matrix $\mathbf{V}$ are nonzero, then

$$\begin{vmatrix} V_1G_{11} - 1 & V_2G_{12} & \cdots & V_MG_{1M} \\ V_1G_{21} & V_2G_{22} - 1 & \cdots & V_MG_{2M} \\ \vdots & & & \vdots \\ V_1G_{M1} & V_2G_{M2} & \cdots & V_MG_{MM} - 1 \end{vmatrix} = 0 \qquad (2.11)$$

if (2.9) is to have nontrivial solutions. Solving (2.11), together with (2.10), enables the perturbed solutions to be found in terms of the known unperturbed ones.

The above treatment, which was originally developed by several workers,[66-68] has been generalized by Koutecký,[2] who made use of the projection operator formalism.[69] Here an association is made between a projection operator T and the space t into which it projects. Since the operator T of the system is made up of the operators T_i of the subsystems, it follows that

$$T = \cup T_i \qquad (2.12)$$

where $\cup$ stands for "union." Diagrammatically, the noninteracting and interacting systems can be represented as shown in Fig. 6. For the noninteracting system, the eigenfunctions of the Hamiltonian operator H^0 form the basis of the function space p^0, whose subspaces are p_i^0. In operator

[65] S. G. Davison and A. T. Amos, *J. Chem. Phys.* **43**, 2223 (1965).
[66] I. M. Lifshitz, *Zh. Eksperim. i Teor. Fiz.* **17**, 1017, 1076 (1947).
[67] G. R. Baldock, *Proc. Cambridge Phil. Soc.* **48**, 457 (1952).
[68] G. F. Koster and J. C. Slater, *Phys. Rev.* **95**, 1167 (1954).
[69] P.-O. Löwdin, *J. Math. Phys.* **3**, 969 (1962).

terminology, suppose

$$P^0 = \bigcup P_i^0 = I, \qquad (2.13)$$

I being the *identity* operator. Thus, the projection capacity of P^0 can be said to be *complete*. Furthermore, for $i \neq j$,

$$P_i^0 P_j^0 = 0, \qquad (2.14)$$

i.e., the projection operators $P_{i,j}^0$ project into the subspaces $p_{i,j}^0$, which have zero overlap, providing $i \neq j$.

For the interacting system, the space p of the eigenfunction of H in (2.1) is given by (Fig. 6)

$$p = p^0 - s; \qquad (2.15)$$

thus, p can be regarded as a subspace of p^0. The removal of the space s from p^0 to form p is exemplified by the removal of an atom from a cyclic chain of atoms, representing an infinite crystal, to form a finite (or semi-infinite) chain of atoms. In general, the space p is the sum of two spaces, namely,

$$p = q + \bar{q} \qquad (2.16)$$

where $\bar{q}$ is the space in which *all* the matrix elements of V are zero and q is the complementary space to $\bar{q}$. Taking the above example further, the formation of a semi-infinite crystal (p) leads to the presence of a surface, which causes the crystal potential to be perturbed in *some* or *all* of the region corresponding to the space q, leaving the remainder $(\bar{q})$ of the crystal *completely* unperturbed. It is now evident that, in operator notation,

$$\bar{Q}V = V\bar{Q} = 0 \qquad (2.17)$$

while from (2.13), (2.15) and (2.16) comes

$$I = S + Q + \bar{Q}. \qquad (2.18)$$

The eigenvalue equation to be solved is

$$H | \psi \rangle = (H^0 + V) | \psi \rangle = E | \psi \rangle \qquad (2.19)$$

which can be written as [cf. (2.7) and (2.9)]

$$| \psi \rangle = GIVI | \psi \rangle \qquad (2.20)$$

where[70] [cf. (2.8) and (2.10)]

$$G = (EI - H^0)^{-1} = \sum_j | \psi_j^0 \rangle \langle \psi_j^0 | /(E - E_j^0), \qquad (2.21)$$

---

[70] A. Messiah, "Quantum Mechanics," Vol. 1, p. 260; Vol. 2, p. 712. North-Holland Publ., Amsterdam, 1962.

$| \psi_j^0 \rangle \langle \psi_j^0 |$ being the projector corresponding to the jth MO. Inserting (2.18) into (2.20) using (2.17) yields[2,32]

$$| \psi \rangle = G(QV - SH^0)Q | \psi \rangle \qquad (2.22)$$

because

$$S | \psi \rangle = 0 \qquad (2.23)$$

and

$$SH | \psi \rangle = SH^0 | \psi \rangle + SV | \psi \rangle = 0. \qquad (2.24)$$

The coefficient equation analogous to (2.9) is

$$\langle \phi_l^0 | \psi \rangle = \langle \phi_l^0 | G(QV - SH^0)Q | \psi \rangle \qquad (2.25)$$

in which

$$\begin{aligned} \langle \phi_l^0 | \psi \rangle &\neq 0, & | \phi_l^0 \rangle &\in p \\ &= 0, & | \phi_l^0 \rangle &\in s. \end{aligned} \qquad (2.26)$$

The advantage of having the term in S in (2.25) is that it enables the initial system to be larger and simpler (e.g., in its symmetry properties) than the system actually being studied. The nontrivial solutions of (2.25) require [cf. (2.11)]

$$\det\{\langle \phi_l^0 | G(QV - SH^0)Q - I | \phi_k^0 \rangle\} = 0 \qquad (2.27)$$

where the order of this determinantal equation is equal to the number l fulfilling the condition $l \in Q$. Finally, it should also be noted that a connection[32] exists between the resolvent and partition[11,71,72] techniques.

b. Crystal Quantum States via the Next-Nearest-Neighbor Approximation

(i) Cyclic Crystal Solutions. Consider a 3-D cyclic monatomic crystal in which the mutually perpendicular elementary translation vectors are $\boldsymbol{\sigma}, \boldsymbol{\tau}, \boldsymbol{\nu}$. For such a crystal, the one-electron wave function can be written in the form

$$| \psi_{\theta \eta \nu}^0(\mathbf{r}) \rangle = N^{-1/2} \sum_{l=0}^{N-1} e^{il\theta} | a_{\eta \nu}^0(\mathbf{r} - l\boldsymbol{\sigma}) \rangle \qquad (2.28)$$

---

[71] P.-O. Löwdin, *J. Chem. Phys.* **19**, 1396 (1951); *J. Mol. Spectry.* **10**, 12 (1963); **14**, 119 (1964).
[72] R. D. Levine and A. T. Amos, *Phys. Status Solidi* **19**, 587 (1967).

where

$$|a_{\eta\nu}^0(\mathbf{r} - l\boldsymbol{\sigma})\rangle = N^{-1} \sum_{m,n=0}^{N-1} e^{i(m\eta + n\nu)} |a^0(\mathbf{r} - l\boldsymbol{\sigma} - m\boldsymbol{\tau} - n\boldsymbol{\nu})\rangle \quad (2.29)$$

are 1-D Wannier functions[68,73,74] or *layer* orbitals[32] and N is the number of unit cells in the direction of an arbitrary elementary translation. The Born–von Karman cyclic boundary conditions lead to

$$\theta, \eta, \nu = 2\pi t/N, \quad 0 \leq t \leq N - 1. \quad (2.30)$$

Comparison of (2.28) with (2.3) shows that

$$c_l^0 = e^{il\theta} \quad (2.31)$$

while the layer orbitals of the 3-D case correspond to the AO of the linear one. If, in addition to $H_{n,n\pm1}$ in (1.7), the next-nearest-neighbor (NNN) matrix elements $H_{n,n\pm2} = \beta_2$ are also nonvanishing, then (1.8) is replaced by the fourth-order difference equation[2,4,68,75–80]

$$2X^0 c_l^0 = (c_{l+1}^0 + c_{l-1}^0) + \rho(c_{l+2}^0 + c_{l-2}^0) \quad (2.32)$$

where $|\rho| = |\beta_2/\beta_1| < 1$ is a measure of the strength of the NNN interaction. From (2.31) and (2.32) it follows that

$$X^0 = \cos\theta + \rho \cos 2\theta. \quad (2.33)$$

The graphs of X^0 vs θ are drawn in Fig. 7 for $0 \leq \theta \leq \pi$ and various values of ρ; $\rho = 0$ corresponding to the tight-binding approximation (TBA). When $|\rho| > \frac{1}{4}$ the curves are no longer monotonic, as in the TBA, but exhibit relative extrema. A plot of the extrema of $X^0(X_M^0)$ against ρ is shown in Fig. 8. As indicated, the band of energy levels can be divided into three zones, depending on the range of ρ. As $\rho \to 0$, zone II degenerates into the normal (TBA) band. In an *exact* solution of the Schroedinger equation, symmetry arguments require that the relative extrema disappear.

(*ii*) *Semi-Infinite Crystal Solutions.* If the N atoms in the cyclic crystal (p^0) are numbered throughout the range $0 \leq n \leq N - 1$, then in the NNN approximation a finite crystal (p) can be created by removing

[73] G. H. Wannier, *Phys. Rev.* **52**, 191 (1937).
[74] G. H. Wannier, "Elements of Solid State Theory," p. 173. Cambridge Univ. Press, London and New York, 1959.
[75] G. R. Baldock, *Proc. Phys. Soc.* **66**, 1 (1953).
[76] M. Tomášek and J. Koutecký, *Czech. J. Phys.* **B10**, 268 (1960).
[77] J. Koutecký, *Czech. J. Phys.* **B11**, 565 (1961).
[78] S. G. Davison and G. L. Derco, *Surface Sci.* **6**, 323 (1967).
[79] W. M. Fairbairn, *Surface Sci.* **9**, 439 (1968).
[80] S. G. Davison and J. Grindlay, *Surface Sci.* **11**, 99 (1968).

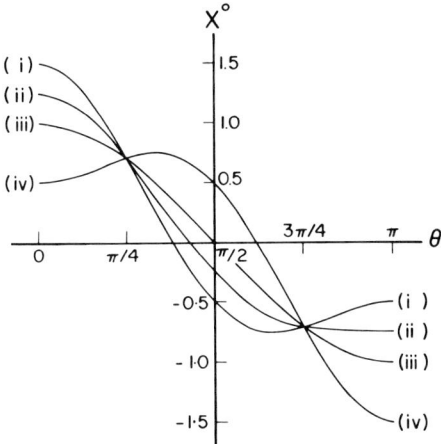

FIG. 7. Graphs of the reduced energy X^0 vs θ. (i) $\rho = 0.5$; (ii) $\rho = 0.25$; (iii) $\rho = 0$; (iv) $\rho = -0.5$.

the $(N - 1)$th and $(N - 2)$th atoms (s), hence,

$$S = \sum_{l=N-2}^{N-1} | a_l^0 \rangle \langle a_l^0 |. \qquad (2.34)$$

For a semi-infinite crystal, $N \to \infty$, and the wave function can be expanded

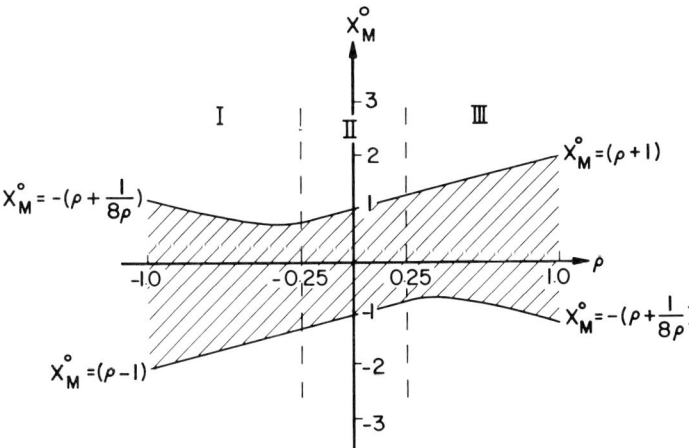

FIG. 8. Variation of the volume band edges (X_M^0) with the NNN parameter (ρ).

as an infinite series of layer orbitals, viz.,

$$|\psi_{\eta\nu}(\mathbf{r})\rangle = \sum_{l=0}^{\infty} c_{\eta\nu;l} |a_{\eta\nu}^{0}(\mathbf{r} - l\boldsymbol{\sigma})\rangle. \tag{2.35}$$

The surface formed at $l = 0$[81] gives rise to a perturbation surface potential V, which penetrates the crystal only to the depth of the first unit cell below the surface plane, whose normal vector is $\boldsymbol{\sigma}$. Thus, the perturbation integral can be written as

$$\langle a_{\eta\nu}^{0}(\mathbf{r} - l\boldsymbol{\sigma}) | V | a_{\eta'\nu'}^{0}(\mathbf{r} - l'\boldsymbol{\sigma})\rangle = A\delta_{\eta\eta'}\delta_{\nu\nu'}\delta_{l0}\delta_{l'0}. \tag{2.36}$$

Although the perturbation potential V is confined to the first atom at $l = 0$, because the approximation being used is the NNN one, the space q must contain the atoms at $l = 0$ and $l = 1$, i.e.,

$$Q = \sum_{l=0}^{1} |a_l^0\rangle\langle a_l^0|. \tag{2.37}$$

Using (2.34)–(2.37) in (2.25) yields[25,82]

$$\sum_k c_k \delta_{lk} = \sum_k c_k \{[G_{l0}(V_{00}\delta_{0k} + V_{01}\delta_{1k}) + G_{l1}(V_{10}\delta_{0k} + V_{11}\delta_{1k})]$$
$$- [G_{l,N-2}(H_{N-2,0}^0\delta_{0k} + H_{N-2,1}^0\delta_{1k})$$
$$+ G_{l,N-1}(H_{N-1,0}^0\delta_{0k} + H_{N-1,1}^0\delta_{1k})]\} \tag{2.38}$$

where

$$\delta_{lk} = \langle a_l^0 | a_k^0 \rangle \tag{2.39}$$

and the other matrix elements are given by

$$\Lambda_{lk} = \langle a_l^0 | \Lambda | a_k^0 \rangle; \quad \Lambda = G, V, H^0. \tag{2.40}$$

With the aid of (2.36), and the definitions of $\beta_{1,2}$, (2.38) reduces to

$$c_l = (AG_{l0} - \beta_1 G_{l,N-1} - \beta_2 G_{l,N-2})c_0 - \beta_2 G_{l,N-1}c_1. \tag{2.41}$$

From (2.21), (2.28), (2.30), (2.39), and (2.40) comes

$$G_{l+k} = G_{l,N-k} = N^{-1} \sum_{\theta} e^{i(l+k)\theta}[E - E^0(\theta)]^{-1} \tag{2.42}$$

which for large N leads to

$$G_l = (2\pi)^{-1} \int_0^{2\pi} e^{il\theta}[E - E^0(\theta)]^{-1} d\theta \tag{2.43}$$

via (2.30). Thus, (2.41) becomes

$$c_l = (AG_l - \beta_1 G_{l+1} - \beta_2 G_{l+2})c_0 - \beta_2 G_{l+1}c_1 \tag{2.44}$$

---

[81] With the crystal being semi-infinite the surface effect at $n = N$ is neglected.
[82] P. Feuer, *Phys. Rev.* **88**, 92 (1952).

for $l \geq 0$. On setting $l = 1$ and 0 in (2.44), the nontrivial solutions for c_0 and c_1 require that

$$\begin{vmatrix} AG_0 - \beta_1 G_1 - \beta_2 G_2 - 1 & -\beta_2 G_1 \\ AG_1 - \beta_1 G_2 - \beta_2 G_3 & -\beta_2 G_2 - 1 \end{vmatrix} = 0 \qquad (2.45)$$

which can also be found from (2.27). Equation (2.45) gives the energy E corresponding to the wave function $|\psi\rangle$ in (2.35).

(a) *Surface state energy and coefficient.* On transforming (2.43) into an integral round a unit circle in the complex plane,

$$G_l = \frac{i}{2\pi\beta_2} \oint \frac{z^{l+1}}{F(z)} dz \qquad (2.46)$$

where by (1.9)

$$F(z) = 2z^2(X^0 - X)/\rho \qquad (2.47)$$

which via (2.31) and (2.32) can be written as

$$F(z) = z^4 + \rho^{-1}(z^3 + z) - 2X\rho^{-1}z^2 + 1. \qquad (2.48)$$

The roots of

$$F(z) = 0 \qquad (2.49)$$

are

$$z_{1,2} = y_1[1 \pm (1 - y_1^{-2})^{1/2}], \qquad z_{3,4} = y_2[1 \pm (1 - y_2^{-2})^{1/2}] \qquad (2.50)$$

where

$$y_{1,2} = (4\rho)^{-1}\{-1 \pm [1 + 8\rho(X + \rho)]^{1/2}\}. \qquad (2.51)$$

Putting

$$y_{1,2} = \cos \theta_{1,2} \qquad (2.52)$$

so that

$$z_{1,2} = \exp(\pm i\theta_1), \qquad z_{3,4} = \exp(\pm i\theta_2) \qquad (2.53)$$

identifies the present treatment[76] with that of Fairbairn,[79] who used the DCO method. The results are in exact agreement.

Using the theory of residues,[64] (2.46) becomes

$$\beta_2 G_l = -\sum_{j=1}^{4} R_j \qquad (2.54)$$

where the residue

$$R_j = [(z - z_j)z^{l+1}/\prod_{j=1}^{4}(z - z_j)]_{z=z_j}. \qquad (2.55)$$

Now, if any of the z roots lie *on* the unit circle, then $X = X^0(\theta_{1,2})$ for some *real* value of $\theta_{1,2}$. However, since $X \neq X^0(\theta_{1,2})$ in (2.46), it follows that

$\theta_{1,2}$ must be *complex*, in which case $|y_{1,2}| > 1$ and $|z_{1,3}| > 1$. Hence, the residues $R_{1,3}$ lie *outside* the unit circle contour and (2.54) with (2.55) yields

$$\beta_2 G_l = (z_4 - z_2)^{-1}(1 - z_1 z_3)^{-1}[z_2{}^l(z_2 - z_1)^{-1} - z_4{}^l(z_4 - z_3)^{-1}] \quad (2.56)$$

where the property

$$z_1 z_2 = z_3 z_4 = 1 \quad (2.57)$$

has also been used.

From (2.45) results

$$\alpha = \frac{A}{\beta_2} = \frac{1 + \beta_1 G_1 + 2\beta_2 G_2 + \beta_2{}^2(G_2 - G_1 G_3)}{\beta_2 G_0 - \beta_2{}^2(G_1{}^2 - G_0 G_2)} \quad (2.58)$$

which, on inserting (2.56) and using

$$1 + \beta_1 G_1 + 2\beta_2 G_2 = (z_1{}^3 z_3{}^3 - z_2 z_4)\Xi$$
$$\beta_2{}^2(G_2{}^2 - G_1 G_3) = z_2 z_4 \Xi, \qquad \beta_2 G_0 = (1 - z_1{}^2 z_3{}^2)\Xi$$
$$\beta_2{}^2(G_2{}^2 - G_1 G_3) = z_2 z_4 \Xi \qquad (2.59)$$
$$\Xi = (1 - z_1 z_3)^{-2}(z_2 - z_1)^{-1}(z_4 - z_3)^{-1}$$

gives

$$\alpha = -z_1 z_3. \quad (2.60)$$

After some tedious manipulations, (2.50), (2.51), and (2.60) lead to

$$2X = (A^2 + \beta_2{}^2)/A\beta_1 + A\beta_1/(A - \beta_2)^2 \quad (2.61)$$

while (2.44) with (2.56) and (2.60) yields

$$e_p/e_0 = (z_4{}^{p+1} - z_2{}^{p+1})/(z_4 - z_2). \quad (2.62)$$

(b) *Surface state existence conditions*. Since the poles at $z_{1,3}$ lie outside the unit circle, it follows immediately from (2.60) that

$$|\alpha| > 1. \quad (2.63)$$

When θ_k ($k = 1, 2$) is complex, y_k can be real or complex, depending on the form of θ_k. For $\theta_k = n_k \pi + i\mu_k$, y_k is real, whereas, if $\theta_k = \zeta_k + i\mu_k$, then y_k is complex.

Equation (2.51) shows that y_k is *real* if

$$8\rho(X + \rho) > -1 \quad (2.64)$$

in which case (2.51) and (2.52) yield

$$|y_2| > |y_1| > 1, \quad (2.65)$$

so (2.50) gives

$$|z_3| > |z_4|, \qquad |z_3| > |z_1| > |z_2|. \quad (2.66)$$

From (2.51) results
$$2\rho y_k^2 + y_k - (X + \rho) = 0 \qquad (2.67)$$
which by (2.65) gives
$$|X - \rho| > 1. \qquad (2.68)$$
Writing (2.61) in the form
$$2X = \rho(1 + \alpha^2)/\alpha + \alpha/\rho(\alpha - 1)^2 \qquad (2.69)$$
and utilizing (2.68) leads to
$$|\rho|(\alpha - 1)^2 > |\alpha|. \qquad (2.70)$$

Although (2.63) is the necessary condition for the existence of surface states, it is sufficient only when y_k is *complex*, which in (2.51) requires
$$8\rho(X + \rho) < -1 \qquad (2.71)$$
so that
$$y_1 = y_2^* \qquad (2.72)$$
and
$$z_1 = z_3^*, \qquad z_2 = z_4^*, \qquad |z_1| > |z_2| \qquad (2.73)$$
in (2.50). Equations (2.69) and (2.71) yield
$$2|\rho|(1 + |\alpha|) > |\alpha|^{1/2} \qquad (2.74)$$
i.e., for $|\rho| < \frac{1}{4}$,
$$\alpha < 1 - (8\rho^2)^{-1}[1 + (1 - 16\rho^2)^{1/2}] \qquad (2.75)$$
which obviously satisfies (2.63). When $|\alpha|$ is large, (2.74) approximates to
$$4\rho^2|\alpha| > 1 \qquad (2.76)$$
which is also obtained by taking ρ small in (2.75). The sum of the y roots in (2.51) gives
$$2\rho(y_1 + y_2) = -1. \qquad (2.77)$$
However, (2.50) yield
$$2y_1 = z_1 + z_1^{-1}, \qquad 2y_2 = z_3 + z_3^{-1}, \qquad (2.78)$$
so (2.77) becomes
$$z_1 + z_3 = \alpha(\rho^{-1} + z_1 + z_3) \qquad (2.79)$$
via (2.60). Moreover, on setting $\alpha < -1$ in (2.60),
$$(z_1 + z_3)^2 > 4z_1 z_3 > 4, \qquad (2.80)$$
so (2.79) gives
$$|\rho| > \tfrac{1}{4}. \qquad (2.81)$$

(c) *Discussion.* Having obtained the existence conditions, the various

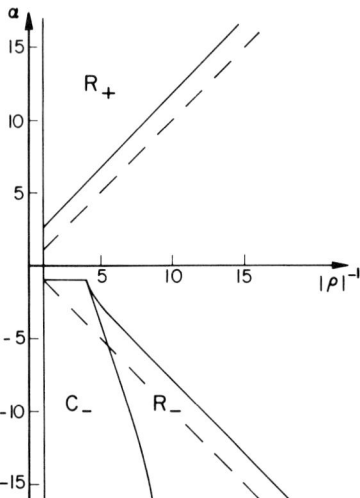

FIG. 9. Mapping of existence conditions in $\alpha \mid \rho \mid^{-1}$ plane. The region R_+ (R_-) corresponds to y_k real and $\alpha > 0$ ($\alpha < 0$), while C_- denotes y_k complex and $\alpha < 0$. The broken lines are for the TBA. Note $\mid \rho \mid^{-1} > 1$ because $\mid \beta_2 \mid > \mid \beta_1 \mid$.

regions satisfying them are mapped in the $\alpha \mid \rho \mid^{-1}$ plane (Fig. 9). In the TBA, the single existence condition for Tamm surface states is[1,2,14]

$$\mid A/\beta_1 \mid = \lim_{\beta_2 \to 0} \mid \alpha\rho \mid > 1 \qquad (2.82)$$

which corresponds to (2.63) and is represented by the broken lines in Fig. 9. As can be seen from the relative positions of the solid and broken lines, for $\alpha > 0$, $\alpha_{\rm NNN} > \alpha_{\rm TBA}$ and, for $\alpha < 0$, $\alpha_{\rm NNN} < \alpha_{\rm TBA}$. Thus, the NNN interaction impedes (facilitates) the occurrence of surface states when $\alpha > 0$ ($\alpha < 0$). Note also that, for $\mid \rho \mid$ small, (2.70) approximates to

$$\mid \rho(\alpha - 2) \mid > 1. \qquad (2.83)$$

For y_k real (i.e., regions $R_\pm$ of Fig. 9), (2.62) can be rewritten as

$$(e_p/e_r)^2 = z_4^{2(p-r)}(1 - \omega^{p+1})^2/(1 - \omega^{r+1})^2 \qquad (2.84)$$

where

$$\omega = z_2/z_4. \qquad (2.85)$$

From (2.57) and (2.66) it follows that $\mid z_4 \mid < 1$, hence, e_p decreases with increasing distance from the surface and the electron is *localized* on the crystal surface. If y_k is complex (i.e., region C_- of Fig. 9), then by (2.53), (2.73), and (2.85)

$$\omega = e^{2i\theta}, \qquad (2.86)$$

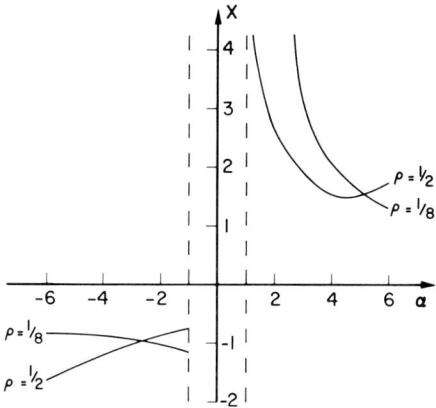

Fig. 10. Variation of X with α for $\rho = \frac{1}{8}$ and $\frac{1}{2}$.

so (2.84) becomes

$$| e_p/e_r |^2 = | z_4 |^{2(p-r)} | \sin(p + 1)\theta/\sin(r + 1)\theta |^2. \qquad (2.87)$$

The presence of the trigonometrical factor means that the coefficients now oscillate as they decrease from the surface.

Returning to the surface energy expression (2.69), the graphs of X vs α, which are subject to (2.63), are drawn in Fig. 10 for $\rho = \frac{1}{8}$ and $\frac{1}{2}$. The curves corresponding to $\rho = -\frac{1}{8}$ and $-\frac{1}{2}$ are obtained by reflecting those of Fig. 10 in the line $X = 0$. The graphs of X vs ρ are displayed in Fig. 11 and the various regions satisfying the inequalities (2.64), (2.68), and (2.71)

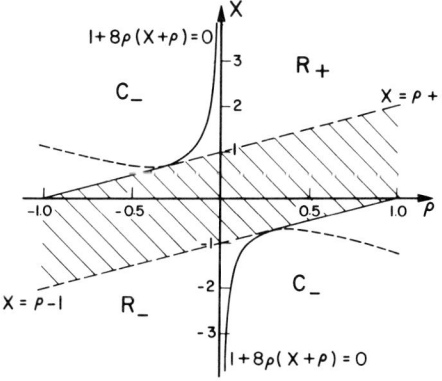

Fig. 11. Complete energy spectrum. Volume (surface) state energies lie inside (outside) broken lines. Pseudostate energies lie in the shaded region.

are labeled using the notation of Fig. 9. The broken lines represent the volume band edges between which all the above inequalities are violated.

In addition to the *multiplicative* combination of the oscillatory and damping terms (2.87), for y_k real, but θ_1 real, $\theta_2 = n_2\pi + i\mu_2$ ($\rho > 0$) or θ_2 real, $\theta_1 = n_1\pi + i\mu_1$ ($\rho < 0$), Fairbairn[79] obtained an *additive* combination of the oscillatory and damping terms in the coefficient expression. For classification purposes, if the damping term is larger (smaller) than the oscillatory one, then these states may be referred to as *pseudo*surface (volume) states.[83] In violation of (2.68), the energy of these pseudostates, which arise as a result of including the NNN interaction, lies in the range $\rho - 1 < X < \rho + 1$, i.e., *embedded*[78] in the volume band as indicated by the shaded region in Fig. 11. Pseudostates can also be called scattering resonances (Section 8).

In the TBA, $\beta_2 \to 0$ and (2.61) reduces to

$$2X_0 = A/\beta_1 + \beta_1/A. \qquad (2.88)$$

Hence,

$$\Delta X = 2(X - X_0) = \rho[2\sigma^2 + \alpha^{-1}(1 + 3\sigma^2)] \qquad (2.89)$$

where

$$\sigma = (\alpha\rho)^{-1}. \qquad (2.90)$$

From (2.89) it follows that

$$|\Delta X|_{\alpha>0} > |\Delta X|_{\alpha<0}. \qquad (2.91)$$

c. Other NNN Work

Baldock,[75] in his DCO calculation of the surface energy of a 2-D monovalent square metallic lattice, discussed the improvements achieved by including the NNN interaction. In the following year, Koster and Slater[68] studied the impurity levels in linear crystals via the TB and NNN approximations. In extending the work to 3-D, a Green function technique was employed and higher-order interactions[83a,b] were taken into account. Koutecký,[77] in his RCO treatment of Shockley[17] surface states of a 1-D chain of equal atoms, joined by bonds of alternating strengths, included the resonance integrals between NN and NNN atoms and showed that surface states occurred if the stronger bond is broken by the surface. Koutecký also discussed the connection between Tamm[14] and Shockley[17] surface states.

Instead of confining the surface perturbation to the first unit cell, Fairbairn[79] allowed it to penetrate to the second unit cell. In this way he

---

[83] M. Fukushima, *Sci. Rept.* **16**, 1 (1967).
[83a] I. Alstrup and K. Johansen, *Phys. Status Solidi* **28**, 555 (1968).
[83b] S. G. Davison and N. F. Taylor, *Chem. Phys. Letters* **3**, 424 (1969).

was able to generalize the earlier work[76,78] and obtain an extra surface state. Furthermore, such a model can be used to examine the effect of the NNN interaction on the electronic process of chemisorbing a foreign atom onto a crystal surface.[84] Fairbairn also pointed out that the attempt by Davison and Derco[78] to simplify their NNN calculation, by neglecting one of the boundary conditions, led to some erroneous results. A detailed discussion of the role of boundary conditions in the TB and NNN approximations, and their effects on the nature of Tamm surface states in semi-infinite and finite linear crystals, was given later by Davison and Grindlay.[80]

IV. Crystal Potential Methods

Suppose that for a bounded crystal, the potential $V(\mathbf{r})$ is known as a function over all space. The solution of the Schroedinger equation then yields all the states of the system, some of which will be surface states. In principle, given $V(\mathbf{r})$, the problem can be solved exactly and all energies obtained. In practice, however, it is convenient to simplify $V(\mathbf{r})$ in the vicinity of the surface. A frequently used technique is to include step discontinuities, which separate one region from another. The limiting use of a step is, of course, an infinitely high wall, which imposes a nodal condition on the wave function.

Kronig and Penney (KP)[13], in their study of the quantum mechanics of electrons in crystal lattices, represented the crystal potential field by a linear array of rectangular wells, which they later transformed into a chain of δ wells subject to the area of each well remaining constant. Saxon and Hutner[85] adopted the δ-well form of the KP potential to formulate the scattering matrix (SM) and Green function methods for investigating the electronic properties of various 1-D crystal models. The crystal field can also be described approximately by a sinusoidal potential variation.[86-89] In this case, the Schroedinger equation is equivalent to the Mathieu equation,[90] whose analytic solution is well known.[91,92]

[84] Cf. Fairbairn's[79] Fig. 1 with Grimley's[1] Fig. 1.
[85] D. S. Saxon and R. A. Hutner, *Philips Res. Rept.* **4**, 81 (1949).
[86] L. Brillouin, *J. Phys. Radium.* **1**, 377 (1930).
[87] P. M. Morse, *Phys. Rev.* **35**, 1310 (1930).
[88] J. C. Slater, *Phys. Rev.* **87**, 807 (1952).
[89] L. Brillouin, "Wave Propagation in Periodic Structures," Chapter 8. Dover, New York, 1953.
[90] É. Mathieu, *J. Math. Pures Appl.* **13**, 137 (1868).
[91] E. T. Whittaker and G. N. Watson, "Modern Analysis," Chapter 19. Cambridge Univ. Press, London and New York, 1958.
[92] N. W. McLachlan, "Theory and Application of Mathieu Functions." Dover, New York, 1964. For surface states, see especially pp. 99–102.

3. Kronig–Penney Model

a. Relativistic Treatment

In order to generalize the classical work of KP[13] and Tamm,[14] a relativistic study is made of the volume and surface states of a monatomic crystal. After describing the Dirac formulation, the SM method[38,85] is used to derive the energy dispersion relation, while the bulk wave function is obtained by solving the Dirac equation directly using Bloch's theorem. The surface problem is then tackled via Tamm's procedure.

Relativistic corrections to energy band calculations are significant for heavy atomic crystals. Up to now, these relativistic calculations[93–98] have involved the solution of the Dirac equation for the so-called "muffin-tin" potential, and have utilized the Foldy–Wouthuysen[99] canonical transformation to decouple the positive and negative energy states and obtain the various relativistic correction terms to $O(\alpha^2)$ in the Hamiltonian, α ($= e^2/\hbar c$) being the fine-structure constant. Moreover, the one-electron crystal wave function is expressed as a linear combination of augmented (or orthogonal) plane waves, the expansion coefficients being determined variationally.

For crystal surface problems, the solving of the Dirac equation for the muffin-tin potential is extremely difficult because of the complexity arising from the boundary conditions. Thus, to make the problem more tractable, it is necessary to resort to a simpler type of potential, such as the KP one chosen here.[100,101]

(*i*) *Basic Theory.* (a) *Dirac formulation.* For an electron of rest mass m_0 moving in a linear potential $V(x)$, the Dirac equation is[102]

$$(i\hbar \, \partial/\partial t - i\hbar c \alpha_x \nabla_x + \beta m_0 c^2 - V)\psi = 0, \tag{3.1}$$

c being the velocity of light, ψ the four-component spinor wave function in

---

[93] L. E. Johnson, J. B. Conklin, Jr., and G. W. Pratt, Jr., *Phys. Rev. Letters* **11**, 538 (1963).
[94] F. Herman, C. D. Kuglin, D. F. Cuff, and R. L. Kortum, *Phys. Rev. Letters* **11**, 541 (1963).
[95] J. B. Conklin, Jr., L. E. Johnson, and G. W. Pratt, Jr., *Phys. Rev.* **137**, A1282 (1965).
[96] P. Soven, *Phys. Rev.* **137**, A1706 (1965).
[97] P. C. Chow and L. Liu, *Phys. Rev.* **140**, A1817 (1965).
[98] T. L. Loucks, *Phys. Rev. Letters* **14**, 693, 1072 (1965); *Phys. Rev.* **139**, A1333 (1965); **143**, 506 (1966). See also "Augmented Plane Wave Method." Benjamin, New York, 1967, for a more extensive survey.
[99] L. L. Foldy and S. A. Wouthuysen, *Phys. Rev.* **78**, 29 (1950).
[100] R. D. Wood and J. Callaway, *Bull. Am. Phys. Soc.* **2**, 18 (1957).
[101] M. L. Glasser and S. G. Davison, *Intern. J. Quantum Chem.* **3S**, 867 (1969).
[102] L. I. Schiff, "Quantum Mechanics." McGraw-Hill, New York, 1955.

(x, t), namely,

$$\psi(x, t) = \begin{bmatrix} \psi_{(1)}(x, t) \\ \psi_{(2)}(x, t) \\ \psi_{(3)}(x, t) \\ \psi_{(4)}(x, t) \end{bmatrix}, \qquad (3.2)$$

and

$$\alpha_x = \begin{bmatrix} 0 & 0 & 0 & 1 \\ 0 & 0 & 1 & 0 \\ 0 & 1 & 0 & 0 \\ 1 & 0 & 0 & 0 \end{bmatrix}, \quad \beta = \begin{bmatrix} 1 & 0 & 0 & 0 \\ 0 & 1 & 0 & 0 \\ 0 & 0 & -1 & 0 \\ 0 & 0 & 0 & -1 \end{bmatrix}. \qquad (3.3)$$

Substituting (3.2) and (3.3) in (3.1) leads to the four simultaneous equations

$$i\hbar \dot{\psi}_{(1)} - i\hbar c \psi_{(4)}' + m_0 c^2 \psi_{(1)} - V\psi_{(1)} = 0 \qquad (3.4a)$$

$$i\hbar \dot{\psi}_{(2)} - i\hbar c \psi_{(3)}' + m_0 c^2 \psi_{(2)} - V\psi_{(2)} = 0 \qquad (3.4b)$$

$$i\hbar \dot{\psi}_{(3)} - i\hbar c \psi_{(2)}' - m_0 c^2 \psi_{(3)} - V\psi_{(3)} = 0 \qquad (3.4c)$$

$$i\hbar \dot{\psi}_{(4)} - i\hbar c \psi_{(1)}' - m_0 c^2 \psi_{(4)} - V\psi_{(4)} = 0. \qquad (3.4d)$$

Inspection of these equations shows that (3.4a) and (3.4d) and (3.4b) and (3.4c) are coupled and can be written in the matrix form

$$i\hbar \dot{\psi} - i\hbar c \sigma_x \psi' + m_0 c^2 \sigma_z \psi - V\psi = 0 \qquad (3.5)$$

where $\sigma_{x,z}$ are the x and z component Pauli spin matrices

$$\sigma_x = \begin{bmatrix} 0 & 1 \\ 1 & 0 \end{bmatrix}, \quad \sigma_z = \begin{bmatrix} 1 & 0 \\ 0 & -1 \end{bmatrix}, \qquad (3.6)$$

and

$$\psi = \begin{bmatrix} \psi_{(1,2)} \\ \psi_{(4,3)} \end{bmatrix} \qquad (3.7)$$

is now a two-component spinor in (x, t). With $V(x)$ being time independent, the space and time coordinates in (3.5) can be separated by writing

$$\psi(x, t) = \phi(x) e^{-iEt/\hbar}. \qquad (3.8)$$

Equations (3.5) and (3.8) give the 1-D time-independent Dirac equation
$$i\hbar c \sigma_x \phi' - m_0 c^2 \sigma_z \phi = (E - V)\phi \qquad (3.9)$$
which on solving yields the stationary state solutions for the linear potential $V(x)$.

(b) *Constant potential solution.* On putting
$$\phi = \begin{bmatrix} \phi_{(1)} \\ \phi_{(2)} \end{bmatrix} \qquad (3.10)$$
and using (3.6), Eq. (3.9) can be expressed as
$$i\hbar c \phi_{(2)}' = [(\epsilon - V) + 2m_0 c^2]\phi_{(1)} \qquad (3.11)$$
$$i\hbar c \phi_{(1)}' = (\epsilon - V)\phi_{(2)} \qquad (3.12)$$
in which
$$\epsilon = E - m_0 c^2. \qquad (3.13)$$
Decoupling (3.11) and (3.12) by differentiation gives
$$\phi_{(j)}'' = -\kappa_v^2 \phi_{(j)}, \qquad j = 1 \text{ and } 2, \qquad (3.14)$$
where
$$\kappa_v^2 = (\epsilon - V)[(\epsilon - V) + 2m_0 c^2]/\hbar^2 c^2. \qquad (3.15)$$
Solving (3.14) for the constant potential V yields the plane wave solutions
$$\phi_{(j)} = \alpha_1^{(j)} \exp(i\kappa_v x) + \alpha_2^{(j)} \exp(-i\kappa_v x), \qquad (3.16)$$
so (3.10) becomes
$$\phi = A_1 \exp(i\kappa_v x) + A_2 \exp(-i\kappa_v x) \qquad (3.17)$$
where
$$A_t = \begin{bmatrix} \alpha_t^{(1)} \\ \alpha_t^{(2)} \end{bmatrix}, \qquad t = 1 \text{ and } 2. \qquad (3.18)$$
From (3.12) and (3.16) it follows that
$$\alpha_t^{(1)} = \mp \gamma \alpha_t^{(2)} \qquad (3.19)$$
where
$$\gamma = (\epsilon - V)/\hbar c \kappa_v, \qquad (3.20)$$
the minus and plus signs corresponding to $t = 1$ and 2, respectively. Hence, (3.17)–(3.19) yield
$$\phi = \begin{bmatrix} -\gamma \\ 1 \end{bmatrix} \alpha_1^{(2)} \exp(i\kappa_v x) + \begin{bmatrix} \gamma \\ 1 \end{bmatrix} \alpha_2^{(2)} \exp(-i\kappa_v x). \qquad (3.21)$$

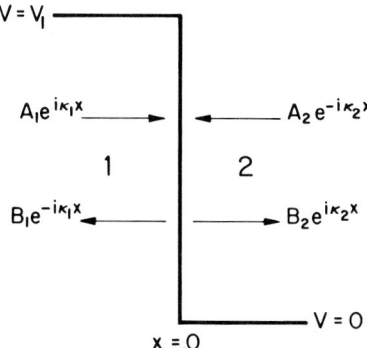

FIG. 12. Step potential scattering process.

(c) *Step potential solution.* The interaction of a free electron with a "step" potential at $x = 0$ can be treated as a scattering problem.[38,85] The incoming and outgoing waves (Fig. 12)[103] are related via the SM equation

$$B = SA \tag{3.22}$$

in which

$$A = \begin{bmatrix} A_1 \\ A_2 \end{bmatrix}, \quad B = \begin{bmatrix} B_1 \\ B_2 \end{bmatrix} \tag{3.23}$$

with

$$A_k = \begin{bmatrix} \alpha_k^{(1)} \\ \alpha_k^{(2)} \end{bmatrix}, \quad B_k = \begin{bmatrix} \beta_k^{(1)} \\ \beta_k^{(2)} \end{bmatrix}, \tag{3.24}$$

and

$$S = \begin{bmatrix} S_{11} & S_{12} \\ S_{21} & S_{22} \end{bmatrix}. \tag{3.25}$$

The waves on either side of the scattering potential are connected by the R matrix in

$$\begin{bmatrix} B_2 \\ A_2 \end{bmatrix} = R \begin{bmatrix} A_1 \\ B_1 \end{bmatrix} \tag{3.26}$$

where

$$R = \begin{bmatrix} R_{11} & R_{12} \\ R_{21} & R_{22} \end{bmatrix}. \tag{3.27}$$

[103] Contrary to Glasser and Davison,[101] a positive step potential is used here.

Using (3.22) and (3.26) the S and R matrices can be written in terms of each other's elements to give

$$S = R_{22}^{-1} \begin{bmatrix} -R_{21} & 1 \\ R_{11}R_{12} - R_{12}R_{21} & R_{12} \end{bmatrix} \qquad (3.28)$$

$$R = S_{12}^{-1} \begin{bmatrix} S_{12}S_{21} - S_{11}S_{22} & S_{22} \\ -S_{11} & 1 \end{bmatrix}. \qquad (3.29)$$

Thus, S and R are infinite when $R_{22} = 0$ and $S_{12} = 0$, respectively, (Section 3,b,iii).

In regions 1 and 2 of Fig. 12, the time-independent spinors can be written as

$$\phi_1 = A_1 \exp(i\kappa_1 x) + B_1 \exp(-i\kappa_1 x) \qquad (3.30)$$

$$\phi_2 = B_2 \exp(i\kappa_2 x) + A_2 \exp(-i\kappa_2 x) \qquad (3.31)$$

where[103]

$$\kappa_1^2 = (\epsilon - V_1)[(\epsilon - V_1) + 2m_0 c^2]/\hbar^2 c^2 \qquad (3.32)$$

$$\kappa_2^2 = \epsilon(\epsilon + 2m_0 c^2)/\hbar^2 c^2. \qquad (3.33)$$

Matching the two-component spinors across the potential discontinuity at $x = 0$ leads to the continuity condition

$$\phi_1(0) = \phi_2(0). \qquad (3.34)$$

The choice of the continuity condition and its effect on the subsequent analysis of relativistic surface and impurity problems has recently been discussed in detail.[103a,b,c] With the aid of (3.19) and (3.24), Eqs. (3.30) and (3.31) in (3.34) give

$$\Gamma(\beta_1^{(2)} - \alpha_1^{(2)}) = \alpha_2^{(2)} - \beta_2^{(2)} \qquad (3.35)$$

$$\alpha_1^{(2)} + \beta_1^{(2)} = \alpha_2^{(2)} + \beta_2^{(2)} \qquad (3.36)$$

where

$$\Gamma = \gamma_1/\gamma_2, \qquad \gamma_k = (\epsilon - V)/\hbar c \kappa_k. \qquad (3.37)$$

From (3.35) and (3.36) result

$$\begin{aligned} \beta_1^{(2)} &= [2\alpha_2^{(2)} - (1 - \Gamma)\alpha_1^{(2)}]/(1 + \Gamma) \\ \beta_2^{(2)} &= [2\Gamma\alpha_1^{(2)} + (1 - \Gamma)\alpha_2^{(2)}]/(1 + \Gamma) \end{aligned} \qquad (3.38)$$

[103a] M. Steślicka, S. G. Davison, and A. G. Brown, *Proc. NBS Mater. Res. Symp., 3rd* (in press) (1969).
[103b] M. Steślicka and S. G. Davison, *Phys. Rev.* **B1**, 1858 (1970).
[103c] S. G. Davison and M. Steślicka, *Intern. J. Quantum Chem.* **4S** (in press) (1970).

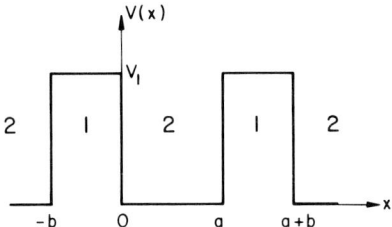

Fig. 13. Kronig–Penney potential.

which, when compared with the corresponding elemental equations obtained from (3.22), show that

$$S_{11} = -(1 - \Gamma)/(1 + \Gamma), \qquad S_{12} = 2/(1 + \Gamma)$$
$$S_{21} = \Gamma S_{12}, \qquad S_{22} = -S_{11}. \tag{3.39}$$

Inserting (3.39) in (3.29) yields

$$R_0 = \begin{bmatrix} \nu_+ & \nu_- \\ \nu_- & \nu_+ \end{bmatrix}, \qquad \nu_\pm = (1 \pm \Gamma)/2, \tag{3.40}$$

for the R matrix at $x = 0$.

(ii) *Bulk Properties.* (a) *Energy relation.* In the KP model,[13] the crystal potential field is represented by a linear array of rectangular wells (Fig. 13). Traveling along the x axis, the translation of the spinor through a distance d, without crossing a discontinuity, is achieved by utilizing the transfer matrix[85]

$$T_d = \begin{bmatrix} e^{i\kappa d} & 0 \\ 0 & e^{-i\kappa d} \end{bmatrix}. \tag{3.41}$$

Thus, the translation by a period $(a + b)$ is accomplished by multiplying the spinor by the matrix

$$T = T_b R_a T_a R_0 \tag{3.42}$$

On taking account of the 1 and 2 regions and setting

$$\chi_1 = \kappa_1 b, \qquad \chi_2 = \kappa_2 a, \tag{3.43}$$

Eqs. (3.40) and (3.41) in (3.42) yield the matrix elements

$$T_{11} = \exp(i\chi_1)[2\Gamma \cos \chi_2 + i(1 + \Gamma^2) \sin \chi_2]/2\Gamma$$
$$T_{12} = i(1 - \Gamma^2) \exp(i\chi_1) \sin \chi_2/2\Gamma \tag{3.44}$$
$$T_{21} = T_{12}{}^*, \qquad T_{22} = T_{11}{}^*,$$

the asterisk denoting the complex conjugate. The T matrix is unimodular, i.e.,

$$\det T = 1 \tag{3.45}$$

and, for the Bloch–Floquet theorem[36,38,85,104] to hold,

$$\cos \theta a = \tfrac{1}{2} \operatorname{tr} T, \tag{3.46}$$

θ being the wave number. From (3.44) and (3.46) comes

$$\cos \theta a = \cos \chi_1 \cos \chi_2 - \Xi \sin \chi_1 \sin \chi_2 \tag{3.47}$$

where

$$\Xi = (\Gamma + \Gamma^{-1})/2. \tag{3.48}$$

If the KP potential is transformed into a linear array of δ potentials in such a way that

$$\lim_{\substack{V_1 \to \infty \\ b \to 0}} V_1 b = p, \tag{3.49}$$

then the corresponding limiting values of χ_1 and γ_1 are

$$\chi_1 = -p/\hbar c, \qquad \gamma_1 = 1. \tag{3.50}$$

Substituting (3.37), (3.43), (3.48), and (3.50) in (3.47), and assuming $p \ll \hbar c$, gives the *relativistic KP transcendental relation* for $\kappa_2 a$, viz.,

$$\cos \theta a = \cos \kappa_2 a + (mp/\hbar^2 \kappa_2) \sin \kappa_2 a, \tag{3.51}$$

$$m = E/c^2 = m_0(1 + v^2/2c^2) \tag{3.52}$$

being the first-order relativistic (1-R) electron mass. In the nonrelativistic (NR) limit, as $c \to \infty$, (3.33) and (3.52) yield

$$\kappa_2 = \kappa = (2m_0 \epsilon)^{1/2}/\hbar, \qquad m = m_0, \tag{3.53}$$

so (3.51) reduces to the usual KP relation

$$f(\kappa a) = \cos \theta a = \cos \kappa a + (m_0 p/\hbar^2 \kappa) \sin \kappa a. \tag{3.54}$$

Hence, comparing (3.51) and (3.54) shows that the Dirac relativistic formulation leads to a KP relation of exactly the same form as that in the NR case, the relativistic contribution appearing in the propagation constant and the electron mass.

The graph of $f(\kappa a)$ against κa is drawn in Fig. 14 for $P = m_0 a p/\hbar^2 = 3\pi/2$. Since $|\cos \theta a| \leq 1$ and the curve extrema lie beyond ± 1, it follows that the various intersections (obtained by varying θ) of the straight lines and the curve, when projected onto the κa axis, lie in the allowed bands

---

[104] E. Merzbacher, "Quantum Mechanics," p. 99. Wiley, New York, 1964.

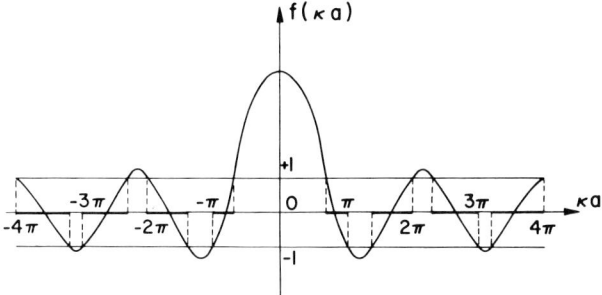

FIG. 14. Graph of $f(\kappa a)$ vs κa for $P = 3\pi/2$. Heavy lines indicate allowed energy bands.

(heavy lines), which are separated by FEG. With P being positive, the upper-band edges are given by[105]

$$\kappa a = n\pi; \quad n = 0, \pm 1, \ldots. \tag{3.55}$$

If $P < 0$, (3.55) locates the lower-band edges.

Equations (3.33) and (3.53) give

$$\kappa_2 = \kappa(1 + \eta^2\chi^2/2) \tag{3.56}$$

where

$$\eta = \hbar/2m_0ac, \quad \chi = \kappa a. \tag{3.57}$$

Similarly, (3.13), (3.52), and (3.53) yield

$$m = m_0(1 + 2\eta^2\chi^2). \tag{3.58}$$

Introducing atomic units,

$$m_0 = \tfrac{1}{2}, \quad \hbar = 1, \quad e^2 = 2, \tag{3.59}$$

with lengths in Bohr radii $a_0 = 0.529$ Å and energies in rydbergs (1 Ryd = 13.6 eV), transforms (3.57) and (3.51) into

$$\eta = (ac)^{-1}, \quad \eta\chi = \kappa/c \tag{3.60}$$

$$g(\chi) = \cos\theta a = \cos\chi(1 + \eta^2\chi^2/2)$$
$$+ \frac{pa(1 + 2\eta^2\chi^2)}{2(1 + \eta^2\chi^2/2)} \sin\chi(1 + \eta^2\chi^2/2). \tag{3.61}$$

By analogy with (3.55), the upper-band edge χ values in the 1-R formulation are found from

$$\kappa_2 a = n\pi \tag{3.62}$$

[105] H. Jones, "The Theory of Brillouin Zones and Electronic States in Crystals." North-Holland Publ., Amsterdam, 1962.

TABLE I. ENERGY SPECTRUM RESULTS FOR $a = 6a_0 > 3$ Å, $pa = 5 \times 10^{-4}, n = 6 \times 10^{-4}$ [a]

n	Band center		Band width		FEG width	
	χ	χ_2	$\Delta\chi$	$\Delta\chi_2$	$\Delta\chi$	$\Delta\chi_2$
1	1.60614	1.60614	3.07090	3.07089	0.07070	0.07070
2	4.71318	4.71316	3.14000	3.13996	0.00159	0.00159
3	7.85438	7.85428	3.14080	3.14069	0.00080	0.00080
4	10.99584	10.99559	3.14106	3.14086	0.00053	0.00053
5	14.13737	14.13684	3.14119	3.14085	0.00040	0.00040
6	17.27892	17.27797	3.14127	3.14077	0.00032	0.00032
7	20.42048	20.41893	3.14133	3.14062	0.00027	0.00027
8	23.56206	23.55967	3.14136	3.14042	0.00023	0.00023
9	26.70364	26.70018	3.14139	3.14018	0.00020	0.00020
10	29.84522	29.84040	3.14142	3.13991	0.00018	0.00018
15	45.55315	45.53609	3.14148	3.13796	0.00011	0.00011
20	61.26110	61.21972	3.14151	3.13516	0.00008	0.00008
25	76.96905	76.88714	3.14153	3.13153	0.00007	0.00007
30	92.67701	92.53427	3.14154	3.12708	0.00006	0.00006

[a] With $V(x)$ being positive, a FEG appears first at $n = 1$.

which by (3.56) leads to the cubic equation

$$\eta^2 \chi^3 + 2\chi - 2n\pi = 0 \tag{3.63}$$

whose solution is[106]

$$\chi = \eta^{-1}[(r + s)^{1/3} + (r - s)^{1/3}] \tag{3.64}$$

where

$$r^2 = s^2 + 8/27, \qquad s = \eta n\pi. \tag{3.65}$$

The positions of the lower-band edges are established by using Newton's approximate method[106] to locate the roots of

$$g(\chi) + (-1)^n = 0. \tag{3.66}$$

As a particular example, the lattice spacing is taken as $a = 6a_0 \simeq 3$ Å. In view of (3.59) and the fine-structure constant

$$\alpha = e^2/\hbar c = 1/137,$$

it follows that $c = 274$. Thus, (3.60) gives $\eta = 6 \times 10^{-4}$. With $pa \ll c$, p is chosen so that

$$pa = 5 \times 10^{-3}.$$

For these parametric values, Table I shows the variation of the band

[106] H. W. Turnbull, "Theory of Equations." Oliver and Boyd, Edinburgh, 1957.

centers, band widths, and FEG widths with the band number n. Since the band widths are close to their maximum value of π, the FEG are extremely small. In fact, for $n \geq 5$, the FEG are so small that the relativistic reduction of the band is greater than the actual width of the FEG.[97] At low (high) energies the NR band width increases (tends to remain constant), whereas in the 1-R case it increases up to a maximum at $n = 4$ and then decreases with increasing n. In contrast, the FEG widths show no relativistic effect and, after an initial decrease, tend to stay constant at high energies. Thus, for n large, the relativistic effect causes the whole spectrum to *shrink*,[98] the shrinkage absorbing the decrease in the band widths, while keeping the FEG widths constant. For large pa (~ 4), the 1-R FEG are wider than the NR ones, particularly at high energies. On varying $\eta(pa)$, keeping $pa(\eta)$ constant, it is found that as $\eta(pa)$ decreases (increases) the maximum 1-R band width occurs at higher values of n. Although the simple KP potential has been used here, these findings are in qualitative agreement with experiment and relativistic augmented plane wave calculations.[98]

(b) *Spinor expression.* According to Bloch's theorem[36-39]

$$\phi_k(x) = e^{i\theta x} u_k(x) \tag{3.67}$$

where $u_k(x)$ is a periodic spinor function, whose period is that of the lattice, and k denotes the potential region. Inserting (3.67) in (3.14) and solving the resulting second-order differential equation gives

$$u_k(x) = A_k \exp[i(\kappa_k - \theta)x] + B_k \exp[-i(\kappa_k + \theta)x], \tag{3.68}$$

A_k and B_k being defined in (3.24). Subjecting $u_k(x)$ to the continuity conditions

$$u_1(0) = u_2(0), \quad u_1(-b) = u_2(a) \tag{3.69}$$

leads to

$$A_1 + B_1 = A_2 + B_2 \tag{3.70}$$

$$A_1 \exp[-i(\kappa_1 - \theta)b] + B_1 \exp[i(\kappa_1 + \theta)b]$$
$$= A_2 \exp[i(\kappa_2 - \theta)a] + B_2 \exp[-i(\kappa_2 + \theta)a] \tag{3.71}$$

which, by (3.43), (3.49), and (3.50) and the assumption that $p \ll \hbar c$, yield

$$A_2\{1 - \exp[i(\kappa_2 - \theta)a]\} = B_2\{\exp[-i(\kappa_2 + \theta)a] - 1\}. \tag{3.72}$$

The second component of this matrix equation gives

$$\lambda = \frac{\beta_2^{(2)}}{\alpha_2^{(2)}} = \frac{1 - \exp[i(\kappa_2 - \theta)a]}{\exp[-i(\kappa_2 + \theta)a] - 1}. \tag{3.73}$$

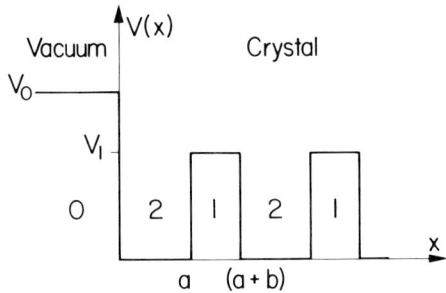

FIG. 15. Potential field for semi-infinite linear crystal with a surface potential of $V_0 > \epsilon$. [From S. G. Davison and M. Stęślicka, *J. Phys. C* **2**, 1802 (1969) (*Proc. Phys. Soc.*). Reproduced by permission of The Institute of Physics and The Physical Society, London.]

From (3.67) and (3.68) comes the bulk spinor

$$\phi_2(x) = A_2 \exp(i\kappa_2 x) + B_2 \exp(-i\kappa_2 x) \qquad (3.74)$$

i.e.,

$$\phi_2(x) = \alpha_2^{(2)} \left\{ \begin{bmatrix} -\gamma_2 \\ 1 \end{bmatrix} \exp(i\kappa_2 x) + \lambda \begin{bmatrix} \gamma_2 \\ 1 \end{bmatrix} \exp(-i\kappa_2 x) \right\} \qquad (3.75)$$

via (3.19) and (3.73).

(*iii*) *Tamm Surface Problem.* For crystals composed of heavy elements, the relativistic shrinkage of the energy spectrum has a drastic effect on the positions of the band edges, especially at high energies. Consequently, with the positions of the band edges being of great importance in surface state calculations, it is clear that relativistic effects should be taken into account.

(a) *Energy relation.* The solution of the Dirac equation in the vacuum region (Fig. 15) is[107]

$$\phi_0(x) = \begin{bmatrix} \gamma_0 \\ 1 \end{bmatrix} \beta_0^{(2)} \exp(l_0 x), \qquad l_0 = -i\kappa_0 > 0 \text{ and real.} \qquad (3.76)$$

The surface continuity condition at $x = 0$ is

$$\phi_0(0) = \phi_2(0) \qquad (3.77)$$

which, together with (3.75) and (3.76), gives

$$\gamma_0 \beta_0^{(2)} + \alpha_2^{(2)} \gamma_2 (1 - \lambda) = 0, \qquad \beta_0^{(2)} - \alpha_2^{(2)} (1 + \lambda) = 0 \qquad (3.78)$$

so, for nontrivial solutions,

$$\gamma_0 (1 + \lambda) + \gamma_2 (1 - \lambda) = 0. \qquad (3.79)$$

[107] S. G. Davison and M. Stęślicka, *J. Phys. C.* **2**, 1802 (1969) (*Proc. Phys. Soc.*).

Equations (3.73) and (3.79) yield

$$e^{i\theta a} = \cos \chi_2 - i\Gamma_0 \sin \chi_2 \qquad (3.80)$$

where

$$\Gamma_0 = \gamma_0/\gamma_2. \qquad (3.81)$$

The existence of surface states is investigated by choosing θ complex,[14] i.e.,

$$\theta = n\pi/a + i\mu, \quad \mu \text{ real} > 0, \qquad (3.82)$$

n being an integer [cf. (1.32) and (3.82)]. The relativistic KP relation (3.51) now becomes

$$(-1)^n \cosh \mu a = \cos \chi_2 + (P'/\chi_2) \sin \chi_2 \qquad (3.83)$$

in which

$$P' = map/\hbar^2. \qquad (3.84)$$

Subtracting (3.80) from (3.83) using (3.82) gives

$$(-1)^n \sinh \mu a = (P'/\chi_2 + i\Gamma_0) \sin \chi_2. \qquad (3.85)$$

With the aid of the identity

$$\cosh^2 \mu a - \sinh^2 \mu a = 1, \qquad (3.86)$$

(3.83) and (3.85) yield the surface state energy expression

$$\chi_2 \cot \chi_2 = \frac{\chi_2^2}{2P'} \left\{ 1 - \frac{(\epsilon - V_0)}{\epsilon \chi_0^2} \left[\chi_2^2 \frac{(\epsilon - V_0)}{\epsilon} - 2i\chi_0 P' \right] \right\} \qquad (3.87)$$

which, to a first-order approximation, can be written as

$$\chi(1 + \eta^2\chi^2/2) \cot \chi(1 + \eta^2\chi^2/2)$$
$$= Q[1 + \eta^2(\Omega_-^2 - \chi^2)] - \Omega_-(1 + \eta^2\Omega_+^2/2) \qquad (3.88)$$

where

$$\Omega_\pm^2 = a^2 q^2 \pm \chi^2, \quad Q = a^2 q^2/2P, \quad q^2 = 2m_0 V_0/\hbar^2. \qquad (3.89)$$

In the NR limit, as $c \to \infty$ and $\eta \to 0$, (3.88) reduces to the well-known Tamm[14] equation

$$\chi \cot \chi = Q - \Omega_-. \qquad (3.90)$$

(b) *Existence conditions.* From (3.37), (3.81), and (3.85) results

$$(-1)^n \sinh \mu a = \frac{\sin \chi_2}{\chi_2} \left[P' + \frac{i\chi_2^2(\epsilon - V_0)}{\epsilon \chi_0} \right] \qquad (3.91)$$

where

$$\chi_2 = \chi + \delta, \quad \delta = \eta^2 \chi^3/2, \qquad (3.92)$$

which is obtained from (3.15), (3.56), and (3.57). In the NR sense, the

$(n + 1)$th FEG occurs in the range $n\pi < \chi < (n + 1)\pi$. However, the presence of the 1-R correction term δ in (3.92) means that the $(n + 1)$th FEG now occurs in the range

$$(n\pi - \delta) < \chi < [(n + 1)\pi - \delta] \qquad (3.93)$$

with $n = 0, 1, \ldots$. In any FEG, the NR counterpart of (3.91) has[108]

$$\operatorname{sign}[(\sin \chi)/\chi] = (-1)^n; \qquad (3.94)$$

thus,

$$P > \Omega_- \qquad (3.95)$$

because $\sinh \mu a > 0$ for $\mu a > 0$. If (3.95) is converted into an equation, then its solutions will give the positions of lower-band edges at the crystal surface.[109] The upper-band edges are found by using (3.55). From the 1-R point of view, (3.94) is valid in the range (a) $[(n + 1)\pi - \delta] > \chi > n\pi$, but, in the range (b) $n\pi > \chi > (n\pi - \delta)$, where $n \neq 0$ since $\chi > 0$, (3.94) must be replaced by

$$\operatorname{sign}[(\sin \chi)/\chi] = (-1)^{n+1}. \qquad (3.96)$$

Consequently, for the (a) and (b) ranges, (3.15), (3.81), (3.84), and (3.91) lead, respectively, to

$$P - \Omega_- \gtreqless \tfrac{1}{2}\eta^2(\Omega_-\Omega_+{}^2 - 4P\chi^2). \qquad (3.97)$$

In the NR limit, as $\eta \to 0$, the upper inequality reduces to Tamm's existence condition (3.95), hence, the surface states satisfying this inequality will be called *relativistic Tamm states* (RTS). However, in the NR limit, the lower inequality simplifies to one which represents the violation of Tamm's condition. Thus, the surface states fulfilling this condition arise solely as a result of the Dirac relativistic formulation and, for this reason, will be referred to as *Dirac surface states* (DSS).

(c) *Energy classification of surface states.* If the inequalities (3.97) are made into an equation, then the χ values given by this equation will correspond to the 1-R lower-band edges at the crystal surface. The positions of the 1-R surface energy levels in the various FEG are given by the solutions of (3.88). A surface state exists, when one of the inequalities (3.97) is satisfied, provided a corresponding solution to (3.88) can be found. However, this proviso only applies to the upper inequality (RTS) in the first FEG, since $n \geq 1$ for the lower inequality (DSS) and solutions of (3.88) always exist in a FEG for which $n \geq 1$.

[108] M. Stęślicka, *Physica* **45**, 506 (1970).
[109] Although, in the NR case, the χ values correspond to the lower band edges at the crystal surface, in the 1-R situation they correspond to the surface energy levels in the FEG.

For the NR lower-band edge values of χ at the crystal surface, the left-hand side (L) of (3.97) vanishes, so

$$P = \Omega_- \tag{3.98}$$

and the right-hand side (R) of (3.97) becomes

$$R = \eta^2 P(\Omega_-^2 - 2\chi^2)/2. \tag{3.99}$$

When $R < 0$, the upper inequality in (3.97) is always satisfied, and

$$V_0/3 < \epsilon < V_0 \tag{3.100}$$

via (3.53), (3.57), and (3.89). If $L = 0$ and $R < 0$, the lower inequality in (3.97) is never satisfied, whereas if $R > 0$, the upper inequality is never satisfied and the lower one is always fulfilled, subject to

$$0 < \epsilon < V_0/3. \tag{3.101}$$

Consider now the violation of the NR Tamm condition (3.95), so that $L < 0$. For $R < 0$, both inequalities in (3.97) can be satisfied mutually exclusively, if (3.100) holds. When $R > 0$, the upper inequality is always violated, but the lower one is always valid, provided (3.101) is fulfilled. If, on the other hand, $L > 0$, then (3.95) holds. Taking $R > 0$ leads to both inequalities being satisfied separately, provided (3.101) holds. Setting $R < 0$, if (3.100) is valid, the lower (upper) inequality is always violated (satisfied).

As shown in Fig. 16, the surface states can be classified according to their NR energy ϵ and the appropriate parametric values of P, q, and a,

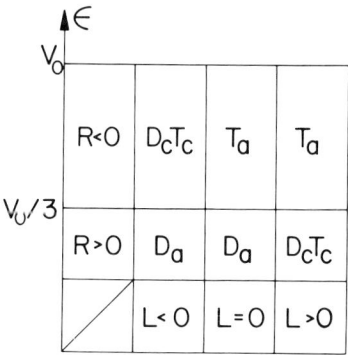

Fig. 16. Classification of surface states according to the range of ϵ. R (L) denotes right- (left-)hand side of (3.97), D and T refer to Dirac and relativistic Tamm states, and the subscript "c" or "a" indicates that a surface state "can" or "always" exist. [From S. G. Davison and M. Stęślicka, *J. Phys. C* **2**, 1802 (1969). Reproduced by permission of The Institute of Physics and The Physical Society, London.]

which make L greater than, equal to, or less than zero. In the case of $L > 0$, i.e., Tamm's NR condition, either a RTS or a DSS can occur when $R > 0$, but when $R < 0$ only a RTS always exists. Thus, like the NR situation, a Tamm-like state can be found over the entire energy range. In contrast, however, a DSS can exist in the lower part of the energy spectrum, when a RTS is not found there. Furthermore, when $R < 0$ ($R > 0$) a RTS is more (less) readily formed than a Tamm state is in the NR case. When $L = 0$, which applies at the NR lower-band edges at the crystal surface, one of either of the relativistic surface states is always found throughout the complete energy spectrum. For this particular value of L, the line $\epsilon = V_0/3$ separates the RTS and DSS energy zones. Of course, in the NR picture, no surface state appears for this L value. For $L < 0$, the NR Tamm condition (3.95) is violated, and it is possible for a DSS to exist throughout the entire energy spectrum. In summary, it is clear from Fig. 16 that a DSS (RTS) always exists for $R > 0$ ($R < 0$) and $L \leq 0$ ($L \geq 0$), and that the D_cT_c zones require both L and R to be positive or negative simultaneously.

b. *Nonrelativistic Treatment*

In contrast to Section 3,a, where the crystal bulk properties were investigated by solving the Dirac equation via the SM method, the Schroedinger equation is now solved by the Green function technique.[85] However, in dealing with the surface problem, the SM approach is again used.

(*i*) *Basic Theory.* For a one-dimensional system the Schroedinger equation can be written as

$$\phi''(x) + \kappa^2\phi(x) = -(2m_0/\hbar^2)V(x)\phi(x) \quad (3.102)$$

where κ is given by (3.53) and $V(x) > 0$. The first step in solving (3.102) is to replace the right-hand side by a Dirac δ function and then seek the solution of the simpler equation

$$G''(x, \tau) + \kappa^2 G(x, \tau) = -\delta(x - \tau) \quad (3.103)$$

which defines the *Green function* $G(x, \tau)$.[102,110]

Equation (3.103) is solved by means of the Fourier transform technique.[64,110] Let $g(k)$ be the Fourier transform of $G(x, \tau)$, so that

$$G(x, \tau) = (2\pi)^{-1/2} \int_{-\infty}^{\infty} g(k) \exp[ik(x - \tau)]\, dk. \quad (3.104)$$

[110] P. M. Morse and H. Feshbach, "Methods of Theoretical Physics." McGraw-Hill, New York, 1953.

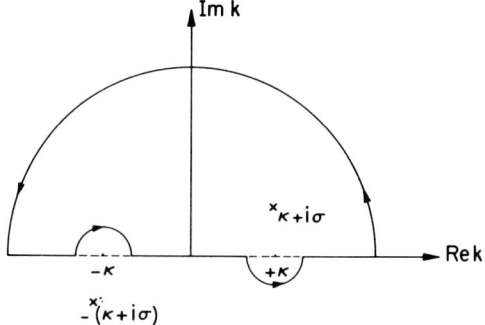

FIG. 17. Outgoing contour of integration for the evaluation of $G(x, \tau)$ in the complex k plane.

Inserting (3.104) in (3.103) yields

$$\int_{-\infty}^{\infty} [(k^2 - \kappa^2)g(k) - (2\pi)^{-1/2}] \exp[ik(x - \tau)] \, dk = 0$$

for all x and τ, which means the integrand must vanish, whence,

$$g(k) = (2\pi)^{-1/2}(k^2 - \kappa^2)^{-1} \quad (3.105)$$

and (3.104) becomes

$$G(x, \tau) = (2\pi)^{-1} \int_{-\infty}^{\infty} (k^2 - \kappa^2)^{-1} \exp[ik(x - \tau)] \, dk. \quad (3.106)$$

The integrand in (3.106) diverges as $k \to \pm \kappa$, so to overcome this difficulty $G(x, \tau)$ is rewritten as

$$G(x, \tau) = (2\pi)^{-1} \lim_{\sigma \to 0} \int_{-\infty}^{\infty} \frac{\exp[ik(x - \tau)] \, dk}{[k^2 - (\kappa + i\sigma)^2]}. \quad (3.107)$$

When $\sigma > 0$, the poles of (3.107) are at $k = \pm(\kappa + i\sigma)$, and the contour of integration in the complex k plane (Fig. 17) is called *outgoing*. The semicircular part of the contour has a very large radius on which k is large and imaginary, making the exponential in the integrand vanish. The calculus of residues gives

$$G(x, \tau) = (2\pi)^{-1} \lim_{\sigma \to 0} [2\pi i \times \text{residue at } (\kappa + i\sigma)]. \quad (3.108)$$

The residue at $(\kappa + i\sigma)$ is obtained by putting $k = (\kappa + i\sigma + \nu)$, and finding the coefficient of ν^{-1} in the Laurent series expansion of the integrand, viz.,

$$\exp[i(\kappa + i\sigma)(x - \tau)]/2(\kappa + i\sigma). \quad (3.109)$$

Hence, (3.108) becomes

$$G(x, \tau) = i \exp(i\kappa |x - \tau|)/2\kappa. \tag{3.110}$$

Having solved (3.103), its solution (3.110) is now used to assist in solving (3.102). Multiplying (3.102) and (3.103) by G and ϕ, respectively, and integrating over all τ before substracting the resulting equations leads to

$$\int_{-\infty}^{\infty} [\phi(\tau) G''(\tau) - G(\tau) \phi''(\tau)] d\tau$$

$$= (2m_0/\hbar^2) \int_{-\infty}^{\infty} V(\tau) G(\tau) \phi(\tau) d\tau - \int_{-\infty}^{\infty} \phi(\tau) \delta(x - \tau) d\tau. \tag{3.111}$$

Since the left-hand side of (3.111) is zero by Green's theorem,[64,110] it follows that

$$\phi(x) = (2m_0/\hbar^2) \int_{-\infty}^{\infty} V(\tau) G(x, \tau) \phi(\tau) d\tau \tag{3.112}$$

where the δ-function property[40,70]

$$\phi(x) = \int_{-\infty}^{\infty} \phi(\tau) \delta(x - \tau) d\tau \tag{3.113}$$

has also been employed. Equation (3.112) shows that, by invoking the use of the Green function technique, the differential Schroedinger equation (3.102) has been converted into an integral equation.

(*ii*) *Volume Solutions.* If the crystal lattice spacing is a, then the integral in (3.112) can be replaced by a sum of integrals over each unit cell, i.e.,

$$\phi(x) = (2m_0/\hbar^2) \sum_{n=-\infty}^{\infty} \int_{na}^{(n+1)a} V(\tau) G(x, \tau) \phi(\tau) d\tau. \tag{3.114}$$

Taking the crystal periodicity into account gives

$$V(\tau + na) = V(\tau) \tag{3.115}$$

and

$$G(x, \tau + na) = i \exp(i\kappa |x - \tau - na|)/2\kappa \tag{3.116}$$

from (3.110). In addition, Bloch's theorem (3.67) yields

$$\phi(\tau + na) = \exp(i\theta na) \phi(\tau). \tag{3.117}$$

Thus, (3.114) can be written as

$$\phi(x) = (im_0/\hbar^2\kappa) \sum_{-\infty}^{\infty} e^{i\theta na} \int_0^a \exp(i\kappa |x - \tau - na|) \phi(\tau) V(\tau) \, d\tau \quad (3.118)$$

or

$$\phi(x) = (im_0/\hbar^2\kappa) \int_0^a F(x, \tau) \phi(\tau) V(\tau) \, d\tau \quad (3.119)$$

where

$$F(x, \tau) = \left[\sum_{-\infty}^{(t/a)} + \sum_{(t/a)+1}^{\infty} \right] \exp(i\theta na) \exp(i\kappa |t - na|) \quad (3.120)$$

with $t = x - \tau$ and (t/a) = integral part of t/a. On setting $t' = t - a(t/a)$, and summing the infinite geometrical series, (3.120) leads to

$$F(x, \tau) = -i \exp[i\theta a(t/a)] \left[\frac{e^{i\theta a} \sin \kappa t' - \sin \kappa(t' - a)}{\cos \kappa a - \cos \theta a} \right] \quad (3.121)$$

after some lengthy manipulation. Consequently, (3.119) becomes

$$\phi(x) = \frac{m_0}{\hbar^2 \kappa} \int_0^a \exp[i\theta a(t/a)] \left[\frac{e^{i\theta a} \sin \kappa t' - \sin \kappa(t' - a)}{\cos \kappa a - \cos \theta a} \right] \phi(\tau) V(\tau) \, d\tau. \quad (3.122)$$

For a 1-D monatomic crystal, the potential field can be described by a linear array of Dirac δ functions, in which case, if the origin is located at one of the lattice sites, then the potential function for the whole system can be expressed as

$$(2m_0/\hbar^2) V(x) = -aU \sum_{n=-\infty}^{\infty} \delta(x - na), \quad (3.123)$$

the parameter $U < 0$ being the strength of the δ potential well at each lattice site. Inserting (3.123) in (3.122) and using (3.113) for the first unit cell, so that $\tau = 0$ and $t' = x'$,

$$\phi(x) = \frac{aU}{2\kappa} \phi(0) \exp[i\theta(x - x')] \left[\frac{e^{i\theta a} \sin \kappa x' - \sin \kappa(x' - a)}{\cos \theta a - \cos \kappa a} \right] \quad (3.124)$$

which, for $x' = x = 0$, immediately reduces to the KP relation [cf. (3.54)]

$$\cos \theta a = \cos \kappa a + (aU/2\kappa) \sin \kappa a. \quad (3.125)$$

(iii) Surface Solutions. (a) *Wave functions.* Consider a semi-infinite linear crystal, with lattice spacing a. Let the potential field of the crystal be represented by (Fig. 18)

$$(2m_0/\hbar^2) V(x) = -aU \sum_{n=0}^{\infty} \delta(x - na), \qquad x > 0$$
$$= -V_0, \qquad V_0 < 0, \qquad x < 0. \qquad (3.126)$$

Inside the crystal ($x > 0$), the general form of the wave function is[111]

$$\phi_1(x) = \{A_1[e^{-i\theta a} \sin \kappa x - \sin \kappa(x-a)]$$
$$+ B_1[e^{i\theta a} \sin \kappa x - \sin \kappa(x-a)]\}/\sin \kappa a \qquad (3.127)$$

by (3.124) and (3.125). For the vacuum region outside the crystal ($x < 0$), the general solution of (3.102) is

$$\phi_0(x) = A_0 \exp(i\kappa_0 x) + B_0 \exp(-i\kappa_0 x) \qquad (3.128)$$

where

$$\kappa_0^2 = \kappa^2 + V_0. \qquad (3.129)$$

(b) *Boundary conditions and S matrix.* The continuity of the wave function at $x = 0$ requires [cf. (3.34)]

$$\phi_0(0) = \phi_1(0). \qquad (3.130)$$

Using (3.113) and integrating the Schroedinger equation over the range $0_- < x < 0_+$ gives[112]

$$\phi'(0_+) - \phi'(0_-) = aU\phi(0). \qquad (3.131)$$

Substituting (3.127) and (3.128) in these boundary conditions yields

$$A_0 + B_0 = A_1 + B_1, \qquad (3.132)$$

$$(\kappa/\sin \kappa a)[A_1(e^{-i\theta a} - \cos \kappa a) + B_1(e^{i\theta a} - \cos \kappa a)]$$
$$- i\kappa_0(A_0 - B_0) = aU(A_0 + B_0), \qquad (3.133)$$

respectively. If Eqs. (3.132) and (3.133) are written in the matrix notation of (3.22), then the elements of the S matrix give[111]

$$R_{22} = e^{i\theta a} - \cos \kappa a - \sin \kappa a(aU - i\kappa_0)/\kappa \qquad (3.134)$$

by (3.28).

---

[111] E. Aerts, *Physica* **26**, 1047 (1960); *in* "Solid State Physics in Electronics and Telecommunications" (M. Désirant and J. L. Michels, eds.), Vol. 1, part 1, pp. 628–638 Academic Press, New York, 1960.

[112] The need for using (3.131) instead of the more familiar condition, viz., $\phi'(0_+) = \phi'(0_-)$, arises from the necessity to attribute a finite value to the depth of the δ-potential well.

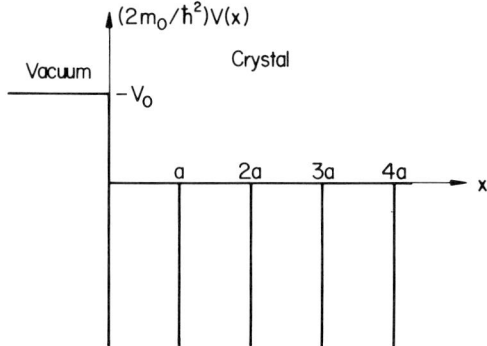

FIG. 18. Dirac δ-potential function for a semi-infinite linear crystal.

(c) *Surface state conditions.* As was already mentioned in connection with (3.28), when $R_{22} = 0$ the S matrix in (3.22) becomes infinite, so B becomes infinitely large compared to A. In this case, if the outgoing waves have θ real, the surface will act as a source of electrons, which is physically unreasonable. On the other hand, if θ is defined by (3.82), then the outgoing waves are attenuated, and the electrons are localized at the surface. Thus, the surface state conditions are

$$R_{22} = 0; \quad \theta = n\pi/a + i\mu, \quad \mu \text{ real} > 0. \tag{3.135}$$

(d) *Energy levels.* From (3.125), (3.134), and (3.135) comes

$$(-1)^n \cosh \mu a = \cos \kappa a + (aU/2\kappa) \sin \kappa a \tag{3.136}$$

$$(-1)^n e^{-\mu a} = \cos \kappa a + \sin \kappa a (aU - i\kappa_0)/\kappa. \tag{3.137}$$

Eliminating $\cos \kappa a$ between these equations gives [cf. (3.85)]

$$(-1)^n \sinh \mu a = (i\kappa_0 - aU/2) \sin \kappa a/\kappa. \tag{3.138}$$

The identity (3.86), together with (3.136) and (3.138), leads to the surface state energy relation[113]

$$\chi \cot \chi + i\chi_0 + V_0/U = 0 \tag{3.139}$$

via (3.57), where χ_0 must be imaginary for surface state energy to be real. Furthermore, since $\exp(-i\kappa_0 x)$ is an exponentially damped wave in the negative x direction, it follows that $i\kappa_0 < 0$. Thus, with the aid of (3.129), and introducing the parameters

$$\Upsilon^2 = -a^2 V_0, \quad \Theta = -a^2 U, \tag{3.140}$$

[113] P. Phariseau, *Physica* **26**, 737 (1960).

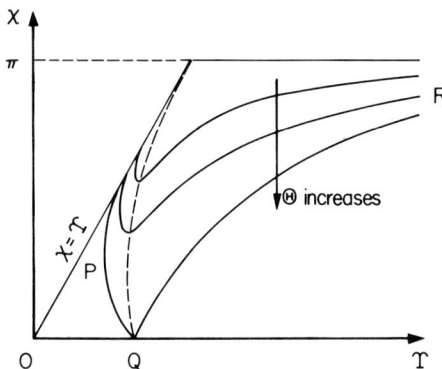

FIG. 19. Graphical solutions of (3.141).

(3.139) becomes [cf. (3.90)]

$$\chi \cot \chi - (\Upsilon^2 - \chi^2)^{1/2} + \Upsilon^2/\Theta = 0. \quad (3.141)$$

The graphical solutions of (3.141) are drawn schematically in Fig. 19. In the $\chi\Upsilon$ plane, the FEG is bounded by the asymptote $\chi = \pi$ and the line $\chi = \Upsilon$, which is present to ensure that the solutions of (3.141) are always real (i.e., $\chi \not> \Upsilon$). The Υ and χ minimum points occur at $P(\Upsilon_p, \chi_p)$ and $Q(\Upsilon_q, \chi_q)$, respectively. The locus of Q, shown as the broken curve in Fig. 19, is given by

$$\Theta = 2(\Upsilon_q^2 - \chi_q^2)^{1/2} \quad (3.142)$$

which is obtained by differentiating (3.141) with respect to Υ and setting $d\chi/d\Upsilon = 0$ at the point (Υ_q, χ_q).

Comparison of Fig. 19 with the KP energy spectrum for a *negative potential*[114] shows that the asymptote at $\chi = \pi$ corresponds to the bottom of the second band, while the point Q corresponds to the top of the first band. This can also be seen from the well-known equation[105] for the top of the energy bands, viz.,

$$\chi \cot \chi/2 = \Theta/2 \quad (3.143)$$

which inserted in (3.141) again leads to (3.142) at the point Q. Note in addition that, when $\Theta = 4$ and $\Upsilon_q = 2$, $\chi_q = 0$ in (3.142), the first band disappears.

(e) *Existence condition.* In the KP relation (3.125),

$$\text{sign}[(\sin \kappa a)/\kappa] = (-1)^{n+1} \quad (3.144)$$

for $(n-1)\pi < \kappa a < n\pi$, $n = 1, 2, \ldots$ being the number of the FEG.

---
[114] Phariseau[113] appears to have incorrectly taken a positive potential in the KP relation, hence, the confusion between the top and bottom of each band.

Therefore, since $\mu > 0$, (3.138) is satisfied only when

$$i\kappa_0 - aU/2 < 0 \qquad (3.145)$$

which by (3.129), (3.140), and the fact that $i\kappa_0 < 0$ yields

$$\Theta < 2(\Upsilon^2 - \chi^2)^{1/2}. \qquad (3.146)$$

Hence, surface states exist only for the solutions of (3.141), which satisfy the inequality (3.146).

It is easy to prove[108] that the existence condition (3.146) is always satisfied if, and only if, $\Upsilon > \Upsilon_q$. Thus, in Fig. 19, surface states are represented by those parts of the curves lying to the right of the broken curve, since it is only these portions which satisfy (3.146). The remaining portions correspond to $\mu < 0$, which gives rise to waves with increasing amplitudes inside the crystal. Each point along the QR part of the curves corresponds to one surface state. Hence, in agreement with Tamm,[14] for the particular values of Θ, Υ, and χ satisfying (3.146), only *one*[115] surface state can occur. When (3.146) is violated, no surface state exists. As demonstrated by Phariseau,[113,114] the QR regions of the curves show that, although it is possible for a surface state to disappear into the top of a band at the Q points, it is impossible for one to attain the bottom of the next band, because of the asymptotic behavior of the curves in the vicinity of R.

(*iv*) *Other Work.* (a) *S-matrix approach.* Besides his work on the surface states of crystals with clean surfaces, Aerts[111] studied the surface states of crystals with impurity contaminated surfaces, via the SM method. He showed that when the impurity potential is weaker (stronger) than that of the bulk atoms, so that an electron is loosely (tightly) bound to the impurity atom, i.e., the impurity can be regarded as a donor (acceptor), the surface level is displaced toward the conduction (valence) band. In this way, Aerts was able to classify the surface conductivity as being of the n (donor) type or p (acceptor) type.

After his investigations of monatomic crystals, Aerts[49] turned his attention to diatomic crystals with clean surfaces, which were discussed in detail in Section 1,b. Once again, the main new feature arising from the presence of two atoms per unit cell was the creation of additional FEG. In the final paper[116] of the series, Aerts analyzed the problem of Shockley surface states, by having a symmetric surface potential. On adsorbing foreign atoms onto the surface, he found that surface states could exist if the impurity potential is less than that of the bulk atoms. By way of contrast, Aerts also computed the energy levels localized at the clean interface

---

[115] This is somewhat in contrast to Aerts,[111] who, by his neglect of the existence condition (3.146), found that it was possible to have a maximum of *two* surface states.

[116] E. Aerts, *Physica* **26**, 1163 (1960).

of two semi-infinite linear crystals composed of different atoms and with different lattice spacings. The interface levels were found to occur in FEG which were common to both crystals. Ghosh and Laskar[117] extended this work to include contaminated interfaces.

(b) *Green function method.* The greatest exponent of the Green function technique has so far been Phariseau. His contributions to the theory of surface states of semi-infinite[113] and finite[118] pure crystals are in agreement with the previous findings. Perhaps one of the more interesting of Phariseau's papers is the one dealing with the energy spectrum of amorphous substances.[119] Utilizing continued fractions, it is shown that, like crystals, such materials have an energy band structure, which arises solely because of the suitable interatomic distances and not from any atomic periodicity. For a system consisting of an amorphous substance embedded in its own crystalline form, extra localized energy levels were found to occur in the FEG. This work laid the foundation for the following article[58] on the subsurface states of deformed crystals (Section 1,c), since the deformed part of the crystal, created by the penetration of the surface perturbation, can be regarded as an amorphous region.

In a similar vein to Aerts' work[49] on AB-type binary crystals, Phariseau[120] has also investigated the surface states of III–V compounds, with particular reference to the zinc blende (or sphalerite) structure, where the directions studied are the 1–D analogs of [100] and [111]. The results for the [100] direction are identical to those of Aerts.[49] However, for the [111] direction, where the atoms are arranged in a sequence of double layers, which are separated by internal and external distances in the ratio of 1:3, the existence of surface states depends, to a large extent, on what type of bond is broken at the surface. In accordance with Koutecky's results,[121] the surface states in the first FEG correspond to the breaking of strong surface bonds. It should be noted that a newly formed surface is naturally more stable if the weaker bonds are broken. According to these concepts, the surface states should then disappear. In the second FEG, the energy levels corresponding to broken weak bonds are associated with lower surface potential barriers than those originating from broken strong bonds.

4. MATHIEU PROBLEM

A simple model of a crystal potential, which is somewhat more realistic for surface state calculations than a δ-function array, is a sinusoidal poten-

---

[117] S. K. Ghosh and A. K. Lasker, *Ind. J. Phys.* **37**, 534 (1963).
[118] P. Phariseau, *Physica* **27**, 590 (1961).
[119] P. Phariseau, *Physica* **26**, 1185 (1960).
[120] P. Phariseau, *Physica* **30**, 608 (1964).
[121] J. Koutecký, *Phys. Status Solidi* **1**, 554 (1961).

tial. The amplitude measures the crystal potential strength in a monatomic crystal, and the ionicity in a diatomic crystal. The crystal is terminated at the surface by a step potential of arbitrary height. The termination can occur at any point in the last unit cell; special cases are terminations at maxima or minima. Because the Schroedinger equation with a sinusoidal potential reduces to the well-known Mathieu equation,[92,122] many results can be obtained quite easily. Statz[21] was the first to suggest the Mathieu problem for surface state computations. Recently, Levine[34] carried out the Mathieu computations for conditions of arbitrary termination. The procedure adopted in the following discussion is: (1) to investigate the properties of the Mathieu equation, (2) to introduce the surface termination, (3) to consider the continuity condition, (4) to treat the finite crystal and hybridization of all states, and (5) to compare the surface state results with those of other models. It will be shown that Tamm states, Shockley states, and ionic surface states can be profitably interrelated by the Mathieu model.

a. Basic Theory

The Schroedinger equation for a 1-D crystal with a sinusoidal potential of amplitude V_2 and a reference potential V_1 is

$$(\hbar^2/2m_0)\psi''(x) + [E - V_1 - V_2 \cos(2\pi x/a)]\psi(x) = 0. \quad (4.1)$$

The potential is periodic in a and, by definition, has maximum (repulsive) value at $x = 0$. (A shift of origin will be considered later.) It is convenient to express this equation in the dimensionless form

$$\psi''(z) + (\alpha - 2q \cos 2z)\psi(z) = 0. \quad (4.2)$$

Here $z \equiv \pi x/a$, α is the dimensionless energy, and q is a measure of the ratio of potential energy amplitude to kinetic energy. The normalized crystalline potential is shown in Fig. 20. Equation (4.2) is the Mathieu equation expressed in a standard form given by McLachlan.[92] (Jahnke and Emde[122] use a slightly different notation.)

In the limiting case of $q = 0$, the crystal potential is flat (free-electron approximation). The standing-wave unnormalized solutions of (4.2) then become $\alpha = m^2$, where $m = 1, 2, 3, \ldots$, and $\psi = +\cos z, +\sin z, +\cos 2z, +\sin 2z, \ldots$. For convenience, the positive sign is always adopted. A solution also exists for $\alpha = m = 0$ and $\psi = 1$.

For $q \neq 0$, the energy α and wave function ψ must be functions of q. These can be expressed in power series expansions. Consider, e.g., the

[122] E. Jahnke and F. Emde, "Table of Functions with Formulae and Curves." Dover, New York, 1945.

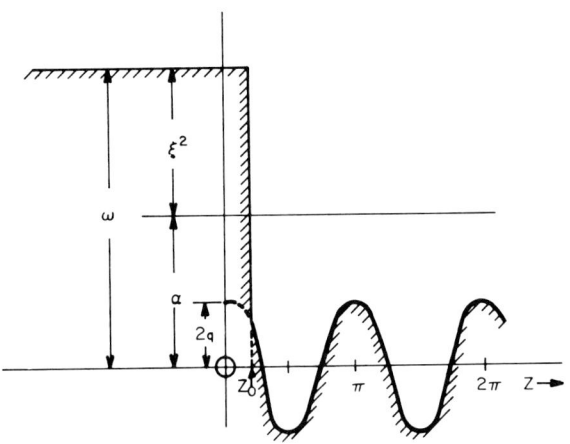

Fig. 20. Surface region of a crystal with sinusoidal potential.

solution which has a limit $\alpha = 1$ and $\psi = \cos z$ when $q = 0$.[123] The expansions are of the form

$$\alpha = 1 + \alpha_1 q + \alpha_2 q^2 + \alpha_3 q^3 + \cdots \quad (4.3)$$

$$\psi = \cos z + qC_1(z) + q^2 C_2(z) + q^3 C_3(z) + \cdots. \quad (4.4)$$

Substituting (4.3) and (4.4) into (4.2), and equating like powers of q equal to zero, gives

$$q^0: \quad \cos z - \cos z = 0 \quad (4.5)$$

$$q^1: \quad C_1'' + C_1 - \cos 3z + (\alpha_1 - 1) \cos z = 0 \quad (4.6)$$

$$q^2: \quad C_2'' + C_2 + \alpha_1 C_1 - 2C_1 \cos 2z + \alpha_2 \cos z = 0. \quad (4.7)$$
$$\vdots$$

These equations can be solved in order, using symmetry conditions and successive integrations.[92] For example, symmetry requires that $\alpha_1 = 1$ in (4.6) and $\alpha_2 = -\frac{1}{8}$ in (4.7). In this manner, all energy coefficients $\alpha_1, \alpha_2, \alpha_3, \ldots$ and all wave function coefficients $C_1, C_2, C_3, \ldots$ can be computed, so that α and ψ in (4.3) and (4.4) can be determined to any degree of accuracy desired. Fortunately, α and ψ have been *tabulated* and *graphed* as functions of q and m.[92,122] The most relevant ψ and α properties are shown in Figs. 21 and 22, respectively.

When $q \neq 0$, the standing wave eigenfunctions ψ of the Mathieu equation (4.2) are called, by definition, the *cosine elliptic functions* $ce_m(z, q)$

---

[123] There is also another solution for $\alpha = 1$ which has $\psi = \sin z$ in the limit when $q = 0$.

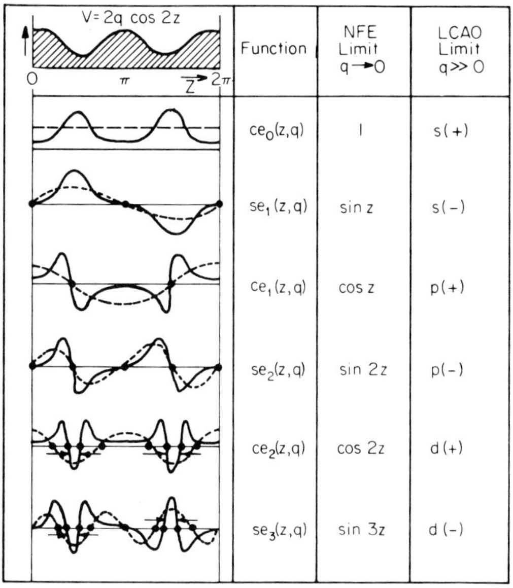

	Function	NFE Limit $q \to 0$	LCAO Limit $q \gg 0$
$V = 2q \cos 2z$			
	$ce_0(z,q)$	1	$s(+)$
	$se_1(z,q)$	$\sin z$	$s(-)$
	$ce_1(z,q)$	$\cos z$	$p(+)$
	$se_2(z,q)$	$\sin 2z$	$p(-)$
	$ce_2(z,q)$	$\cos 2z$	$d(+)$
	$se_3(z,q)$	$\sin 3z$	$d(-)$

FIG. 21. Some properties of Mathieu functions. $(+)$ indicates a bonding orbital; $(-)$ indicates an antibonding orbital.

and the *sine elliptic functions* $se_m(z, q)$. They have properties qualitatively shown in Fig. 21. In the free-electron limit, there is a flat lattice potential and $q = 0$. Then, by definition, $ce_m(z, q) \to \cos mz$ and $se_m(z, q) \to \sin mz$. These are the free-electron wave functions shown as dashed curves in Fig. 21. The nearly free electron (NFE) limit is defined by $q \approx 0$, for which $ce_m(z, q) \approx \cos mz$ and $se_m(z, q) \approx \sin mz$. As q increases, the attractive potential wells centered at $z = \pi/2$ (also $z = 3\pi/2$) get deeper. This causes the eigenfunction amplitudes and nodes to approach closer to $z = \pi/2$, to take advantage of the attractive potential there. For very large q, the eigenfunctions become more like LCAO functions. They are schematically shown by the solid curves in Fig. 21. They resemble s, p, d, etc. bonding and antibonding atomic orbitals. To be specific, the $ce_0(z, q)$ and $se_1(z, q)$ functions become s-like (no nodes near $z = \pi/2$), the $ce_1(z, q)$ and $se_2(z, q)$ functions become more p-like (one node at $z = \pi/2$), the $ce_2(z, q)$ and $se_3(z, q)$ functions become d-like (two nodes near $z = \pi/2$), etc. Here, m is the appropriate FEG index. Thus, the Mathieu functions display a smooth conceptual transition between the NFE and LCAO limits.

The corresponding standing wave eigenvalues $\alpha = a_m$ or $\alpha = b_m$, for arbitrary values of q, are known and reproduced as solid lines in Fig. 22a. Shaded areas of the diagram indicate allowed Bloch-like states of the type

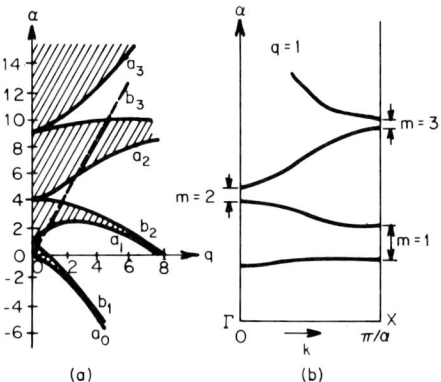

FIG. 22. (a) Energy bands versus sinusoidal amplitude. (b) Energy bands versus wave vector for $q = 1$.

$e^{\pm i\mu z}\phi(\pm z, q)$. The unshaded areas of the diagram are the FEG, where the wave functions have the forms $e^{\mp \mu z}\phi(\pm z, q)$. From the figure it can be seen that for $q = 0$ there is no FEG; this is the free-electron limit. As q increases, the degeneracy is lifted. This is strongest at the first FEG $m = 1$, where the potential and the wave function happen to have the same periodicity. For very large q, the allowed bands become narrower (more atomic-like) and the narrow-band energies are pulled down lower to take advantage of the greater attractive potential near $z = \pi/2$.

For completeness, the α vs k-Brillouin zone representation of Fig. 22a is constructed schematically in Fig. 22b for $q = 1$. The bottom of the first allowed band is the a_0 state. The first FEG occurs at the edge X of the Brillouin zone (BZ) and is bounded by the b_1 and a_1 states. The second FEG occurs at the center Γ of the BZ and is bounded by the b_2 and a_2 states. The FEG at the edge of the BZ correspond to $m =$ odd, while FEG at the center of the BZ correspond to $m =$ even. Since CdS, ZnS, and many other II–VI and III–V crystals have their main FEG in the center Γ of the 3-D BZ,[124] an appropriate 1-D Mathieu analog might be the FEG $m = 2$. In general, then, attention will be given to all FEG m, not merely the most obvious first FEG, $m = 1$.

This concludes the brief review of the properties of the Mathieu equation and the bulk states of the sinusoidal potential. The eigenfunctions and eigenvalues mentioned above, for arbitrary q, will now be used as a basis for discussing the types and energies of surface states. Essentially, the surface states will be composed of the band-edge states described above, which have been analytically continued by using a complex-wave vector.

[124] See, e.g. M. L. Cohen and T. K. Bergstresser, *Phys. Rev.* **141**, 789 (1966).

b. The Surface

Consider the following model of a surface, schematically shown in Fig. 20. The surface is located at an arbitrary position z_0, which may, or may not, be at a maximum ($z_0 = 0$) or a minimum ($z_0 = \pi/2$) of the unit cell. In the figure, this location is chosen, e.g., to be $z_0 = \pi/8$. Ionic and covalent prototypes for terminations $z_0 = 0, \pi/2, \pi/4$, and $3\pi/4$ will be given subsequently. Also, let the crystal be of infinite extent for $z \geq z_0$; the consequences of presuming a large, but finite, slab of crystal are trivial and will be discussed later.

Inside the crystal ($z \geq z_0$) the potential is sinusoidal. An allowed surface state wave function inside the crystal satisfies Mathieu's equation (4.2) and has the form

$$\psi_i(z, q) = \phi(z, q) \exp[-\mu(z - z_0)] \tag{4.8}$$

and the eigenvalue is α. Here μ is positive by definition and $\psi_i(z, q) \to 0$ as $z \to \infty$.

Outside the crystal ($z \leq z_0$) the potential has a step of height ω (in dimensionless units) with respect to the mean of the sinusoidal crystal variations. The wave function outside the crystal has the form

$$\psi_0(z, q) = \text{const} \exp[-\xi(z_0 - z)] \tag{4.9}$$

and the same eigenvalue α ($\alpha = \omega - \xi^2$). Here ξ is positive, by definition, and $\psi_0(z) \to 0$ as $z \to -\infty$. Continuity of the wave function and its first derivative at the surface $z = z_0$ gives

$$\phi(z_0, q)/\phi'(z_0, q) = (\xi + \mu)^{-1}. \tag{4.10}$$

For many physical situations, the quantity on the right will be sufficiently small to be treated as a perturbation. For the extreme case of an infinitely high step potential ($\xi \to \infty$), (4.10) reduces to a very simple nodal condition $\phi(z_0, q) = 0$.

The most general form of $\phi(z, q)$ will be a linear combination (or hybrid) of all the bulk standing wave states $\text{se}_r(z, q)$ and $\text{ce}_r(z, q)$ for all FEG r. A discussion of such a general form will be given later. An excellent approximate form of $\phi(z, q)$ can be realized by noting that, within the mth FEG, the surface state will be a hybrid mainly of the $\text{ce}_m(z, q)$ and $\text{se}_m(z, q)$ functions which border the FEG. To be specific, the $\text{se}_m(z, q)$ function corresponds to the eigenvalue b_m, which lies at the bottom of the FEG; also, the $\text{ce}_m(z, q)$ function corresponds to the eigenvalue a_m, which lies at the top of the FEG. (The special case $m = 0$ will be treated later). Accordingly, the surface state wave function of (4.8) is taken to be (aside from a normalizing constant)

$$\psi_i(z, q) = [\lambda \, \text{ce}_m(z, q) - \text{se}_m(z, q)]e^{-\mu z} \tag{4.11}$$

where the term in brackets is $\phi(z, q)$, μ is positive by definition, and λ is a hybridizing parameter. This approximation is valid when the FEG is sufficiently narrow compared to band widths ($q < \alpha/2$ as shown by the dashed line in Fig. 22a).

For completeness, it will be useful to consider explicitly the *unallowed* surface state wave function $\psi_i^*(z, q)$ which also satisfies Mathieu's equation. This is obtained by setting $z = -z$ in (4.11) to obtain

$$\psi_i^*(z, q) = [\lambda \operatorname{ce}_m(z, q) + \operatorname{se}_m(z, q)]e^{\mu z} \qquad (4.12)$$

where μ is defined positive, as before. This function *increases* exponentially with z as $z \to \infty$ and, therefore, must be *rejected* for a semi-infinite crystal.

The hybridizing parameter λ can be computed in terms of bulk properties q, m, and α by direct substitution of (4.11) into Mathieu's equation (4.2), which, on following McLachlan's method,[34,92] gives

$$\phi'' - 2\mu\phi' + \mu^2\phi + (\alpha - 2q \cos 2z)\phi = 0. \qquad (4.13)$$

Combining (4.11) and (4.13) and incorporating the orthogonality integrals[34,92,122]

$$\int_0^{2\pi} \operatorname{ce}_m^2 (z, q)\, dz = \int_0^{2\pi} \operatorname{se}_m^2 (z, q)\, dz = \pi \qquad (4.14a)$$

$$\int_0^{2\pi} \operatorname{ce}_m (z, q) \operatorname{se}_m (z, q)\, dz = 0 \qquad (4.14b)$$

$$\int_0^{2\pi} \operatorname{ce}_m (z, q) \operatorname{ce}_m' (z, q)\, dz = 0 \qquad (4.14c)$$

$$\int_0^{2\pi} \operatorname{se}_m (z, q) \operatorname{se}_m' (z, q)\, dz = 0 \qquad (4.14d)$$

$$\int_0^{2\pi} \operatorname{ce}_m (z, q) \operatorname{se}_m' (z, q)\, dz = K_{mm} \approx m\pi \qquad (4.14e)$$

$$\int_0^{2\pi} \operatorname{se}_m (z, q) \operatorname{ce}_m' (z, q)\, dz = -K_{mm} \approx -m\pi \qquad (4.14f)$$

where the approximations for K_{mm} in (4.14e) and (4.14f) are valid for the range of interest, leads to

$$(\alpha + \mu^2 - a_m)\lambda + 2\mu m = 0 \qquad (4.15a)$$

$$-2\mu m\lambda + (\alpha + \mu^2 - b_m) = 0. \qquad (4.15b)$$

For a sufficiently narrow FEG, and for the energy lying within the FEG, $b_m \leq \alpha \leq a_m$, it can be shown that (4.15a) and (4.15b) can be approximated by the very simple matrix equation

$$\begin{bmatrix} \alpha - a_m & 2\mu m \\ -2\mu m & \alpha - b_m \end{bmatrix} \begin{bmatrix} \lambda \\ 1 \end{bmatrix} = 0 \qquad (4.16)$$

which will be used throughout until Section 4,e.

The secular equation of this matrix yields the energies

$$\alpha_{\pm} = \tfrac{1}{2}\{a_m + b_m \pm [(a_m - b_m)^2 - 16m^2\mu^2]^{1/2}\}. \qquad (4.17)$$

If there were no surface, then there would be no surface state, and μ would vanish; the energies would be simply the band-edge energies $\alpha_+ = a_m$ and $\alpha_- = b_m$. The presence of a surface, however, allows $\mu > 0$. This acts as a surface "perturbation," which *attracts* α_+ and α_- toward each other and toward the center of the FEG. Such an attraction arises because of the minus sign inside the square root in (4.17). The minus sign, in turn, is a consequence of the antisymmetric matrix in (4.16). Equation (4.17) is very cumbersome as it stands, partly because of its double roots. It can be inverted to yield

$$\mu = (2m)^{-1}[(a_m - \alpha)(\alpha - b_m)]^{1/2} \qquad (4.18)$$

which can also be obtained by inspection of (4.16). By definition, to have a surface state, only the $(+)$ square root of μ need be taken.

The hybridization parameter λ can also be obtained from (4.16):

$$\lambda = (\alpha - b_m)/2\mu m. \qquad (4.19)$$

Here, λ *must be positive, because, from the definition of a surface state, μ is positive and $b_m < \alpha < a_m$.* This is equivalent to a "surface state selection rule" $\lambda > 0$, whose important consequences are shown later. Combining (4.18) and (4.19) yields

$$\lambda = [(\alpha - b_m)/(a_m - \alpha)]^{1/2}. \qquad (4.20)$$

Further, (4.18) and (4.20) can be rearranged to yield α and μ as single-valued functions of λ:

$$\alpha = (b_m + \lambda^2 a_m)/(1 + \lambda^2) \qquad (4.21a)$$

$$\mu = \lambda(a_m - b_m)/2m(1 + \lambda^2). \qquad (4.21b)$$

Let us now investigate the consequences of (4.21a) and (4.21b), considering λ as an independent variable. Later λ will be fixed by detailed surface considerations. When $\lambda = 0$, then $\alpha = b_m$ and $\mu = 0$. The energy is at the bottom of the FEG and there is no surface state. When $\lambda = 1$, $\alpha =$

$(a_m + b_m)/2$ and $\mu = (a_m - b_m)/4m$, and a surface state occurs whose energy is in the middle of the FEG, where the bulk damping μ (or localization) is maximum. When $\lambda = \infty$, then $\alpha = a_m$ and $\mu = 0$. Near the top and bottom of the band edges, the energy varies as the square of the damping wave vector μ as seen from (4.17) or (4.18). All these results seem reasonable and extremely convenient from an analytical point of view. They apply to all values of m and q, which characterize the bulk. In Section 4,e, it will be shown that these results should be modified somewhat, when the hybridization of other bands is taken into account.

For completeness, the situation for λ negative should be investigated. From (4.19) and (4.21b) it is evident that μ must also be negative. This situation (negative λ, negative μ) defines precisely the unallowed wave function $\psi_i^*(z, q)$ of (4.12). Thus, negative λ values correspond to a physically unreal situation and must be rejected. Finally, for λ pure imaginary, α lies in the allowed bands (4.20), and μ is pure imaginary. A Bloch wave is then formed which is not a surface state, and can be ignored here.

c. The Continuity Condition

To have a surface state, it is necessary to have the correct bulk crystal energy α and *also* to have the correct continuity condition on the wave function. Here, the continuity condition will be investigated in some detail. The combination of (4.10) and (4.11) yields the general continuity condition

$$\lambda = \frac{\mathrm{se}_m(z_0, q) - (\xi + \mu)^{-1} \mathrm{se}_m'(z_0, q)}{\mathrm{ce}_m(z_0, q) - (\xi + \mu)^{-1} \mathrm{ce}_m'(z_0, q)} . \quad (4.22)$$

This relates λ to the band gap index m, the crystal surface location z_0, the amplitude $2q$ of the crystal potential, and the height of the step potential ξ, as shown in Fig. 20. Recall from Section 4,b that a surface state occurs only if λ is positive. *Therefore, the condition for a surface state is that the right-hand side of (4.22) be positive.* In general, the relation between m, z_0, ξ, μ, q, and λ is complicated. For certain special examples, however, it can be considerably simplified, even without numerical analysis.

For the first example, consider the case where $z_0 = 0$, which corresponds to the termination of the crystal at a *maximum* with respect to the unit cell. Substituting $z_0 = 0$ in (4.22), it can be shown that λ is always negative.[34] Thus, there can be no surface states for the sinusoidal potential terminated at a potential maximum. The connection of this conclusion with Shockley[17] states will be pointed out in Section 4,g.

In a similar manner, consider the case where $z_0 = \pi/2$. The lattice is terminated at a *minimum*. Substituting $z_0 = \pi/2$ in (4.22), it can be shown that there will be no surface states for $m =$ even, but there will be surface states for $m = $ odd. The notations $m = $ even and $m = $ odd directly correspond to the center and edge, respectively, of the 1-D BZ, as shown in Fig. 22b. The states will be closer to the top of the FEG, the higher the ratio of ξ/m. For an infinite step potential, $\xi \to \infty$, so $\lambda \to \infty$, and the surface state disappears into the top of the FEG. The connection with Tamm states will be pointed out in Section 4,f.

For other terminations z_0, a more general treatment must be carried out. The simplest possible case of this is where $q \sim 0$ and $\xi \sim \infty$, which corresponds to a nearly empty lattice and an infinitely high surface potential. In this case, (4.22) reduces approximately to

$$\lambda = \tan mz_0. \tag{4.23}$$

The right-hand side of (4.23) is positive only when

$$0 < mz_0 < \pi/2, \quad \pi < mz_0 < 3\pi/2, \quad 2\pi < mz_0 < 5\pi/2, \ldots . \tag{4.24}$$

These are the conditions necessary for surface states to appear. The energy is then obtained from (4.21a) and (4.23), viz.,

$$\alpha = b_m \cos^2 mz_0 + a_m \sin^2 mz_0. \tag{4.25}$$

It is convenient to depict these results schematically in Fig. 23. Here α is plotted against z_0 in the range $0 < z_0 < \pi$. The surface states are shown as sinusoidal solid curves in the various FEG. For example, consider the

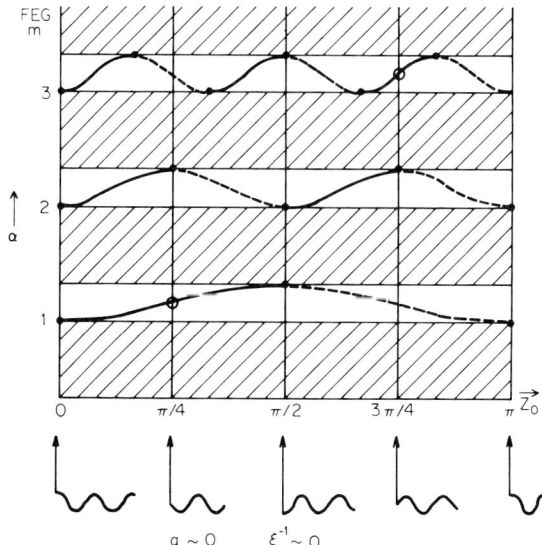

Fig. 23. Surface state energies versus z_0 for a nearly empty lattice and an infinite surface potential.

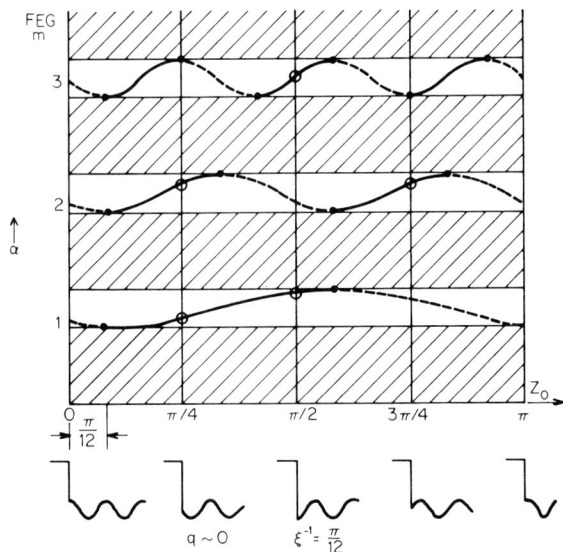

Fig. 24. Surface state energies versus z_0 for a nearly empty lattice and a finite surface potential. Note phase shift of ξ^{-1}.

first FEG ($m = 1$). Then (4.24) allows surface states provided $0 < z_0 < \pi/2$. One such state is shown by the open circle at $z_0 = \pi/4$. The energy increases smoothly from the bottom to the top of the FEG as z_0 is "turned" on to $\pi/2$. But for $z_0 > \pi/2$, λ becomes negative, so there is an *unallowed* surface state; this is indicated by the dashed line. This *unallowed* state has the correct crystal eigenvalue α, the correct continuity condition of the wave function, but the *incorrect* behavior at $z \to \infty$. Nevertheless, it will be very convenient, for symmetry reasons, to always include the unallowed states as dashed lines in all the diagrams. In a similar manner, the second FEG ($m = 2$) has surface states in two regions; for $0 < z_0 < \pi/4$ and $\pi/2 < z_0 < 3\pi/4$. Generally, the mth FEG has m different regions of surface states. At $z_0 = 0$, all energies α lie at the bottom band edges b_1, b_2, b_3, b_4, etc. At $z_0 = \pi/2$, the energies α alternate a_1, b_2, a_3, b_4, etc. The most striking feature of Fig. 23 is its intricate structure. It is difficult to explain this structure in simple classical terms; it arises mainly from the quantum-mechanical continuity condition.

A more general case can be investigated where $q \sim 0$ and ξ is finite. This corresponds to a nearly empty lattice and a finite step barrier. Equation (4.22) then becomes

$$\lambda = \frac{\sin mz_0 - [m/(\xi + \mu)] \cos mz_0}{\cos mz_0 + [m/(\xi + \mu)] \sin mz_0}. \tag{4.26}$$

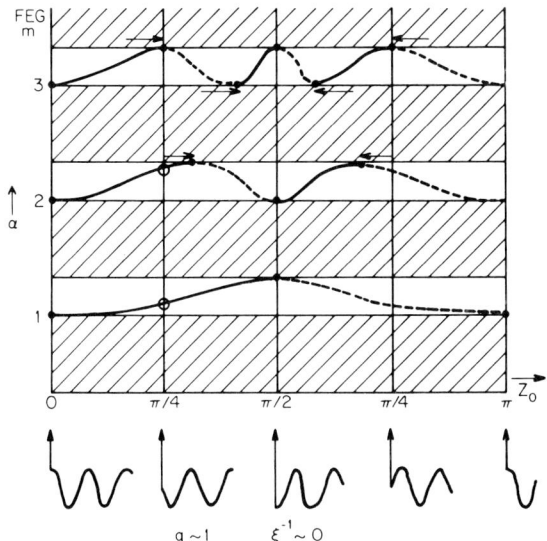

FIG. 25. Surface state energies versus z_0 for a strongly periodic lattice and an infinite surface potential. Note compression toward $\pi/2$.

For $\xi \gg \mu$ and $m/\xi \ll 1$, which corresponds to a large, but finite barrier, the new terms can be considered as a perturbation to the previous results. Using a trigonometric identity, (4.26) reduces to

$$\lambda = \tan m(z_0 - \xi^{-1}). \tag{4.27}$$

The only distinction between this equation and (4.23) is a slab of constant thickness ξ^{-1} subtracted from the crystal (essentially a phase shift).

The surface states corresponding to (4.27) can be immediately inferred by translating Fig. 23 uniformly to the right a distance ξ^{-1}. This is carried out in Fig. 24, where ξ^{-1} is taken as $\pi/12$. The higher bands for $m > 2$ are not accurately displayed, since the inequality $m/\xi \ll 1$ does not hold. One important result here is that for ξ finite, surface states appear at $z_0 = \pi/2$, as shown by the open circles on the diagram. They appear only for $m =$ odd; this is in agreement with our previous calculations at $z_0 = \pi/2$. Notice also that surface states are not present at $z_0 = 0$, in agreement with our previous calculation at $z_0 = 0$. The states at $m = 2$, $z_0 = \pi/4$ and $m = 2$, $z_0 = 3\pi/4$ will be called "ionic surface states," for reasons given later.

A crystal with $q > 0$ can also be easily handled in a similar manner. Consider first the case where $q > 0$ and $\xi \to \infty$; this corresponds to a crystal with a strong periodic potential and an infinite surface potential. The NFE method no longer applies. Instead, the properties of the Mathieu functions are used directly. In this case, (4.22) reduces to

$$\lambda = \mathrm{se}_m(z_0, q)/\mathrm{ce}_m(z_0, q). \tag{4.28}$$

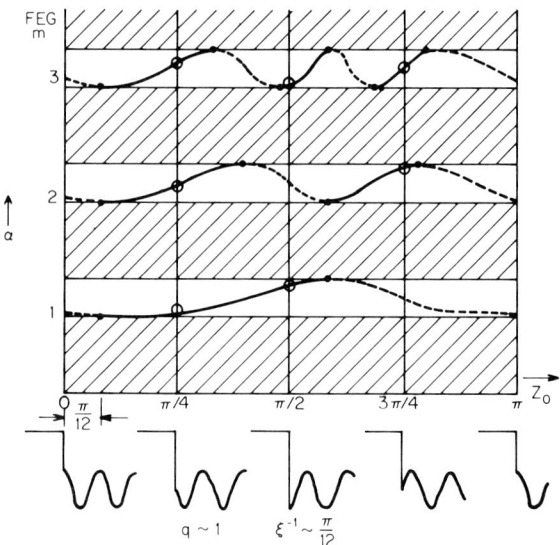

FIG. 26. Surface state energies versus z_0 for a strongly periodic lattice and a finite surface potential. Covalent states are at m = odd, $z_0 = \pi/2$, ionic states are at $m = 2$, $z_0 = \pi/4, 3\pi/4$.

It was shown in Fig. 21 that the main effect of increasing q was to draw the $ce_m(z, q)$ and $se_m(z, q)$ amplitudes and nodes closer to $z = \pi/2$, where the attractive potential was greatest. Thus, the locations of z_0, where $\lambda = 0$ or ∞, will be pulled closer to $z_0 = \pi/2$ as q increases. In this way, Fig. 23 can be simply "compressed" to construct Fig. 25. The shift of nodes is shown schematically in Fig. 25 by arrows. In this figure $q = 1$.

For the final, and most significant, example consider the more general case where $q > 0$ and ξ is finite. Equation (4.22) can now be expanded without further simplification. For m = odd[34,92]

$$\lambda = \frac{\sum_{r=0}^{\infty} B_{2r+1}^{(2n+1)} \left[\sin(2r+1)z_0 - (2r+1)(\xi+\mu)^{-1} \cos(2r+1)z_0 \right]}{\sum_{r=0}^{\infty} A_{2r+1}^{(2n+1)} \left[\cos(2r+1)z_0 + (2r+1)(\xi+\mu)^{-1} \sin(2r+1)z_0 \right]}. \quad (4.29)$$

When ξ is sufficiently large, so that $\xi \gg \mu$ and $(2r+1)/\xi \ll 1$ for all r of interest (applicable for $m/\xi \ll 1$), then

$$\lambda = se_{2n+1}\left[(z_0 - \xi^{-1}), q\right]/ce_{2n+1}\left[(z_0 - \xi^{-1}), q\right]. \quad (4.30)$$

Similar relations apply for m = even. Thus, this case *superposes* the effects of shifting Fig. 23 to the right by an amount ξ^{-1} and compressing it toward the center by an amount proportional to q. The result is shown in Fig. 26 for $\xi = 12/\pi$ and $q = 1$. It should be noted that at $z_0 = 0$ there are still no surface states, while at $z_0 = \pi/2$ there are still surface states for

$m =$ odd. Moreover, at $m = 2$, $z_0 = \pi/4$, the "ionic surface state," first indicated in Fig. 24, has moved closer to the *lower* band edge, and that at $m = 2$, $z_0 = 3\pi/4$, the other "ionic surface state," has moved closer to the *upper* band edge.

In summary, the continuity condition has imposed severe conditions upon z_0, q, ξ, and m in order to have a surface state; the general picture is shown in Fig. 26. Fortunately, using the properties of Mathieu functions one can decompose Fig. 26 into a *translation* (ξ^{-1}) and a *compression* ($\sim q$) of the more basic Fig. 23. The physically important surface states lie at $z_0 = 0$, $\pi/4$, $\pi/2$, and $3\pi/4$, and at $m = 1$, 2. The connection between these states and the Tamm, Shockley, and ionic surface states will be pointed out in Sections 4,f, 4,g, and 4,h, respectively.

d. The Finite Crystal

Surface states of a finite crystal should have somewhat different properties from those of the semi-infinite crystal treated previously. Physically, it is reasonable to expect that, for a very large finite crystal, the semi-infinite model would be a very accurate description of the system. Here, these qualitative effects are expressed on a quantitative basis. Using first-order perturbation theory, it can be shown that[34] the energy shift $\Delta \alpha$, caused by a finite lattice of N cells rather than an infinite one, is given by

$$\Delta \alpha \sim \omega \exp(-2\pi N\mu). \tag{4.31}$$

Thus, for sufficiently large N, $\Delta \alpha \to 0$ and the previous results are recovered. For thin crystals (N small), low damping (μ small), and high surface barriers (ω big), $\Delta \alpha$ can no longer be neglected, and a different calculation is required.

e. Hybridization of All States

Consider the most general surface state $\psi_i(z, q)$, which is a hybrid of *all* the analytically continued bulk standing-wave states, i.e.,[34,92,122]

$$\psi_i(z, q) = \sum_{r=0}^{\infty} [C_r \operatorname{ce}_r(z, q) - S_r \operatorname{se}_r(z, q)] e^{-\mu z} \tag{4.32}$$

where C_r and S_r are functions of q for each FEG r. Suppose the surface state is in the mth FEG. The most important terms in the sum are C_m and S_m, which have already been considered ($\lambda = C_m/S_m$). The purpose of this discussion is to consider the contributions from the states in the other FEG. The assumption of a narrow FEG is no longer required, and the results apply to *all* of Fig. 22a. Following the same procedure used to obtain (4.16) for the general case gives (4.33), where

$$\begin{bmatrix} \alpha'-a_0 & 0 & 0 & h_{20} & 0 & 0 & 0 & h_{40} & \cdots \\ 0 & \alpha'-b_1 & -H_{11} & 0 & 0 & 0 & 0 & 0 & \cdots \\ 0 & H_{11} & \alpha'-a_1 & 0 & 0 & 0 & -h_{13} & 0 & \cdots \\ -h_{20} & 0 & 0 & \alpha'-b_2 & -H_{22} & 0 & 0 & 0 & \cdots \\ 0 & 0 & 0 & H_{22} & \alpha'-a_2 & 0 & 0 & h_{42} & \cdots \\ 0 & 0 & -h_{31} & 0 & 0 & \alpha'-b_3 & -H_{33} & 0 & \cdots \\ 0 & h_{13} & 0 & 0 & 0 & H_{33} & \alpha'-a_3 & 0 & \cdots \\ -h_{40} & 0 & 0 & 0 & -h_{42} & 0 & 0 & \alpha'-b_4 & \cdots \\ \vdots & \vdots & \vdots & \vdots & \vdots & \vdots & \vdots & \vdots & \end{bmatrix} \begin{bmatrix} C_0 \\ S_1 \\ C_1 \\ S_2 \\ C_2 \\ S_3 \\ C_3 \\ S_4 \\ \vdots \end{bmatrix} = 0 \qquad (4.33)$$

$$\alpha' = \alpha + \mu^2 \qquad (4.34\text{a})$$

$$H_{rr} = 2\pi\mu^{-1}K_{rr} \qquad (4.34\text{b})$$

$$h_{rt} = 2\pi\mu^{-1}K_{rt}. \qquad (4.34\text{c})$$

It can be shown that $K_{rt} < K_{rr}$. In particular, when $q \to 0$, $h_{rt} = 0$, but $H_{rr} \neq 0$.

The fact that $h_{rt} = 0$, for $r + t = $ odd, has an interesting physical implication. A surface state located at Γ in the BZ (even-numbered FEG $m = 2, 4, 6, \ldots$) is a linear combination of the unperturbed standing-wave states, all of which are located at Γ; the unperturbed standing-wave states at X (odd-numbered FEG $m = 1, 3, 5, \ldots$) have vanishing matrix elements h_{rt}. Similarly, a surface state located at X in the BZ is formed from the unperturbed standing-wave states at X; the unperturbed states at Γ do not contribute. Note that the dotted boxes along the diagonal in (4.33) include the dominant terms. The previous analysis is equivalent to neglecting all the smaller coupling terms h_{rt}. The determinant of (4.33) then reduces to a simple product of 2×2 determinants, one for each m, except $m = 0$. The neglect of h_{rt}, therefore, uncouples all the FEG, and the surface state problem for the mth FEG can be solved independently of the other gaps. Furthermore, each 2×2 determinant is of the form (4.16) in the narrower FEG limit when $\alpha' \sim \alpha$; i.e., when μ^2 in the diagonal elements of (4.33) can be neglected. The effect of including the μ^2 term is to include the effect of parabolicity on the band structure. In essence, each electron effective mass becomes less than each hole effective mass. But the lowest level is an exception; the μ^2 term must be included in the zeroth approximation.[125]

The main effect of the antisymmetric off-diagonal terms H_{rr} was shown to *attract* the FEG edges toward each other, except for the ground level a_0, which is lowered by the μ^2 term in α'. This is schematically shown in Figs. 27a and 27b. It should be noted that the energies in Fig. 27 are only the *possible* energies obtained by analytic continuation. The corresponding wave functions must also have positive μ and satisfy the continuity condition at $z = z_0$, in order for the possible energies shown there to be true surface state energies.

The coupling terms h_{rt} perturb the independence of the FEG. Without carrying out detailed calculations, it is possible to estimate the qualitative aspects of introducing h_{rt}, by recognizing that the matrix is antisymmetric, so the coupling terms tend to *attract* energy levels; a special case of this was explicitly shown in (4.16) for one isolated FEG. The levels, as shifted by the attractive interaction, are shown in Fig. 27c.

---

[125] The statement by Levine[34] that the a_0 level becomes a metallic surface state lying within the first allowed band is erroneous. In that reference, the μ^2 term in the diagonal element of (4.33) was neglected in the treatment of a_0.

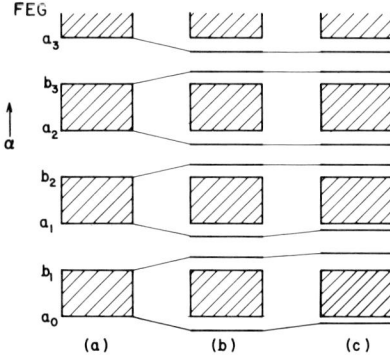

Fig. 27. Schematic energy diagram of: (a) bulk states; (b) possible surface states in narrow-band approximation; (c) possible surface states including interband attractive terms.

The level most affected by this type of perturbation would be the a_0 level, which lies at the bottom of the first allowed band. This will be shifted to a higher energy due to the h_{02} coupling term as shown in Fig. 27c. There is no term pulling in the opposite direction. The a_1 and b_1 hybrids are also affected in a similar manner. They are attracted by the h_{13} and h_{31} coupling terms and their energy is increased somewhat as shown in Fig. 27c. The second and higher bands are probably not greatly affected, since the attractive perturbation tends to act in opposite directions.

Few essentially new results occur as a result of using the complete matrix (4.33) instead of the 2×2 matrix (4.16). Therefore, Figs. 23–26 are essentially unchanged.

f. Tamm States

Tamm states[14] are regarded by most workers to constitute a special category of surface states. They are usually considered to appear when there is a strong surface "perturbation," and to disappear when there is no surface "perturbation." This definition is hazy, and sometimes misleading, since the very presence of a surface indicates some significant perturbation.

A Mathieu analog of a *strong surface "perturbation"* in a monatomic crystal would be the termination at, say, a potential *minimum*, as shown by $z_0 = \pi/2$ in Figs. 20 and 26. The atoms in the monatomic crystal are presumed to be located at each potential minimum: $z = \pi/2, 3\pi/2, 5\pi/2, \ldots$. As shown in Fig. 26, the expected Tamm surface states appear at $z_0 = \pi/2$, but only in the odd-numbered FEG. The surface "perturbation" seems strong only for certain FEG, according to the Tamm picture. In other words, the Tamm states for a sinusoidal potential appear only at X in the 1-D BZ, as shown in Fig. 22b. The conclusion that Tamm states appear in the odd-numbered FEG for the Mathieu problem was inferred first by Statz,[21] by analogy between Shockley's band-crossing diagram[17] and the Mathieu α vs q diagram (Fig. 22a). Statz did not consider a continuity condition, however, or derive an energy level as in (4.21a) or (4.22). The fact that a Tamm surface state should appear in the FEG $m = 1$ can also be predicted by the NFE method, following Maue[15] and Goodwin.[16] According to this method, a surface state will appear in the mth FEG if the mth Fourier coefficient is negative (mth potential harmonic is terminated at a minimum). Translating this language into the Tamm states of the Mathieu problem gives $z_0 = \pi/2$, $V_1 < 0$, $V_2 = V_3 = V_4 = \cdots = 0$, and a Tamm state is expected to appear at $z_0 = \pi/2$ in the first FEG $m = 1$. There is such a state as shown in Fig. 26. The NFE method is silent regarding the problem of Tamm surface states in FEG $m > 1$, since there is no harmonic higher than the fundamental in the Mathieu problem. In other words, the α vs q diagram of Fig. 22a shows that α varies linearly with q for $m = 1$, but α varies quadratically with q for $m > 1$. The linear part can be accommodated by the NFE theory, but the quadratic part requires a higher-order perturbation theory. Square wells are rich in higher harmonics and the expected surface states are indicated by Statz.[21]

On the other hand, a Mathieu analog of *no surface "perturbation"* in a monatomic crystal would be the termination at a potential maximum $z_0 = 0$. (The surface step is still, strictly speaking, a nonvanishing perturbation.) Thus, Tamm states should then be *absent* at a termination near $z_0 = 0$, as shown in Fig. 26. Statz[21] had anticipated this result, and it is also the result expected for $m = 1$ using the NFE method of Goodwin. As outlined above, the potential harmonics evaluated at $z_0 = 0$ are $V_1 > 0$, $V_2 = V_3 = V_4 = \cdots = 0$. Since $V_1 > 0$, no surface states with $z_0 = 0$ can appear in the first FEG.

At this juncture, it may be instructive to compare the results of the Mathieu problem with those obtained by the TB-CO methods as discussed in Part III. To do this, it is necessary to relate the terminologies. On one hand, the surface is characterized by a termination location z_0 within the last unit cell, a phase shift ξ^{-1} related to the step height at the surface, and a damping constant μ within the crystal. In the LCAO representation, the surface is characterized by a perturbed Coulomb integral α', different from the bulk Coulomb integral α, and a damping constant μ (LCAO). The way to compare terminologies, then, is to equate the μ's. For the case of the surface states near the bottom of the FEG ($\lambda \sim 0$) in the Mathieu problem

$$\mu = \tfrac{1}{2}(a_m - b_m)(z_0 - \xi^{-1}) \qquad (4.35)$$

while for the single-band LCAO problem, as given by Goodwin,[16]

$$\mu(\text{LCAO}) = \pi^{-1}[\ln(\alpha' - \alpha) - \ln \beta]. \quad (4.36)$$

Thus, the correspondence[34] between the two terminologies is

$$\alpha - \alpha' = \exp[\pi z_0(a_m - b_m)/2] \quad (4.37)$$

$$\beta = \exp[\pi(a_m - b_m)/2\xi]. \quad (4.38)$$

In physical terms, the Coulomb energy difference $\alpha' - \alpha$ in LCAO terminology is related to the termination z_0, as seen in (4.37). Positive $\alpha' - \alpha$ corresponds to a greater surface repulsive potential on the last cell, which is the same effect as positive z_0, as seen from Fig. 26. Also, the threshold energy ($\alpha' - \alpha \geq \beta$ for surface states) is related to the step height ξ at the surface (i.e., phase shift) as seen in (4.38). These results are valid for all m. Somewhat similar arguments can be applied to the case of surface states near the top of the FEG ($\lambda \to \infty$).

g. Shockley States

Shockley states[17] are considered by most workers to constitute another special category of surface states, complementary to Tamm states. Shockley states are usually considered to appear when there is no surface perturbation, and to disappear when there is a strong surface perturbation, provided the bands are "crossed." As mentioned before, this definition is hazy and sometimes misleading. In 1-D problems involving periodic potentials, a shift of one-half cell (a redefinition of where the atomic nuclei are located) sometimes causes Tamm states to become Shockley states and vice versa. This feature was clearly stated by Lippmann.[24]

A Mathieu analog of a very weak surface perturbation in a monatomic crystal would be the termination at, say, a potential *maximum*, as shown by $z_0 = 0$ in Figs. 20 and 26. How can one tell if the bands in the Mathieu problem are "crossed"? As shown in Fig. 21, the functions $ce_m(z, q)$, $se_m(z, q)$ smoothly become atomic functions as q increases. These functions are ordered by energy in the normal way: bonding s, antibonding s, bonding p, antibonding p, bonding d, ..., etc. Thus, one would suspect that in the α vs q diagram of Fig. 22a there is no band "crossing" and, hence, Shockley states should not appear. As seen in Fig. 26, there are no "Shockley" states at $z_0 = 0$.

There is no guarantee, however, that the termination at a potential maximum is the most appropriate termination for simulating a 1-D covalent lattice (i.e. Si, Ge, diamond) as Shockley tacitly assumed. First, Kleinman and Phillips[126] computed that the valence–electron pseudo-

---
[126] I. Kleinman and J. C. Phillips, *Phys. Rev.* **125**, 809 (1962).

potential has a *relative minimum in between the NN atoms of a covalent lattice*. This minimum causes a localization of the covalent bond *in between* the covalent atoms, as can be anticipated from simple chemical arguments. Thus, the more appropriate place to terminate a covalent "dangling bond" seems to be at the potential minima.[5,127] In the Mathieu problem, this would correspond to a termination at $z_0 = \pi/2$. There are indeed surface states here, but they have already been considered under the category of Tamm states. This underscores the kind of confusion that surrounds the Tamm–Shockley categorization. Second, it is now known that covalent crystals have reconstructed surfaces according to LEED, as shown in Part VII. Apparently, freshly formed dangling bonds are so unstable that the surface atoms spontaneously reconstruct to form a more stable entity. In this sense, one cannot speak of dangling bonds in the usual way, since they are saturated to some unknown extent by the atomic reconstruction. Finally, the transformation $q \rightarrow -q$ changes the order of the bulk wave functions from ce_0, se_1, ce_1, se_2, ce_2, ce_3, ... to ce_0, ce_1, se_1, se_2, ce_2, ce_3, ...[92,122] without changing the energies. This would seem to be an operation which simultaneously shifts the origin by $\pi/2$ (adds a half-cell to the crystal) and also "crosses" the bands. Thus, many possibilities for confusion exist in this definition of Shockley states. The situation seems to be somewhat simpler in dealing with the CO methods. There, Shockley states are supposed to appear if the surface Coulomb and resonance integrals are unperturbed, and if the bands are crossed.[2,16,22,128]

Since the classification of surface states appears to depend on the type of material (i.e. ionic, covalent, metallic crystals, etc.) being considered, it is perhaps wise to refrain from making any general classification, such as Tamm or Shockley states, and to just treat each material on its merits. In this way, one may avoid the confusion which arises as a result of awkward definitions. An example of such a possible reclassification is given in the following section.

h. Ionic Surface States

Ionic and partially ionic crystals, such as the III–V, II–VI, and I–VII compounds, seem to have a unique distribution of intrinsic surface states (Section 17,c). These crystals are of the class AB, where A is the metallic cation and B is the nonmetallic anion. The AB notation is useful, even though the crystals have generally a fractional charge. The detailed discussion of ionic surface states in Sections 1,b, 10, and 17,c shows that the

---

[127] J. D. Levine (unpublished). See P. Mark *in* "Clean Surfaces: Their Preparation and Characterization for Interfacial Studies." (G. Goldfinger, ed.). Dekker, New York, 1969.

[128] P. Mourad, *Phys. Status Solidi* **12**, 817 (1965).

bulk conduction (valence) band can be considered composed primarily of the A (B) ions, and the FEG width is a measure of the electrostatic or Madelung energy. Levine and Mark[30] computed the Madelung energies of the surface ions and thereby deduced a "pair-like" distribution of surface states. An A- (B)-like surface state seems to appear a few tenths of an electron volt below (above) the conduction (valence) band, provided the surface has both A and B ions in the surface plane. This "pair-like" distribution of surface states, shown schematically in Fig. 34e, will be conveniently called "ionic surface states" for the rest of the review. It is felt that this terminology is very suggestive and much more useful than other possible terminologies involving subclasses of Tamm (or Shockley) states for the separate A- and B-like surface species.

In the Mathieu model, ionic surface states also appear with similar properties. They can be found in Fig. 26, by invoking the following two considerations. First, an AB crystal can be simulated in the Mathieu model by assuming the A (B) ions to be located at the sinusoidal potential maxima (minima). This correspondence indicates that the A (B) ions are electropositive (electronegative), in agreement with usual chemical arguments. Accordingly, a termination at an A (B) ion would be at $z_0 = 3\pi/4$ ($\pi/4$) in Fig. 26. These locations are neither at potential maxima or minima, but are midway between the ions. Second, the valence-conduction FEG for many III–V and II–VI crystals is at Γ ($\mathbf{k} = 0$) in the BZ. This corresponds to the *second* FEG in the Mathieu model ($m = 2$), since the first FEG (at X in the 1-D BZ) separates the s and p bands of the filled B ions, in a simple chemical picture. The A (B)-like surface states are indeed found at $m = 2$ in Fig. 26 with $z_0 = 3\pi/4$ ($z_0 = \pi/4$). Furthermore, it can be shown that, as the ionicity is increased (i.e. q is increased), the A (B) states retreat into their respective conduction (valence) bands.

It is illuminating to compare the ionic surface state properties deduced above with those of the band-edge LCAO model (Section 1). There the ionic crystal is simulated by placing s (p) orbitals on the A (B) ions. In 3-D, three orthogonal p orbitals have to be used. The analogy between the methods can best be seen by comparing the energy matrices, rewritten here in the Mathieu model notation[34]

$$\begin{bmatrix} a_m - \alpha & -2\pi^{-1}K_{mm\mu} \\ 2\pi^{-1}K_{mm\mu} & b_m - \alpha \end{bmatrix} \begin{bmatrix} \lambda \\ 1 \end{bmatrix} = 0 \qquad (4.39)$$

and the band-edge LCAO model notation[33]

$$\begin{bmatrix} \alpha_A - E & -2\beta \sinh \mu \\ 2\beta \sinh \mu & \alpha_B - E \end{bmatrix} \begin{bmatrix} R \\ 1 \end{bmatrix} = 0. \qquad (4.40)$$

In (4.39), m is the FEG index, which for ionic surface states is equal to 2. Since $a_2 = \alpha_A$ and $b_2 = \alpha_B$, the diagonal elements are the same in (4.39) and (4.40). The off-diagonal elements of (4.39) contain the $\mathbf{k}\cdot\mathbf{p}$-like integral in dimensionless form, viz.,

$$K_{mm} = \int_0^{2\pi} \mathrm{ce}_m(z, q) \, (d/dz) \, \mathrm{se}_m(z, q) \, dz. \qquad (4.41)$$

This integral couples the $\mathrm{ce}_m(z, q)$ and $\mathrm{se}_m(z, q)$ functions, which are the band-edge wave functions of the top and bottom, respectively, of the mth FEG. By comparison, the off-diagonal term in (4.40) contains an integral of the resonance type with strength β. It also couples the band edges, but in a somewhat different way. In any case, the off-diagonal term β gives the curvature to the band edges, as shown in Fig. 2, and can thus be determined from effective-mass experiments.

In the Mathieu problem, the λ parameter hybridizes the $\mathrm{se}_m(z, q)$ and $\mathrm{ce}_m(z, q)$ functions, while in the LCAO problem the R parameter hybridizes the ψ_A and ψ_B functions.

There are also similarities with regard to the surface-matching equations. Both λ and R are uniquely determined by these equations, and the existence conditions for surface states are essentially $\lambda > 0$ and $R > 0$.[33] This similarity holds even though the Mathieu problem requires equating the values and slopes of ψ at the surface z_0, while the LCAO problem simply removes atoms to the left side of the surface ($a_0 = 0$, $a_{-1} = 0$, $a_{-2} = 0$, etc.).

It is also illuminating to relate this band-edge LCAO method to the Madelung problem.[30] There the band-edge energies are simply taken as α_A and α_B and the surface state energies are simply taken as α_A' and α_B'. This simplicity is a consequence of the lack of a quantum-mechanical foundation in the Madelung problem. Also, the classical Madelung problem contains no bulk bandwidth; this is a consequence of the absence of off-diagonal terms containing β.

V. Three-Dimensional Calculations

Up to now, only 1-D models have been considered for reasons of simplicity.[129] In this section, more realistic models are to be treated, which take into account the other two-dimensions parallel to the surface. Three types of complications arise: first, experimental evidence that certain surfaces are reconstructed or mildly rearranged; second, from the lack of knowledge of the crystal potential variation in the vicinity of the surface and, third,

---

[129] In the RCO method, a 3-D system is reduced to a 1-D analog by using the so-called "layer orbitals," which are constructed so that they contain the complementary symmetry of the virtual 1-D crystal.

from the computational difficulties which are inherent in any 3-D calculations.

The surface atoms are frequently displaced both laterally and normally to the surface (Sections 1,c and 15). Since it is necessary to know the atomic positions for accurate calculations, one must refer to experimental data, to take into account such geometrical effects. If the potential were known completely, for a 3-D finite crystal, the Schroedinger equation could be solved, in principle, exactly and the bulk and surface states would appear naturally, without the need for using artificial boundary or matching conditions. However, in practice, one must resort to idealizations of the surface regions. Two simplifying assumptions include a step-potential termination in 1-D and a planar discontinuity termination in 3-D.[130] From the computational point of view, 3-D calculations of bulk band structure are now standard.[98,105] However, when a surface is present, the complexity arising from the boundary conditions describing the surface effect makes the problem an order of magnitude more difficult.

5. Direct Extension of One-Dimensional Picture

The simplest way of looking at a real crystal in 3-D is to take the 1-D models and incorporate the other two dimensions by superposition if possible. For example, in the crystal orbital approach (Part III), the 3-D bulk reduced energy expression for an idealized monatomic cubic crystal is[131]

$$X = \cos\theta_x + \cos\theta_y + \cos\theta_z \tag{5.1}$$

where θ_z is the dimensionless wave vector normal to the surface and $\theta_{x,y}$ are those parallel to it. The corresponding wave function has the form

$$\psi = \psi(\theta_x)\psi(\theta_y)\psi(\theta_z). \tag{5.2}$$

On taking the surface into account, $\theta_z = i\mu_z$ or $(\pi + i\mu_z)$, so (5.1) becomes[1,67]

$$X = \cos\theta_x + \cos\theta_y \pm \cosh\mu_z \tag{5.3}$$

and the wave function (5.2) is now exponentially damped in the z direction. For $\theta_{x,y,z} = 0$ and π, $X_0 = 3$ and $X_\pi = -3$, so the volume bandwidth is

$$\Delta X_v = X_0 - X_\pi = 6.$$

In (5.3) the value of μ_z is determined by the surface state existence conditions, the minimum value being $\mu_z = 0$[33], which corresponds to a band edge. In this case, on taking the negative sign in (5.3) and setting $\theta_{x,y} = 0$ and

[130] V. Heine, *Proc. Phys. Soc.* **81**, 300 (1963).
[131] N. F. Mott and H. Jones, "The Theory of the Properties of Metals and Alloys." Dover, New York, 1958.

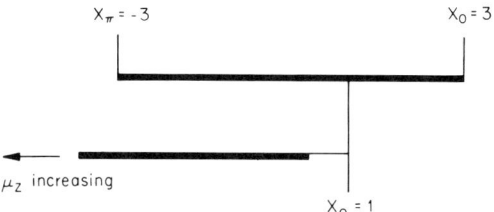

Fig. 28. Overlapping of volume band ($\Delta X_v = 6$) and surface band ($\Delta X_s = 4$).

π, then $X_0 = -3$, so the surface state bandwidth is

$$\Delta X_s = 4$$

and the volume and surface bands are completely overlapped. As μ_z increases, the surface band moves to the left (Fig. 28) and starts to separate from the volume band. Although this separation depends on μ_z, the bandwidths do not, thus $X_v = 6$ and $X_s = 4$. It should be pointed out that, in the more general 3-D problem, μ_z is a function of $\theta_{x,y}$. A further discussion of volume and surface band overlapping, and its relation to scattering resonance, is given in Section 8.

6. Analytical Continuation

The simple procedure used above can, in principle, also be applied to more complicated situations. As shown by Heine,[130,132] Boudreaux and Heine,[132a] Blout,[133] Jones,[134] and Stern et al.,[135] exponentially damped surface states can be obtained by *analytic continuation* of the bulk band structure. Here the bulk energies $E_n(\theta_x, \theta_y, \theta_z)$ and wave functions $\psi_n(\theta_x, \theta_y, \theta_z)$ for the nth band are evaluated for complex θ_z. There still remains the problem of matching the boundary conditions, where it is necessary to use a linear combination of a number of analytically continued bulk wave functions, frequently supplemented by other, more localized wave functions. It should be noted, of course, that analytic continuation is independent of the representation, e.g., augmented plane wave (APW)[136], $\mathbf{k} \cdot \mathbf{p}$,[137] tight-binding (TB),[2,100] pseudopotential (PP),[139] etc.

[132] V. Heine, *in* "Solid Surfaces" (H. Gatos, ed.). North-Holland Publ., Amsterdam, 1964.
[132a] D. S. Boudreaux and V. Heine, *Surface Sci.* **7**, 426 (1967).
[133] E. I. Blout, *Solid State Phys.* **13**, 305 (1962).
[134] R. O. Jones, *Proc. Phys. Soc.* **89**, 443 (1966).
[135] R. M. Stern, J. J. Perry, and D. S. Boudreaux, *Rev. Mod. Phys.* **41**, 275 (1969).
[136] P. Phariseau, *Phys. Status Solidi* **3**, 2391 (1963).
[137] C. M. Chaves, N. Majlis, and M. Cardona, *Solid State Commun.* **4**, 271 (1966).
[138] D. Pugh, C. Rhys-Roberts, and R. H. Tregold, *Proc. Phys. Soc.* **80**, 534 (1962).
[139] R. O. Jones, *Phys. Rev. Letters* **20**, 992 (1968).

7. Complications at the Boundary Region

Even the best modelistic band structure picture, suitably analytically continued, gives no guarantee that the derived surface states are realistic, because of the uncertainties inherent in the approximations associated with the boundary regions. In most APW, PP, $\mathbf{k}\cdot\mathbf{p}$ and NFE[16] methods, the surface region is represented by a planar discontinuity. As shown in Fig. 29, the three major objections to such an approach are: (a) The locus of the constant energy points in the surface zone is not, as usually assumed, a plane, but reflects the "atomic graininess," which is indicated by the broken lines.[5] (b) Each of these loci have different profiles for different energies as depicted in the figure. For an accurate calculation, the wave function inside and outside the crystal must be matched along the undulating constant energy contours, for each energy. (c) Ignoring the above structure, it is common practice to idealize the energy profile by a planar discontinuity,[16,136,137,139] where the wave functions are matched only at a few symmetry points in the surface unit cell. Moreover, the appropriate location of such a planar discontinuity is unknown.

An attractive method, which seems to reduce the above shortcomings to a minimum, is the CO–TBA (Part III). Here the surface region is represented by a few *perturbed* Coulomb and resonance integrals. These can sometimes be estimated via the simple concepts of dangling bond (Section 9), classical electrostatics (Section 10), and quantum mechanics.

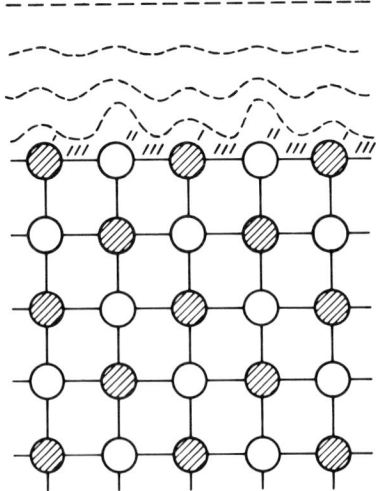

Fig. 29. Schematic potential energy profile of the surface of a diatomic lattice. The dashed curves represent the loci of constant potential energy points. The primes denote various perturbed Coulomb and resonance integrals in the CO–TBA method.

It is conceded that the CO methods are not the best methods for bulk band structure calculations. In other words, the use of a few AO's does not adequately yield a complete description of the E vs k curves over an energy range of, say, 20 eV. On the other hand, if one is interested only in surface states within a particular FEG of, say, 2 eV, then the use of the CO method is more justified. It is possible to simulate the known band-edge diagrams by the use of the "band-edge" method.[33,140] The basic idea is to assign Coulomb integrals to the well-known band edges and to assign resonance integrals to the well-known band curvatures (electron and hole effective masses). This particular method has been successfully applied to III–V and II–VI compounds,[140] where the valence–conduction FEG is direct and lies at the center Γ of the BZ.[124] By combining this method with experimental measurements of surface states, the location of the A and B surface atoms has recently been inferred by Levine and Freeman.[140]

8. Density of States

A feature which is absent from 1-D calculations is the energy distribution of intrinsic surface states, the so-called *density of states*. The number of surface states can be of the order of one per surface atom. In 3-D, the surface states interact among themselves and, consequently, a broadening of the surface band occurs. Both the energy and density of states are, in principle, experimentally observable (Part VII). Some simple examples of possible surface state band structures and energy distributions are schematically shown in Fig. 30. The energy is plotted against the wave vector parallel to the surface plane for both bulk and surface states. The bulk states have an additional normal wave vector $\mathbf{k}_n$, so that the bulk bands become a quasi continuum with a central band gap, as is found for most III–V and II–VI compounds.[124]

In case A (Fig. 30), a large number ($\sim 4 \times 10^{14}$ cm^{-2}) of surface levels are concentrated in a very narrow energy range ($\sim 10^{-1}$ eV). The density of states is, therefore, $\sim 4 \times 10^{15}$ cm^{-2} eV^{-1}, which is the situation in Ge and Si (Part VII). For cases B and C, the surface bands are gently perturbed from the bulk band edges, so that a large overlap can exist between surface and bulk bands as illustrated. Case C differs from case B in that it crosses the bulk band edge, where the states become scattering resonances,[140] instead of the usual localized type. Curve B resembles C in that it has a similar curvature at $k_{||} = 0$ (i.e. similar effective mass). The density of states for an effective mass ratio of 0.2 yields $\sim 10^{14}$ cm^{-2} eV^{-1} (Part VII). This is two orders of magnitude below that of case A. In agreement with experiment, 3-D computations[140] indicate that curves B and C (not A) are more

---

[140] J. D. Levine and S. Freeman, *Bull. Am. Phys. Soc.* **14**, 787 (1969); *Phys. Rev.* (in press) (1970).

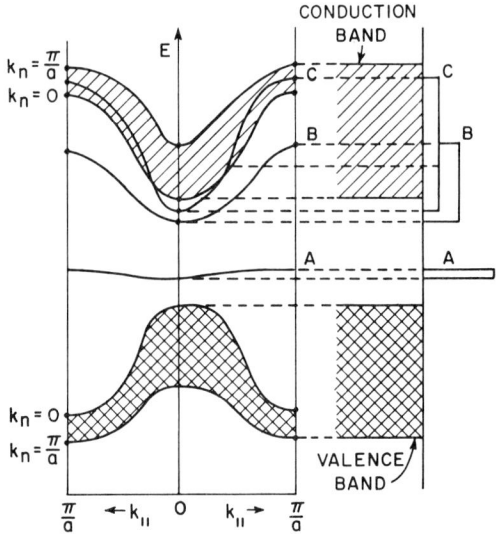

FIG. 30. Energy versus wave vector and density of states for some possible surface state models.

appropriate for zinc blende. From the density of states point of view (right-hand side of Fig. 30), the surface levels appear to merge into the conduction band, however, this can be deceiving, since discrete levels are always possible providing no band edge crossing has occurred in **k**-space.

VI. Semiclassical Methods

As described above, the surface state problem is extremely difficult to solve using quantum theory in 3-D. For this reason, it is desirable to introduce simple semiclassical analogs, where applicable. These are crude and inaccurate compared to the corresponding quantum theory models, but their strength and usefulness lies in their simplicity and in their ties to familiar concepts in structural chemistry and classical electrostatics. Furthermore, certain integrals needed in the quantum theory approach can be estimated from the semiclassical models, as will be shown later.

The semiclassical model of a covalent crystal is the tetrahedral bond. In this model the surface is represented naturally by a broken or dangling bond. On the other hand, the semiclassical model of an ionic crystal is an array of ions. In this model the surface ions have a reduced coordination number, which causes a reduced surface Madelung potential. These two somewhat complementary models will be reviewed *very briefly below*. Of course, for III–V and II–VI compounds neither model is very satisfactory, so one must somehow extrapolate between the two extremes of covalent and ionic prototypes.

9. Dangling Bond Approach

In the conventional ball-and-stick model of a covalent crystal, a free surface is represented by a plane of broken bonds (also called dangling bonds). The bulk atoms have four tetrahedral bonds, which are doubly occupied, because of electron spin, to yield the closed shell of eight electrons. The surface atoms may have one or two broken bonds, depending on the crystal face. Three broken bonds is probably an unstable situation.

Geometrical calculations will now be carried out for the diamond structure. Properties include the crystal face, the number of broken bonds per atom, the atom packing density, and the broken-bond density. These are given in Table II for Ge. Here S is the fcc crystal edge ($S = 5.65$ Å for Ge and 5.42 Å for Si). Table II indicates that the surface density of broken bonds is least for (111), greater for (110), and greatest for (100). Handler[141] has utilized this table to relate Ge surface state data; Huff et al.[29] have used it to describe InSb surface states; and Allen and Fowler[142] have employed it to correlate work function measurements on different faces of Ge. A main deficiency of the table is that the surfaces undergo atomic reconstruction, according to LEED data, i.e., the dangling bonds rearrange to partially saturate themselves. This effect is discussed in Part VII, and is not taken into account in this table.

The dangling-bond concept was the first semiclassical explanation offered for the existence of surface states. (The first quantum explanation was made by Tamm[14]). Each broken bond was supposed to represent one, or maybe more, electronic surface states. Handler[143] and Huff et al.[29] recognized that the surface states were perturbed bulk states, formed out of the valence and conduction bands. But no attempt was made to carry out computations of surface state energy levels using the dangling-bond concept.

Dangling bonds are frequently used to explain the *structural* and *chemical* properties of Si and Ge (see Section 17), and the polar surfaces of partially ionic zinc blende and wurtzite semiconductors. Consider, e.g., the zinc blende compound GaAs. The {111} crystal polar direction consists of

---

[141] P. Handler, in "Semiconductor Surface Physics" (R. H. Kingston, ed.), p. 23. Univ. of Pennsylvania Press, Philadelphia, Pennsylvania, 1957.
[142] F. G. Allen and A. B. Fowler, J. Phys. Chem. Solids **3**, 107 (1957).
[143] P. Handler, J. Phys. Chem. Solids **14**, 1 (1960).

alternating layers of A and B atoms. The ideal (111) or A face of GaAs terminates in a plane of Ga atoms, while the ideal ($\bar{1}\bar{1}\bar{1}$) or B face terminates in a plane of As atoms. Experimentally it is found that the A face has different etching and oxidizing properties from the B face. This was explained by Gatos and Lavine[28,144] in the following way. The bulk atoms in GaAs have eight shared electrons as shown in Fig. 31a. Upon the formation of a (111) surface by breaking one bond, as shown schematically in Fig. 31b, the possibility exists that there will be seven electrons each for the surface atoms. But from chemical arguments, Gatos and Lavine claim another configuration is more probable, as shown in Fig. 31c. Here Ga (atomic valence 3) loses its valence electron to the As (atomic valence 5) on the other free surface of the crystal to complete a shell of eight electrons.

TABLE II. IDEALIZED PROPERTIES OF UNRECONSTRUCTED Ge SURFACES[a]

Face	Broken bonds per atom	Density of surface atoms	Bonds cut/cm^2	Relative number bonds cut/cm^2
(111)	1	$4/\sqrt{3}S^2 = 7.3 \times 10^{14}$ cm^{-2}	$4/\sqrt{3}S^2$	1.0
(110)	1	$2\sqrt{2}/S^2 = 8.92 \times 10^{14}$ cm^{-2}	$2\sqrt{2}/S^2$	1.23
(100)	2	$2/S^2 = 6.3 \times 10^{14}$ cm^{-2}	$4/S^2$	1.73

[a] Actually, the surfaces of Ge and Si are reconstructed.

The surface Ga has only six shared electrons. There will be a tendency for sp^2 planar hybridization and the Ga will be relatively chemically inert. On the other hand, the surface As has eight shared electrons, and there will be a tendency toward keeping the sp^3 tetrahedral bonds. The dangling electron pair on As makes it a chemically active donor, and it should be most active in oxidizing environments. In this manner, many observed chemical differences between the A and B faces were explained[145]. Similar results were obtained with the polar A and B faces of wurtzite. Early studies on spontaneous bending of thin (111) III–V crystals were also interpreted by this broken-bond model[146]; but they have been criticized,[147] because surface-polishing work damage was not etched away, and residual stresses and/or

[144] H. C. Gatos and M. C. Lavine, *J. Phys. Chem. Solids* **14**, 169 (1960); see also H. C. Gatos, *J. Appl. Phys.* **32**, 1232 (1961).
[145] A. N. Mariano and R. E. Hanneman, *J. Appl. Phys.* **34**, 384 (1962).
[146] M. C. Finn and H. C. Gatos, *Surface Sci.* **1**, 361 (1964).
[147] A. Taloni and D. Haneman, *Surface Sci.* **8**, 323 (1967).

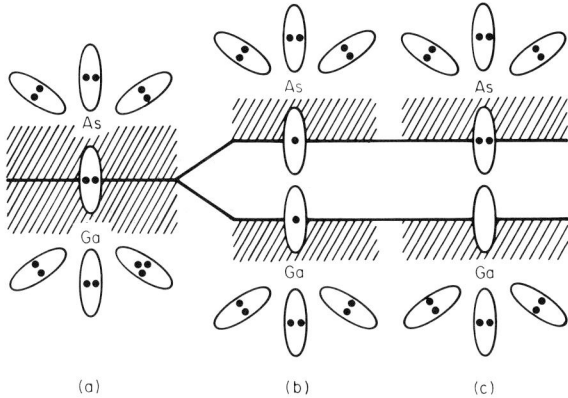

Fig. 31. Dangling bonds on {111} GaAs.

surface oxides probably caused most of the bending. Dangling bonds are frequently used in theories of crystal nucleation and growth.[148]

These dangling-bond approaches are thrown somewhat into doubt by the LEED observations that the supposedly ideal (111) polar faces of zinc blende are, in fact, reconstructed [or sometimes form (110) unreconstructed microfacets]. Haneman[149,150] proposed a model in which the LEED information is taken into account by having $\frac{1}{4}$ of the surface A atoms slightly raised and $\frac{3}{4}$ of the surface A atoms slightly lowered on the A face of (111) zinc blende. Essentially, the raised atoms were assumed hybridized mainly s type, while the lowered atoms were assumed hybridized mainly p type. Haneman[149,150] and Holt[151] considered the charges on the atoms of the A and B faces to be $6 + \frac{3}{4}$ and $6 + \frac{5}{4}$, respectively. MacRae[152,153] has also used certain dangling bond concepts in interpreting LEED patterns on (110) and polar faces of zinc blende.

In conclusion, therefore, it can be said that dangling-bond ideas have proved useful in the structure and chemistry of intrinsic surfaces, but have been singularly unsuccessful in correlating the energies and energy distributions of intrinsic surface states, apart from the crude postulate that every dangling bond yields of the order of one surface state.

[148] R. G. Sangster, *in* "Compound Semiconductors" (R. K. Willardson and H. L. Goering, eds.), Vol. 1, p. 241. Reinhold, New York, 1967.
[149] D. Haneman, *J. Phys. Chem. Solids* **14**, 162 (1960).
[150] D. Haneman, *Phys. Rev.* **121**, 1093 (1961).
[151] D. B. Holt, *J. Appl. Phys.* **31**, 2231 (1960).
[152] A. U. MacRae and G. W. Gobeli, *in* "Semiconductors and Semimetals" (R. K. Willardson and A. C. Beer, eds.), Vol. 2. Academic Press, New York, 1966.
[153] A. U. MacRae, *Surface Sci.* **4**, 247 (1966).

10. Madelung Potential Method

The ionic prototype of a crystal is an array of point charge ions. It is well known that most ionic crystal structures[154] and bond energies[155] have been explained using this ionic prototype; even though covalency is also present to some extent. The electronic levels of ionic crystals and partially ionic crystals (NaCl, ZnO, ZnS) were computed semiclassically by Seitz[156] and Mott and Gurney.[157] They used simple electrostatics and Madelung sums (rather than quantum mechanics or dangling bonds). Even though a Madelung sum is an idealization, because of electron–electron screening, it is very useful for certain simple applications.[158]

Levine and Mark[30] have extended this semiclassical method to treat ionic surface states. Their approach is idealized and approximate, but serves the purpose of conveniently categorizing bulk and surface states on all ionic crystals and on all their crystal faces. Such a simple convenient categorization has not proved possible, as yet, for covalent or metallic crystals. The Madelung potential method will first be reviewed for bulk states, and will then be extended to surface states. The main numerical results have been computed in the form of ratios of various surface and bulk properties; these ratios are expected to be more accurate than differences between absolute magnitudes.

Consider an ionic crystal of the form AB where A is the cation and B is the anion. In the bulk crystal, the anions interact to form the valence band, which is occupied at $0°K$. Also the cations interact to form the conduction band, which is unoccupied at $0°K$. In the discussion to follow, the valence and conduction bands will be idealized as discrete levels, and the percentage of ionic character will be idealized at 100%. Departures from these idealizations will be considered later.

Experimental evidence that the valence band is anion-like for III–V, II–VI, and I–VII crystals has been given by Swank,[159] as reviewed in Part VII. His plot of photoelectric threshold versus electronegativity *of the anion* yields an accurate straight line for eighteen crystals. LEED evidence also indicates that the ionic component of these crystals dominates, since the electrostatically neutral (checkerboard-like) surfaces are not reconstructed (see Section 15).

---

[154] A. F. Wells, "Structural Inorganic Chemistry." Oxford Univ. Press, London and New York, 1962.
[155] F. Seitz, "Modern Theory of Solids," p. 76. McGraw-Hill, New York, 1940.
[156] F. Seitz,[155] pp. 408 ff and 447 ff.
[157] N. F. Mott and R. A. Gurney, "Electronic Processes in Ionic Crystals," p. 86. Dover, New York, 1964.
[158] N. F. Mott, *Proc. Phys. Soc.* **49**, 258 (1937).
[159] R. K. Swank, *Phys. Rev.* **153**, 844 (1967).

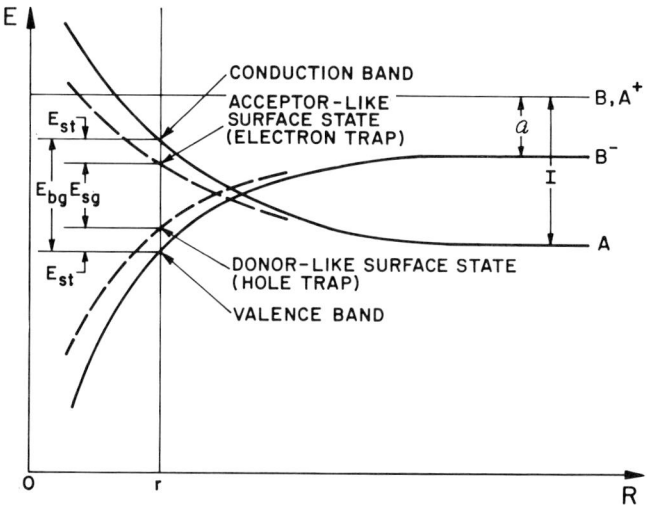

FIG. 32. Energy levels as a function of interionic distance R for an AB ionic crystal.

According to Seitz,[156] the energy levels vary with internuclear distance R (lattice constant) in a manner as shown in Fig. 32. For $R \to \infty$ the levels of the separated species are A and B^-. These are related to the ionization potential I and electron affinity α, as shown. Since A is lowest, the separated atoms are stable as *neutral* A and B species. As R is imagined to be uniformly reduced, there is a crossover, due to the Madelung potential, which is symmetric about the midgap position $(I + \alpha)/2$. At the equilibrium distance r, the B^- level is lowest, so the ionic crystal species are stable as A^+ and B^- ions. The bulk bandgap E_{bg} (valence-to-conduction band) is given by

$$E_{bg} = 2V_b - I + \alpha \qquad (10.1)$$

where V_b is the bulk Madelung potential. The greater the Madelung potential, the greater the bandgap; this is an essential feature. In a similar manner,[30] the *surface state band gap* E_{sg}, shown in Fig. 32, becomes

$$E_{sg} = 2V_s - I + \alpha. \qquad (10.2)$$

Here I and α are assumed to be the same as in the bulk, but the surface Madelung potential V_s is assumed different ($V_s \neq V_b$). From Fig. 32 the surface state trap depth E_{st} is defined:

$$E_{st} = (E_{bg} - E_{sg})/2. \qquad (10.3)$$

In this simple electrostatic model, *the surface states appear in symmetric A and B pairs with the same* E_{st}. Formally, E_{st} is related to the Coulomb integral difference $\alpha_A - \alpha_A'$ and $\alpha_B' - \alpha_B$ of Part III. Since ratios are more accurate then absolute magnitudes, (10.1)–(10.3) can be transformed into

$$E_{st}/E_{bg} = (1 - \gamma)/2(1 - \mu). \tag{10.4}$$

Here a geometric ratio γ and a material ratio μ are introduced and defined by

$$\gamma = V_s/V_b \tag{10.5}$$

$$\mu = (I - \mathcal{C})/2V_b. \tag{10.6}$$

The main general conclusion that can be drawn from (10.4) is that the surface state traps are relatively deeper for crystals with *smaller* γ and for *larger* μ. These ratios will now be evaluated, in turn. The normal displacements of a few percent in the surface atoms will be ignored in this simple treatment.

The relevent Madelung potentials needed to compute γ are defined as

$$V_l = c_l ez/r, \quad l = b \text{ or } s, \tag{10.7}$$

where $c_{b(s)}$ is the bulk (surface) Madelung constant, e is the unit electric charge, z is the valence (possibly fractional), and r is the anion–cation distance. The quantities r and z are assumed, for lack of other knowledge, to be the same for both the bulk and surface species. Values of c_b for rocksalt, CsCl, zinc blende, and wurtzite crystals are known to be[160] 1.748, 1.763, 1.638, and 1.641, respectively. These values are obtained from the usual Coulomb sum

$$c_b = \sum_{i,j,k} q_{ijk}/R_{ijk} = \sum_{i,j,k} Q \tag{10.8}$$

where q_{ijk} is ± 1, depending on the sign of the ion located at index position (i, j, k) and R_{ijk} is the distance measured from $(0, 0, 0)$ to (i, j, k), expressed in terms of the nearest cation–anion distance.[160] The point $(0, 0, 0)$ is excluded from the sum. As one example, $c_b = (6/1 - 12/\sqrt{2} + 8/\sqrt{3} - 6/\sqrt{4} + \cdots)$, for NaCl.

In a similar manner, the *surface* Madelung constant can be evaluated[30] from

$$c_s = \sum_{i \geq 0, j, k} Q. \tag{10.9}$$

Here the index i is taken normal to the surface plane $i = 0$, and the j and k indices project in the surface plane. As an example, consider the (100) face of NaCl, for which (ideally) $c_s = (5/1 - 8/\sqrt{2} + 4/\sqrt{3} - 5/\sqrt{4} + \cdots)$. The computational procedure in calculating such a poorly converging sum

---

[160] L. Pauling, "The Nature of the Chemical Bond," p. 508. Cornell Univ. Press. Ithaca, New York, 1960.

can be considerably simplified, *if* c_b and c_s can be written in expanded forms

$$c_b = \sum_{i=0} Q + \sum_{i>0} Q + \sum_{i<0} Q = \sum_{i=0} Q + 2 \sum_{i>0} Q \qquad (10.10)$$

$$c_s = \sum_{i=0} Q + \sum_{i>0} Q = \tfrac{1}{2} c_b + \tfrac{1}{2} \sum_{i=0} Q \qquad (10.11)$$

where the sums over the indices j, k are implied, just as in (10.8) and (10.9). This separation is possible only for certain crystal faces, which have appropriate symmetry about $i = 0$. For example, in wurtzite, $(11\bar{2}0)$ is appropriate, but $(10\bar{1}0)$ is not. Also, the sum $\sum_{i=0} Q$ must converge. This rules out all the layered surfaces such as zinc blende (111), wurtzite (0001), and rocksalt (111). According to LEED, these layered surfaces are frequently reconstructed anyway, as shown in Part VII. Therefore, if (10.10) and (10.11) apply, the ratio $\gamma = c_s/c_b$ becomes

$$\gamma = \tfrac{1}{2} + \tfrac{1}{2} c_b^{-1} \sum_{i=0} Q. \qquad (10.12)$$

γ is a purely geometric property of a crystal surface, since it is independent of the anion and cation elemental constituents. The problem of computing γ is then reduced to computing the Madelung constant for the desired 2-D plane (sum over $i = 0, j, k$) since c_b is known. The range of γ is thus $\tfrac{1}{2} < \gamma < 1$. Values of γ have been computed[30] for a variety of crystal faces which are electrostatically neutral. For example, $\gamma = 0.96$ for rocksalt (100). The large value of γ implies that rocksalt (100) surface states are very similar to rocksalt bulk states. Another two examples are $\gamma = 0.92$ for wurtzite $(11\bar{2}0)$ and $\gamma = 0.91$ for zinc blende (110).[161] Thus, the wurtzite and zinc blende surface states should be relatively closer to midgap than the rocksalt surface states. The ratio ρ of the surface coordination number to the bulk coordination number is also significantly different for the two cases considered above. To be specific, $\rho = \tfrac{5}{6}$ for rocksalt (100) and $\rho = \tfrac{3}{4}$ for wurtzite $(11\bar{2}0)$ or zinc blende (110). Thus, the greatest discrepancy between surface ions and bulk ions occurs for wurtzite and zinc blende, rather than for rocksalt. In general, by using a computer, the sum in (10.9) can be performed even when the expansions (10.10) and (10.11) are invalid; this has been done by Nosker for wurtzite $(10\bar{1}0)$. The result[161] is $\gamma = 0.88$.

The computation of μ in (10.6) requires a knowledge of I, $\mathcal{A}$, z, c_b, and r. Since I and $\mathcal{A}$ depend on the fractional charge z, μ can be written as $\mu(z)$, and can be evaluated at integral values of z, using the first and second ionization potentials and the electron affinities and extrapolating in between.

---

[161] R. Nosker's computer-generated γ values (to be published in *Surface Sci.*) are given; these are probably more accurate than those of Levine and Mark.[30]

For example, consider CdS. Taking $z = 2$, and using $I(2)$ and $\alpha(2)$, gives $\mu(2) = 0.58$. Taking $z = 1$, and using $I(1)$ and $\alpha(1)$, yields $\mu(1) = 0.41$. A linear extrapolation to Birman's suggested fractional charge[162] $z \sim 0.5$, gives $\mu(0.5) = 0.33$. Levine and Mark[30] computed μ values for 46 halides, oxides, and sulfides. Generally, the greater the ionicity, the smaller the value of μ. For example, μ for NaCl ($z = 1$) is $\mu(1) = 0.06$. In principle, there is nothing to prevent E_{st}/E_{bg} in (10.4) from becoming $\gtrsim 1$, especially for large μ. This may be the case, if the fractional charge z is very small ($z \sim 0.1$), as in certain narrow bandgap ($E_{sg} < 1$ eV) III–V compounds.[163]

Finally, the valence and conduction bands are idealized here as discrete levels. Actually, both the bulk states and the surface states will be broadened into bands. A portion of the surface state bands will certainly overlap the bulk bands and be outside of the bandgap; this portion of the surface state distribution will *not* be easily detectable. See Part VII, especially Fig. 34e.

Historically, this semiclassical Madelung potential method was the first method to indicate the crude symmetry of A and B surface states. It had the obvious disadvantage that it was not based on the solution of the Schroedinger equation. For this reason, subsequent quantum-mechanical methods were employed. The DCO or band-edge LCAO method (Section 1) and the Mathieu problem approach (Section 4) showed that the nature of ionic surface states is very general and independent of the method employed for calculation.

Experimental evidence of the ionic surface states on II–VI and III–V crystals is reviewed in Section 17. A very favorable comparison of experiment and ionic surface state theory is given there. Mark has extended the formalism to the study of crystal morphology[127] and to chemisorption states.[164]

VII. Comparison between Theory and Experiment

Only in the past ten years has a great amount of surface state data been gathered which are characteristic of atomically clean surfaces. The main reason for this advance has been the recent availability of ultrahigh vacuum (UHV) equipment in the range $p = 10^{-11}$ to 10^{-10} Torr. At 10^{-6} Torr, one monolayer of residual gas arrives at a free surface in roughly 1 sec; but at 10^{-10} Torr the arrival time is roughly 10^4 sec or >1 hr. Thus, experiments can be carried out in UHV with a minimum fear of ambient contamination. For certain substrate and adsorbate combinations, the sticking coefficient is low; this helps to keep the surface clean for a longer time.

[162] J. L. Birman, *Phys. Rev.* **109**, 810 (1958).
[163] G. A. Wolff, *in* "Compound Semiconductors" (R. K. Willardson and H. G. Goering, eds.), Vol. 1. Reinhold, New York, 1967.
[164] P. Mark, *J. Phys. Chem. Solids* **29**, 689 (1968).

In UHV, one can observe the *intrinsic states* of a free surface, rather than *extrinsic states*, due to the presence of foreign atoms. Examples of extrinsic states are those due to chemisorbed atoms, physisorbed atoms, a surface oxide, electrolytes, residue from etchants, and metallic deposits. Intrinsic surface states are the main subject of this review and the relevent data and key experiments will be presented here in some detail. Extrinsic states will be discussed very briefly in Section 17,d. Complications sometimes appear from the surface preparation, such as p-type conversion due to wall outgassing; this will be considered separately in Section 17. Other complications sometimes appear, because of *edge states* due to surface dislocations or cleavage steps and surface atom reconstruction; these will also be considered separately in Sections 15 and 16.

Metals, insulators, and semiconductors have free surfaces and, therefore, may in principle have surface states. Can they be ever detected experimentally? In metals, there is no bandgap, and the density of bulk electrons at the Fermi level is high. Metallic surface states are thus difficult to distinguish experimentally from the bulk states. However, metallic surfaces have other characteristic properties which can be measured. These include (1) surface plasma losses,[165] (2) surface paramagnetic properties,[166] (3) surface scattering properties,[167] (4) wave function distribution as detected by Auger methods,[168] (5) dangling bonds, as detected by brightness of facet edges on metallic field emission,[169] and (6) participation in chemisorption.[170] But none of these seems to be directly connected with the intrinsic surface states as computed in the previous theoretical sections, so they will not be considered further here.

Insulators, such as the alkali halides, silver halides, aluminum oxide, and glass, are frequently full of electronic traps (for both electrons and holes) within the bulk. In alkali halides, these traps have well-defined names and properties: F centers, U centers, V centers, etc., and each corresponds to a particular point or cluster defect. Because of these bulk traps, the surface states, if present, would not be easily distinguished. Also, in these materials, ionic conduction frequently predominates over electronic conduction. For these reasons, no data have yet been accumulated regarding intrinsic surface states on large-bandgap insulators. Some AgCl data has

---

[165] R. H. Ritchie, *Surface Sci.* **4**, 497 (1965); H. Raether, *ibid.* **8**, 233 (1967).
[166] R. W. Zuehlke, *J. Chem. Phys.* **45**, 411 (1966).
[167] J. W. Geus, *Surface Sci.* **2**, 48 (1964); H. J. Juretschke, *ibid.* **5**, 111 (1966).
[168] H. D. Hagstrum, Y. Takeishi, G. E. Becker, and D. D. Pretzer, *Surface Sci.* **2**, 26 (1964).
[169] Z. Knor and E. W. Müller, *Surface Sci.* **10**, 21 (1968); T. T. Tsong, *ibid.* **10**, 303 (1968).
[170] J. D. Levine and E. P. Gyftopoulous, *Surface Sci.* **1**, 171 (1964); E. P. Gyftopoulous and J. D. Levine, *J. Appl. Phys.* **23**, 67 (1961).

been obtained,[171] but not in UHV. Silver halide–dye interface states have been most extensively studied in photographic research.[172]

Semiconductors seem to have just the right properties for observation of surface states (and interface states). They have forbidden gaps where a surface state, if present, can be easily detected. In some cases, a resolution of 10^{10} states cm^2 is available from electronic measurements (Section 17,d); this corresponds to $\sim 10^{-5}$ of a monolayer. Semiconductor technology is such that crystals can be grown with a relatively small number of bulk traps. Thus, their response to fields, light, ambients, and other probes for surface states can be relatively easily determined. It is clear, then, that surface state theory and experiment can be compared best for semiconductors. This comparison is carried out in Section 17,d, for Si, Ge, III–V, and II–VI compounds. Surface state theory and experiment seem to be in better agreement for the unreconstructed ionic faces of III–V and II–VI semiconductors than for the reconstructed Si and Ge surfaces.

11. Basic Phenomena

To understand semiconductor surfaces, it is necessary to consider the physics of the space-charge layer. Because of the scarcity of free electrons and holes in the semiconductor bulk, the presence of some surface states can dominate the space-charge layer behavior. *This "amplification effect" makes it possible to easily detect the semiconductor surface states.* The space-charge layer arises from the solution of the 3-D Poisson equation

$$\nabla^2 V(x, y, z) + \rho(x, y, z)/K\epsilon_0 = 0 \qquad (11.1)$$

where V is the local potential (volts), K the local dielectric constant, ϵ_0 the permittivity of free space (8.85×10^{-12} farads/m) and ρ is the charge density of free and trapped charge (coulombs/m^3). Surface states can be included in ρ by the introduction of appropriate δ functions. Of course, if $\rho = 0$, there are no charge inhomogeneities, and thus no surface states. Although considerable progress has been made in defining the 3-D aspects of (11.1), with respect to the bulk[173] and surface[174] states, it has become traditional to idealize (11.1) by assuming a *1-D model*; i.e., the dimensions parallel to the free surface are presumed unimportant. This idealization rules out, at the start, all many-body interactions[175] between surface state electrons. The idealization also treats the bulk as a homogeneous solid,

---

[171] A. M. Goodman and G. Warfield, *Phys. Rev.* **120**, 1142 (1960).
[172] C. E. K. Mees and T. H. James, eds., "The Theory of the Photographic Process." Macmillan, New York, 1966.
[173] R. Seiwatz and M. Green, *J. Appl. Phys.* **29**, 1034 (1958).
[174] J. D. Levine, *Surface Sci.* **10**, 313 (1968).
[175] D. Pines, "Elementary Excitations in Solids." Benjamin, New York, 1964.

even though bulk donors and acceptors are discrete centers. Furthermore, the idealization assumes that the positions of the states, relative to the band edges, are independent of V or the surface state occupancy; i.e., the states move with the bands. For bulk states, this has proved acceptable in the past, but for surface states it may fail. The reason is that uncompensated charged surface states can cause a considerable macroscopic field in the bulk. Thus, surface states can yield greater influence than a similar number of randomly distributed bulk states. The occupation statistics of surface states is, at present, assumed to be similar to the Fermi statistics of bulk states, although it has been proposed by Haneman[176] that this may have to be altered in the future, because of Coulomb interactions between surface state electrons. Finally, the image potential at a free semiconductor surface is never explicitly considered in the traditional solution of (11.1); instead it is implicitly incorporated (very crudely) into the electron affinity χ of the crystal, as a boundary condition. This omission is not serious for bulk states somewhat near the surface, but it can become serious for surface states.

The sometimes formidable mathematics involved in solving (11.1), even with the simplifications described above, are now well established. For mathematical details the reader is referred to Many et al.[7] and Frankl.[8] Equations and figures are available there, which relate all the relevant variables, such as potentials and charge densities. For the purpose of this review, these equations and figures will *not* be derived or repeated. Rather, only the terminology will be introduced, and the most relevant results stated in qualitative form. This will aid the reader to understand *what* was measured, and *how* it was interpreted by the experimenter. Commentaries on some experimental work of others (up to 1966) are given by Many et al.[7] and Frankl.[8] These are helpful, but frequently are intertwined with a great amount of detail, of little direct relevance to the surface state theorist, or to the experimentalist interested in intrinsic surface state distributions. The purpose of this review is to omit most of the detail, to include the later references (through mid-1969), and to spotlight the essential features of experimentally measured intrinsic surface state distributions. For details, the reader is referred to the original references.

The basic energy quantities used by experimenters to describe surface state properties are indicated schematically in Fig. 33. The figure corresponds to a p-type crystal ($p_b > n_b$) with a depletion layer ($p_b > p_s$ and $p_s > n_s$). Here p_b, n_b, p_s, n_s refer to the bulk hole density, bulk electron density, surface hole density, and the surface electron density, respectively. One can also consider similar diagrams for a p-type crystal with an accumu-

---

[176] D. Haneman, *Proc. Intern. Conf. Phys. Semiconductors Exeter, England* (A. C. Strickland, ed.), p. 842. Academic Press, New York, 1962.

lation layer ($p_b < p_s$ and $p_s > n_s$), or a p-type crystal with an inversion layer ($p_b > p_s$ and $p_s < n_s$). Analogous n-type crystals with depletion, inversion, or accumulation layers can also be considered. The energy definitions and relations described below follow from Fig. 33, but are largely independent of any of the possibilities enumerated above. The diagram includes a metallic field plate to the right of the semiconductor. If this is present, it can be used to apply an external field to the semiconductor surface. The gap between semiconductor and field plate is understood to be a vacuum in the study of clean surfaces (it can be an oxide in the study of interface states). The relevent energies are: vacuum reference level E_{vac}, conduction band edge E_c, valence band edge E_v, intrinsic level $E_i \approx \frac{1}{2}(E_c + E_v)$, energy gap $E_g = E_c - E_v$, Fermi energy E_F, work function

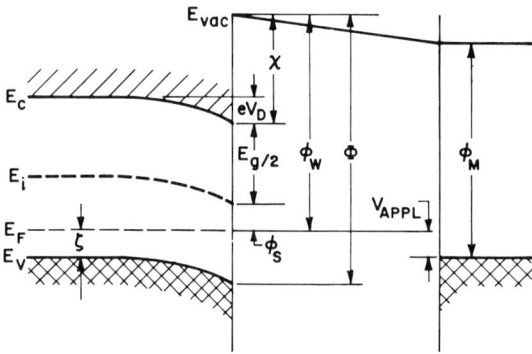

FIG. 33. Terminology relating to the space-charge layer in a p-type semiconductor. In the absence of an applied voltage and surface states, the band bending eV_D vanishes.

$\phi_w = E_{vac} - E_F$, photoelectron threshold (cube law extrapolation) Φ which frequently corresponds to the valence band edge at the surface, as shown (more precisely, Φ can be called the crystal ionization potential), electron affinity χ, doping parameter $\zeta = E_F - E_v$, band bending eV_D, surface potential ϕ_s, work function of metallic field plate ϕ_M, if present, and applied voltage V_{appl}, if present. The doping parameter for an extrinsic p-type semiconductor is given approximately by

$$\zeta = kT \ln N_v/N_A. \qquad (11.2)$$

Here N_A is the acceptor density (cm^{-3}) and N_v is the valence-band effective density of states (states within $\sim kT$ of the band edge). As usual, k is Boltzmann's constant and T is the temperature. Several relations between the above energies for p-type samples (assuming Φ corresponds to the valence-band edge) are

$$\Phi = \chi + E_g \tag{11.3}$$

$$\phi_w = \chi - eV_D + E_g - \zeta \tag{11.4}$$

$$\phi_s = \tfrac{1}{2}E_g - eV_D - \zeta. \tag{11.5}$$

In agreement with Fig. 33, V_D is taken as a positive quantity in (11.3) to (11.5). For the case of no surface states and no applied fields, $V_D = 0$ and the bands are flat up to the surface. For the case of detectable photoelectric emission from surface states, Φ does not correspond to the valence-band edge and (11.3) becomes $\Phi \leq \chi + E_g$. It is known that χ is dependent on the crystal face.[142] It has even been proposed that χ is dependent on doping for strongly degenerate p-type silicon.[177–179] Nevertheless (11.3)–(11.5) are frequently used equations, at least for preliminary studies on surface states. Eliminating χ from (11.3) and (11.4) gives

$$eV_D = \Phi - \phi_w - \zeta. \tag{11.6}$$

Thus, by measuring the photoelectric threshold, work function, and doping, the band bending is obtained in the absence of applied fields (subject to the limitations outlined above). Of course, for n-type samples, the above equations must be changed to account for the fact that ζ in n-type samples is measured from the conduction-band edge. The substitution $\zeta \to (E_g - \zeta)$ in (11.3) to (11.6) gives the appropriate equations for n-type samples. (There are a number of subtleties involved in obtaining Φ from a plot of photoemission yield *versus* photon energy; these will be considered in Section 13,c).

If $V_D \neq 0$, this indicates that there are surface states present. Different ways of computing V_D, other than by using (11.6), are given below. In the most direct method, V_D can be determined by *photovoltage* experiments[159]; essentially, light of energy $h\nu > E_g$ and intensity I is introduced to create a large concentration of free electron-hole pairs at the surface. This tends to flatten the bands. The usual plot is ΔV (from Kelvin probe or capacitance measurements) versus I. At saturation $\Delta V = V_D$. One must take care to subtract out the Dember voltage, which occurs because electrons and holes created by light near the surface region diffuse with different mobilities in towards the bulk.

Alternately, V_D can be inferred from *conductivity* measurements made parallel to the surface. The major assumption here is that surface state

---

[177] G. W. Gobeli and G. F. Allen, *Phys. Rev.* **127**, 141 (1962).
[178] F. G. Allen and G. W. Gobeli, *Phys. Rev.* **127**, 150 (1962).
[179] F. G. Allen and G. W. Gobeli, *Proc. Intern. Conf. Phys. Semiconductors, Exeter, England* (A. C. Strickland, ed.), p. 818. Academic Press, New York, 1966.

charges are relatively immobile and that the conductivity occurs in the bulk region. This is supported for Si and Ge by other experiments.[180,181] Also a crystal can be cleaved in UHV. If there is a change in surface conductance per square $\Delta\sigma$ (mho/square) monitored during cleavage, the change in surface space charge $\Delta p/e$ (cm^{-2}) can be estimated from an equation whose simplest form (for p-type conductivity) is

$$\Delta\sigma = e\,\Delta p\,\mu_{\text{ps}} \qquad (11.7)$$

where μ_{ps} is the hole mobility at the surface (cm^2/V-sec). Generally, the surface mobility is less than the bulk mobility and the scattering mechanism (specular, diffuse, mixed) must be estimated to compute μ_{ps}. The most widely quoted relation is the one by Schrieffer.[182] Once Δp is known from (11.7), the standard relations between Δp and V_D can be used to infer V_D. The problem is usually complicated by the inclusion of a term $\Delta n e \mu_{\text{ns}}$ in (11.7), which is related to the change in electron conduction at the surface, and by the inclusion of quantized "light holes," if a strong inversion layer is present.[183]

If $\Delta\sigma$ cannot be monitored *during* cleavage as described above, then V_D can still be inferred, but by a less satisfactory and more indirect procedure. Suppose a crystal is cleaved, but its bulk conductivity is not known, because of uncertainties in doping or crystal geometry. Then one cannot immediately tell how much of the conductivity σ is due to the surface layer, $\Delta\sigma$, and how much to the bulk, σ_0, since the conduction paths are in parallel. The bulk contribution can be subtracted out by finding $\sigma_{\min}$, which is the minimum conductivity obtained after the bands have been swung by an applied electric field or by the use of p- or n-type adsorbates. The quantity $\sigma - \sigma_{\min}$ then has a known relation to V_D.[184]

The importance and measurement of the band bending V_D has been stressed above. Generally, if a measurement on a free surface gives band bending in the absence of absorbates or applied fields, then there *must* be surface states present. To determine just how many surface states are present and to fix their distribution in energy, more complicated analyses and experimental techniques must be used. These will be briefly outlined below.

12. Surface State Distributions

Suppose it has been found that $V_\text{D} \neq 0$, so that some surface states must be present. How can one deduce the surface state distributions from

[180] D. E. Aspnes and P. Handler, *Surface Sci.* **4**, 353 (1966).
[181] A. Kobayashi, K. Sugiyama, H. Arata, and Z. Oda, *J. Phys. Soc. (Japan)* **16**, 2481 (1961).
[182] J. R. Schrieffer, *Phys. Rev.* **97**, 641 (1955); see also Frankl,[8] p. 110 ff.
[183] P. Handler and S. Eisenhour, *Surface Sci.* **2**, 64 (1964).
[184] See Frankl,[8] p. 107.

the data? There is no clear-cut answer to this question, because of the peculiarities of the Fermi statistics. That is, only those states at, or near, the Fermi level can be detected easily. For example, the electronic states a few electron volts above the conduction band in a metal or semiconductor cannot be probed easily, except by optical absorption or hot electron scattering. Thus, the problem is generally open-ended. *If* the Fermi level could be moved at will throughout the entire energy band structure and if experiments could be carried out over a large temperature range, the problem would then be in closed form; the surface state structure would be uniquely determined. It is precisely because the Fermi level, in practice, can be moved only by a restricted amount, and the temperature range is limited, that the data obtained are restricted, and the surface state information obtained is incomplete. Another consequence of the limited traverse of the Fermi level, is that the total number (charged and uncharged) of surface states must frequently be assigned a lower limit. To help clarify these points, some useful surface state models are reviewed below.

Surface states, like bulk impurity states, are generally considered as either acceptor-like or donor-like. By convention, acceptor-like states are neutral when empty, and negatively charged when filled with one electron. The donor-like states are positively charged when empty and neutral when filled with one electron. For the more insulating semiconducting crystals ($E_g > 2$ eV), the terminology is sometimes different.[30] An acceptor-like state is equivalent to an electron trap (neutral when it has no electrons and negatively charged when it has one electron) and a donor-like state is equivalent to a hole trap (neutral when it has no hole, and positively charged when it has one hole). Both terminologies "donor and acceptor" and "electron and hole traps" will be used whenever appropriate. It seems to be a property of surface states that the acceptor levels (electron traps) most frequently lie *above* the donor levels (hole traps). In this respect, they resemble the trap behavior in the more insulating crystals.[185] Those workers who are familiar only with the shallow extrinsic donors and acceptors in Si and Ge may find this behavior unusual, since the shallow donors lie just below the conduction band and the shallow acceptors lie just above the valence band. Intrinsic surface states (and bulk trap states) seem to have the opposite behavior; the acceptor (donor) states generally lie below (above) the conduction (valence) band. From this fact alone, one can conclude that surface states are *not* akin to shallow donors and acceptors. (As mentioned above, it is always assumed in this section that *intrinsic* surface states are considered. The above conclusions will, of course, be modified if *extrinsic* surface states are present.)

[185] A. Rose, "Concepts in Photoconductivity and Allied Problems." Wiley (Interscience), New York, 1963.

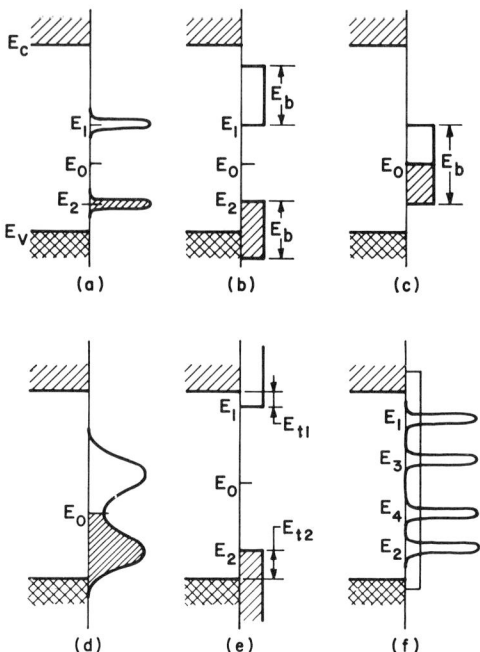

Fig. 34. Some useful intrinsic surface states models: (a) two discrete levels, (b) two bands, (c) merged bands, (d) cosh-like bands, (e) ionic-like bands, (f) possibilities of extrinsic distributions.

It may occur that some surface states are located outside of the forbidden gap. This case was previously discussed in Section 8, and it was shown to exist whenever there was no crossing of the surface and bulk energy curves in **k** space.

As a simple example of a possible surface state distribution, consider a *two discrete level* model as shown in Fig. 34a. Let the number of available surface states of the acceptor (donor) type be N_1 (N_2), and the negative (positive) charge due to the occupied (unoccupied) acceptors (donors) be $-Q_1$ (Q_2). The energy locations are E_1 and E_2, respectively. The total surface state charge $Q_{ss} = Q_1 + Q_2$, using Fermi statistics and unit statistical weight, is

$$Q_{ss} = -eN_1[1 + \exp(E_1 - E_F)/kT]^{-1} + eN_2[1 + \exp(E_F - E_2)]^{-1}. \tag{12.1}$$

For simplicity, assume $N_1 = N_2 = N$ and define

$$E_0 = \tfrac{1}{2}(E_1 + E_2), \quad \Delta E_F = E_F - E_0, \quad \tfrac{1}{2}E_{sg} = E_1 - E_0 = E_0 - E_2. \tag{12.2}$$

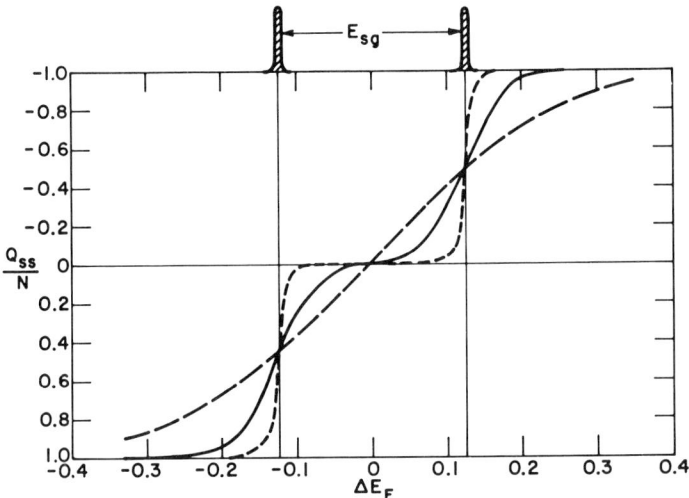

FIG. 35. Two discrete level model. Charge in surface states as a function of the Fermi level and temperature. (—, $T = 300°K$; ···, $T = 75°K$; – –, $T = 1200°K$.)

The new quantities E_0 and ΔE_F define, respectively, the neutral energy, where $Q_{ss} = 0$, and the location of the Fermi energy with respect to this location. Equation (12.1) now becomes

$$-Q_{ss}/eN = [1 + \exp(\tfrac{1}{2}E_{sg} - \Delta E_F)/kT]^{-1}$$
$$- [1 + \exp(\tfrac{1}{2}E_{sg} + \Delta E_F)/kT]^{-1} \quad (12.3)$$

which can be rearranged into

$$-Q_{ss}/eN = \sinh(\Delta E_F/kT)/[\cosh(E_{sg}/2kT) + \cosh(\Delta E_F/kT)]. \quad (12.4)$$

The function $-Q_{ss}/eN$ represents the normalized negative charge in the surface states. Essentially, this function is sinh-like for $\Delta E_F < \tfrac{1}{2}E_{sg}$, and saturates for $\Delta E_F > \tfrac{1}{2}E_{sg}$. A plot of Q_{ss}/N vs ΔE_F is given in Fig. 35 for $T = 300°K$ (solid curve), $T = 75°K$ (dotted curve), and $T = 1200°K$ (dashed curve). The temperatures are room temperature, $\tfrac{1}{4}$ room temperature, and 4 times room temperature. The band gap between the surface states E_{sg} is presumed independent of temperature (a simplification) and is taken to be 0.25 eV in the figure. This value seems to be characteristic of Si. The specific temperatures and energies can be generalized by recognizing that ΔE_F and $\tfrac{1}{2}E_{sg}$ always appear with temperature in the ratios $\Delta E_F/kT$ and $\tfrac{1}{2}E_{sg}/kT$. In this manner, the band gap can be specified in terms of temperature: E_{sg} is $40kT$ at 75°K, $10kT$ at room temperature, and $2.5kT$ at 1200°K.

Examination of Fig. 35 shows that, even for this simple *discrete pair* distribution of surface states, there are considerable energy and temperature variations in the measurable charge present Q_{ss}. Basically, this occurs because the Maxwellian exponential tails extend beyond the energy level of the surface state. In the high temperature case, Q_{ss} varies linearly with ΔE_F and seems to bear no relation to the discreteness of surface states, from which the charges arose. The slope at E_0 increases with increasing T; it can be obtained from (12.4) as

$$-(\partial Q_{ss}/\partial \Delta E_F)_{E_0} = (eN/kT)/[1 + \cosh(E_{sg}/2kT)]. \quad (12.5)$$

The slope is a quantity which is closely related to the measured field effect mobility, which will be described later. Figure 35 differs from Fig. 5.4 of Many et al.,[7] which has abcissa and curves labeled by *normalized* energies of the form E/kT; in that form it is difficult to comprehend the temperature dependence.

As a second simple example, consider a *two band* model of surface states, as shown in Fig. 34b. Assume both bands have a constant *energy density* of states $\bar{N}$ (states/cm² eV). Then

$$Q_{ss} = -e\bar{N} \int_{E_1}^{E_1+E_b} [1 + \exp(E - E_F)/kT]^{-1} dE + e\bar{N}$$

$$\times \int_{E_2-E_b}^{E_2} [1 + \exp(E_F - E)/kT]^{-1} dE. \quad (12.6)$$

Here E_b is the "width" of the acceptor and donor bands. (It has been assumed for simplicity that $\bar{N}_1 = \bar{N}_2 = \bar{N}$, and $E_{b1} = E_{b2} = E_b$). Also assume, for simplicity, that $\bar{N}$ can be estimated for a two-dimensional free-electron gas with spin degeneracy 2, characterized by an appropriate isotropic effective mass m_s^*. In this case

$$\bar{N} = 4\pi m_s^* h^{-2} = 4.2 \times 10^{14} m_s^*/m_0 \quad \text{states/cm}^2 \text{ eV}. \quad (12.7)$$

Obviously, the approximation in (12.7) must break down away from the surface state band edge; nevertheless, it is customary to consider (12.7) to apply over "most" of the surface state bands.

The integral in (12.6) can be solved in closed form, contrary to the three-dimension band problem, to give

$$Q_{ss}/e\bar{N}E_b = (kT/E_b) \ln\{[1 + \exp(\tfrac{1}{2}E_{sg} + E_b - \Delta E_F)/kT]$$
$$\times [1 + \exp(\tfrac{1}{2}E_{sg} + \Delta E_F)/kT]\}$$
$$- (kT/E_b) \ln\{[1 + \exp(\tfrac{1}{2}E_{sg} - \Delta E_F)/kT]$$
$$\times [1 + \exp(\tfrac{1}{2}E_{sg} + E_b + \Delta E_F)/kT]\} \quad (12.8)$$

which is plotted for $E_b = E_{sg} = 0.25$ eV and the three temperatures 75°K,

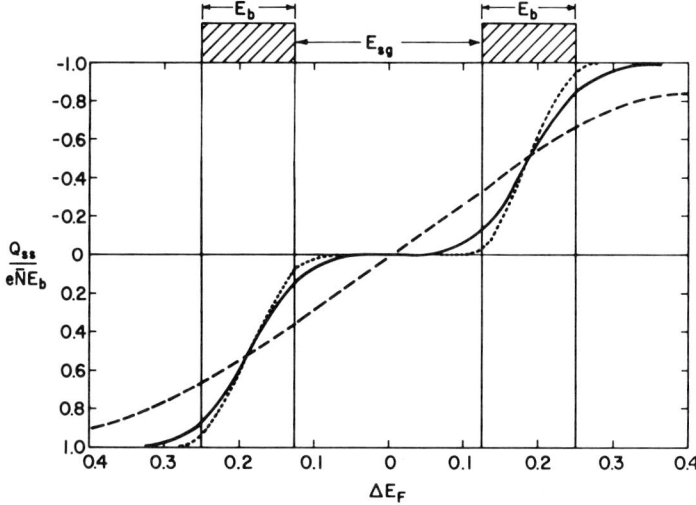

FIG. 36. Two band model. (—, $T = 300°K$; ..., $T = 75°K$; --, $T = 1200°K$.)

300°K, and 1200°K in Fig. 36. On inspection, this function has the following properties. For a Fermi level far above the acceptor surface state band $Q_{ss} \to -e\bar{N}E_b$. This is the total charge of filled acceptors, $-eN_A$. If it is assumed that these are *intrinsic* surface states, then N_A is $\sim 4 \times 10^{14}$ cm²; this defines the surface state effective mass and bandwidth since, by definition,

$$E_b = N_A/\bar{N} \simeq 1.0 m_0/m_s^*. \qquad (12.9)$$

A ratio $m_s^*/m_0 \sim 10$ gives a surface-state bandwidth of ~ 0.1 eV; this seems to be typical of Si and Ge. By contrast, a different ratio ~ 0.2 gives a surface-state bandwidth ~ 5 eV; this is more typical of II–VI compounds, as will be shown later. In a similar manner, it can be shown that for $\Delta E_F \to -\infty$, $Q_{ss} \to +e\bar{N}E_b$, and the donor states are fully charged (positively). In the range where ΔE_F lies within a surface state band, (12.8) becomes $Q_{ss} \simeq e\bar{N}(\Delta E_F - \frac{1}{2}E_{s_k})$. In other words, Q_{ss} varies nearly linearly with ΔE_F in this degenerate range. In the range within the surface state gap, (12.8) becomes sinh-like, just as in (12.4). At the band edge, $Q_{ss} = -e\bar{N}kT \ln 2$. The figure shows that, even though $\bar{N}$ is constant, Q_{ss} varies with T *within* the surface state band. This effect occurs because the Maxwellian tails must be taken into account. Of course, for $E_b \to 0$, the previous *two discrete level* distribution is the appropriate limit.

For a third simple example, consider the case where the pair of acceptor and donor surface states have formed a *merged band* of surface states as shown in Fig. 34c. The neutral level E_0 is in midband. If the energy density of states is $\bar{N}$ and the bandwidth is E_b, as above ($E_b = 0.25$ eV in the

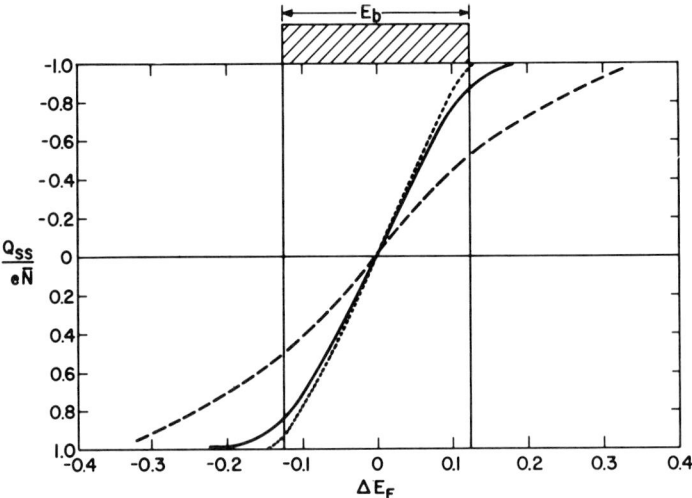

FIG. 37. Merged band model. (—, $T = 300°K$; ..., $T = 75°K$; ---, $T = 1200°K$.)

figure), then

$$Q_{ss} = -e\bar{N} \int_{-\frac{1}{2}E_b}^{\frac{1}{2}E_b} [1 + \exp(E - \Delta E_F)/kT]^{-1} dE$$

$$+ e\bar{N} \int_{-\frac{1}{2}E_b}^{\frac{1}{2}E_b} [1 + \exp(\Delta E_F - E)/kT]^{-1} dE \qquad (12.10)$$

and the integrals yield

$$\frac{Q_{ss}}{e\bar{N}E_b} = \frac{2kT}{E_b} \ln \frac{\cosh[(\frac{1}{2}E_b - \Delta E_F)/2kT]}{\cosh[(\frac{1}{2}E_b + \Delta E_F)/2kT]}. \qquad (12.11)$$

This is plotted for the same three temperatures in Fig. 37.

By comparing Figs. 35–37, the characteristics of the various distributions can be best seen. For the 1200°K case, all three distributions yield Q_{ss} essentially linear with ΔE_F. This is because the bandgap (or bandwidth) is $\sim 2.5kT$ wide. To distinguish between these distributions it is necessary to carry out experiments at a much lower temperature (i.e. eliminate the "thermal noise"). The effect is easily scaled. For example, suppose $E_{sg} = 0.06$ eV, then $E_{sg} = 2.5kT$ at room temperature. This seems to be the case for Ge; one cannot uniquely determine this surface state distribution unless experiments are carried out at temperatures much lower than room temperature.

It is instructive to formally consider the most general distribution of surface states. For this purpose, let $\bar{N}_A(E)$ and $\bar{N}_D(E)$ be the acceptor and the donor energy density of states, respectively. E is the energy measured from $-\infty$ to $+\infty$. In practice, $\bar{N}_A(E)$ and $\bar{N}_D(E)$ will be nonzero only in restricted regions near the bulk bandgap. Then

$$Q_{ss}/e = -\int_{-\infty}^{\infty} f(E - E_F, T)\bar{N}_A(E) \, dE + \int_{-\infty}^{\infty} f(E_F - E, T)\bar{N}_D(E) \, dE \tag{12.12}$$

where $f(X, T)$ is the Fermi function $f(X, T) \equiv [1 + \exp(X/kT)]^{-1}$. Donors can be related to acceptors, since $f(X, T) = 1 - f(-X, T)$, so that (12.12) becomes

$$Q_{ss}/e = -\int_{-\infty}^{\infty} f(E - E_F, T)[\bar{N}_A(E) + \bar{N}_D(E)] \, dE + N_D \tag{12.13}$$

in which

$$N_D = \int_{-\infty}^{\infty} \bar{N}_D(E) \, dE. \tag{12.14}$$

A convenient analysis of (12.13) is made possible by splitting the Fermi function $f(X, T)$ into two parts

$$f(X, T) \equiv f_0(X, 0) + f_1(X, T) \tag{12.15}$$

where f_0 is the step function

$$f_0(X, 0) = \begin{cases} 1, & X < 0 \\ 0, & X > 0. \end{cases} \tag{12.16}$$

Both f_0 and f_1 are discontinuous at $X = 0$. Combining (12.13) and (12.15) gives

$$Q_{ss}/e = -\int_{-\infty}^{E_F} [\bar{N}_A(E) + \bar{N}_D(E)] \, dE$$

$$- \int_{-\infty}^{\infty} f_1(E - E_F, T)[\bar{N}_A(E) + \bar{N}_D(E)] \, dE + N_D. \tag{12.17}$$

This can be differentiated to yield

$$\frac{\partial Q_{ss}/e}{\partial E_F} = -[\bar{N}_A(E_F) + \bar{N}_D(E_F)]$$

$$- \int_{-\infty}^{\infty} \frac{\partial f_1(E - E_F, T)}{\partial E_F} [\bar{N}_A(E) + \bar{N}_D(E)] \, dE. \tag{12.18}$$

The first term on the right in (12.18) is the charge contribution due to the sum of the donor and acceptor states actually present at E_F; the second term is the contribution due to the Maxwellian tails of states located some distance from E_F. At $T = 0$, the second term must vanish. This analysis shows that it is erroneous to determine $[(\bar{N}_A(E_F) + \bar{N}_D(E_F)]$ by a simple differentiation of Q_{ss} as one might first expect. The second term in (12.18) may frequently be significant, especially for smaller values of E_{sg}/kT or E_b/kT. This is shown dramatically in Fig. 35–37.

Although $\bar{N}(E)$ is generally thought to be independent of temperature, the charge present $Q_{ss}(E_F, T)$ is not. Thus, all experiments which have been interpreted with a model of a constant $\bar{N}(E)$ over a narrow energy range could be repeated at a lower temperature to detect the structure, if any, present in $\bar{N}(E)$. Essentially, the distribution $\bar{N}(E)$ is probed experimentally by a "spectrometer" whose "resolution" is a few kT wide. It should also be noted that Figs. 36 and 37 show temperature dependences, even when the surface states are degenerate (E_F lying inside a surface state band).

13. Experimental Methods

The various methods which have been used to determine the energy distribution of surface states will now be briefly reviewed. The experimental methods used for studying *extrinsic* surface states and *interface* states are interesting, but are not really relevant to intrinsic surface states, which are of concern here. The methods discussed below fall into four categories: doping, applied fields, light and all others. These methods will now be described *very* generally. Certain subtleties and details will be reserved for Section 17, where the data are actually given.

a. Doping

A very useful method consists in performing the same experiments on crystals of different doping, ranging from degenerate n type to degenerate p type. The doping is related to ζ by (11.2). Sometimes a crystal is used which is not homogeneously doped, as in a p–n junction; then measurements at different locations on this crystal give the doping variations. Sometimes it is experimentally impossible to vary the doping by great amounts; this occurs for the II–VI and some III–V compounds. For example, CdSe is only n type and ZnTe is only p type. This may be changed in the future by ion-implantation techniques.[186]

V_D can be directly obtained at each doping by measuring ϕ_w and Φ and

---

[186] E. R. Pollard and J. H. Hartke, *Bull. Am. Phys. Soc.* **14**, 115 (1969).

using (11.6). From this, the surface potential ϕ_s can be obtained from (11.5) as a function of ζ. The surface potential gives the location of the Fermi level at the surface with respect to midgap (more precisely, E_i). Knowing ϕ_s, the space-charge density Q_{sc}/e (cm^{-2}) can be constructed theoretically. Since there are no external fields present $Q_{sc} = -Q_{ss}$, so the surface state charge Q_{ss} can be found as a function of ϕ_s ($\phi_s = E_F - E_i$) for various values of ζ. The plot Q_{ss} versus ϕ_s frequently resembles a sinh-like curve as in Figs. 35 and 36. By appropriate curve fitting, the possible surface state distributions can be determined.[177-179]

It is less satisfactory to measure limited data. Measurements of photoemission (not ϕ_w) versus doping[187a-c] yield some information about the band bending, if it is severe and if the bands bend down at the surface. This occurs because the bulk electrons in the space-charge layer contribute to the *value* and *shape* of the photoemission yield curve, Y vs $h\nu$. It may occur that just ϕ_w (not Φ) is measured[142] as a function of ζ. For crystals with a variation of doping from degenerate n type to degenerate p_type, the work function *change* $\Delta\phi_w$ is expected to be E_g, in the absence of surface states. If it turns out that $\Delta\phi_w \ll E_g$, the bands can be said to be "clamped" at the surface by the surface states. An inverse measure of the "clamping" strength is the ratio of $\Delta\phi_w/E_g$. It will be shown later that this ratio is $\sim 6\%$ for Ge, and $\sim 20\%$ for Si.

V_D can also be measured as a function of doping without measuring either ϕ_w or Φ. For example, V_D can be obtained by measuring the change of conductivity upon cleavage,[180,188] using appropriate models of surface scattering. In fact, any method which yields V_D can be done, in principle, for different dopings. Thus, the Fermi level is moved through the surface states; it acts as a probe of the surface state distribution as seen in Figs. 35-37.

b. Applied Fields

Instead of varying the doping, one can vary the applied field for a crystal of specific doping. This is done by having a (usually metallic) field plate parallel to the surface plane, as shown in Fig. 33.

First consider the DC field effect measurements. The appropriate equation (Gauss's law) relating Q_{sc} and the applied field E, in the absence of surface states, is

$$Q_{sc} = K\epsilon_0 E. \tag{13.1}$$

[187a] J. van Laar and J. J. Scheer, *Philips Res. Rept.* **17**, 101 (1962).
[187b] J. van Laar and J. J. Scheer, *Proc. Intern. Conf. Phys. Semiconductors Exeter, England* (A. C. Strickland, ed.), p. 827. Academic Press, New York, 1962.
[187c] J. van Laar and J. J. Scheer, *Surface Sci.* **3**, 189 (1965).
[188] P. Handler, *Appl. Phys. Letters* **3**, 96 (1963).

For a dielectric constant $K = 10$ and a field $E = 10^5$ V/cm, $Q_{sc} = 8.85 \times 10^{-8}$ C/cm². This corresponds to a number of charges $Q_{sc}/e = 5.5 \times 10^{11}$ cm⁻². In practice, 10^5 V/cm is the maximum one can expect for a clean surface (with no solid spacers, such as Mylar, mica, or an oxide layer). This would correspond to, say, 10^3 V across a vacuum gap 10^{-2} cm wide. Thus, if there are intrinsic surface states present, which can accommodate a much greater number of surface states than 5×10^{11} cm², the DC field effect method cannot affect or adequately probe the states. For this reason, the number of surface states would be reported as "at least as great as 5×10^{11} cm⁻² (0.1% of a monolayer)." The variation in doping to degeneracy, by comparison, can probe up to 10% of a monolayer.[177-179] [The situation is considerably different for *real* (usually contaminated) surfaces with Mylar or mica spacers. There the maximum field is $\sim 2 \times 10^6$ V/cm² and $Q_{ss}/e = 10^{13}$ cm⁻² (2.5% of a monolayer). This field is close to the breakdown field of most solids.] In the case of intrinsic surface states on Si, Ge or the III–V and II–VI compounds, it may be that the surface states are partly located outside the bandgap. Neither the field effect nor the doping variations can directly probe this part of the distribution.

Because of the inability of the DC field effect to effectively probe surface states on Si or Ge, a perturbation method is very frequently used. This method consists in measuring the *differential* change of surface conductivity $\Delta\sigma$ with respect to the *differential* charge Q_f applied at the field plate. A "field-effect mobility" is defined by

$$\mu_{fe} \equiv -d\,\Delta\sigma/dQ_f. \qquad (13.2)$$

This is not a true mobility, because not all of the Q_f charges are mobile. The surface state charges Q_{ss} are presumed immobile, as mentioned earlier. For the case where surface states dominate and the surface is strongly accumulating but not degenerate, it can be shown that[189]

$$dQ_{ss}/du_s \sim \tfrac{1}{2}\,\Delta\sigma/\mu_{fe} \qquad (13.3)$$

where u_s is the normalized surface potential ($u_s = \phi_s/kT$). It is conventional to interpret the results in terms of a constant energy density of states $\bar{N}$, within a few kT of E_0, as in Fig. 37. Then $dQ_{ss}/du_s = e\bar{N}kT$ and (13.3) becomes, in this *merged band model*,

$$\bar{N} = -\tfrac{1}{2}\,\Delta\sigma/ekT\mu_{fe}. \qquad (13.4)$$

For Si,[190] $\Delta\sigma$ is 0.59 μmho/square, $\mu_{fe} = -1$ cm²/V sec and $kT = 0.025$ eV; this gives $\bar{N} = 1 \times 10^{14}$ states/cm² eV. If this were due to a *merged* band of intrinsic surface states as in Fig. 37, then the effective mass according to

---

[189] See Many *et al.*,[7] p. 427 ff.
[190] M. Henzler, *Phys. Status Solidi* **19**, 833 (1967).

(12.7) would be $m_s^*/m_0 \sim 0.24$, and the bandwidth according to (12.9) would be $E_b \sim 5$ eV.

On the other hand, if a model of *two discrete levels* were present as in Fig. 35, and if the Fermi level were locked near midsurface-state gap $\Delta E_F = 0$, then (12.5) can be used together with (13.3) to get a different interpretation of the same field effect data:

$$N = -(\tfrac{1}{2}\Delta\sigma/e\mu_{fe})[1 + \cosh(\tfrac{1}{2}E_{sg}/kT)]. \qquad (13.5)$$

When $E_{sg} = 0.25$ eV and $kT = 0.025$ eV, the term in brackets becomes 75 and $N = 2 \times 10^{14}$ states/cm², which is near that expected for one surface state per surface atom. The *two discrete level* model seems more correct than the *merged band* model, as found from other experiments on Si. Clearly, this calculation shows that the field-effect mobility is of limited validity in defining *distributions* of surface states, which is because it is only a point measurement. From Figs. 35–37, however, it can be seen that additional measurements of temperature or doping dependence can help to distinguish the details of surface state structure. Another model used in conjunction with the interpretation of field-effect mobility is that of a *single discrete band* located at E_F as shown in Fig. 37 with $E_b = 0$. This yields a number[189] of surface states $N = 4kT\bar{N}$, where $\bar{N}$ is the energy density of states in (13.4). Frankl[8,191,192] has considered the field-effect mobility criteria in great detail and has found that in some cases, where data has been reported and analyzed by others, the results are self-inconsistent; they contradict the properties of the Fermi function f. That is, it must always be true that $f \geq df/du_s$ and $df/du_s \leq \tfrac{1}{4}$.

Pulsed field-effect measurements[193–195] are used to determine the dynamic response of the emptying and filling of surface states. By analyzing the time constant as a function of $1/T$, an effective Arrhenius energy level (i.e. $E_2 - E_v$ or $E_c - E_1$ in Fig. 34a) and capture cross section can be inferred. Acceptor (donor) states are usually found closer to the conduction (valence) band, as in Fig. 34a. Sinusoidal field-effect measurements have been performed with varying frequency to help distinguish between fast and slow states. It is now established[196] that there are no slow states on clean surfaces; the slow states only appear at oxides or otherwise contaminated surfaces of semiconductors. They are called "slow states" because their communication with the interface is poor.

[191] D. R. Frankl, *Surface Sci.* **4**, 201 (1966); see also Frankl,[8] p. 213.
[192] D. R. Frankl, *Surface Sci.* **6**, 334 (1967); see also Frankl,[8] p. 212.
[193] B. A. Nesterenko and O. V. Snitko, *Soviet Phys. Solid State* **6**, 2321 (1965).
[194] B. A. Nesterenko and O. V. Snitko, *Surface Sci.* **5**, 380 (1966).
[195] B. A. Nesterenko, O. V. Snitko, and V. T. Rozumnyuk, *Surface Sci.* **9**, 407 (1968).
[196] Y. Margoninski, *Phys. Rev.* **132**, 1910 (1963).

If the DC field effect is extremely large, field emission occurs, and sometimes surface states can be found. This has been carried out for Ge field-emission tips by Shepherd.[196a] See also Arthur.[196b]

c. *Light*

Light is a frequently used probe for the investigation of surface states. Certain effects caused by light have already been briefly mentioned. They are the photovoltage effect and photoemission into vacuum. The photovoltage effect measures only V_D and does not directly measure the interaction of surface states with light. Photoemission can, in principle, activate electrons from filled surface states into the vacuum. If the filled surface states lie above the valence band, then the photoelectric yield Y would have a "tail" at low energy, which can be detected experimentally. The magnitude of the surface state yield is impossible to estimate at present, since the cross section is unknown. One of the main problems in photoemission measurements is to distinguish the surface state contribution from the various bulk contributions. Experimentally the quantum yield Y (electrons emitted/electrons absorbed) frequently has the form[197]

$$Y = C_1(h\nu - \Phi)^{r_1} + C_2(h\nu - \Phi_2)^{r_2} \tag{13.6}$$

where Φ and Φ_2 are the lower and upper energies, respectively, and r_1 and r_2 are integers or half intergers $1, \frac{3}{2}, 2, \frac{5}{2}, 3$, etc. Sometimes there seems to be three terms in (13.6), as for Ge. Other times, the contributions are considered tangential,[198] not additive, i.e., both terms are present with both Y and its first derivative continuous at $h\nu_0$. Theoretical estimates[199-201] have been made for the powers r_1, r_2, assuming different mechanisms. Direct bulk transitions correspond to $r_2 = 1$. Sometimes there is no linear portion, as with bombarded and annealed Si.[202] The low-energy tail may have a power $r = 3$, which is difficult to distinguish experimentally from $r = \frac{5}{2}$. This "cube law" or "$\frac{5}{2}$-power law" portion seems to be the case for all materials.[159,203,204] Kane[199] theoretically computed the type of power law to be

---

[196a] W. B. Shepherd, *J. Vacuum Sci. Technol.* **6**, 908 (1969).
[196b] J. R. Arthur, Jr., *J. Appl. Phys.* **36**, 3221 (1965).
[197] G. W. Gobeli and F. G. Allen, in "Semiconductors and Semimetals" (R. K. Willardson and A. C. Beer, eds.), Vol. 2. Academic Press, New York, 1966.
[198] T. E. Fischer, *Phys. Rev.* **142**, 519 (1966); *Helv. Phys. Acta* **41**, 827 (1968).
[199] E. O. Kane, *Phys. Rev.* **127**, 131 (1962); **147**, 335 (1966).
[200] N. B. Kindig and W. E. Spicer, *Phys. Rev.* **138**, A561 (1965).
[201] D. Brust, *Phys. Rev.* **139**, A489 (1965).
[202] F. G. Allen and G. W. Gobeli, *J. Appl. Phys.* **35**, 597 (1964).
[203] G. W. Gobeli and F. G. Allen, *Phys. Rev.* **137**, 245 (1965).
[204] J. J. Scheer and J. van Laar, *Phys. Letters* **3**, 246 (1963).

expected from eleven different mechanisms, and none of these have $r = 3$. The mechanisms he considered include various combinations of volume and surface transitions, direct and indirect transitions, and scattered and unscattered transitions. However, since most workers today use the cube law ($r = 3$) as a convenient way to parametrize their photoemission data near threshold, the same practice will be followed here. A theoretical derivation of the photoelectric yield involves the matrix element between the initial and final states, as well as the energy distributions of all these states.[200,201] Thus, the band structure of a crystal near the energy corresponding to E_{vac} in Fig. 33 enters into the determination of the form and r values to be used in (13.6). Fischer[205] reviews the photoemission processes and concludes that surface state photoemission can, in principle, be detected, but only as a departure from the cube-root law in the photoemission tail. This would require pure samples which are not amorphous.[206] Energy distributions of photoelectron emissions have been obtained using a retarding potential.[202] The analysis of this distribution is a valuable asset in interpretation of the emission mechanisms. Yet, at present, surface state emissions have still not been unambiguously identified.[159,203,205]

Surface states can also interact with light of less than bandgap energy $h\nu \leq E_g$. An example would be the photoionization of occupied surface acceptors ($h\nu > E_c - E_1$ in Fig. 34a) or photoionization of surface donors ($h\nu \geq E_2 - E_v$ in Fig. 34a). These and similar transitions have been measured using internal reflection spectroscopy,[207–209] photovaporization,[210] and photodesorption.[211,212]

d. Other Methods

Temperature variations can be used to probe the surface states distribution, since the doping parameter ζ is temperature dependent and since Q_{ss} is temperature dependent as shown in Figs. 35–37. Filled surface state traps can be thermally emptied[213] and "thermally stimulated current" can be measured in photoconductors.[214]

[205] T. E. Fischer, *Surface Sci.* **13**, 30 (1969).
[206] A. I. Gubanov, "Quantum Electron Theory of Amorphous Conductors." Consultants Bureau, New York, 1965.
[207] G. Samoggia, A. Nucciotti, and G. Chiarotti, *Phys. Rev.* **144**, 749 (1966).
[208] N. J. Harrick, "Internal Reflection Spectroscopy." Wiley (Interscience), New York, 1967.
[209] G. Chiarotti, G. Del Signore, and S. Nannarone, *Phys. Rev. Letters* **21**, 1170 (1968).
[210] G. A. Somorjai and J. E. Lester, *J. Chem. Phys.* **43**, 1450 (1965).
[211] H. Moesta, *Surface Sci.* **2**, 267 (1964).
[212] R. Williams and A. Willis, *J. Appl. Phys.* **39**, 3731 (1968).
[213] P. V. Gray and D. M. Brown, *Appl. Phys. Letters* **8**, 31 (1966).
[214] P. Mark, *J. Phys. Chem. Solids* **26**, 959 (1965).

Indications of surface state populations have been given by measurements of electron spin resonance on freshly cleaved, crushed, or annealed crystals.[215-218] A properly calibrated sample can yield the number of spins/cm^2, which are related, in turn, to the number of *unpaired* surface states.

Following the analogy of impurity bands, known to exist in highly doped semiconductors, conductivity within the surface state bands[180,181,188,219] is possible. This has been investigated for Si and Ge, but no surface state conduction was found, for reasons yet unknown. The channel effect[220,221] on n–p–n and p–n–p junctions has also been used to study surface states.

Recombination of excess bulk carriers (i.e. created by light) is enhanced, if *certain types* of surface states are present. These act as recombination centers. Basically, an effective surface recombination center must trap *both* holes and electrons, with a reasonable cross section for each. Also, the band bending and bulk mobilities must be such as to allow excess electrons and holes easy access to the surface. It stands to reason that not all surface states will be good recombination centers; thus, the recombination method can probe only a limited part of the surface state distribution.

14. Surface Preparation

It is to little avail if precise experiments are made on ill-defined surfaces. Intrinsic surface states may be easily masked by extrinsic surface states, surface oxides, metallic adsorbates from etchants, etc. For this reason, special attention must be given to clean surface preparation.

A good way to prepare a clean surface is by cleavage in UHV. Only certain crystal planes of certain crystals cleave easily. Some of the most important of these are

(111)   IV crystals (diamond type), Si, Ge
(110)   III–V and II–VI compounds (zinc blende), GaAs, CdTe
(11$\bar{2}$0)   II–VI compounds (wurtzite), CdS, CdSe
(100)   IV–VI and I–VII compounds (rocksalt), PbS, AgBr.

Sometimes, there are important exceptions. For example, ZnO (wurtzite) in needle form has {10$\bar{1}$0} prism faces and cleaves across the needle in the

---

[215] D. Haneman, *Phys. Rev.* **170**, 705 (1968).
[216] M. F. Chung and D. Haneman, *J. Appl. Phys.* **37**, 1879 (1966).
[217] D. Haneman, M. F. Chung, and A. Taloni, *Phys. Rev.* **170**, 719 (1968).
[218] M. F. Chung, *Bull. Am. Phys. Soc.* **14**, 787 (1969).
[219] A. Kobayashi, Z. Oda, S. Kawaji, H. Arata, and K. Sugiyama, *J. Phys. Chem. Solids* **14**, 37 (1960).
[220] H. Statz, G. A. de Mars, L. Davis, Jr., and A. Adams, Jr., *Phys. Rev.* **101**, 1272 (1956).
[221] D. R. Palmer, S. R. Morrison, and C. E. Dauenbaugh, *Phys. Rev.* **129**, 608 (1963).

{0001} direction (cleavage plane ⊥ to the long c axis).[222] The cleavage perfection depends on the type of cleavage geometry. Some cleavages are better than others, as judged both by the optical and electron microscope[202] and the Y vs $h\nu$ plot.[202] Good cleavage may require a special geometry of the crystal and cutter. Poorer cleaved (or broken) surfaces sometimes have an upward curvature in Y vs $h\nu$ plots.[187a–c,203] Surfaces have been prepared which were cleaved in relatively inert liquid N_2 [223] or in an organic liquid[224]; these surfaces have somewhat questionable cleanliness, because of the possibility of extrinsic surface states being present. Of course, surfaces cleaved in *air* and not reprocessed are about worthless for studying *intrinsic* surface states.

Another way to obtain an acceptable clean surface (free from foreign atoms) is to "bombard and anneal." That is, a desired crystal face may be cut, ground, polished, etched, and mounted in ultrahigh vacuum. The surface is then subjected to ion bombardment (usually argon), followed by an anneal, and the cycle is repeated a number of times. Such a procedure was first developed by Farnsworth and co-workers.[225] The bombardment and anneal method removes work damage, oxides, etchants, etc., and usually leaves an acceptable clean surface, as judged by LEED.

Heating of an air-exposed crystal in UHV *rarely* yields an intrinsic surface. In fact, surface contaminants can be *added* to the initial air-induced contaminants (1) by outgassing and diffusion of crystal bulk impurities to the surface, and (2) by outgassing of the chamber walls in the vicinity of the crystal sample. Such surface contaminants may be removed by cycles of ion bombardment and anneal, but rarely by heating alone.

For these reasons, it is concluded that the only two methods which can be somewhat trusted to yield intrinsic surfaces are the cleavage method and the bombardment-and-anneal method. (Even these difficult and time-consuming methods have their own special problems, such as cleavage perfection and p-type film formation, as will be seen in Section 17). About 5% of surface state studies use these two methods and, thus, they deal with *intrinsic* surface states; this will form the body of Section 17. The remaining 95% of surface state studies do not use these two methods of surface preparation, for reasons of convenience; they are extensively reviewed elsewhere.[7,8]

---

[222] G. Heiland and P. Kuntzmann, *Surface Sci.* **13**, 72 (1969).
[223] F. Proix and P. Handler, *Surface Sci.* **5**, 81 (1966).
[224] H. K. Henisch and W. P. Noble, Jr., *Surface Sci.* **4**, 486 (1966).
[225] See, for example, H. E. Farnsworth, R. E. Schlier, G. Burger, and R. M. Burger, *J. Appl. Phys.* **26**, 252 (1955); R. E. Schlier and H. E. Farnsworth, *in* "Semiconductor Surface Physics" (R. H. Kingston, ed.), p. 3. Univ. of Pennsylvania Press, Philadelphia, Pennsylvania, 1957.

A recent, and useful, tool for determining low concentrations of surface impurities (<5% of a monolayer) is the method of Auger spectroscopy.[226] Field ion emission[169] gives the geometric locations of surface species, including impurities, but it has not been applied to semiconductors, because of their weak mechanical strength under the high electric field required. An up-to-date review of the many different methods of surface analysis has been compiled by Mark and Levine.[226a]

15. Characterization by LEED

LEED is a powerful tool for assigning the *structure* of surface atoms. It is much more relevant to surface state properties than, e.g., optical methods. Reconstruction and microfacets of <100 Å have been seen in CdS[227] by LEED; these are invisible to the best optical microscope.

In LEED, a collimated monoenergetic electron beam is incident on a surface, and the elastically scattered electrons are measured on a fluorescent screen or in a Faraday cage. The beam penetrates only a few atomic layers so it "sees" the surface rather than the bulk. A review of the LEED method and results is available.[228] Basically, there are three types of information in most LEED studies. In most LEED patterns, there is an array of spots, as in an x-ray Laue pattern. The spot locations indicate the lateral atomic periodicity of the lattice. A surface with an atomic periodicity equal to that of the bulk will have a set of spots at the expected location in the LEED pattern; but a surface with, say, double the atomic periodicity will have an additional set of spots in between the expected spots. Second, the relative intensity from one spot to another can tell the relative weight of the different structures. Third, the variation of a single spot intensity as a function of incident electron energy can tell a great deal about the structure, via Bragg's law. Unfortunately, there are complications in LEED, not seen in x-ray diffraction. That is, the electrons interact among themselves giving multiple scattering; self-consistent theoretical methods[229] have been derived which yield fractional order Bragg peaks, i.e. $n = 1, 1\frac{1}{3}, 2, 2\frac{2}{3}, \ldots$, etc. Also, one can only estimate the scattering factor and the surface potential; the problem is especially acute for interpreting LEED patterns on binary crystals.

According to LEED results, all Si and Ge surfaces seem to be *recon-*

[226] See, e.g. R. E. Weber and W. T. Peria, *J. Appl. Phys.* **38**, 4355 (1967).
[226a] P. Mark and J. D. Levine, eds., "Modern Methods of Surface Analysis." North-Holland Publ., Amsterdam (to be published) (1970).
[227] B. D. Campbell and H. E. Farnsworth, *Surface Sci.* **10**, 197 (1968).
[227a] B. D. Campbell, C. A. Hauge, and H. E. Farnsworth, *in* "The Structure and Chemistry of Solid Surfaces" (G. A. Somorjai, ed.), p. 33–1. Wiley, New York, 1969.
[228] Proc. of Symposium on the Structure of Surfaces. *Surface Sci.* **8**, 1 (1967).
[229] See, e.g. E. G. McRae, *Surface Sci.* **8**, 14 (1967).

structed[54,153,230]; i.e., they have strong extra spots, uncharacteristic of an ideally terminated lattice. In diamond, the reconstruction spots are very weak[55] (<1% relative intensity). It also appears that reconstruction is common for "polar" faces, such as zinc blende (111), ($\bar{1}\bar{1}\bar{1}$)[152,153] and wurtzite (0001) and (000$\bar{1}$).[227,227a] These faces all have one species of atom in the surface layer and are unstable from simple electrostatic arguments.[30,231] The detailed reason for the reconstruction is not fully understood at present. Vaguely, one can say that the dangling bonds distort via atomic movement, to saturate themselves, or that the electrostatic charges shift to obtain a more neutral surface. More explicit atomic pictures have been proposed for the {111} faces of Si, Ge, GaAs, and GaSb by Lander and Morrison,[54,230] MacRae,[153] Seiwatz,[232] and Haneman.[150,233] The Lander and Morrison and MacRae models presume that certain atoms are removed; the Seiwatz model presumes conjugation as in organic molecules; and the Haneman model presumes a warped or corrugated surface, where some of the *surface* atoms are raised by <1 Å and some are lowered by <1 Å. Recent evidence on Na adsorption,[234] regrowth after cleavage,[235] bond energetics,[236] and electron spin resonance[215–217] seems, at present, to favor the Haneman model for Si and Ge. But *each* pattern of each crystal face of each crystal (sometimes at each temperature too) invites a controversial explanation for the pattern. For this reason no attempt will be made here to assign atomic structure to the LEED patterns.

The "ionic" or "compensated" surfaces of partly ionic compounds seem to be unreconstructed.[152,237,238,238a] That is, there are no new spots. This is probably due to their electrostatic neutrality in the surface plane (equal number of A and B ions in the surface plane of an AB crystal).[30,231] The intensity analysis seems to indicate that, e.g., in a (110) surface of GaAs, the Ga sublattice is rigid, but slightly shifted laterally or normally (<1 Å) with respect to the As sublattice.[152,153] The same seems to be observed for all the III–V zinc blende compounds[152,238] as well as in (10$\bar{1}$0) CdS (wurtzite).[238a] (These crystal faces, if ideally terminated, would naturally

---

[230] J. J. Lander and J. Morrison, *J. Appl. Phys.* **34**, 1403 (1963); H. E. Farnsworth, R. E. Schlier, and J. A. Dillon, Jr., *J. Phys. Chem. Solids* **8**, 116 (1959).
[231] R. Nosker, P. Mark, and J. D. Levine, *Surface Sci.* **19**, 291 (1970).
[232] R. Seiwatz, *Surface Sci.* **2**, 473 (1964).
[233] N. R. Hansen and D. Haneman, *Surface Sci.* **2**, 566 (1964).
[234] P. W. Palmberg and W. T. Peria, *Surface Sci.* **6**, 57 (1967).
[235] D. Haneman, W. D. Roots, and J. T. P. Grant, *J. Appl. Phys.* **38**, 2203 (1967).
[236] A. Taloni and D. Haneman, *Surface Sci.* **10**, 215 (1968).
[237] I. Murkland and S. Anderson, *Surface Sci.* **5**, 197 (1966).
[238] L. P. Feinstein and D. P. Shoemaker, *Surface Sci.* **3**, 294 (1965).
[238a] G. K. Zyrianov and A. T. Hoang, *Bull. Leningrad Univ., Phys. Chem.* No. 10, p. 69 (1969).

have a right-hand and a left-hand side.) Other surfaces are not altered at all in the lateral direction. One such surface is the (100) face of NaCl.[237] There is a lattice dilation of a few percent, in the normal direction,[239] but this is a second-order effect compared to the reconstruction and mild lateral shifts described above. Finally, *reconstructed surfaces* obtained by cleaving have phase changes upon heating,[54,230] and the spot picture changes. Each surface investigation can be assigned a label, e.g., Si (111)-7. There seems to be no phase changes with temperature for the unreconstructed surfaces.[152]

16. LINE DEFECTS

Among the weaknesses of LEED is its inadequacy to detect certain surface features less than, say, 15% of the nominal surface atoms.[153] For example, *line defects* due to edges of terraces or edges of microfacets (or dislocations or grain boundaries) would not appear in LEED, if there were <1% of the surface atoms affected. To estimate this contribution, consider that the best-cleaved Si (111) crystals have terraces 2000 ± 1000 Å apart and >50 Å high, as obtained from electron microscopy.[202] The number of surface atoms in the line defects are

$$N_l \approx \left(\frac{1 \text{ line}}{0.2 \times 10^{-4} \text{ cm}}\right)\left(\frac{1 \text{ atom/line}}{3 \times 10^{-8} \text{ cm}}\right)$$

$$\approx 1.6 \times 10^{12} \quad \text{atoms/cm}^2. \tag{16.1}$$

For comparison, there are 8×10^{14} atoms/cm^2 in the normal surface. Thus, the edge atoms occupy about 0.2% of a monolayer. Similar figures have been obtained for the best cleavages on III–V crystals where the terraces are 7000 ± 3000 Å apart and 50–200 Å high.[152] Suppose that the edge atoms had surface states associated with them. The charges in these states would be compensated by a nearly cylindrical space-charge sheath, and the surface would appear "patchy." This patch effect would, supposedly, show up in electrical or optical measurements. It is useful to contrast these supposed "terrace" or "edge" states with edge dislocation states known to exist in *bulk* Ge. Pearson *et al.*[240] displayed and explained the acceptor properties of bulk edge dislocations states and Shockley[241] predicted that these form an impurity band, and are responsible for low-temperature conduction in bulk Ge. By contrast, surface states do not seem to con-

---

[239] G. C. Benson, P. I. Freeman, and E. Dempsey, *J. Chem. Phys.* **39,** 320 (1963); H. Tokutaka and M. Prutton, *Surface Sci.* **11,** 216 (1968).
[240] G. L. Pearson, W. T. Read, Jr., and F. J. Morin, *Phys. Rev.* **93,** 666 (1954); W. T. Read, Jr., *Phil. Mag.* **45,** 775 (1954); **46,** 111 (1955).
[241] W. Shockley, *Phys. Rev.* **91,** 228 (1953).

duct.[180,219,241a] Igras[242] has measured edge dislocation states using etch pits and the "electron mirror" technique.

It seems that the edge atoms do *not* affect the LEED pattern[54,230] or the Y vs $h\nu$ plots for these well-cleaved crystals.[177,178,202] Furthermore, the charges *measured* in the surface states in Si amount to $>10\%$ of a monolayer[178] ($Q_{ss}/e > 0.4 \times 10^{14}$ states/cm^2), which is >50 times the density of edge atoms. Also, Swank found similar Φ and ϕ_w results on II–VI crystals with different perfections of good cleavages.[159]

17. Data and Interpretation

As seen in the previous sections, data acquisition on intrinsic surface states is, in fact, difficult to carry out experimentally and complicated to analyze and interpret. Each experimentalist presents his raw data, his interpretation of it, and his own procedure for obtaining a clean surface. There are pros and cons for each step taken. For example, cleaved surfaces have more reproducible photoemission properties than bombarded and annealed surfaces,[202] but the former sometimes have less reproducible LEED patterns than the latter.[243] It is easy, therefore, for any experiment concerned with a surface state distribution to be criticized on various grounds. Sometimes the criticism can be mild, as in an energy shift of an entire distribution, and sometimes it is severe, as in the use of a mica spacer in intimate contact with the surface. To reproduce here all the raw data, even on clean surfaces, would therefore be folly.

Instead, only those experiments will be reported and evaluated in this section, *if the work is necessary from the point of view of the theoretical models of intrinsic surface states*. Other experimental work regarding adsorbates, interface states, and the physical, chemical, and structural aspects of surfaces will be purposely eliminated for clarity. Emphasis will be on the electrical properties of free intrinsic surfaces.

For simplicity in presentation, the zero of energy will be defined to be near midgap $E_i = 0$. In particular, the location of the Fermi level at the surface will be $E_F = 0$ (intrinsic surface), $E_F > 0$ (n-type surface), and $E_F < 0$ (p-type surface). Whenever possible, the results of each experimenter will be presented, not the detailed steps each took to get the results. This procedure saves considerable space and time, and makes it easy for a new worker to comprehend the field. This section will treat silicon, ger-

---

[241a] Studies on cleaved ZnO at 90°K were made by G. Heiland and P. Kuntzmann.[222]

[242] E. Igras, *Proc. Intern. Conf. Phys. Semiconductors Exeter, England* (A. C. Strickland, ed.), p. 832. Academic Press, New York, 0962.

[243] H. E. Farnsworth, J. B. Marsh, and J. Toots, *Proc. Intern. Conf. Phys. Semiconductors, Exeter, England* (A. C. Strickland, ed.), p. 836. Academic Press, New York 1962.

manium, and III–V and II–VI crystals, in that order. A very small section is added at the end, containing aspects of *extrinsic* surface states, for purposes of comparison.

a. Silicon

Of all materials, Si has probably the best known intrinsic surface-state distribution. It was essentially determined in 1962 by Allen and Gobeli.[177–179] Their pioneering efforts were followed by many other publications, but the distribution has not been changed significantly. Instead, the many electrical, optical, and electron paramagnetic resonance (EPR) experiments generally support the distribution, or at least do not strongly contradict it.

Silicon investigations are complicated, because the surface is reconstructed. After cleavage in UHV at room temperature, the LEED pattern is called either Si (111)-1 $\times$ $\sqrt{3}$ or Si (111)-1 $\times$ 2, depending on whether[215] or not[54,230] oblique axes are used as reference axes. After cleaving and heating, this pattern changes irreversibly to Si (111)-7; this is apparently the same structure as that obtained after ion bombardment and annealing. For more details about these structures and an intermediate structure Si (111)-5 see Lander[54] and Lander and Morrison.[230] It is also difficult to cleave Si; some cleavage geometries are *far* better than others.[177–179,202] The theoretical analyses of Si surface states are not trivial, because of the *indirect* bulk band gap ($E_g = 1.09$ eV at room temperature). According to these aspects, Si does not seem to be a good choice for carrying out basic surface state measurements. It has five other assets, however. Its commercial importance, its controlled purity, its control of n and p dopings, its freedom from bulk traps, and its simplicity of bonding in the bulk (pure covalent binding). By contrast, the ionic faces of the III–V and II–VI compounds have direct bandgaps and unreconstructed surfaces, but these compounds have more complications in bulk preparation and doping than does Si (or Ge). It is worth repeating that only *intrinsic* surface states are considered in this section. The vast literature regarding the Si–oxygen adsorbate system and the Si–SiO_2 interfaces will *not* be considered until Section 17,d.

Consider the qualitative properties of cleaved Si surfaces, as described by Allen and Gobeli,[177–179] Van Laar and Scheer,[187a–c] Handler,[188] Aspnes and Handler,[180] Henzler,[190] Henzler and Heiland,[244] Haneman,[215] and Fischer.[245] The surface states seem to be arranged in acceptor–donor pairs as suggested in the *two discrete level* model in Fig. 34a, the *two band model* ($E_b \sim 0.2$ eV) in Fig. 34b, or a *cosh-like model* in Fig. 34d. The Fermi level

---

[244] M. Henzler and G. Heiland, *Solid State Commun.* **4**, 499 (1966).
[245] T. E. Fischer, *Surface Sci.* **10**, 399 (1968).

moves only ~0.2 eV at the surface over the range from degenerate n-type to degenerate p-type doping, thus, $E_{sg} \sim 0.2$ eV. The data do not permit, at this time, further selection between these three models or similar models. The main evidence for these models is that at a plot of Q_{ss} vs E_F follows a sinh-like curve[178] as predicted theoretically for these models: see (12.4), (12.8), and (12.17). That is, the slope dQ_{ss}/dE_F at the center E_0 of the surface-state distribution is the minimum slope. Thus, a merged band model of Fig. 35c can be ruled out for cleaved Si: see (12.11). The surface state distributions for cleaved-and-heated or bombarded–annealed Si are much less definitive; the reason is that the relevant experiments were not carried out over a range of dopings.

Are the observed surface states intrinsic, or are they due to surface defects, such as cleavage steps, thermal-etch pits, and edge dislocations? The question can be resolved in two ways. First, the maximum number of states reported, $N \geq 10^{14}$ cm^{-2}, seems to be much greater than the number of surface steps, etch pits, or dislocations, as discussed in Section 16. Second, the presence of the inequality sign in most data analyses *does not contradict* the supposition of intrinsic surface states $2N = 8 \times 10^{14}$ cm^{-2}. As seen from Figs. 35–37, the magnitude of the surface states can be determined only if the Fermi level can be passed *through* the states. Since this has not been done for Si, the experimenter is forced to use the $\geq$ sign. The problem is especially serious for the field-effect mobility measurements, where the number of surface states, within $\sim \pm kT$ of the Fermi level, is the maximum number measured. In this sense, the importance of the field-effect mobility[7,8] has been somewhat overrated. Besides, with the postulated distributions for cleaved Si, a field-effect mobility measurement carried out near the density-of-states minimum ($E_F = E_0$) would yield a gross underevaluation of the surface states present. This would occur for a 200 Ω-cm p-type Si crystal. There is one case where the opposite inequality $\leq$ has been used; it has been offered tentatively by Fischer[245] as an explanation for insensitivity to temperature dependence; this will be discussed in more detail later. The surface states on Si seem to have unusual dynamic properties. They photoemit when there are electrons in the states, but *only* for n-type samples[177,178] (on a cleaved surface). They do not seem to conduct electricity,[180] even though Q_{ss}/e can be as much as 10% of a monolayer for degenerate n-type samples.[177,178] The EPR signal after cleavage is $\sim 10^{14}$ cm^{-2}, indicating intrinsic surface states[215]; it *decreases* by an order of magnitude or more[216,217] upon heating in UHV. It is interesting that the surface state photoemission of p-type samples *increases* abruptly after heating in UHV.[202]

Thus, the qualitative aspects of Si surface states are: (1) a distribution as in Fig. 34a, b, or d, not c; (2) a density high enough to be related to intrinsic surface states; and (3) unusual dynamic properties. The location of E_0 is a matter of some dispute. This matter and others will be considered below, where the individual papers are briefly reviewed.

The main contribution of Allen and Gobeli[177–179] was to examine ϕ_w and Φ (cube law) over a *continuous* range of dopings on cleaved Si, from degenerate n-type to degenerate p-type. This doping range was able to move the Fermi level by ~ 0.2 eV, which is only about 20% of the band gap. This is the extent of the "clamping action" of the surface states. To estimate the distribution of surface states, they assumed $\Phi = 5.15$ eV *constant* for all dopings and located at the valence-band edge at the surface. Φ is generally different from Φ (apparent). The observed decrease in Φ (apparent) for highly doped p-type samples was attributed to the yield from subsurface levels (escape depth is ~ 25 Å), plus a decrease in χ for very degenerate samples. The observed decrease in Φ (apparent) for highly doped n-type samples was attributed to the photoemission of filled acceptor-like surface states; the corresponding Y vs $h\nu$ distribution was exponential, rather than a power law as predicted by Kane.[199] (All power laws approach a vertical asymptote on a semilog plot.[159])

Only in the range -0.4 eV $< E_F < 0.0$ eV was $\Phi = 5.15$ eV, without corrections. Using Φ and ϕ_w for different dopings, they computed Q_{ss} vs E_F; the plot was sinh-like. To be specific, the curve fitted to the data was

$$Q_{ss}/e = -2.44 \times 10^{12} \sinh[(E_F/kT) + 9.7] \text{ cm}^{-2}.$$

E_0 is located at $-9.7kT = -0.23$ eV. From the fact that the data did not depart greatly from the sinh-like curve, over the range in E_F investigated, they concluded that the states must lie outside the range investigated. This is in accordance with the theoretical plots in Figs. 35 and 36. For the *two discrete level* model the parameters are (see Figs. 34a and 35)

$$E_0 = -0.23 \text{ eV}, \quad N_1 = N_2 \geq 1.8 \times 10^{14} \text{ cm}^{-2}, \quad E_{sg} \leq 0.30 \text{ eV}.$$

For the *two band model* the parameters are (see Figs. 34b and 36)

$$E_0 = -0.23 \text{ eV}, \quad \bar{N}_1 = \bar{N}_2 \geq 2.6 \times 10^{15} \text{ cm}^{-2} \text{ eV}^{-1}$$

$$E_{sg} = 0.2 \text{ eV}, \quad E_b \leq 0.16 \text{ eV}.$$

If one ignores the inequalities in $\bar{N}_1$ (worst case situation), then $N_1 = \bar{N}_1 E_b$, and

$$N_1 = N_2 = 4 \times 10^{14} \text{ cm}^{-2}, \quad m_s^* = 6.5 m_0.$$

The large surface state effective mass means that there is a large density of states per eV interval. By comparison, the bulk has a small effective mass and a comparably larger bandwidth as seen from (12.9). A third model was considered, which followed a cosh-like distribution as in Fig. 34d, with a minimum $\bar{N}(E_0)$ of 0.98×10^{14} cm^{-2} eV^{-1}. This distribution was obtained by differentiating the sinh-like curve of Q_{ss} vs E_F, as in (12.18), but ignoring the temperature-dependent terms on the right-hand side. As seen from Fig. 37, even the degenerate distributions of surface states can have considerable temperature dependence of Q_{ss}. The cosh-like distribution could not extend beyond ± 0.13 eV from E_0, or else the intrinsic surface states are exceeded. The location of the center of surface state distribution, E_0, is a matter of some dispute, as seen below, but it is not significant as far as the distribution *shape* is concerned. (An attempt to measure V_D directly by photovoltage experiments was unsuccessful.) There is general agreement among all workers, and for all Si preparations, that $0 < E_0 < -0.4$ eV; in other words, a nondegenerate p-type surface. The maximum number of surface states reported by Allen and Gobeli has been questioned by Frankl[246]; he claims that if the degenerate bulk dopings are eliminated from the analysis, the number would be less.

Other studies of the properties of cleaved Si will now be reviewed. Van Laar and Scheer[187a-c] measured Φ for different dopings ranging from degenerate n type to degenerate p type. Unfortunately, ϕ_w was not measured, so the surface properties had to be inferred from the *shape* of Y vs $h\nu$ plots on p-type samples. Because the bands bend upward, electrons in the bulk (<30 Å from the surface) can contribute to Y. This effect changes the shape of Y vs $h\nu$ plots. By appropriate mathematics and curve fitting, and by analysis of their indium adsorption data, they concluded that, near E_F, $\bar{N} \geq 10^{15}$ states/cm^2 eV. By applying a $\frac{3}{2}$-power law, which they observe from photoemission yields, they conclude that the surface state distribution must be cosh-like about a neutral level E_0 which, they claim, is near midgap $E_0 = 0$. They account in this manner for intrinsic surface states 8×10^{14} cm^{-2}. In an earlier paper, the stabilization of the Fermi level by ~ 0.1 eV for various p-type dopings gave them a lower limit of $N \geq 2 \times 10^{13}$ donor states within a "narrow energy range."[187a] The low-energy tails (cube law or $\frac{5}{2}$ law) were first attributed by van Laar and Scheer[187a] to strained parts of the cleaved (practically broken) crystal; they included other data at the time, which showed how the low-energy photoemission tail on Si (111) was increased by sandblasting or etching. This behavior has also been observed by others.[247] In later papers,[187b,c] they attributed this tail to surface state emission, even for p-type samples; this is in marked contrast to the assumption of Allen and Gobeli.[177-179] Most other evidence, reviewed below, seems to favor Allen and Gobeli's assumption that Φ (cube law) corresponds to the valence-band emission for p-type samples.

---

[246] See Frankl,[8] p. 201.
[247] N. I. Vitrikhovskii and M. V. Kurik, *Surface Sci.* **3**, 415 (1965); M. T. Thomas, G. Shimaoka, and J. A. Dillon, Jr., *ibid.* **6**, 261 (1967).

Aspnes and Handler[180] measured the change in conductivity occurring upon cleavage of p-type samples of Si: one gold-doped and one boron-doped. By determining the degree of conductivity increase or decrease upon cleavage and making assumptions regarding scattering, they found that $E_0 = -0.27 \pm 0.08$ eV, independent of assumptions regarding Φ. The result of this electrical measurement is in good agreement with Allen and Gobeli, but not with van Laar and Scheer. A better experiment would be to cleave the crystals of different doping and find the one which has *null* conductivity change upon cleavage. This was done by Handler[188] in a previous paper, the result being that E_0 agrees with the Allen and Gobeli value $E_0 = -0.23$ eV. Aspnes and Handler carried out cleavages at temperatures as low as $T = 105°K$, in an attempt to detect conduction in surface states. Before cleavage a typical sample had 11×10^{-12} mho conductance; but there was no increase in conduction upon cleavage. With continuous *merged bands* or *cosh-like* models of surface states, they compute a mobility $\mu_{ss} < 10^{-8}$ cm^2/V-sec and $\bar{N} > 1.8 \times 10^{12}$ cm^{-2} eV^{-1}. With the *two band* or *two discrete level* models, they can account for the experimental results, because of the semiconductor nature of the surface states. But at 300°K, they would then expect a measurable surface state conductivity, due to thermally generated free carriers, of $\sim 10^{12}$ cm^{-2} in the surface state band; they did not see this. That is, their conductivity data (on two samples) seems self-consistent, without including the extra surface state conduction. On conductance grounds alone, they find it impossible to select a distribution of surface states. It seems that the surface states on Si (and also Ge as shown in the following section) simply do not conduct; this goes against a major assumption of 3-D surface state theory; the reason for this discrepancy is not clear. Possibly the surface states *are* conducting, but an appropriate ohmic contact has not been made.

Field-effect measurements were made by Henzler[190] and Henzler and Heiland[244] on a p-type crystal. They took special steps to shield the crystal and to measure the field effect for various distances of the field electrode. The data were $E_F = -0.36 \pm 0.02$ eV, $\mu_{fe} = -1$ cm^2/V-sec (p type) and $\Delta\sigma = 0.59 \pm 0.12 \times 10^{-6}$ mho/square. Assuming a *merged band* model and using (13.4), they computed $\bar{N} = 1 \times 10^{14}$ cm^{-2} eV^{-1}. But, as mentioned in Section 13,b, the same data can be interpreted with a *two discrete level* model, using (13.5). The result then becomes $2N = 4 \times 10^{14}$ cm^{-2} and $E_{sg} = 0.25$ eV. The latter model gives excellent agreement with Allen and Gobeli's assumed *distribution* of intrinsic surface states, as well as the general location of the neutral surface state level E_0. Similar field-effect measurements were made on n-type Si by Kawaji and Takishima.[248] They found

---

[248] S. Kawaji and Y. Takishima, *Surface Sci.* **1**, 119 (1964).

that $E_F = -0.30$ eV, but they made no estimates of the surface state distributions.

Haneman[215] measured the EPR signal of a cleaved 0.7 cm² single crystal surface; the signal was taken out of the noise by an accumulation technique. In this manner, he found a total of 0.8×10^{14} spins/cm². These correspond to the number of *unpaired* surface state or space-charge electrons; thus, this figure is a lower limit. Haneman also found no hyperfine structure and no anisotropy of the g tensor. He took these results to mean that a percentage of electrons in surface states were not localized on specific sites and that they were essentially free electrons. He also made the following suppositions, some of which are based on his previous studies[150,233,235]: (1) the surface atoms were split into two groups; (2) some of these are raised (called s-like); (3) some of these are lowered (called p-like); (4) the two types are arranged in adjacent furrows; (5) conductivity occurs only down the furrows, not transverse to them; and (6) cleavage steps exist parallel to the furrows and perpendicular to the contacts. In this manner, Haneman explains the negligible conductivity[180] found for surface state bands. He also speculates that the surface state distributions of Allen and Gobeli (i.e. Fig. 34a) can be accounted for by separate bands due to the s-like and p-like surface atoms. This interpretation is not unique. Other possible speculations for the origin of these distributions can be (1) the Jahn–Teller effect as proposed by Ducros[249]; and (2) the spin correlations of surface states as studied by Davison and Amos[65] and Tomášek[250]: essentially, the surface state bands are characterized by the lower and upper bands having $\uparrow \downarrow$ spins, respectively. Chung[218] carried out electron paramagnetic resonance studies on crushed Si powder of various dopings, and found a maximum signal for the intrinsic samples.

Fischer[245] carried out an analysis of photoemission energy distributions for an n-type and a degenerate p-type crystal. The analysis was performed entirely without measuring ϕ_w; instead the threshold and saturated points of the photoemission energy distribution were given prime emphasis. The measured temperature dependences in the range $80°K < T < 300°K$ were found to be slight. Further analysis showed that Q_{ss} was essentially independent of T for both samples. This independence indicated electron degeneracy, which was shown to provide *lower limits* to $\bar{N}$ in the surface state *two band* model. That is, $\bar{N} \leq 4 \times 10^{14}$ cm⁻² eV⁻¹, which is about one order of magnitude lower than that of Allen and Gobeli, $\bar{N} \geq 26 \times 10^{14}$ cm⁻² eV⁻¹. Reference to Fig. 36 (also Figs. 35 and 37) shows that true temperature independence occurs only at the *centers* of the surface state bands; this is a more severe restriction than the simple result predicted by

[249] P. Ducros, *Surface Sci.* **10**, 295 (1968).
[250] M. Tomášek, *Surface Sci.* **4**, 471 (1966).

the simplified equations used by Fischer [his Eq. (10)], namely, that temperature independence is guaranteed, provided E_F lies within the surface state band. It seems probable that the *cosh-like* distribution proposed by Allen and Gobeli has the least temperature dependence and, hence, is in greatest accord with the results of Fischer. This conclusion is obtained from investigating (12.17), and also from noting the closeness of the curves in Fig. 37 for $T = 300°K$ and $T = 75°K$. Aspects of Fischer's analysis, in close agreement with the work of Allen and Gobeli, include the locations of the Fermi energy with respect to the band edges at the surface for both n- and p-type samples.

The field-effect measurements of Palmer *et al.*[251] seemed to infer that $E_0 > 0$. The work was later reinterpreted by Handler[252] who showed that (1) part of it was essentially consistent with the results of Allen and Gobeli, and (2) the other part was spurious, due to the strong capacitative coupling with the uncleaved surface. This completes the most relevant studies carried out on cleaved Si.

Experiments on cleaved and heated Si were carried out by Allen and Gobeli and reported in a later paper.[202] However, they measured ϕ_w and Φ (including photoelectron energy distribution) for only one doping, so they could not determine a surface state distribution. It seems that the surface states photoemit after heating, since Φ (cube law) was assigned to the Fermi level located within a surface state band. The location of E_F was proposed to be $E_F = -0.23$ eV, the same as E_0 before heating. In the heating process, Φ was assumed to be decreased by 0.23 eV. Chung and Haneman[216] studied the EPR of crushed Si powder in UHV (supposedly the major facets were {111}). An anneal in UHV at $\sim 1100°K$ caused the EPR signal to become $\sim 2\%$ that of the unannealed crushed Si. Some explanations were offered involving the new surface structure. Similar results were reported later by Haneman *et al.*[217] on air-crushed and mechanically polished Si crystals. It is interesting that annealing in UHV causes both a great *increase* in surface state photoemission and a great *decrease* in the EPR signal.

Bombarded and annealed Si was studied by Allen and Gobeli,[202] Law,[253] Heiland and Lamatsch,[254] Nesterenko and Snitko,[194,195] and Dillon and Farnsworth.[255] In spite of this great effort, the surface state distribution is still unknown, primarily because a range of dopings was not used. Allen and

[251] D. R. Palmer, S. R. Morrison, and C. E. Dauenbaugh, *Phys. Rev. Letters* **6**, 170 (1961).
[252] P. Handler, *Phys. Rev.* **126**, 971 (1962).
[253] J. T. Law, *J. Appl. Phys.* **32**, 848 (1961).
[254] G. Heiland and H. Lamatsch, *Surface Sci.* **2**, 18 (1964).
[255] J. A. Dillon, Jr., and H. E. Farnsworth, *J. Appl. Phys.* **29**, 1195 (1958).

Gobeli find similar results $E_0 \sim -0.23$ eV for this treatment as for the cleaved-annealed sample. The bombarded–annealed surface differed, however, in the following respects: (1) there was no linear portion of the Y vs $h\nu$ plot (instead there was only a curved portion), and (2) there was a slight p-type surface film formed (presumably due to surface contamination). This occurred even though the sputtering was carried out *before* the first heating. It is in contradiction to Dillon and Oman,[256] who found that the p-type film can be removed, if the sample is ion-bombarded before the first heating. By comparison, the p-type film did not appear on the cleaved-and-heated samples. The p-type film is a source of trouble to many workers who use the bombardment-annealed method, as described below.

Dillon and Farnsworth[255] measured ϕ_w and Φ for the (111), (110), and (100) faces of p-type Si. They found the following sequence in the magnitude of the work functions: $\phi_w(100) > \phi_w(110) > \phi_w(111)$. This is the same order as the surface density of cut bonds on the respective faces (Table II, Section 9).[141] They also found that the photoemission threshold Φ was smaller than ϕ_w, in contradiction to all expectation; they attributed this to oxygen impurities in the Si samples. Thus, it is impossible to draw conclusions from this data regarding the surface distribution. Law[253] was aware of the p-type film contamination problem and took steps to prevent it. By carrying out conductance measurements (junction characteristics) of a p–n junction, and related photoconduction measurements, he found that $E_F = -0.37$ eV on both n and p sides; if a p-type film were present, then E_F would be even closer to midgap ($E_F = 0$).

Heiland and Lamatsch[254] were also aware of the contamination problem and took steps to prevent it. They studied the conductance properties along a 1000 Ω-cm p-type Si single crystal filament (the circumference has different crystal faces). By bombardment and annealing *in stages*, they could drive the surface through a conductance minimum. Apparently, bombardment without annealing leaves an n-type surface. They found that $E_F = -0.25$ eV. Field-effect mobility measurements gave $\mu_{fe} = -0.01$ cm^2/V-sec and $\bar{N} = (1.5 \pm 0.5) \times 10^{15}$ cm^{-2} eV^{-1} at E_F, assuming a *merged band* model, or $N \geq 10^{14}$ cm^{-2}, assuming a *single discrete level* model, with the level located at E_F. Thus, these results support the assertion of Allen and Gobeli that, for cleaved-and-heated or bombarded-and-annealed Si, E_F lies inside the surface state band and the photoemission threshold arises from surface state photoemission. Nesterenko and Snitko[193] and Nesterenko *et al.*[194] measured the pulsed field effect and the field-effect mobility, for both n-type and p-type (111) samples. A mica plate was used as a spacer for the field effect, but this was kept from contacting the Si

[256] J. A. Dillon, Jr., and R. M. Oman, *Proc. Intern. Conf. Semiconductor Phys.* Czech. Acad. Sci., Prague, 1961. See also Many *et al.*,[7] p. 441.

surface by means of nonetched protrusions on the Si. The results were consistent with the *two band* model, as in Fig. 36. It is clear from the dynamics that the upper band (beginning at $E_1 \sim 0.1$ eV) is an electron acceptor and the lower band (beginning at $E_2 \sim -0.1$ eV) is a hole acceptor (electron donor). The neutral level was presumed to be at midgap $E_0 = 0$. Furthermore, $N_1 = N_2 = 10^{14}$–10^{15} cm^{-2} and $\bar{N}_1 = \bar{N}_2 = 1 \times 10^{15}$ cm^{-2} eV^{-1}. One word of caution is in order, however. They found *no* effect due to monolayer adsorption of O_2, Au, or Sb; this may be due to an insulating oxide layer on the Si, which has not been fully removed by their procedure (supposedly the same as the Farnsworth procedure).

As mentioned earlier, the great variety of experiments carried out on clean Si (including field emission tips[257]) has yielded comparatively scant information on the surface state distribution. This is mainly because most experiments were not done with a large range of dopings. For Si, an applied field simply cannot swing the bands very much, due to the clamping action of the surface states. Also it is difficult to get reproducible Si surfaces because (1) Si is hard to cleave satisfactorily, (2) bombarded–annealed Si has a tendency toward formation of a p-type contamination layer, and (3) Si surfaces have different reconstructed surface phases, depending on processing.

Theoretical 3-D analyses applicable to diamond, Si, and Ge surface states have been carried out by Koutecký and Tomášek,[26] Chaves *et al.*[137] Jones,[139] Koutecký,[258] Pugh,[259] and Bartoš.[260] These analyses and results will be briefly reviewed below.

Koutecký and Tomášek[26] chose to represent the bulk of a diamond-like crystal by a tight-binding scheme, where the atomic orbitals are selected to be sp^3 tetrahedral orbitals. The (111) surface termination was accomplished by removing atoms (i.e. requiring the expansion coefficients to vanish), leaving the surface Coulomb and resonance integrals unchanged. That is, they looked for the so-called Shockley states. They found (1) a bulk band structure with a conduction band, a valence band, and two valence bands, whose energies are independent of **k**, (2) a *merged band* of surface states as in Fig. 34c, whose bandwidth E_b is much less than the bulk bandgap $E_b \lesssim 0.2 E_g$, and (3) that the neutral level E_0 was located some distance above the midgap, depending on their choice of parameters.

In a later study, Koutecký[258] considered the effects of perturbing the surface Coulomb integral (i.e. Tamm states). When this perturbation was sufficiently strong, the surface state merged band could be moved below

[257] F. G. Allen, *J. Phys. Chem. Solids* **8**, 119 (1959).
[258] J. Koutecký, *Surface Sci.* **1**, 281 (1964).
[259] D. Pugh, *Phys. Rev. Letters* **12**, 390 (1964).
[260] I. Bartoš, *Surface Sci.* **15**, 94 (1964).

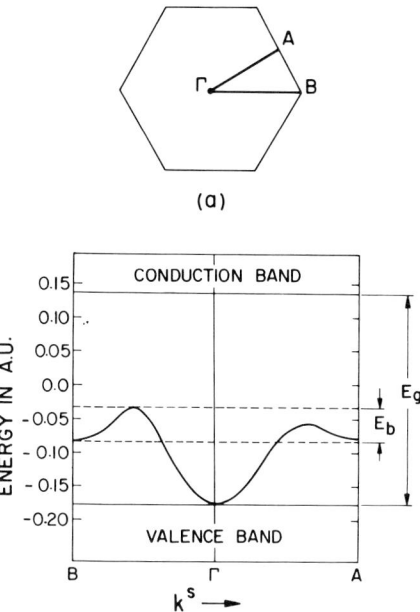

FIG. 38. Theoretical surface state band structure of unreconstructed (111) diamond; computed by D. Pugh, *Phys. Rev. Letters* **12**, 390 (1964).

midgap. A certain perturbation of the surface resonance integral was later considered, and this also moved the surface state band to the lower-half of the bandgap (E_0 became negative). In these studies, only the bandwidth of the surface state band was indicated, not the **E** vs **k** diagram, or the density of states.

Pugh[259] carried out a more elaborate calculation, using the same sp³ orbitals as basis states, and computing some 33 necessary diatomic molecular integrals between first, second, and third nearest neighbors in the surface and bulk regions. He also computed E vs **k** for the 2-D hexagonal zone of the surface state band (Figs. 38a and 38b). Its minimum is at Γ and its maxima were not on symmetry points. The bulk bandgap is E_g. In addition, Pugh computed the *merged band* of surface states for diamond (111) lying below midgap, with over 90% of the states located in a very narrow bandwidth, shown by E_b in the figure.

Chaves *et al.*[137] used a **k·p** formalism to yield an accurate representation of the band structure of bulk Si and Ge. They computed surface state energies *at a selected point* of the BZ, assuming a step discontinuity at a location midway between atom layers. Their results showed the surface

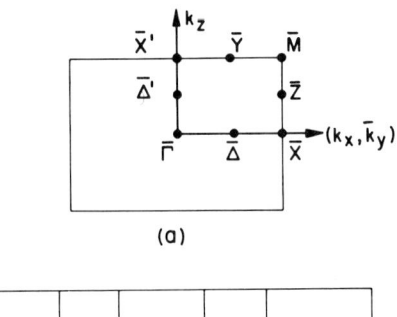

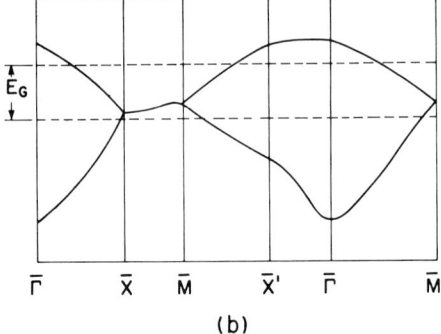

FIG. 39. Theoretical surface state band structure of unreconstructed (110) Si; computed by R. O. Jones, *Phys. Rev. Letters* **20**, 992 (1968).

state at X_1 in the BZ to be located slightly below (0.07 eV) the conduction-band edge for Si, and slightly above for Ge.

Bartoš[260] recently extended this **k·p** method to the (100) faces of Si and Ge, for different locations of an assumed surface potential discontinuity. He found that at the selected point Γ in the 2-D BZ, the surface state was located ∼0.20 eV above midgap for Ge, and was absent entirely for Si. These results were apparently insensitive to the discontinuity location.

Jones[139] carried out a pseudopotential computation of the (110) face of Si and constructed an E vs **k** diagram for the surface state band (Fig. 39). He found a continuous band of surface states across the band-gap, as in Fig. 34f, and many of the states were located outside the bandgap.

Since the above theories do not take into account surface reconstruction, which is known to exist from LEED measurements on Si and Ge, it is not surprising that the comparison between surface state theory and experiment shows little correlation, at the present time. In particular, some of the theoretically computed surface states are located near the conduction band, which is at variance with experiment. Moreover, these calculations do not predict the *pair-like* donor–acceptor states with E_0 negative, as in Figs. 34a, b, and d.

b. Germanium

The surface state distribution for Ge still seems to be controversial, even though many workers have been concerned with this topic. There are probably three main reasons for this. First, the surface states are frequently thought to lie close to the valence-band edge, and in many cases lie *within* the valence band; this makes them difficult to distinguish and analyze. Second, Ge has an indirect bandgap of 0.67 eV, about half that of Si. Because of this feature, the Fermi level clamping in Ge can be achieved with much lower surface state densities.[261] Alternately, given the same surface state densities in Ge and Si, the Fermi level clamping in Ge will be much greater than that in Si; this seems to be the case. Third, the great majority of papers on the subject deal with the field-effect mobility; this has inherent limitations for surface state distribution analysis, as mentioned in the Si section above, unless other parameters are simultaneously measured.

The LEED patterns of Ge are strikingly similar to those of Si.[54,230] That is, cleavage at room temperature gives a reconstructed surface Ge (111)-$1 \times \sqrt{3}$. Upon heating to 470°K, Ge becomes Ge (111)-8. Further heating above 570°K yields Ge (111)-12.[61] The (111) bombarded-and-annealed surface has, essentially, the same LEED pattern as that of the cleaved-and-heated surface, namely, Ge (111)-12. There are subtle differences, however, which can be resolved by the Farnsworth method of obtaining LEED patterns using a Faraday cage. With this method, Farnsworth et al.[243] found that there is a higher degree of perfection in the bombardment–annealed surface than in the cleaved-and-heated surface. The (100) and (110) surfaces have other specific LEED patterns (few popular terminologies as yet).

First consider the properties of cleaved Ge, as measured by Gobeli and Allen,[261] Henzler,[61,262] and Palmer et al.[221,263] Gobeli and Allen attempted to determine the surface state distribution of Ge, following the same procedure they successfully used for Si. That is, they measured ϕ_w and Φ (cube law extrapolation) for various dopings of Ge ranging from degenerate p type to nearly degenerate n type. The striking result was that $\phi_w = \Phi$ *independent* of doping, within experimental error of ±0.02 eV. Thus, the Fermi level is very strongly clamped at the surface; it can swing *at most* by 0.04/0.67 or ~6% of the bandgap. By contrast, for Si the swing is 0.2/1.1 or ~20% of the bandgap. They considered various possible models to

---

[261] G. W. Gobeli and F. G. Allen, *Surface Sci.* **2**, 402 (1964).
[262] M. Henzler, *J. Appl. Phys.* **40**, 3756 (1969).
[263] D. R. Palmer, S. R. Morrison, and C. E. Dauenbaugh, *J. Phys. Chem. Solids* **14**, 27 (1960).

explain the data, and concluded that, in the absence of other information, the best model would be the *two discrete level* model with the center of the distribution located at the valence-band edge:

$$E_0 = -0.33 \pm \text{few } kT, \quad E_1 = E_0 + \text{few } kT, \quad E_2 = E_0 - \text{few } kT$$

$$E_{\text{sg}} \sim \text{few } kT, \quad N_1 = N_2 \geq 1.4 \times 10^{13} \text{ cm}^{-2}$$

Thus, at least 1/50 of the expected Ge intrinsic surface states, 7×10^{14} cm^{-2}, are accounted for in the inequality. The distribution was considered to be similar to that of Si, except for two differences: (1) in Ge, E_0 lies practically at the valence-band edge, while in Si, it lies $\sim$0.3 eV above the valence-band edge; and (2) E_{sg} is much narrower in Ge than in Si. The surface state occupation Q_{ss} for Ge will then be similar to the 75°K curve in the *two discrete level* model of Fig. 35. (The statistics are approximately scaled by $E_{\text{sg}}/kT = $ const.) It is seen from this figure that, for a band gap only a few kT wide, there is hardly any evidence of surface state structure in Q_{ss}. In fact, the curve of Q_{ss} vs E_F closely resembles the 300°K curve in the *merged band* model of Fig. 37. Only upon studying the low temperature dependence of Q_{ss} can the surface state structure be extracted from the "thermal noise."

To help decide whether Φ (cube law) relates to surface state emission or to valence-band emission, additional experiments were carried out using polarized light.[264] They detected the directional effect in the distribution of the photoemitted electrons, which persisted into the cube-root tail. This means that there was *not* complete random scattering and the threshold process could not be assigned to an indirect process, as they claimed previously for Si.[177-179] It was called, instead, a "quasi-direct" process, where $k_{\parallel}$ is still preserved, but $k_{\perp}$ is not during the act of absorption. It is not generally realized that the data only show $\sim$40% directional (unscattered) effect in the tail section; thus, other scattering processes contributing to the yield *cannot* be ruled out. Allen and Gobeli recognize that it is possible, though not probable, to have surface states give this directional effect. Thus, the problem of assignment of the photoelectric threshold Φ is still not definitely settled, even with the polarization experiments.

Henzler[61] measured the conductance and field-effect mobility of six samples of Ge with the same p-type doping. The result was $\mu_{\text{fe}} = -100$ cm^2/V-sec, which indicated that the surface was p-type in agreement with Allen and Gobeli. But the Fermi level location at the surface was $E_F = -0.23$ eV (nondegenerate). Henzler also computed $\bar{N}$ at E_F, based on a *merged band* model, and found that $\bar{N} = 2 \times 10^{14}$ cm^{-2} eV^{-1}. As usual, with

---

[264] G. W. Gobeli, F. G. Allen, and E. O. Kane, *Phys. Rev. Letters* **12**, 94 (1964); *Proc. Intern. Conf. Phys. Semiconductors* p. 937. Dunod, Paris, 1964.

this type of analysis, assumptions must be made regarding the surface scattering mechanism. Henzler used the surface mobility data of Handler and Eisenhour,[183] derived from magnetoconductive measurements on bombarded and annealed Ge surfaces.

Palmer et al.[221] attempted to measure the surface state distribution by using cleaved p–n–p junctions and monitoring the channel effect. They found that a freshly cleaved sample could not be affected by the bias voltage, because the excess hole density Δp, due to the acceptor surface states, was much larger than the change in ionized donors with applied field. In other words, the surface state density was too high to measure. However, they concluded that $N > 1.5 \times 10^{12}$ cm^{-2}. By selectively dosing the sample with O_2, they could decrease Δp, decrease the surface states, and measure their distribution in energy across the gap, using the current–voltage channel characteristics. This observation and technique is interesting, but not particularly relevant to the surface states on *clean* Ge. The work is a continuation of their previous studies,[263] where they cleaved both n-type and p-type Ge samples. This gave virtually identical results. They measured field-effect mobility, by using one face of the cleaved sample as the field plate, since it was located only ~ 0.04 cm away. The result was $\mu_{\text{fe}} = -(50 - 150)$ cm^2/V-sec and no change was observed for at least 100 hr at 10^{-9} Torr pressure. They estimated that $E_F = -0.10 \pm 0.02$ eV and $N \geq 10^{11}$ cm^{-2}. In both works,[221,263] however, there were measured pressure bursts of 10^{-8} to 10^{-7} Torr upon cleavage (probably greater near the crystal); this complicates the interpretation of the results in terms of clean surfaces, as acknowledged by the authors.

Banbury et al.[265] and Barnes and Banbury[266] have reported results which are definitely out of line with the above. That is, $E_F = -0.05$ eV. This work has been questioned by Many[7] on two grounds: there was strong capacitance to the uncleaved parts of the crystal,[266] and there was a mica spacer in intimate contact with the Ge surface of interest.[265]

Very important experiments on *cleaved-and-heated* Ge were made by Henzler.[61] He computed $\bar{N}$ from the field-effect mobility as a function of heating temperature T up to 650°K. The results were summarized in his plot of $\bar{N}$ vs T, which showed a monotonic decrease from $\bar{N} = 2 \times 10^{14}$ cm^{-2} eV^{-1} at $T = 300°$K to $\bar{N} = 2 \times 10^{13}$ cm^{-2} eV^{-1} at $T = 650°$K. $\bar{N}$ is, of course, determined at E_F, whose location at the surface increased upon heating from $E_F = -0.23$ eV to $E_F = -0.15$ eV. The most striking aspect of the results was the lack of structure (i.e. local maxima or minima)

[265] P. C. Banbury, E. A. Davies, and G. W. Green, *Proc. Intern. Conf. Phys. Semiconductors, Exeter, England* (A. C. Strickland, ed.), p. 813. Academic Press, New York, 1962.
[266] G. A. Barnes and P. C. Banbury, *Proc. Phys. Soc.* **71**, 1020 (1958).

in the curve $\bar{N}$ vs T. This was interpreted to mean that the measured surface states are *not* located in the grain boundaries or at defects of other kinds. The reasoning was as follows: (1) From LEED experiments as a function of T, it is found that the surface has stable phases in certain ranges of T, but becomes highly disordered during structural transitions. (2) If the measured surface states were due to the disorder in the grain boundaries or other defects, then the states should reach a local maximum during structural transitions. (3) This was not observed. Henzler considers possible reasons for surface states on Ge and concludes from his experiments that the "termination of the crystal potential at the surface" is to be rejected in favor of the "special arrangement of atoms in the uppermost layers, different from bulk arrangement."

Henzler[262] has carried out more recent field-effect mobility experiments on cleaved Ge, after various stages of annealing $300°K < T < 600°K$, while *simultaneously* monitoring the LEED pattern in a vacuum of $<5 \times 10^{-10}$ Torr. His previously reported relation[61] between surface conductivity and field-effect mobility for different T was essentially repeated. A detailed analysis of the data was carried out, however, assuming a variety of models, including the *two discrete level* model of Fig. 34a and the *merged band* model of Fig. 34c. Two distributions had to be used, since the LEED data showed that the initial 2×1 phase changed gradually to the 2×8 phase at higher temperatures. By curve fitting the data to the simple selected models, Henzler found what he considered the best distributions. These are reproduced in Fig. 40. A two discrete level set appears just after cleavage; it corresponds to the 2×1 structure, and

$$E_0 = -0.30 \quad \text{eV}, \quad E_1 = -0.36 \quad \text{eV}, \quad E_2 = -0.21 \quad \text{eV}$$

$$E_{sg} = 0.15 \quad \text{eV}, \quad 2N = 0.6 \times 10^{14} \quad \text{cm}^{-2}.$$

After a 150°C anneal, another two discrete level set appears; it corresponds to the 2×8 structure (more precisely called a "weak 8 structure"), and

$$E_0 = -0.21 \quad \text{eV}, \quad E_1 = -0.25 \quad \text{eV}, \quad E_2 = -0.17 \quad \text{eV}$$

$$E_{sg} = 0.08 \quad \text{eV}, \quad 2N = 0.3 \times 10^{13} \quad \text{cm}^{-2}.$$

At an intermediate annealing temperature, both sets are present. The reported numbers N are really lower limits, since the states present far from E_1 and E_2 will not be counted by the Fermi statistics. Finally, his results are found to be consistent with a recent optical study of surface states on cleaved Ge surfaces.[209]

Gobeli and Allen[261] measured ϕ_w and Φ for one cleaved-and-heated p-type crystal. They concluded that E_F changed from $E_F = -0.33$ eV, before heating, to $E_F = -0.22$ eV after heating. The general trend to a

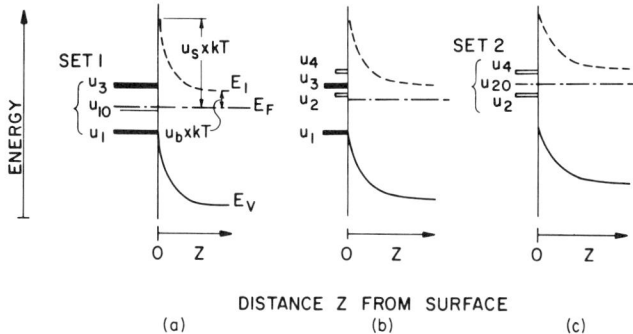

Fig. 40. Henzler's distribution of surface states for cleaved Ge (111): (a) 2 × 1 structure after cleavage, (b) Mixed after 80°C, (c) weak 2 × 8 structure after 150°C. [From M. Henzler, *J. Appl. Phys.* **40**, 3756 (1969). Reproduced by permission of The Institute of Physics and The Physical Society, London.]

more intrinsic (though still p type) surface upon heating agrees with Henzler.[61,262] This completes the experiments on cleaved and cleaved-and-heated Ge.

Experiments on bombarded-and-annealed Ge have been carried out by Allen and Fowler,[142] Forman,[267] Handler and Portnoy,[268] Margoninski,[196] Nesterenko and Snitko,[193] and Kobayashi and co-workers.[181,219] The results are somewhat conflicting, except for the general observations that the surfaces are all p type (possibly degenerate) and that (100) is more p type than (111). Allen and Fowler,[142] as early as 1957, measured ϕ_w for four different n- and p-type dopings for the crystal faces (111), (110), (100). They noted that Ar bombardment cleans the surface quickly and effectively, but cannot produce a surface that remains clean on subsequent heatings, until the bulk material has been thoroughly outgassed. They also considered the possibility that contamination films may form on the surface, during the annealing, from trace impurity gases present in the Ar or on the chamber walls. Evidence that this was *not* the case in their experiments was the insensitivity of ϕ_w as the Ge sample was cooled slowly from 1000°K anneals. Also, there was no p-type contamination film, since the n-type samples remained n type. Part of the disagreement among workers (see below) may be due to the lack of control of these complications. Allen and Fowler's results[142] showed that $\phi_w(100) > \phi_w(110) > \phi_w(111)$ and $\phi_w(100) - \phi_w(111) = 0.06$ eV. Furthermore, these values were insensitive to doping, in agreement with later work.[261] From the insensitivity to temperature during the cooling operation, and the photovoltage experiments of Garrett

[267] R. Forman, *Phys. Rev.* **117**, 698 (1960).
[268] P. Handler and W. M. Portnoy, *Phys. Rev.* **116**, 516 (1959).

and Brattain,[269] they concluded that there was band bending at the surface of $eV_D > 0.1$ eV and that the surface state density must be $\bar{N} > 10^{14}$ cm^{-2} eV^{-1}. Finally, they referred to Smoluchowski's theoretical calculations,[270] based on electron smoothing at a surface, which gave an order of work functions $\phi_w(111) > \phi_w(110) > \phi_w(100)$; this is just the opposite of that observed experimentally. They concluded, therefore, that dangling bonds and surface rearrangement, not included in Smoluchowski's analysis, were responsible for the discrepancy.

Handler,[141] in an excellent review article on bombarded–annealed Ge, prior to 1958, carried the study of Allen and Fowler[142] a little further. He noted that the order $\phi_w(100) > \phi_w(110) > \phi_w(111)$ is the *same* as that of the dangling-bond surface density (Table II, Section 9). This interpretation is now open to question, because of the reconstruction known from subsequent LEED measurements. From the analysis of many measurements carried out by himself and others, in the years 1955–1956, Handler concluded that $N > 1 \times 10^{13}$ cm^{-2} and $\bar{N} \sim 4 \times 10^{14}$ cm^{-2} eV^{-1}.

Forman[267] measured field-effect mobility and conductivity on n-type (111) and (100) samples, and found that the (100) surface is more highly p type than the (111) surface, in line with the above ϕ_w measurements. Thus, if the (111) surface is characterized by the Fermi level within one kT of the valence-band edge, as computed by Forman, then the (100) surface must be even more degenerate p type. These studies have been criticized by Frankl[192] on the grounds that, for self-consistency, either the specularity factor is less than one, or some donor states must be present as well as acceptor states. Frankl[192] concluded that the appropriate surface state distribution for (111) is a *merged band* model, as in Fig. 37, but with zero width. The properties of this distribution are estimated to be $E_0 = -0.25$ eV at $2N = 2 \times 10^{13}$ cm^{-2}; i.e., considerably less than the *expected* density of surface states $2N \sim 7 \times 10^{14}$ cm^{-2}. It should be stressed that Frankl's arguments are based strongly on data relevant to oxygen adsorption on Ge, so his conclusions with respect to the *clean* surface are questionable.

A measurement similar to Forman's was carried out independently by Handler and Portnoy,[268] except that results on a (100) n-type sample were obtained over the temperature range $77°K < T < 300°K$. They found temperature-independent properties and concluded that the (100) surface must be degenerate p type, in agreement with Forman. They interpreted their results in terms of a *two band* model of surface states, as in Fig. 34b, but with E_1 lying deep within the valence band. From the temperature independence of μ_{fe} and $\Delta\sigma$, they obtained an *upper limit* to the surface state density: $\bar{N} \leq 4 \times 10^{13}$ cm^{-2} eV^{-1}. [Their assumed quantization of

[269] C. G. B. Garrett and W. H. Brattain, *Phys. Rev.* **99**, 376 (1955).
[270] R. Smoluchowski, *Phys. Rev.* **60**, 661 (1941).

bulk holes (later called light holes) in the strong accumulation layer is irrelevant to the above conclusions.] Frankl[191] has interpreted this work along different lines. By computing nine different surface properties at the temperatures 300°, 200°, and 100°K, he was able to show (assuming completely specular reflection) that the data were consistent with $E_0 = -0.25$ eV. That is, a p-type surface but not degenerate. Assuming a *merged band* model of zero width, Frankl obtains $N \geq 1.2 \times 10^{13}$ cm^{-2}.

Margoninski[196] also carried out similar experiments. He measured field-effect mobility, surface conductivity, and surface recombination velocity on n-type (111) and (100) samples. He computed (assuming completely specular reflection) that the (111) surface is strongly p type, but not degenerate, while the (100) surface is p-type degenerate. To be specific, for (111): $E_F = -0.25$ eV and $\bar{N} = 2.4 \times 10^{14}$ cm^{-2} eV^{-1}; and for (100): $E_F = -0.35$ eV and $\bar{N} = 5 \times 10^{14}$ cm^{-2} eV^{-1}. From the surface recombination experiments, he concluded that acceptor-like surface recombination centers were present near E_F with a total number $N \sim 10^{14}$ cm^{-2}. He claims that the surface recombination centers may be connected with the surface states, as observed from the conductivity measurements. This is in general agreement with the earlier surface recombination velocity measurements of Wang and Wallis.[271]

More recently, Nesterenko and Snitko[193] carried out similar experiments on p-type (111) surfaces. They found a strongly degenerate p-type surface, and used the same surface state distribution model as Handler and Portnoy.[268] The results were $E_F = -0.38$ eV and $\bar{N} = 10^{15}$ cm^{-2} eV^{-1}, which may differ from the others, because the Ge surfaces were bombarded with He, rather than Ar, and because quartz filaments were used to isolate the field plate from the Ge surface. They also observed $\mu_{fe} = 1400$ cm^2/V-sec, which was about an order of magnitude greater than the other workers, and close to the bulk mobility $\mu_b = 1900$ cm^2/V-sec. Noise spectra were measured. (They were lowest for a free surface and increased tenfold upon oxidation.)

An attempt was made by Kobayashi and co-workers[181,219] to find surface state conductivity of the type looked for in Si by Aspnes and Handler.[180] They measured σ and the Hall effect of Ge (111) in the range 1.7°K $< T <$ 300°K, but their samples were "thermally converted." They found a conductance, which did not disappear at the cryogenic temperatures, and naturally attributed this to conductance in the 2-D surface state band. The sample was prepared by heating, without bombardment and anneal. In the following year, Kobayashi and co-workers[181] carried out the same experiment; this time after bombardment and anneal. They found that the low

---

[271] S. Wang and G. Wallis, *J. Appl. Phys.* **30**, 285 (1959).

temperature conductance disappeared, and ascribed their previous "anomalous" results[219] to p-type impurities, "probably boron transferred from the glass bulb."

As seen from the above discussion, most surface state properties of Ge seem to be conflicting at present. To further emphasize this point, Table III is constructed. In the first column, the Ge surfaces are catagorized according-

TABLE III. NATURE OF CLEAN Ge SURFACES

Preparation	References	Degenerate p type	Very degenerate p type	Nondegenerate p type
Cleaved (111)	a	×		
	b			×
	c			×
	d	×		
Cleaved and heated (111)	b			×
	a			×
Bombarded and annealed (111)	e	×		
	f; g, p. 202			×
	h			×
	i		×	
	d	×		
Bombarded and annealed (100)	e		×	
	j		×	
	g, p. 213; k			×
	h	×		
	d		×	

[a] G. W. Gobeli and F. G. Allen, *Surface Sci.* **2**, 402 (1964).

[b] M. Henzler, *Surface Sci.* **9**, 31 (1968); *Bull. Am. Phys. Soc.* **13**, 1375 (1968); *Phys. Rev.* (to be published).

[c] D. R. Palmer, S. R. Morrison, and C. E. Dauenbaugh, *J. Phys. Chem. Solids* **14**, 27 (1960).

[d] A. Many, Y. Goldstein, and N. B. Grover, "Semiconductor Surfaces," p. 487. North-Holland Publ., Amsterdam, 1965.

[e] R. Forman, *Phys. Rev.* **117**, 698 (1960).

[f] D. R. Frankl, *Surface Sci.* **6**, 334 (1967).

[g] D. R. Frankl, "Electrical Properties of Semiconductor Surfaces." Pergamon Press, Oxford, 1967.

[h] Y. Margoninski, *Phys. Rev.* **132**, 1910 (1963).

[i] B. A. Nesterenko and O. V. Snitko, *Soviet Phys. Solid State* **6**, 2321 (1965).

[j] P. Handler and W. M. Portnoy, *Phys. Rev.* **116**, 516 (1959).

[k] D. R. Frankl, *Surface Sci.* **4**, 201 (1966).

ing to preparation: cleaved (111), cleaved and heated (111), bombarded and annealed (111), and bombarded and annealed (100). The second column lists the references, including the experimenters and the relevant commentaries. The other three columns are labeled degenerate p-type, very degenerate, and not degenerate; one of these is crossed for each reference. Examination of the table shows the considerable scatter in results; the number of crosses being 5, 4, and 7 for the three columns. Furthermore, there seems to be no statistical correlation (equally weighed references). A further disturbing fact is that similar LEED patterns are obtained on cleaved-and-heated (111) and bombarded–annealed Ge (111)[54,230]; yet the table shows no such similarity.

If the Ge surface is either degenerate or very degenerate p type (probability $\sim$50% from the table), or has E_1 near the valence-band edge, this has important theoretical implications. (1) There must be surface states lying *within* the valence band. (2) These states must be *localized* (rather than just being a scattering resonance). (3) The theoretical 3-D arguments in Section 8 of no-band-crossing in **k**-space can explain these states. (4) The possibility exists for the surface state *on other materials* to be similarly continued from the forbidden band into the allowed valence (and possibly conductance) bands. This seems to be the case for the II–VI and III–V crystals, as will be shown in the next subsection.

Additional evidence that a Ge surface is either degenerate or very degenerate p type has been obtained by Proix and Handler,[223] who cleaved Ge under liquid N_2. Although the surface is covered with liquid N_2 (not intrinsic), there is probably only a slight interaction, due to the inertness of N_2 and the low temperature. However, Henzler's recent work[262] seems to point to a nondegenerate p-type surface on Ge (111), both with and without heating. The surface state distribution seems similar to that found for Si by Allen and Gobeli,[261] except for the smaller E_{sg} and the observation that E_1 lies near the valence-band edge.

Some 3-D theoretical calculations applicable to Ge have been reviewed previously at the end of Section 17,a. It is shown there that the comparison is poor, at the present time, probably because reconstruction has not been taken into account.

c. III–V and II–VI Crystals

These crystals differ from Si and Ge because they are partially ionic. For example, ZnS and GaAs have estimated effective charges $\pm 0.5e$[162] and $\pm 0.2e$,[163] respectively. Crystal modifications include zinc blende and wurtzite; these have tetrahedral bonds as in Si and Ge, but differ with respect to the placement of next-nearest neighbors. The rocksalt modifications of BaO, CdO, ... have apparently not yet been studied for their surface state properties. It is easier to cleave zinc blende or wurtzite[197] than Si or Ge. Hence, the intrinsic surface preparation used by most workers to date is simply cleavage in UHV without heating. Certain cleaved surfaces (zinc blende (110), wurtzite (10$\bar{1}$0)) seem to be unreconstructed (no extra spots in the LEED pattern), compensated (equal numbers of A and B ions in the surface plane), and thermally stable (no changes in the LEED spot pattern upon heating). One exception[227a] seems to be ZnO, whose needles can be cleaved to expose the {0001} polar faces,[222,272] which seem to be unreconstructed.[272] It will be shown below that ZnO also has exceptional surface state properties. The intrinsic polar surfaces, zinc blende {111} and wurtzite {0001}, *nominally* consist of layers of either all A ions or all B ions. They can be formed using Ar bombardment-and-anneal techniques.[273] Electrostatic calculations indicate that these nominal layers should be unstable,[30] but can be stabilized[231] by allowing vacancies in $\frac{1}{4}$ of the surface atoms (consistent with MacRae's interpretation[229] of LEED data), or by heating, so as to form microfacets of compensated surfaces.[231] It has been found experimentally that heated zinc blende {111} forms (110) microfacets,[152] and heated wurtzite (0001) forms (30$\bar{3}$8) microfacets.[227] A major disadvantage in studying the surface state properties of these semiconductors is the inability to control the doping over a large range, due to self-compensation. For example, CdS can only be grown n type and ZnTe can only be grown p type.[274] It has been reported that heating CdS crystals in UHV after cleavage can introduce an inhomogeneity in the donor concentration.[275] The surface layers, $d < 50$ Å, sometimes have a donor density ~ 10 times that of the bulk.

Surface state data on the *compensated surfaces*, zinc blende (110) and wurtzite (11$\bar{2}$0), have been obtained for CdS, CdSe, CdTe, ZnO, ZnS, ZnSe, and ZnTe by Swank[159]; for InP by Fischer[198]; for AlSb by Fischer[276]; for InAs by Fischer *et al.*[277]; and for GaAs, GaSb, InAs, and InSb by Gobeli and Allen.[197,203] All these data were taken in a similar manner: by measuring work function ϕ_w and/or photoelectric threshold Φ on cleaved, room-temperature samples. In many cases additional information was obtained from the energy distributions of photoemission, or by photovoltage experiments, all carried out *in situ*.

[272] K. Müller *in* "The Structure and Chemistry of Solid Surfaces" (G. A. Somorjai, ed.), p. 35–1. Wiley, New York, 1969.
[273] Sputtering rates seem to be equal for both A and B components of an AB compound. See S. P. Wolsky, E. J. Zdanuk, and D. Shooten, *Surface Sci.* **1**, 110 (1964).
[274] M. Aven and J. S. Prener, eds., "Physics and Chemistry of II–VI Compounds." North-Holland Publ. Amsterdam, 1964.
[275] C. Sebenne and M. Balkanski, *Surface Sci.* **5**, 410 (1966).
[276] T. E. Fischer, *Phys. Rev.* **139**, 1128 (1965).
[277] T. E. Fischer, F. G. Allen, and G. W. Gobeli, *Phys. Rev.* **163**, 703 (1967).

TABLE IV. BAND-BENDING ASPECTS OF CLEAVED III–V AND II–VI CRYSTALS

Crystal	Bandgap E_g (eV)	E_F pinned at valence band	Band bending	Reference
ZnS	3.6	No	n ↑	a
ZnO	3.3	No	n ↑	a
ZnSe	2.7	No	n ↑	a
CdS	2.5	No	n ↑	a
ZnTe	2.2	No	p ↓	a
CdSe	1.7	No	n ↑	a
AlSb	1.6	No	p ↓	b
CdTe	1.5	No	n ↑	a
GaAs	1.4	No	n ↑	c
InP	1.3	No	n ↑	d
Si	1.09	No	n ↑ p ↓	e
GaSb	0.7	Yes		c
Ge	0.67	Yes		f
InAs	0.35	Yes		g
InSb	0.18	Yes		c

[a] R. K. Swank, *Phys. Rev.* **153**, 844 (1967).
[b] T. E. Fischer, *Phys. Rev.* **139**, 1128 (1965).
[c] G. W. Gobeli and F. G. Allen, *Phys. Rev.* **137**, 245 (1965).
[d] T. E. Fischer, *Phys. Rev.* **142**, 519 (1966); *Helv. Phys. Acta* **41**, 827 (1968).
[e] F. G. Allen and G. W. Gobeli, *Phys. Rev.* **127**, 150 (1962).
[f] G. W. Gobeli and F. G. Allen, *Surface Sci.* **2**, 402 (1964).
[g] T. E. Fischer, F. G. Allen, and G. W. Gobeli, *Phys. Rev.* **163**, 703 (1967).

Is there any pattern to the data obtained? Gobeli and Allen[197,203] proposed a correlation in 1965 that the antimonides, GaSb and InSb, are similar to Ge and the arsenides, GaAs and InAs, are similar to Si, in the sense that the former class had the Fermi level pinned at or near the valence band at the surface, while the latter class did not. But in 1967, a more complete study of InAs by Fischer *et al.*[277] showed that E_F was pinned at the valence band, in contradiction to the previous result.[197,203] This showed that the above proposed correlation[197,203] was no longer useful.

In a search for a new correlation, while writing this review, it was anticipated that, since these crystals are known to be partially ionic, a correlation involving the bulk bandgap E_g, which is a crude measure of the ionicity, might apply.[278] Table IV is constructed to show that E_g does, in fact, correlate many aspects of the data. Here thirteen III–V and II–VI

[278] J. D. Levine, *Bull. Am. Phys. Soc.* **14**, 787 (1969); *J. Vacuum Sci. Technol.* **6**, 549 (1969); invited paper, *2nd Combined Symp. Modern Aspects Vacuum and Thin Film Sci.*, Bell. Labs., 1969; invited paper, *Gordon Conf. Surface Phys., Meriden, New Hampshire*, 1969).

crystals are arranged according to E_g, from 0.18 eV for InSb to 3.6 eV of ZnS and the experimenters are referenced in the last column. Whether or not the Fermi level is pinned at (or near) the valence-band edge at the surface is indicated in the third column by *yes* or *no*. *Yes* is indicated for those crystals with $E_g < 1$ eV and *no* for those with $E_g > 1$ eV. (Si and Ge are also included in the table for comparison.) Thus, the correlation seems to apply. The correlation can be taken even further, i.e., one can investigate whether the bands bend up or down if the crystal has n- or p-type doping. The results for $E_g > 1$ eV are given in the fourth column of Table IV, and pictorially indicated in Figs. 41 and 42. (Si and Ge, as studied by Allen and Gobeli, are also included for comparison.) The notation n↑ (p↓) indicates that n- and (p)-type crystals have bands which bend up (down) at the surface. It is seen from the table and the figures that n↑, p↓ are present, but n↓, p↑ are absent. This observation indicates that these crystals have the expected "ionic surface state" behavior indicated in Fig. 34e and described previously in Sections 1, 4, and 10. Jones[139] has theoretically inferred similar conclusions by considering an ionic perturbation to the Si surface state structure. That is, the perturbation splits the degeneracy along $\bar{X}$–$\bar{M}$ in Fig. 39. Experimentally, the acceptor-like intrinsic surface states below the conduction band cause the bands to bend up at the surface for an n-type crystal. Also, the donor-like states above the valence band cause the bands to bend down at the surface for a p-type crystal. The n↑ p↓ correlation appears to break down when $E_g < 1$ eV, since E_F is pinned at or near the valence band. However, this inadequacy is, in fact, expected, since those crystals with very small bandgaps are normally considered to be hardly ionic. Thus, certain qualitative features of the intrinsic surface states of the compensated surfaces of III–V and II–VI compounds seem to be readily explained using simple ionic concepts. Koutecký and Tomášek[279] have computed the surface state properties of the polar surfaces (111) and ($\bar{1}\bar{1}\bar{1}$) of zinc blende, known to be reconstructed. They used an extension of their previous models,[26,258] including the effect of ionicity as alterations in the A and B Coulomb integrals α and resonance integrals γ. Levine and Freeman[140] have computed the surface state band structure of zinc blende (110), for various locations of surface A and B atoms. Both A- and B-like surface state bands can appear, depending on the surface atom locations and Coulomb energies.

Quantitative evaluations of surface states on all these crystals have not been carried out before. To do this it is necessary to carry out the following steps. First, obtain V_D for all crystals mentioned above and for each doping parameter ζ. The relation between ζ and the donor or acceptor uncompensated concentration N_D or N_A is given by an equation of the form of

[279] J. Koutecký and M. Tomášek, *Surface Sci.* **3**, 333 (1965).

TABLE V. SURFACE STATE ANALYSIS OF SOME III–V CRYSTALS ($E_g > 1$ eV)

	E_g (eV)	eV_D (eV)	ζ (eV)	$L \times 10^{-6}$ (cm)	F_s	$Q_{sc}/e \times 10^{10}$ (cm^{-2})	m^*/m_0	$\Delta E_{1,2}$ (eV)	E_{t1} (eV)	E_{t2} (eV)
CdS	2.47	0.07	0.15	5.1	2.0	5.4	0.2	0.090	0.13	—
CdSe	1.67	0.12	0.15	5.1	2.8	7.6	0.2	0.082	0.20	—
CdTe	1.50	0.23	0.16	6.2	4.1	9.0	0.2	0.077	0.31	—
ZnO	3.25	0.01	0.10	1.9	0.4	3.0	0.2	0.102	0.01	—
(ZnS)	(3.6)	(1.2)	(0.30)	(0.01)	(9.7)	(0.3)	(0.2)	(0.125)	(1.38)	—
ZnSe	2.73	0.58	0.15	5.1	6.6	18	0.2	0.060	0.67	—
ZnTe	2.23	−0.18	0.15	5.1	3.5	−9.4	0.2	0.075	—	0.26
GaAs	1.40	0.59	0.05	1.4	6.7	6.7	0.07	0.057	0.58	—
InP	1.29	0.13	0.11	4.7	2.8	7.8	0.07	0.055	0.18	—
AlSb	1.62	−0.24	0.12	0.83	4.1	−65.0	0.9	0.067	—	0.29

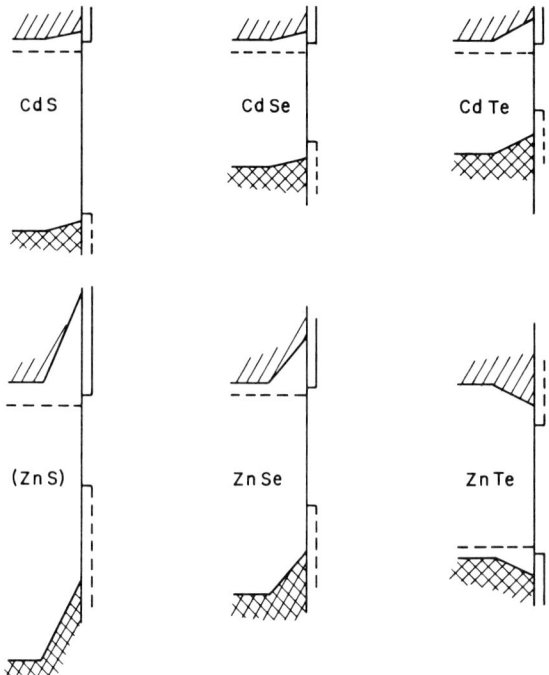

Fig. 41. Schematic diagram of band-bending and intrinsic surface states in some II–VI crystals.

(11.2). A separate analysis carried out later shows that the energies of the surface states are *insensitive* to exact effective mass m^* or dielectric constant K. These quantities affect the energies in the form $\sim kT \ln(m^*/K^2)^{1/4}$. In Fischer's work on InP[198] and AlSb,[276] V_D is not explicitly given, but it can be computed easily from the doping concentration and the effective mass (necessary to compute N_c or N_v). To be specific: for InP, N_D is given as 6×10^{15} cm^{-2} and $m^* = 0.073 m_0$ [280]; this yields $N_c = 4.8 \times 10^{17}$ cm^{-3} and $\zeta = 0.11$ eV. Using Fischer's Fig. 1, one finds that $V_D = 0.13$ eV. Also, for AlSb, N_A is given as 2×10^{17} cm^{-3} and $m^* = 0.9 m_0$ [276] (effective density-of-states mass for holes); this yields $N_v = 2.1 \times 10^{19}$ cm^{-3} and $\zeta = 0.12$ eV. Using Fischer's Fig. 2, it is found that $eV_D = -0.24$ eV. These values $eV_D = 0.13$ eV, 0.24 eV are small (as acknowledged by Fischer), but certainly not negligible. Using $m^* \sim 0.2 m_0$ for all II–VI compounds

[280] D. Long, "Energy Bands in Semiconductors," p. 107. Wiley (Interscience), New York, 1968.

(accurate within a factor of ~ 2), the doping levels can be computed from ζ ($\zeta = \Delta\phi_s - eV_D$ from Swank's[159] Table 1); there is some self-inconsistency with Swank's doping ranges for ZnTe.

For convenience, the values of V_D and ζ are given explicitly for all crystals with $E_g > 1$ eV in Table V, together with the bandgaps (always measured at room temperature). The corresponding band-bending diagrams are shown in Figs. 41 and 42. Because of the lack of other information, the electron and hole surface states (surface traps) are assumed equal; dashed

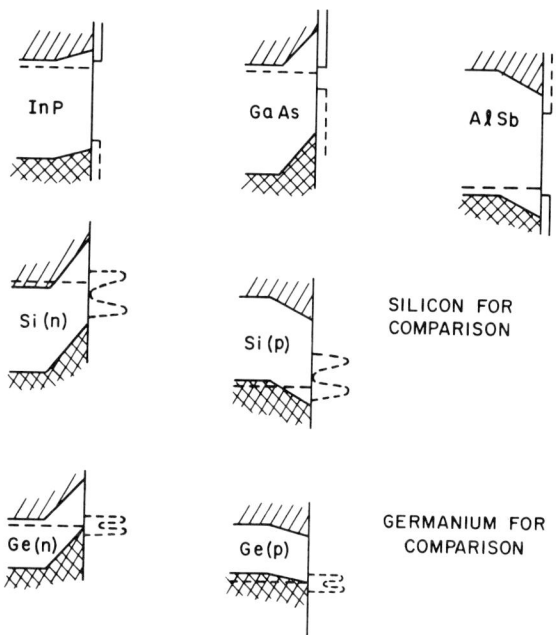

FIG. 42. Schematic diagram of band-bending and intrinsic surface states in some III–V crystals, Si, and Ge.

lines indicate this uncertainty. The diagram for ZnO is not included, since the bands are apparently flat ($eV_D = 0.01$ eV). The results for ZnS are questionable according to Swank,[159] so ZnS is written as (ZnS), a more recent assessment of the ZnTe data[282] gives $eV_D \sim -0.1$ eV; and two dopings of Si[178] and Ge[94] are included for comparison in Fig. 42. The next step is to compute the space charge Q_{sc} from V_D and ζ. From Q_{sc}, the locations of the surface states can be assigned. Since all surfaces considered here have mild depletion layers, the formulas are simple. They are (for p-type samples)

$$Q_{sc} = -eN_A LF_s \tag{17.1}$$

$$L = (KkT/4\pi e^2 N_A)^{1/2} \tag{17.2}$$

$$F_s = \sqrt{2}[|v| + \exp(-|v|) - 1]^{1/2} \tag{17.3}$$

$$v = eV_D/kT. \tag{17.4}$$

For n-type samples, replace N_A by N_D and change the sign of Q_{sc}. Also, K is taken as ~ 10 for all materials, which is accurate within a factor of 2. The Debye lengths L, space-charge functions F_s, and space-charge Q_{sc} for each crystal are given in Table V.

From Section 12 it is seen that a unique characterization of the surface state distribution is not possible unless the Fermi level and the temperature are varied by considerable amounts. Nevertheless, it is possible to try an assumed distribution in accordance with that postulated in Part VI, i.e., for the ionic or partially ionic crystals, there should be a band of acceptor-like states below the conduction band and a band of donor-like states above the valence band. Suppose the states had the *"ionic-like"* distribution of Fig. 34e, where the electron and hole surface trap depths are assumed to be equal, in the absence of other information. This is similar to Fig. 34b, except that the surface state distributions merge with the corresponding valence and conduction bands. Whether that portion of the surface states *within the valence and conduction* bands are localized or virtual, depends on the crossing-behavior in **k** space as outlined in Section 8. However, this aspect is irrelevant to the present discussion, since only those surface states within the bulk bandgap are easily measurable. Let the upper level of the donor surface state band be E_2 as in Fig. 34e and suppose the Fermi level E_F lies above it (surface states not degenerate). Then the Maxwellian approximation of a 2-D electron gas gives

$$Q_{ss} = e\bar{N}kT \exp[-(E_F - E_2)/kT] \tag{17.5}$$

where $\bar{N}$ is given by (12.7). Suppose, as in Part VI, that for large bandgap materials, m_s^* of the surface states and m^* of the bulk states, from which they arise, are sufficiently comparable (say, within a factor of 2).[278,140,281] This is only possible for fairly well-separated acceptor-like and donor-like surface states[140]; fortunately this seems to be the case for large bandgap materials $E_g > 1$ eV, as seen in Figs. 41 and 42. Thus, $\bar{N}$ is adequately fixed by m^*, and since there are no external electric fields

$$Q_{ss} + Q_{sc} = 0. \tag{17.6}$$

[281] S. Freeman (to be published).

Hence, $E_F - E_2$ can be computed from

$$E_F - E_2 = kT \ln(-Q_{sc}/e\bar{N}kT). \tag{17.7}$$

For n-type crystals $E_F - E_2$ becomes $E_1 - E_F$. In general, one can define $\Delta E_1 = E_1 - E_F$ and $\Delta E_2 = E_F - E_2$, as shown in Fig. 34e. The smaller of these will be called ΔE; it is given in Table V. It will be shown below that ΔE varies weakly with m^* and K, since they appear as cube and square roots in the logarithm. This insensitivity somewhat justifies the many approximations made above. Specifically, (17.7) can be rearranged to yield

$$\Delta E = \tfrac{1}{2}\zeta + kT \ln(8\pi e^2)^{1/2}(2\pi m^* kT)^{1/4}/(hKF_s^2 kT)^{1/2}. \tag{17.8}$$

Thus, ΔE is roughly equal to $\tfrac{1}{2}\zeta$, apart from a weakly varying quantity which, in the cases of interest, practically vanishes. This is supported by inspection of Table V. Also, $\Delta E \gtrsim 2kT$, so the assumed Maxwellian statistics are justified at room temperature. Finally, the trap depths with respect to the band edges can be defined as

$$E_{t1} = E_c - E_1 \tag{17.9a}$$

$$E_{t2} = E_2 - E_v. \tag{17.9b}$$

The trap depths are included as the last item in Table V.

By combining these results, it can be shown that, to a first approximation,

$$E_{t1} \simeq |eV_D| + \tfrac{1}{2}\zeta, \quad \text{n type} \tag{17.9c}$$

$$E_{t2} \simeq |eV_D| + \tfrac{1}{2}\zeta, \quad \text{p type} \tag{17.9d}$$

where V_D is the *measured* band bending. For n-type (p-type) samples the bands bend up (down) at the surface, as shown in Figs. 41 and 42. Since E_{t1} and E_{t2} are supposed to be independent of doping, (17.9c) and (17.9d) show that crystals with the larger dopings (smaller ζ) should have the larger band bending V_D. Data are not available at this time to check this prediction. Alternately, if m_s^* and E_{t1} are assumed to be two unknowns, then at least two pieces of data (on differently doped crystals) would be necessary to determine these. The above quantitative results were obtained by fixing $m_s^* \approx m^*$, within a factor of 2. A DC field-effect experiment, carried out in UHV would probably be a very useful method for probing the surface state energies. This is possible in the III–V and II–VI compounds (but impossible for Si and Ge) because only a few charges are present $Q_{ss}/e \approx 10^{10}$–10^{11} cm^{-2} as seen from Table V. This estimate is obtained by simply recognizing that for $m^* \approx 0.2 m_0$, room temperature, and $\Delta E \sim 0.1$ eV, Q_{ss}/e is

$$Q_{ss}/e \approx 4.2 \times 10^{14}(0.2)(0.025)e^{-4}$$

$$\approx 4 \times 10^{10} \quad \text{cm}^{-2}$$

from (17.5). A DC field-effect experiment carried out to make the surface state band just degenerate ($\Delta E = 0$, $E_F = E_t$) would require a greater charge:

$$Q_{ss}/e \approx 4.2 \times 10^{14}(0.2)(0.025)$$

$$\approx 2 \times 10^{12} \quad cm^{-2}$$

and if $E_t \approx 0.15$ eV, a flat-band condition would require an extreme charge

$$Q_{ss}/e \approx 4.2 \times 10^{14}(0.2)(0.15)$$

$$\approx 1 \times 10^{13} \quad cm^{-2}$$

which may be difficult to obtain, in practice.

At this point, a detailed review will be given of the surface state results in Table V and Figs. 41 and 42.

(1) For the carefully studied series of Cd compounds, CdS, CdSe, and CdTe, the surface state trap depths E_{t1} are 0.13, 0.20, and 0.31 eV, respectively. The trend is that the more ionic the crystal, the smaller E_{t1}.

(2) The series of Zn compounds, ZnO, ZnS, ZnSe, and ZnTe, is more complicated to interpret. ZnO has $eV_D \sim 0.01$ eV, so this is clearly an exception among all the III–V and II–VI crystals studied. ZnS has a reported eV_D which is now known to be too high.[282] ZnTe is p type, so that, at present, a definitive trend in surface state properties cannot be established for the Zn series.

(3) The magnitude of the surface state traps is of the order of a few tenths of an electron volt, in agreement with numerical predictions based on the Madelung method[30] (Section 10).

(4) It can be postulated that Q_{ss} is really caused by edge states, due to steps and ridges, which appear upon cleavage. But Swank[159] found similar band-bending results on crystals with different perfections of cleavage, and Fischer[283] found similar photoemission results on cleaved and broken GaAs. While the edge state postulate cannot be conclusively ruled out at this point, the intrinsic surface state models seem more than adequate to explain the data on a variety of different crystals and dopings, without invoking edge states.

(5) Workers in metal–semiconductor contacts have used a rule of thumb[284–286] that *metal–semiconductor interface states* are more pronounced in those crystals with smaller bandgaps (or lower ionicity). This rule of

[282] R. Swank (private communication, 1969).
[283] T. E. Fischer, *Surface Sci.* **10**, 274 (1968).
[284] C. A. Mead, *Solid State Electron.* **9**, 1023 (1966).
[285] W. Ruppel, in "II–VI Semiconducting Compounds" (D. G. Thomas, ed.), p. 1260. Benjamin, New York, 1967.
[286] S. Kurtin and C. A. Mead, *J. Phys. Chem. Solids* **29**, 1865 (1968).

thumb seems to apply crudely to intrinsic surface states as shown above; the two big exceptions being ZnO and ZnS. Experiments on ZnO and ZnS, performed on a few samples, could bear repeating. Present indications are that the E_t value of ZnS is much smaller than that previously reported.[282] Turner and Rhoderick[287] have demonstrated that metal–semiconductor interface states are *greatly* dependent on whether or not a surface is atomically clean before metal deposition.

(6) The acceptor-like surface states on different crystal faces of CdS seem to be similar, since Swank[159] found $eV_D = 0.07$ eV for cleaved ($11\bar{2}0$) CdS, while Campbell and Farnsworth[227] found $eV_D = 0.02$ to 0.06 eV for bombarded-and-annealed (0001) CdS (Cd face).

The data upon which the above conclusions are based will now be briefly reviewed. Swank[159] carried out experiments by cleavage in $\sim 10^{-11}$ Torr and measured Φ by a cube-root extrapolation. He also noted a residual tail, which could be due either to surface state emission or to bulk impurity state emission. The same conclusion was reached later by Fischer.[198] Separate proof that Φ (cube root) corresponds to the valence-band edge (as assumed by Swank), was a series of photovoltage experiments he carried out *in situ*. The photovoltage V_D for CdSe and CdTe was measured to be 0.10 and 0.29 eV, respectively. These agree fairly well with 0.12 and 0.23 eV, respectively, as obtained by using Φ and ϕ_w. CdS was an anomaly in the photovoltage experiments for reasons pointed out by Swank. If Φ were taken from a linear plot instead of a cube-root plot, then V_D, as obtained by the different methods, would differ by >1 eV. Thus, Swank's interpretation seems to be reliable. There were twelve crystals tested for CdS, five for ZnSe, a few for CdTe and ZnO, and one each for CdSe, ZnS, and ZnTe.[282] Swank suggested that his data can be interpreted by the semiclassical Madelung model.[30] He also correlated the values of Φ obtained by different workers on III–V, II–VI, and I–VII crystals, with electronegativity of the anion. This correlation is reproduced in Fig. 43. Notice that there is a good straight line through the data points of the seventeen crystals (ZnO is again an exception; it falls outside of the graph); this indicates that the valence band of all these crystals is anion-like, in accordance with the Madelung,[30] LCAO,[33] and Mathieu[34] models. Fischer *et al.*[277] carried out similar measurement techniques using ϕ_w, Φ, and the photoelectron energy distributions. Fischer prefers not to use the extrapolated cubic threshold. Instead, he suggests the use of energy distribution measurements for the determination of semiconductor surface properties.[205] Gobeli and Allen[197] present an excellent review of photoemission properties including data up to 1965. Van Laar and Scheer[187c] have studied Φ for different dopings of GaAs, but they do not measure ϕ_w. Hence, they cannot obtain the band bending for any doping. They note, however, that the Fermi level

[287] M. J. Turner and E. H. Rhoderick, *Solid State Electron.* **11**, 291 (1968).

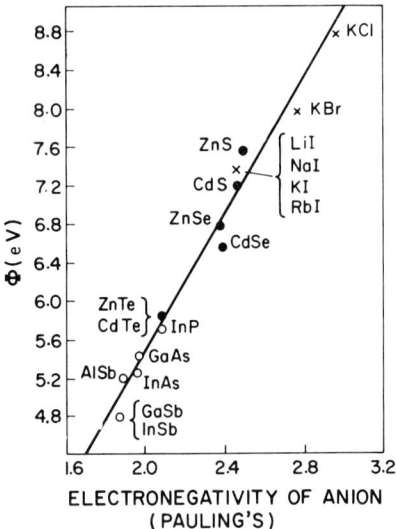

FIG. 43. Photoelectric threshold Φ versus electronegatives of nonmetallic constituents. [From R. K. Swank, *Phys. Rev.* **153**, 884 (1967).]

can be shifted ~ 1 eV ($E_g = 1.4$ eV) by going from n-type to p-type material. They seem to use the terms "stabilized" and "unstabilized" to be mutually exclusive classes, instead of the gradual transition as implied in Figs. 41 and 42. A variety of data on Φ for various different materials, with one doping each, has been obtained for different materials,[204] but this is not useful information for surface state determination, unless it is supplemented with measurements of ϕ_w and/or energy distributions of photoemission.

Additional and independent evidence of the ionic-like surface states, as shown in Fig. 34e has been obtained by Somorjai and Lester.[210] They studied the *evaporation rate* of CdS at high temperatures in high vacuum, and found that the rate can be increased up to a factor of 5 by simultaneously shining greater-than-bandgap light on the sample. Since the free surface was continuously evaporated and reformed, it qualifies as a suitably *clean surface*, for the investigation of intrinsic surface states. They found that the rate-limiting step in evaporation was the capture of free electrons and holes at the surface. Using the above proposed model of ionic-like surface states, their data can be explained in the following terms. In simple classical ionic theory, CdS exists as Cd^{++} and S^{--} ions. To evaporate neutral cadmium, two electrons must be gained to form neutral Cd, which is the species detected in the vapor phase. That is, the reaction must be

$$Cd^{++} + 2e \rightarrow Cd \uparrow .$$

The two electrons can be found in the acceptor-like surface state traps at the surface as shown in Fig. 34e. Thus, the rate limiting step for Cd evaporation is the capture of these two electrons at the surface, in agreement with experiment. The same argument applies for sulfur evaporation, since

$$S^{--} - 2e \rightarrow S \uparrow .$$

Here the loss of two electrons is the same as the gain of two holes. These holes can be found in the donor-like surface state traps at the surface as shown in Fig. 34e. For CdS evaporation, both donor and acceptor states must be present, according to the data.

Finally, mention should be made of the IV–VI semiconductors, i.e., PbS, PbSe, PbTe, etc. These have applications as infrared photoconductors. Apparently, only a few surface studies have been made on a *clean* surface. Oman[288] and Oman and Priolo[289] measured ϕ_w and Φ for (100) PbS, cleaved in UHV. The surface state results are unclear, because of the poor quality of the cleavages.

d. *Extrinsic Surface States*

For completeness, it is desirable to compare intrinsic surface states with extrinsic surface states. As mentioned previously, most surface state data reported are implicitly extrinsic surface state data, because of experimental convenience. In this short section, extrinsic surface states will be simply defined as all those surface states not intrinsic to a free surface. Surfaces which fit into the extrinsic category are:

(1) Surfaces with adsorbates of unknown species. It is known that intrinsic surface states can be passivated by adsorbates. An example is CdS handled in air[290] and not cleaned in UHV by argon-bombardment-and-anneal procedures. Heating in UHV without argon bombardment rarely purifies the surface. The heating problem is especially serious for binary crystals, since the two components have different vapor pressures.

(2) Surfaces with adsorbates (chemisorbed or physisorbed) of known species. An example is Si freshly cleaved in UHV and exposed to a small amount of cesium[245] or oxygen.[216] Adsorption may, or may not, be reversible. Also the adsorption may, or may not, yield a LEED pattern, characteristic of a well defined 2-D surface phase.

(3) Surfaces with adsorbates of one or two monolayers. There is no clear distinction between surface states and interface states, since 3-D

---

[288] R. M. Oman, *J. Appl. Phys.* **36**, 2091 (1965).
[289] R. M. Oman and M. J. Priolo, *J. Appl. Phys.* **37**, 524 (1966).
[290] A. Many and A. Katzir, *Surface Sci.* **6**, 279 (1967); A. Many, J. Shappir, and U. Shaked, *ibid.* **14**, 156 (1969); J. Shappir and A. Many, *ibid.* **14**, 169 (1969).

phases are often patchy and not well developed. An example would be a Cu crystal with a few monolayers of oxygen.[291]

(4) Interfaces between two solid materials. This would include semiconductor–oxide and semiconductor–metal interfaces.

(5) Interfaces between solid and liquid phases. An example of this would be Ge cleaved while immersed in pure liquid nitrogen[223] or in a dielectric liquid.[224] Another example would be ZnO in contact with an electrolyte.[292]

One of the general features of extrinsic surfaces is that their physical, chemical, and structural properties *cannot* easily be characterized. Effects are present which are absent in intrinsic surface states. Results generally depend on (1) the adsorbate species, pressure, exposure time, and surface temperature; (2) the history of processing and growth; e.g., steam-grown oxides and dry plasma-grown oxides have different electrical properties; (3) the magnitude and frequency of applied voltage; sometimes there are fast states, slow states, and drifts of impurity ions in oxide layers; and (4) the strains built in at the solid–solid interface. An adequate review of the vast amount of extrinsic surface state data will not be attempted here. Instead, certain selected results regarding the fast states will be treated below. The fast states are located, by definition, at, or very near, the interface of interest.

Sometimes experimenters unfortunately confuse *intrinsic* surface states with *extrinsic* surface states. For example, this has been done for CdS by Many et al.,[290] and for InSb by Huff et al.[29] In both cases, the experimental data were taken on surfaces prepared either by cutting, polishing, and etching or by vapor phase growth. They were not cleaved in UHV, nor cleaned by argon-bombardment-and-anneal. In the CdS studies,[290] there was a Mylar or mica spacer in physical contact with the surface. Yet the respective authors presumed that the surface states they measured are intrinsic, since they compared their experimental results with theories of intrinsic surface states. In fact, their derived surface state distributions in these two cases were very different from those previously observed for *intrinsic* CdS[159] and InSb.[203] To be specific, Many et al.[290] found *no* surface states on the $(11\bar{2}0)$ and polar faces of extrinsic CdS. On the other hand, Swank[159] found intrinsic surface states on CdS $(11\bar{2}0)$ and Campbell and Farnsworth[227] found intrinsic surface states on CdS (0001), as mentioned in the previous section. Also, Huff et al.[29] found a surface state distribution on InSb (110), as in Fig. 34f, with the neutral level near midgap and with a superimposed constant density of states across the entire band gap. By contrast, Allen and Gobeli[203] found the surface Fermi level pinned at the valence band for intrinsic InSb (110). These CdS and InSb examples were

[291] R. N. Lee and H. E. Farnsworth, *Surface Sci.* **3**, 461 (1965).
[292] T. Freund and S. R. Morrison, *Surface Sci.* **9**, 119 (1968).

included only to show that considerable confusion can easily arise, unless sharp distinctions are made between *intrinsic* and *extrinsic* surface states.

An extensive effort has been made to study the Si–SiO$_2$ interface states. It now seems that, with care, this interface can be near-perfect, in the sense that, surface states within the *entire* band gap $E_v < E_F < E_c$ cannot be detected within experimental resolution.[213,292a] The *upper* limit is $N < 10^{10}$ states/cm^2, using MOS diagnostics. This perfection is achieved by taking great pains to eliminate H$_2$O and trace metallic impurities in oxide growth, and by proper annealing procedures. The observed *intrinsic* surface states of $\sim$1 per surface Si atom (Section 17,a) seem to disappear with proper oxide "passivation."

Metal–semiconductor interface states have also been extensively studied.[284–286] It appears that there are some qualitative relations between intrinsic surface states and interface states even though the environments of the surface atoms are considerably different. Those crystals with the greater number of intrinsic surface states *within the bandgap* seem to have the greater number of metal–semiconductor interface states. This effect is detected by observing how the metal–semiconductor barrier height varies with the work function of the metal. More data are necessary to improve on these crude empirical relations. Heine[293] has proposed a theory for these contacts, which takes into account the penetration of the metal wave functions into the semiconductor.

Extrinsic surface states (fast states) frequently fall into a pattern. That is, the surface states frequently appear in pairs, and the acceptor states generally lie *above* the donor states as in Figs. 34a, b, e, and f. This causes the n-type (p-type) samples to have bands which bend up (down) at the surface as in Figs. 41 and 42. Sometimes the states are considered discrete as in Fig. 34a, banded as in Fig. 34e, or continuous across the entire gap as in Fig. 34c, or combinations of these, which are schematically indicated in Fig. 34f. The above pattern has been found for extrinsic surface states on Si,[213,294,295] Ge,[294,223] AgCl,[171] InSb,[29,296,297] CdS,[214,298,299] and GaAs.[300,301]

Experiments conducted in such a way that no clear pattern has emerged

---

[292a] A. G. Revesz and K. Zaininger *RCA Rev.* **29**, 22 (1968).

[293] V. Heine, *Phys. Rev.* **138A**, 1689 (1965).

[294] See Many *et al.*,[7] pp. 347–425 and Frankl,[8] pp. 220–268 for data on extrinsic Si and Ge surface state data prior to 1967.

[295] R. Memming and G. Schwandt, *Surface Sci.* **5**, 97 (1966).

[296] H. Pagnia, *Surface Sci.* **10**, 239 (1968).

[297] J. L. Davis, *Surface Sci.* **2**, 33 (1964).

[298] P. Mark, *J. Phys. Chem. Solids* **25**, 911 (1964); **26**, 1767 (1965).

[299] K. Sawamoto and H. Toyoda, *Rev. Elect. Commun. Lab.* **14**, 177 (1966) (Nippon Telegraph and Telephone Public Corporation, Japan).

[300] S. Kawaji and H. C. Gatos, *Surface Sci.* **1**, 407 (1964).

[301] I. Flinn, *Surface Sci.* **10**, 32 (1968).

include GaP [302] and GaAs.[187c] Adsorption states are clearly indicated in ZnO.[212,222] A clear exception seems to be donor-like surface states which sometimes can lie *above* the Si conduction band edge in Si–SiO$_2$ interfaces.[292a]

Extrinsic surface state observations sometimes lead to conclusions valid for intrinsic surface states. For example, Mark[214,298] studied the photoconductive and surface-dominated trapping behavior of $(11\bar{2}0)$ CdS platelets, which were reproducibly "cleaned" by flushing with one atmosphere of pure N$_2$. This type of "cleaning" does not guarantee intrinsic surfaces as mentioned above; in retrospect, they were probably extrinsic surfaces. As a consequence of these experiments, Mark was the first to clearly postulate the existence of surface state pairs on CdS. This "pair" concept formed the basis of a new class of theories which, as shown above, now seems valid for most intrinsic surface states on ionic or partially ionic crystals.

The theoretical models applicable to extrinsic surface states are extensive; they extend to the related fields of point defects, chemisorption, physisorption, catalysis, etc. and no attempt will be made to consider them here. Besides, as mentioned above, the related experimental data frequently are not characterized well enough for a meaningful comparison to be made with theory.

VIII. Concluding Remarks

The aim of this review article has been fivefold: (a) to solve several different surface problems by a wide range of different techniques, and so provide a useful interrelated compendium for surface scientists; (b) to investigate surface states via a variety of complementary approaches; (c) to enable the nonspecialists to understand the basic concepts and the mathematical techniques in the simplest possible way; (d) to summarize the experimental methods used in determining surface state properties; and (e) to survey the data obtained on a variety of different crystals, and compare the results with theory.

It is found experimentally, for example, that on Si and Ge all surface state distributions seem to appear in pairs of donor-acceptor bands, whereas the theoretical 3-D calculations predict only a single band. Perhaps this result is not surprising, since theoretical calculations have not yet included surface reconstruction effects. By contrast, certain surfaces of III–V and II–VI compounds are unreconstructed, and the comparison between theory and experiment is satisfactory. In particular, an acceptor- (donor)-like surface state band appears a few tenths of an electron volt below (above) the conduction (valence) band. The surface and bulk bands overlap con-

---

[302] E. J. M. Kendall, *Surface Sci.* **7**, 486 (1967).

siderably and the effective masses are approximately the same. Experimental observations on large bandgap crystals ($E_g > 1$ eV) indicate that the n- (p)-type samples have bands which bend up (down) at the surface, in agreement with the composite theoretical picture of ionic surface states.

In the present survey, no attempt has been made to treat the dynamical properties, such as transport phenomena, many-body aspects, optical effects, and the interaction with surface phonons, magnons, and excitons, because it is only now that the static properties are somewhat understood.

Future trends in surface state theory may include more widespread use of computers, more realistic and sophisticated models to include the dynamical effects mentioned above, and more precise treatment of the surface region, and reconstruction, if present. On the experimental side, future programs may include crystals having a large variation in doping concentrations, measurements performed at cryogenic temperatures to detect surface state distribution, and systematic studies of new classes of materials.

List of Abbreviations

AO	atomic orbital	MOS	metal-oxide-semiconductor
APW	augmented plane wave		
BZ	Brillouin zone	NFE	nearly free electron
CO	crystal orbital	NN	nearest neighbor
DC	direct current	NNN	next-nearest neighbor
DCO	direct crystal orbital	NR	nonrelativistic
DSS	Dirac surface states	PE	potential energy
EPR	electron paramagnetic resonance	PP	pseudopotential
		RCO	resolvent crystal orbital
FEG	forbidden energy gap	RTS	relativistic Tamm states
1-R	first-order relativistic	SM	scattering matrix
KP	Kronig–Penney	TB	tight binding
LCAO	linear combination of atomic orbitals	TBA	tight-binding approximation
LEED	low-energy electron diffraction	UHV	ultrahigh vacuum
		ZE	zero-point energy

ACKNOWLEDGMENTS

The authors wish to thank Drs. T. E. Fischer, S. Freeman, A. M. Goodman, P. Mark, M. Stęślicka, R. K. W. Swank, and K. Wronsci for their many helpful comments and useful suggestions. The authors are also indebted to Miss Arlene Grier for her patient typing of preliminary drafts, and to Mrs. Prudence Davison for her excellent typing of the final manuscript and assistance in proof reading.

Crystal Growth Mechanisms: Energetics, Kinetics, and Transport

R. L. PARKER

National Bureau of Standards, Washington, D.C.

Introduction	152
I. Thermodynamics of Phase Changes: Phenomenology of the Problem	155
1. Introduction	155
2. Thermodynamics of Bulk Phase Equilibrium	156
3. Thermodynamics of Interfaces	158
4. Nonequilibrium Thermodynamics	160
II. Statistical Mechanics of Crystal Growth: Theoretical Physics of the Problem	161
5. Equilibrium Statistical Mechanics	161
6. Applications of Equilibrium Statistical Mechanics to Phase Changes	162
7. Nonequilibrium Statistical Mechanics	170
III. Stefan Problem: Mathematical Physics of Crystal Growth Transport Processes	170
8. Introduction	170
9. Stefan Problem without Complications: Various Shapes	172
10. Stefan Problem with Complications: Various Shapes	185
11. Some Approximate Solutions to the Stefan Problem	195
12. Some Questions Regarding Stefan Problems	197
IV. Nucleation	197
13. Introduction	197
14. Classical Nucleation	198
15. Some Modern Developments in Nucleation	204
V. Interface Structures and Interface Kinetics	208
16. Interface Structures in Equilibrium	208
17. Vapor Phase Growth Interface Kinetics: Theoretical and Experimental	226
18. Solution Growth Interface Kinetics	239
19. Melt Growth Interface Kinetics	245
20. Kinematic Theory of Crystal Growth	250
VI. Morphological Stability	253
21. Introduction	253
22. Morphological Stability Problem without Complications: Various Shapes	255
23. Stability Problem with Complications	262
24. Experimental Studies of Morphological Stability	268

VII. Effects of Impurities on Crystal Growth Mechanisms.................... 272
    25. Introduction... 272
    26. Growth of Binary Chains.. 273
    27. Poisoning of Kinks during Layer Growth by Step Motion around Screw
        Dislocations... 276
    28. Sucrose Growth from Solution: Step Pinning........................ 278
    29. Step Pinning (Continued)... 280
    30. Kinematic Waves and Impurities in Dissolution: Etch Pits............ 281
    31. Macroscopic Particles as the Impurity............................. 283
VIII. Fluid Flow Effects.. 284
    32. Introduction... 284
    33. Flow past a Flat Plate... 286
    34. The Rotating Disk and the Czochralski Apparatus................... 290
    35. The Sphere in Fluid Flow.. 294
    36. Thermal or Natural Convection and Crystal Growth in Horizontal Boats. 295
    37. Fluid Flow Resulting from Crystal/Liquid Density Differences......... 298

Introduction

GENERAL

My aim in this review has been to survey and organize, from the point of view of the physicist, the extremely diverse phenomena coming under the heading of crystallization, and in particular to attempt to provide a physics framework for understanding their mechanisms. This framework consists of the application of statistical mechanics, thermodynamics, kinetic theory, and especially transport theory to crystallization problems. Emphasized here is the quantitative approach on well-defined systems, both for experiment and theory. Deemphasized are theories having so many adjustable parameters as to permit fitting nearly any observation, and purely qualitative observations and observations of poorly defined systems, in both of which there are a great many. Crystal growth techniques and apparatus and technique-oriented research are also deemphasized. Also excluded, but for different reasons, namely that they are subfields too vast in themselves, are solid–solid phase transformations and the crystallization of polymers.

The subject of the growth of crystals is an interdisciplinary one, in the sense that contributions to this field have been and continue to be made by scientists and engineers from many professional fields: solid state physicists, mineralogists, crystallographers, physical chemists, mathematicians, chemical engineers, metallurgists, and probably many others. A similar diversity exists in the wide range of media of publication of the research, and in the departments of universities and research organizations in which the work is performed. This review will inevitably reflect this diversity of interests, but it would be desirable in principle to begin with the most general possible point of view.

The review is arranged in eight sections in each of which crystallization problems are discussed by analyzing them from the point of view of a particular discipline of physics. It begins with the more general disciplines, i.e. thermodynamics and statistical mechanics and proceeds to more specialized ones, i.e. mass transport and fluid flow. Within each section, where appropriate, the discussion proceeds from the simplest shape or morphology treated under the simplest conditions to more complex shapes and more elaborate conditions.

It has become customary to refer to crystallization as more an art than a science. The rate at which the art part is changing to science has greatly increased during the past 15 years and it is hoped that this review will show this. There still appears, however, to be a certain lack of connectedness between the various parts of the subject, a certain inability to treat crystallization as an overall, coupled process; this is reflected in this review and appears to be an area where further work is most needed.

CONTINUOUS AND DISCONTINUOUS TRANSFORMATIONS

One can at least imagine the possibility that all the molecules throughout a liquid, as the temperature is lowered, could gradually, uniformly (homogeneously), and continuously lose their liquid character–long range disorder, and gradually, uniformly (homogeneously) and continuously achieve a crystalline solid character–long range order. A similar continuous process could also be imagined for crystallization from vapor or from solution.

On the contrary, all observation and experience of the crystallization process tells us that the process takes place only in a very small fraction of the space occupied by the system—namely at the interface between crystal already present (either seed or nucleus) and mother medium. The process is discontinuous or heterogeneous in the spatial sense in that the crystallizing system is composed of two contiguous and different parts—fluid mother medium and solid crystal.

This discontinuous phase change is consistent with the thermodynamic description of freezing, condensation, and precipitation as being first order processes—i.e. having a change of volume and of heat content (existence of latent heat). The latent heat corresponds to an abrupt increase of entropy S on melting—as we know an increase of disorder—and hence an abrupt change of the slope of the Gibbs free energy G with temperature T at constant pressure p since $(\partial G/\partial T)_p = -S$. Thus there is another discontinuity, namely a mathematical discontinuity in the slope of the Gibbs function, in fact formed by the intersection of the Gibbs functions for the two phases, resulting from the spatial discontinuity of the two phases having different entropies and specific volumes (see also Part I).

A thermodynamically continuous phase change (first order derivatives of G continuous, second order phase change) may occur continuously in the spatial sense (homogeneously), although it need not. According to Turnbull,[1] examples of the former (i.e. spatially continuous) exist in some order–disorder transitions, and superconductivity is an example of the latter.

Transport; Shape Stability

Since the crystallization process is spatially discontinuous and takes place only in certain parts of the crystallizing system, namely at the interfaces, the crystallizing substance must travel from its position in the mother fluid medium to the interface. Or in the case of melt growth, the heat of freezing must travel away from the interface. Such effects are termed transport effects and, taken together with the interface process itself, the two constitute the entire process of crystal growth on seed crystals or nuclei. Discussions of the mechanism of crystal growth must consider both, as well as the nucleation process itself. As Mott[2] pointed out, the transport process may usually be treated as macroscopic, involving the processes of diffusion of heat or matter. On the other hand, the interface process involves the detailed atomic structure of the surface.

Depending on the system studied, either the transport or the interface effects can be much the slowest and hence rate controlling, or both may control the rate together and indeed, interact with each other. In melt growth, heat of freezing may be removed through the crystal or through the undercooled liquid. The former thermal arrangement is stable to shape perturbations and, if interface kinetics are not sluggish, the latter thermal arrangement is unstable to perturbations, since a perturbation experiences more driving force for growth the more it extends into the supercooled liquid. In solution growth, from supersaturated solution, again for fast kinetics, the interface tends to be unstable, for similar reasons. The instabilities often develop others and the resultant branched system is termed dendritic growth.

Flat Faces

Flat faces are often, but certainly not always, observed on crystals grown from various mother media. As Frank[3,4] has pointed out, the existence

[1] D. Turnbull, *in* "Advances in Materials Research in NATO Nations," Agardograph 62, p. 55. Pergamon Press, Oxford, 1963.
[2] N. F. Mott, *Discussions Faraday Soc.* **5,** 11 (1949).
[3] F. C. Frank, *in* "Growth and Perfection of Crystals" (R. H. Doremus, B. W. Roberts, and D. Turnbull, eds.), p. 3. Wiley, New York, 1958.
[4] F. C. Frank, *in* "Encyclopedic Dictionary of Physics" (J. Thewlis, ed.), Vol. 2, p. 208. Pergamon Press, Oxford, 1961.

of flat faces on a polyhedral crystal growing in a supersaturated environment cannot be understood unless one introduces growth steps proceding laterally from a growth center, for in a diffusion field a portion of crystal surface near an edge or corner must experience a higher supersaturation than a portion near the middle, and yet it grows no faster, so that no "local" boundary condition can be strictly correct. The most typical step source center is a dislocation or group of them. Part V treats this in detail. However, solving the coupled volume transport–interface kinetics problem for such growth center shape-controlled faces without a local boundary condition would be a formidable problem and has not been done. The crystal remains flat and not dendritic presumably due to a sufficient degree of interface control.

On the other hand, many substances and systems grow with nonflat faces and one may assume that a local boundary condition is then more suitable. We discuss this growth in detail in Part III on the Stefan problem.

CRYSTAL MORPHOLOGY

As may be surmised already, understanding the crystal form or morphology is an extremely complex problem, involving, in general, interactions between transport effects and surface kinetic effects, with the added complication of morphological instability (Part VI). The fantastic variety of snow crystals[5] is an illustration of the complexity.

Crystal form can of course reflect crystal structure and interatomic bonding, but it does this through the interface effects which must be taken together with transport effects if morphology is to be understood. Traditional equilibrium (minimum free energy) shapes, often discussed, are often irrelevant for the nonequilibrium processes influencing shape mentioned above, primarily because the Gibbs–Thomson effect becomes relatively negligible for greater than micron-sized crystals, in usual supersaturation environments. Furthermore, the usual assumptions for equilibrium shape calculations give shapes which vary considerably for very slight changes in these assumptions, as Herring[6] pointed out.

I. Thermodynamics of Phase Changes: Phenomenology of the Problem

1. INTRODUCTION

I believe that statistical mechanics, which deals with the prediction of the behavior of systems of many molecules, is the broadest useful discipline that can serve as a foundation for crystal growth problems. However, we

[5] U. Nakaya, "Snow Crystals." Harvard Univ. Press, Cambridge, Massachusetts, 1954.
[6] C. Herring, *Phys. Rev.* **82**, 87 (1951).

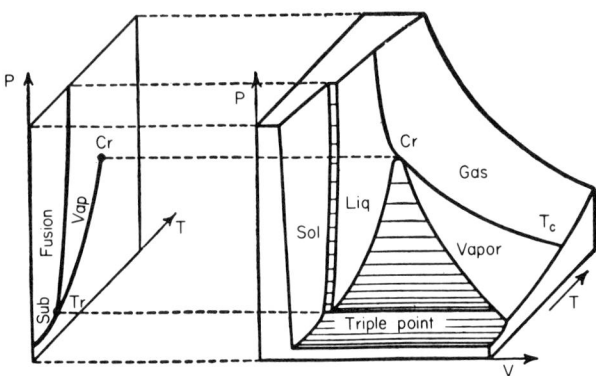

FIG. 1. p–V–T (pressure–volume–temperature) surface, schematic [M. W. Zemansky, "Heat and Thermodynamics." McGraw-Hill, New York, 1943].

discuss the thermodynamics first: it helps us to classify the subject of crystal growth and to define driving forces for crystal growth. Equilibrium thermodynamics cannot, of course, tell us what we really wish to know: how fast do molecules add on to a crystal under particular driving forces, and where do they add, i.e., into what shape does the crystal grow? This is part of transport theory and of interface kinetic theory.

Nor does thermodynamics explain (i.e. predict, using no undetermined coefficients) the existence or origin of the phase transformation. As Uhlenbeck[7] has noted, phase diagrams showing the melting curve, vaporization curve, sublimation curve, triple point, and critical point are so familiar that we think they need not be explained, forgetting that they have not been explained from first principles. That is, no one can predict, from first principles, the melting point of a substance. The absence of this sort of successful theoretical foundation seems to be one reason that crystal growth is still primarily a subject composed of many fragments not well tied together (Widom[8] discusses the current state of these predictions for certain rare gases; see also Part II).

2. Thermodynamics of Bulk Phase Equilibrium

Phase equilibrium thermodynamics is that part of thermodynamics dealing with phase changes of substances, such as sublimation, fusion, and vaporization. The p–V–T surface (pressure–volume–temperature) for a substance (Fig. 1) is determined experimentally for real substances, and in

[7] G. E. Uhlenbeck and G. W. Ford, "Lectures in Statistical Mechanics." Am. Math. Soc., Providence, Rhode Island, 1963.
[8] B. Widom, *Science* **157**, 375 (1967).

it the ruled surfaces are regions where two phases are in equilibrium. The thermodynamics (first and second laws) of such diagrams tells us how changes in such functions as the internal energy U, the Gibbs free energy G, and volume are related during a reversible process. $G = H - TS = U + pV - TS$ (where H is enthalpy and S entropy) is particularly useful for phase changes[9] since

$$dU = T\,dS - p\,dV + \sum \mu_j\,dn_j \quad \text{or} \quad dG = V\,dp - S\,dT + \sum \mu_j\,dn_j \tag{2.1}$$

when the external variables are p, T, and μ_j, where μ_j is the chemical potential of the jth component having n_j moles present; for a reversible, isothermal, isobaric process involving one component, $dG = 0$ or $G = $ constant: a minimum. Hence, the equations for the sublimation, fusion, and vaporization curves (intersections of ruled surfaces with planes of constant volume) are given by setting equal the molar Gibbs functions of the appropriate phases. The slopes of the sublimation, fusion, and vaporization curves are given by the Clausius–Clapeyron equation derived from Eq. (2.1) as

$$dp/dT = \Delta H / T\,\Delta V \tag{2.2}$$

where ΔH is the difference in molar enthalpy and ΔV the difference in molar volume of the two coexisting phases.

The phase changes indicated are of the first order i.e. there is a change of volume and of heat content (latent heat exists). The first order derivatives of G: $(\partial G/\partial T)_p = -S$, $(\partial G/\partial p)_T = +V$, change discontinuously. There is an abrupt increase in disorder, e.g., in melting a crystal to a liquid. Gibbs phase rule may be used to show that there is only one point where the three phases solid, liquid, and vapor coexist.

The fact that the phase transition at constant pressure is sharp, that it takes place at constant temperature, is "explained" by noting that $dG = 0$ is a minimum principle—the material takes up that state which has the minimum G. For a given pressure, for $T < T_M$, the melting temperature, the solid has the lower value of G; for $T > T_M$, the liquid does (Fig. 2). If the melting process is carried out reversibly, the material must melt at $T_M(p)$, the temperature at which the two G's are equal. [Note that $S = -(\partial G/\partial T)_p$.] This illustrates the concept of phase stability, i.e. that phase with the lowest free energy is stable. One asks the thermodynamic question, at low temperatures why is the crystal more stable (lower G) than the liquid? This is equivalent to asking why are phase diagrams the way they are. We discuss this briefly in Part II; see also Turnbull[9] and Temperley.[10]

---

[9] D. Turnbull, *Solid State Phys.* **3**, 225 (1956).
[10] H. N. V. Temperley, "Changes of State." Cleaver-Hume, London, 1956.

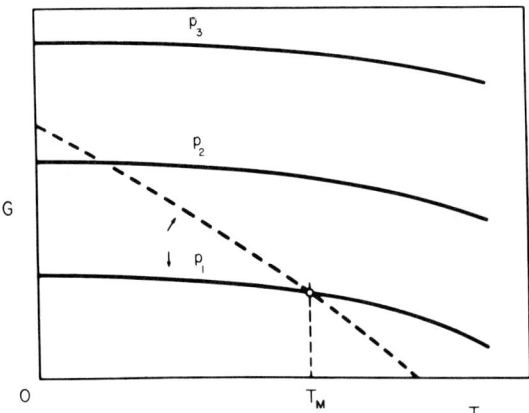

Fig. 2. Gibbs free energy G vs temperature T for solid (—) and liquid (– – –) (including supercooled liquid), for three pressures $p_1 < p_2 < p_3$. The intersection gives the melting point T_M at p_1 [P. M. Morse, "Thermal Physics." Benjamin, New York, 1964].

The role of adding additional chemical components is introduced in Eq. (2.1) via the chemical potential μ_i. Phase diagram analysis is then more complicated.[9] At equilibrium the chemical potential of each component is uniform throughout the system.

3. Thermodynamics of Interfaces

a. Classical Part

The formation of one phase within or adjoining another involves surfaces or interfaces whose energies can be a significant part of the problem. This application of thermodynamics has seen some significant developments in recent years, although like that for bulk substances, it was substantially developed by Gibbs in the 19th century. The classical results may be summarized by noting that the reversible work done, by a fluid–fluid interface, between phase α and phase β having interfacial tension γ, on expanding its area by an amount dA is $-\gamma\, dA$ and hence Eq. (2.1) becomes

$$dU = T\, dS - p_\alpha\, dV_\alpha - p_\beta\, dV_\beta + \sum \mu_i\, dn_i + \gamma\, dA. \qquad (3.1)$$

From this one can show[9] that for an interface between α and β, having principle radii of curvature R_1 and R_2, the pressures in each p_α and p_β are related by

$$p_\beta - p_\alpha = \gamma(1/R_1 + 1/R_2); \qquad (3.2)$$

and finally that the chemical potential inside a spherical drop of radius r

is $\mu_{\beta,r} = \mu_{\beta,\infty} + V_\beta(2\gamma/r)$ if V_β is assumed constant (where V_β is the molar volume of phase β), i.e. is higher than the value in the bulk $\mu_{\beta,\infty}$ by the amount $V_\beta(2\gamma/r)$.

For the fluid–fluid case, the surface tension γ (work per unit area to form the interface) is numerically equal to the surface stress (force per unit length), but not for the solid–fluid case. For a faceted crystal of size l and volume V, whose surface tension may vary with orientation, the counterpart to the above equation is

$$\mu_{\beta,l} = \mu_{\beta,\infty} + [d\sum(\gamma A)/dV_\beta]V_\beta \tag{3.3}$$

(Turnbull,[9] Herring[11]) where $\sum \gamma A$ is the sum of the surface tension of each face times that area A of the face (see Part V for exact definition of γ). If the crystal is in its *equilibrium* shape (having $\sum \gamma A$ a minimum for fixed volume), then Wulff's theorem[11] gives the geometrical construction of this shape.

b. Surface Stress

In solids the surface stress f_{ij} and the surface tension are related by

$$f_{ij} = \delta_{ij}\gamma + (\partial\gamma/\partial\epsilon_{ij}), \qquad i,j = 1, 2 \tag{3.4}$$

(see, e.g. Mullins[12]) where the ϵ_{ij} are the strains in the solid and δ_{ij} is the Kronecker δ. In fact $(\partial\gamma/\partial\epsilon_{ij})$ and f_{ij}, unlike γ, may be either positive or negative. Among more recent advances in the thermodynamics of surfaces are those described in the following reviews and papers: (1) Mullins' excellent review[12] which gives a very clear discussion of surface stresses, and of the effect of temperature on surface energies and surface structures; (2) Cabrera and Coleman[13] and Cabrera[14] discuss thermal faceting in regard to the Wulff plot, and surface and volume stresses in small crystals; (3) Frank[15] discusses the geometry of polar diagrams of the surface tension; (4) Cahn and Hilliard[16] on the thermodynamics of diffuse interfaces involving a "gradient energy" term; and (5) Chernov[17] on the stability of crystal

---

[11] C. Herring, in "Structure and Properties of Solid Surfaces" (R. Gomer and C. S. Smith, eds.), p. 5. Univ. of Chicago Press, Chicago, Illinois, 1953.
[12] W. W. Mullins, in "Metal Surfaces: Structure, Energetics, and Kinetics," p. 17. Am. Soc. Metals, Cleveland, Ohio, 1963.
[13] N. Cabrera and R. V. Coleman, in "The Art and Science of Growing Crystals" (J. Gilman, ed.), p. 3. Wiley, New York, 1963.
[14] N. Cabrera, in "Solid Surfaces" (H. C. Gatos, ed.), p. 320. North-Holland Publ., Amsterdam, 1964.
[15] F. C. Frank, in "Metal Surfaces: Structure, Energetics and Kinetics," p. 1. Am. Soc. Metals, Cleveland, Ohio, 1963.
[16] J. W. Cahn and J. E. Hilliard, *J. Chem. Phys.* **28**, 258 (1958).
[17] A. A. Chernov, *Soviet Phys. Usp. (English Transl.)* **4**, 116 (1961).

vertices and stepped surface profiles. Several of these subjects are discussed in Part V.

4. Nonequilibrium Thermodynamics

The transport processes involved in crystal growth are taken up in subsequent parts. The differential equations describing heat flow, diffusion of matter, and hydrodynamics are well known and have a firm experimental foundation. They may also be regarded as part of nonequilibrium thermodynamics[18] which, unlike equilibrium thermodynamics, is concerned with predicting rates of change during steady state irreversible processes.

The basis of this part of thermodynamics is the notion of generalized fluxes J_m (e.g., heat flow) and forces F_n (e.g., temperature differences) and of a set of linear relations connecting them:

$$\mathbf{J}_m = \sum_n L_{mn} \operatorname{grad} F_n. \qquad (4.1)$$

The relations are assumed to be linear, if the system is sufficiently near equilibrium. The coefficients L_{mn} are to be determined by experiment. If the forces and fluxes are properly chosen (by a process involving consideration of the rate of entropy flow in the system—this is not always an easy formulation—see Christian[19]), then Onsager[20] proved that $L_{mn} = L_{nm}$, which is really new information, not contained in the original transport equations. For the Onsager relations to be useful, at least two forces or fluxes must be operative and there must be some interaction[19] between the effects—i.e. a diffusion of matter driven by both concentration and temperature gradients.

The theorem of minimum entropy production rate[18] is another part of the general theory that is of interest in crystallization processes. It requires that the rate of entropy production be a minimum in the stationary state. Such a minimum, variational principle could be quite useful as a nonequilibrium analogy to the useful principle that the Gibbs free energy is a minimum in phase equilibria. Unfortunately this "principle," although used by several workers in crystal growth,[21,22] can be shown to give wrong results in treating quite simple systems,[23] (discussion of Kirkaldy[22]). The transport

---

[18] S. R. deGroot and P. Mazur, "Non-Equilibrium Thermodynamics." North-Holland Publ., Amsterdam, 1962.

[19] J. W. Christian, "The Theory of Transformations in Metals and Alloys." Pergamon Press, Oxford, 1965.

[20] L. Onsager, *Phys. Rev.* **37**, 405 (1931); **38**, 2265 (1931).

[21] W. A. Tiller, *J. Appl. Phys.* **34**, 3615 (1963).

[22] J. S. Kirkaldy, *in* "Decomposition of Austenite by Diffusional Processes" (V. F. Zackay and H. I. Aaronson, eds.), p. 39. Wiley (Interscience), New York, 1962.

[23] J. W. Cahn and W. W. Mullins, discussion appended to Kirkaldy,[22] p. 123.

differential equations, on the other hand, can be put in variational form and do give the correct results.

II. Statistical Mechanics of Crystal Growth: Theoretical Physics of the Problem

As noted above, the purpose of equilibrium statistical mechanics is to derive the equilibrium properties of a macroscopic molecular system from the laws of molecular dynamics (see e.g. Huang[24]). Calculating crystal growth rates involves, of course, departures from equilibrium statistical mechanics. Such rate calculations are made in transport theory or kinetic theory or in nonequilibrium statistical mechanics.

Nevertheless[9] an adequate theory of phase changes should predict (a) the relative stability of different phases of the same substance and (b) the formation of one phase in another. Or as Christian[19] says, (a) the reasons for transformation and (b) the mechanisms of the transformation. Of course, the transformation will go in a direction and at a rate motivated by the Gibbs free energy difference between the actual system and the equilibrium one. And these free energies are often known from experiment. But, speaking fundamentally, which phase is the more stable? How can this be predicted, given the forces between the atoms?

5. Equilibrium Statistical Mechanics

The application of equilibrium statistical mechanics to a system can be reduced to the problem (which may be very difficult) of calculating the partition function of the system. The partition function for a canonical ensemble is defined as

$$Z_N = (1/N!h^{3N}) \int \exp -(E(N)/kT) \, d\Gamma(N) \qquad (5.1)$$

where N is the number of (identical) particles in the system, h is Planck's constant, k is Boltzmann's constant, $E(N)$ is the total (mechanical) energy of the system in classical mechanics, and $d\Gamma(N)$ a volume element of phase space.

The pressure is given by

$$p = kT \, \partial \ln Z / \partial V \qquad (5.2)$$

and the Gibbs free energy by

$$G = -kT \ln Z + pV. \qquad (5.3)$$

For a system of discrete energy levels the integral in (5.1) is replaced by the sum over energy levels of the system, $\sum_i \exp(-E_{N,i}/kT)$.

[24] K. Huang, "Statistical Mechanics." Wiley, New York, 1963.

6. Applications of Equilibrium Statistical Mechanics to Phase Changes

There have been a great many studies published in which statistical mechanics is used in examining phase transformations. We give here only a few examples. Temperley[10] reviews the literature up to 1955; Uhlenbeck[7,25] gives later references; see also Widom[8] and Cohen.[26]

a. The Condensation of Vapors/Gases to Liquids

(i) The earliest attempt to explain quantitatively the deviations from ideal gas ($pV = NkT$) behavior at high gas densities was that of van der Waals. As we know, the van der Waals equation (see e.g., Widom,[8] Kahn[27])

$$p = NkT/(V - Nb_1) - a_1(N/V)^2 \qquad (6.1)$$

includes two terms neglected in an ideal gas: $-Nb_1$ representing an effect due to the finite volume of the molecules, and $a_1(N/V)^2$ representing attractive forces between the molecules, a and b being constants for a particular gas. Plotting p–V isotherms of this equation (Fig. 3) we see that for isotherms below the critical temperature $T_c{}^V$ (isotherm with horizontal slope at one point) the equation predicts a "wiggle," a nonphysical decrease in volume with decreasing pressure. Maxwell's equal area construction (horizontal line) is then interpreted as giving the correct behavior of the system in this region, when both the vapor phase and the liquid phase are in equilibrium. Although this equation qualitatively represents many properties of the vapor–liquid equilibrium near $T_c{}^V$ fairly well (i.e. algebraic degree of the critical isotherm, form of the coexistence curve (dashed curve), heat capacity), it suffers in an exact quantitative comparison with experiment. It is possible to derive it from an approximate partition function, which may be obtained by considering only the effect of molecular interaction on the translational motions of each molecule (neglecting rotations and internal excitations) (see e.g. Morse[28]).

(ii) We next consider another example, the two-dimensional lattice gas or Ising model. The Ising model is a square two-dimensional array of magnetic atomic dipoles. The dipoles can only point up or down, and the

---

[25] G. E. Uhlenbeck, N. Rosenzweig, A. J. F. Siegert, E. T. Jaynes, and S. Fujita, "Statistical Physics." Benjamin, New York, 1963.
[26] E. G. D. Cohen, ed., "Fundamental Problems in Statistical Mechanics," Vol. II. Wiley, New York, 1968.
[27] B. Kahn, *in* "Studies in Statistical Mechanics" (J. deBoer and G. E. Uhlenbeck, eds.), Vol. III, p. 289. North-Holland Publ., Amsterdam, 1965.
[28] P. M. Morse, "Thermal Physics." Benjamin, New York, 1964.

nearest-neighbor interaction energy is zero when parallel, and $\varphi/2$ when antiparallel. This simple system shows the ferromagnetic phase transition, i.e. for temperatures less than $T = T_c{}^I$ a spontaneous magnetization exists; for $T > T_c{}^I$ it disappears. Onsager[29] rigorously solved this system, obtaining the partition function and finding the transition temperature $T_c{}^I$. The Ising model is believed to be the only exactly solved model in two or more dimensions showing a phase transition.

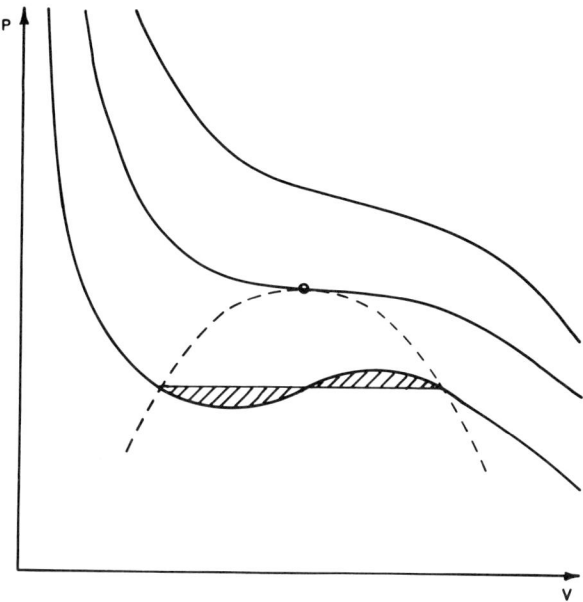

FIG. 3. p–V isotherms of van der Waals' equation, showing phase transition, critical isotherm (T_c), equal-area construction, and dashed coexistence curve.

Lee and Yang[30] devised a simple model of a fluid based on the Ising model of ferromagnetism. Their model, called the two-dimensional lattice gas, consists of a square, two-dimensional array of cells. Each cell may be either vacant or occupied by an atom. If two molecules are in neighboring cells, their potential energy of interaction is taken to be -2ϵ; if they are in the same cell, it is ∞; and if farther apart than a first neighbor cell, it is zero. As Widom[8] points out, this scheme resembles a kind of square-well potential with hard spheres.

[29] L. Onsager, *Phys. Rev.* **65**, 117 (1944).
[30] T. D. Lee and C. N. Yang, *Phys. Rev.* **87**, 410 (1952).

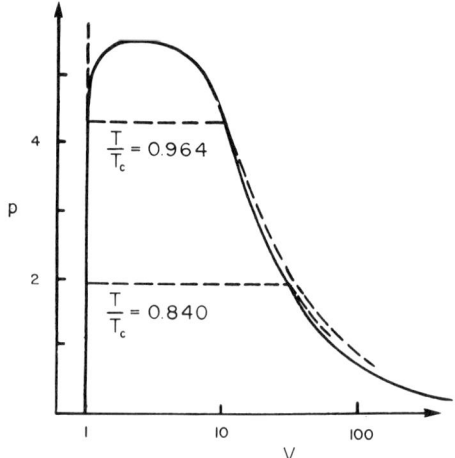

FIG. 4. p-V isotherms of two-dimensional lattice gas, showing phase transition (isotherms, ---; coexistence curve, —) [T. D. Lee and C. N. Yang, Phys. Rev. **87**, 410 (1952)].

For this system Onsager[29] (see Huang[24]) had calculated the partition function per molecule Z^I exactly, in a very complex derivation. The result is

$$\ln Z^I = \ln(2\cosh 2\beta\epsilon) + (2\pi)^{-1} \int_0^\pi d\varphi \ln \tfrac{1}{2}[1 + (1 - \kappa^2 \sin^{+2}\varphi)^{1/2}] \quad (6.2)$$

where $\kappa = (e^{2\beta\epsilon} - e^{-2\beta\epsilon})/(e^{2\beta\epsilon} + e^{-2\beta\epsilon})^2$, $\beta = 1/kT$. From this Lee and Yang[30] derived the equation of state for the two-dimensional lattice gas. A plot is shown in Fig. 4. The T_c^I is exactly given by

$$\exp(-\epsilon/kT_c^I) = \sqrt{2} - 1. \quad (6.3)$$

As is seen from the dashed isotherms, the system has a phase transformation generally similar to the van der Waals gas, but in the present case the result follows exactly from the theory, with no nonphysical wiggles, and no need to use Maxwell's construction. The coexistence curve, critical isotherm, and heat capacity are quantitatively different from the van der Waals prediction.

(iii) The three-dimensional lattice gas problem has never been exactly solved, but through numerical calculations many of its properties near the critical point have been determined.[31] Certain of these are found to agree quite closely with experiment, in particular the coexistence curve, critical isotherm, and specific heat.[8]

[31] C. Domb, Advan. Phys. **9**, 149 (1960).

b. *Freezing of Liquids and Melting of Crystals*

We emphasize once again that equilibrium theories of freezing and of condensation do not concern themselves with the kinetics and mechanisms (such as the rates) of the transformation process. This is the justification for ignoring, at this stage, the notions of nucleation and growth which characterize the rate processes of many phase transformations. Instead, the equilibrium notion of a "cooperative" process is emphasized, i.e. a process in which the particles dealt with in statistical mechanics could not be treated as independent particles, or as nearly independent ones, with the interactions regarded as a perturbation, but on the contrary their interactions, as in the Ising model, are central and must be taken into account at the start. Condensation and freezing are both examples of cooperative processes.[31] The entire assembly is viewed as "cooperating" in the phase transformation, which is in turn taken to be homogeneous. (Of course, in the nucleation and growth picture, only a small fraction of the molecules—those in the interface—are involved in the rate processes of the phase change at any one time.)

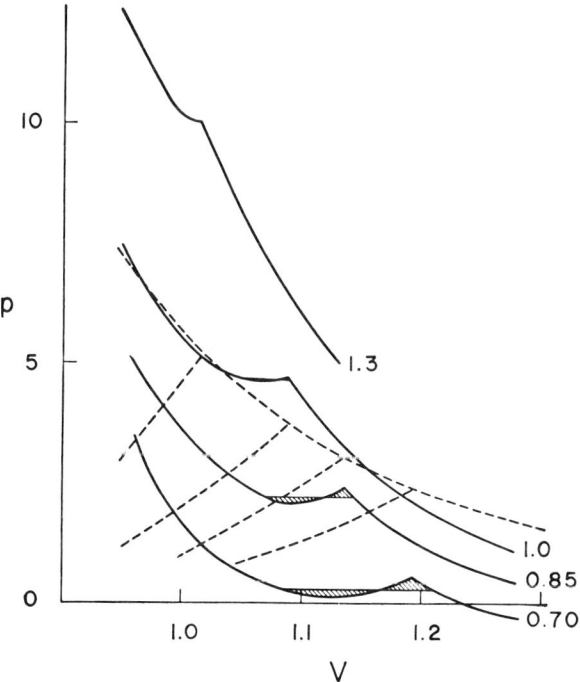

Fig. 5. p–V isotherms of Lennard-Jones and Devonshire melting theory, showing phase transition and equal area construction [J. deBoer, *Proc. Roy. Soc.* **A215,** 4 (1952)].

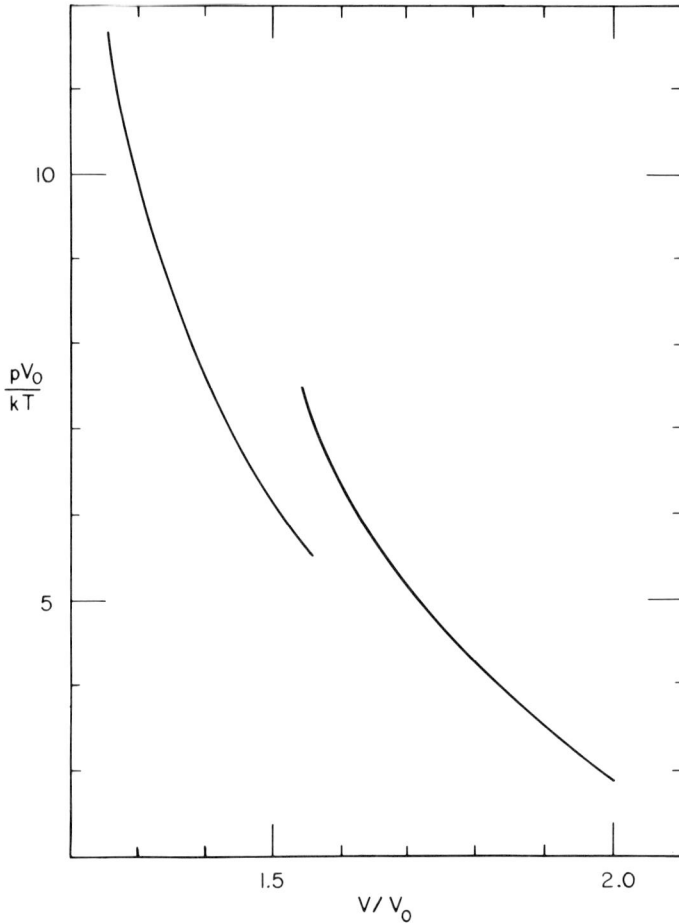

Fig. 6. Equation of state of hard spheres, showing separate, overlapping branches, as calculated by B. J. Alder and T. E. Wainwright [*J. Chem. Phys.* **27**, 1208 (1957)] for a system of 108 molecules. V_0 is close-packed volume.

There have been many equilibrium theories of the structure of liquids and the freezing of liquids.[32] For reviews of the latter we mention Temperley,[10] Ubbelohde,[33] Brout,[34] and Widom.[8] We confine ourselves here to discussing two of these theories of freezing.

[32] T. J. Hughel, ed., "Liquids: Structure, Properties, Solid Interactions." Elsevier, Amsterdam, 1965.
[33] A. R. Ubbelohde, "Melting and Crystal Structure." Oxford Univ. Press (Clarendon), London and New York, 1965.
[34] R. Brout, "Phase Transitions." Benjamin, New York, 1965.

(i) *The Theory of Lennard-Jones and Devonshire.*[35] In this model of the melting of crystals, all the atoms are on their proper lattice sites in the crystal (assumed to be fcc). At higher temperatures some move to interstitial sites, increasing the disorder and entropy. The energy required to so displace an atom is assumed to decrease as the number of such displaced atoms is increased. The liquid is assumed to exist when there is maximum disorder (equal occupation of both regular and interstitial sites). The interatomic potential used in the calculations was $1/r^{12}$ repulsive, $1/r^6$ attractive, where r is the distance between two atoms, and the results were applied to the case of argon. Figure 5 shows the resulting isotherms.[36] The phase change is clear. The theory also gives good agreement with the entropy and volume changes (for argon) on melting, and the melting curve (T_M vs p). But see also Henderson and Barker,[37] and the remarks by Frank and by Seeger in Rudman *et al.*[38]

(ii) *Machine Calculations of Alder and Wainwright.*[39] This study of the equation of state of hard spheres (i.e. no attractive forces, billiard ball collisions) was made by actually solving the simultaneous classical equations of motion of 32 and 108 particles in a box with periodic boundary conditions. The volume was varied from about 1.2 to 2.0 times V_0, the close-packed volume. The pressure was evaluated by the rate of change of momentum of the particles colliding with the wall. The results are shown in Fig. 6, and consist of two separate and overlapping branches. On the lower branch the particles were found to be confined by their neighbors (i.e. extended short-range order) while on the upper branch the particles were able to exchange with the surrounding particles (liquid). These curves suggest that this system has a phase transition between the less dense and the more dense phase.

c. *The Statistical Mechanics Calculation of Surface Structures*

(i) *Surface Melting, Roughness, and Ising Model.* Burton and Cabrera[40] (see also Burton *et al.*[41]) as did Lee and Yang,[30] recognized that the Onsager solution of the Ising model for a ferromagnet could be adapted to

---

[35] J. E. Lennard-Jones and A. F. Devonshire, *Proc. Roy. Soc.* **A169,** 317 (1938); **A170** 464 (1939).
[36] J. deBoer, *Proc. Roy. Soc.* **A215,** 4 (1952).
[37] D. Henderson and J. A. Barker, *Mol. Phys.* **14,** 587 (1968).
[38] P. S. Rudman, J. Stringer, and R. I. Jaffee, eds., "Phase Stability in Metals and Alloys." McGraw-Hill, New York, 1967. See pp. 554, 572.
[39] B. J. Alder and T. E. Wainwright, *J. Chem. Phys.* **27,** 1208 (1957). See also W. W. Wood and J. D. Jacobson, *ibid.* **27,** 1207 (1957).
[40] W. K. Burton and N. Cabrera, *Discussions Faraday Soc.* **5,** 33 (1949).
[41] W. K. Burton, N. Cabrera, and F. C. Frank, *Trans. Roy. Soc.* **A243,** 299 (1951).

the treatment of phase transformations of two-dimensional arrays of molecules. Burton and Cabrera wished to calculate how rough a surface (ideally close-packed initially) of a crystal might become as the temperature is raised. A simple cubic crystal (100) surface is considered, and is assumed flat (no vacancies) at $T = 0°K$. The potential energy of interaction between two nearest-neighbor atoms is φ, or $\varphi/2$ per atom. If, however, one atom of the surface is higher than the other by one or more lattice distances, a first neighbor bond is broken, and an extra energy $\varphi/2$ per atom is associated with this arrangement. Defining the surface roughness $S_r = (U - U_0)/U_0$ where U_0 is the surface potential energy per atom of a perfectly flat surface ($= \varphi/2$) and U is that of the actual surface, it is clear that $S_r = 0$ for a flat surface and $S_r > 0$ is a measure of the roughness. This two-dimensional, two-level problem corresponds exactly to the Ising model, and hence suffers a phase change ("surface melting") at $T = T_c^I$ where, for a square lattice, T_c^I is given by Eq. (6.3) as $kT_c^I/\varphi = 0.57$. (Note $\varphi/2 = \epsilon$.) [We note that bulk melting is a first order phase change, the Ising "surface melting" is a second order one.] The surface roughness S_r is given by differentiation of Eq. (6.2); for $T \ll T_c^I$, $S_r = 4 \exp(-2\varphi/kT)$. For an adatom (adsorbed atom), four bonds are unpaired (compared to a surface atom), with an energy $\varphi/2$ per bond. Thus $\exp(-2\varphi/kT)$ is the probability of exciting single atoms to the surface. [For an adatom pair, there are six unpaired bonds and S_r is proportional to $\exp(-3\varphi/kT)$.] Thus, for $T \ll T_c^I$, there are only single atoms, not clusters of them, or steps, upon the surface. The T_c^I was shown to be generally much greater than the experimental (bulk) melting point of the solid, and hence the general conclusion was that one does not have thermodynamically produced steps on close-packed crystal faces even near the melting point; possible effects of the neighboring medium are not included.

(ii) *Surface Roughness of a Solid in Contact with Its Melt.* Jackson[42,43] studied a problem similar to that of Burton and Cabrera, namely the degree of roughness of a surface, but in addition considered the nature of the medium adjacent to the surface. The model used is a plane surface in contact with either its own vapor or its own liquid. Single atoms are imagined to be added to this surface, from the medium, and the change in free energy is calculated as a function of the number of already occupied sites, using the Bragg–Williams method and dealing with nearest-neighbor bonds only. This is analogous to the free energy of mixing calculated on

---

[42] K. A. Jackson, *in* "Growth and Perfection of Crystals" (R. H. Doremus, B. W. Roberts, and D. Turnbull, eds.), p. 319. Wiley, New York, 1958.

[43] K. A. Jackson, *in* "Liquid Metals and Solidification," p. 174. Am. Soc. Metals, Cleveland, Ohio, 1958.

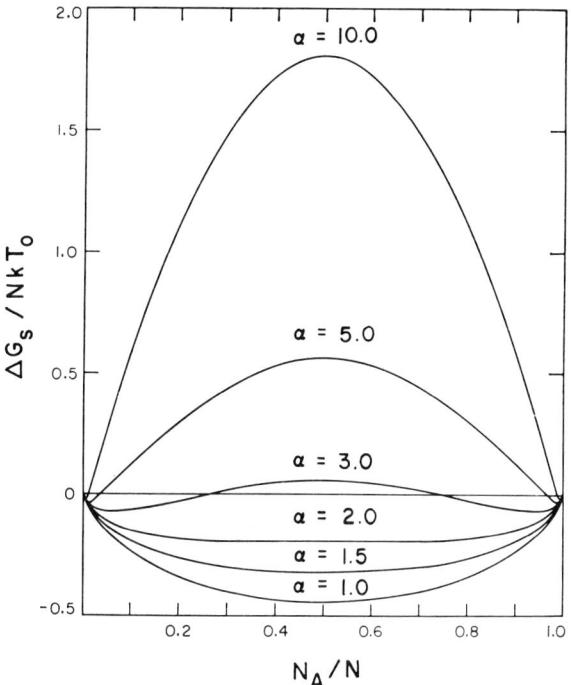

Fig. 7. Surface roughness theory of Jackson, showing free energy change $\Delta G_s/NkT_0$ resulting from addition of N_A molecules to N possible sites vs N_A/N (α is defined in the text) [K. A. Jackson, *in* "Liquid Metals and Solidification," p. 174. Am. Soc. Metals, Cleveland, Ohio, 1958].

"regular solution" theory (see e.g. Christian[44]). The result is

$$\frac{\Delta G_s}{NkT_0} = \frac{\alpha N_A}{N^2}(N - N_A) - \ln\left[\frac{N}{N - N_A}\right] - \frac{N_A}{N}\ln\left[\frac{N - N_A}{N_A}\right] \quad (6.4)$$

where $\Delta G_s/NkT_0$ is the normalized free energy change per atom added to the surface, T_0 the equilibrium temperature, N the total number of sites on the surface, and N_A the actual number of these occupied. The parameter $\alpha \cong (L/kT_0)f_k$ depends on the latent heat L of the process and f_k ($f_k \leq 1$) is a crystallographic factor, i.e. the fraction of all first neighbors lying in a plane parallel to the crystal face considered. Figure 7 shows Eq. (6.4) for several values of α. It is seen that for $0 \leq \alpha \leq 2$ the free energy is a minimum only when the surface is half populated with adatoms—i.e. is rough. Whereas for $2 \leq \alpha \leq \infty$ the reverse is true—the surface, apart from a few

[44] See Christian,[19] pp. 174 and 204.

adatoms is smooth. For a vapor medium α is generally large (large L) but for a melt medium α is often less than 2. Hence the existence of "surface melting" depends critically on the medium adjacent and may exist at the (bulk) melting point. (See also Part V.)

7. Nonequilibrium Statistical Mechanics

The transport laws of the conduction of heat, of diffusion of matter, and of the flow of fluids are essential as aids in understanding the nonequilibrium processes of crystal growth. These laws are often stated as phenomenological relations deduced from experiment (i.e. Fick's law) with purely experimentally determined coefficients. However, they can be derived from the Boltzmann transport equation of nonequilibrium statistical mechanics (see e.g. Huang[24]) and also the transport coefficients (diffusivity, viscosity, conductivity) can be rigorously derived, at least for a gas not too far from equilibrium, using the notions of collisions and mean free path. Both the thermal conductivity and viscosity can be shown, for a gas having molecular mass m, to be approximately $(mkT)^{1/2}/a^2$, "a" being the molecular diameter.[24] Calculations of these coefficients for fluids are discussed by Rice.[45]

III. Stefan Problem: Mathematical Physics of Crystal Growth Transport Processes

8. Introduction

As noted earlier, all experience and observation indicate that the crystallization process is heterogeneous, not homogeneous. The molecules do not, in unison, all gradually convert from being molecules of a fluid to becoming molecules of a solid (although this is rather like the theoretical picture often treated as in Part II). Instead, here and there, groups of molecules will become solid through the nucleation process. These groups or nuclei may then grow to substantial size, by transport processes. The molecules which condense upon the interface of the nucleus have been transported from some more distant part of the solution or vapor; or particularly, in the case of freezing, the heat given up by the molecules on condensing must be transported away from the interface. In Part III we discuss these transport processes; in Part IV, nucleation processes. (Of course, if a seed crystal is already present in the system, then the transport process may take place at once upon it without the need for the bulk nucleation process just mentioned.) Either the nucleation process or the

---

[45] S. A. Rice, *in* "Liquids: Structure, Properties, Solid Interactions" (T. J. Hughel, ed.), p. 51. Elsevier, Amsterdam, 1965.

transport process can be rate-controlling in a particular case. Yet a third process which can be rate-controlling is that known as the interface kinetic process. This is the detailed molecular mechanism whereby molecules proceed from a position in the fluid medium, just outside the crystal–fluid interface, to their (relatively) permanent resting place in the crystal surface. We discuss the various mechanisms of interface kinetics in Part V, although in the present part we will mention the use of interface kinetic laws as boundary conditions for the transport process.

A problem treated by Stefan[46] is illustrative of the general class of transport problems involving the diffusion differential equation with a moving or free boundary, which are called Stefan problems. Stefan studied the rate of growth in thickness of the ice layer in polar seas. In this case the heat of crystallization had to pass through the solid layer of ice to be carried away by the cold air, which was assumed to be at a temperature below freezing. The problem is to determine the unknown position of the boundary as a function of time. It is not a steady state one, in that the thicker the ice becomes the more difficult it is to extract the heat through it and hence the slower the process. It appears to be useful to regard all problems in crystal growth as being of the Stefan type, whether steady state or not, and whether or not the boundary motion is believed known. Thus, as we have just mentioned, heat flow from superheated melts (Bridgman, Czochralski, Chalmers process for crystal growth; freezing of ice; castings, solidification); heat flow into undercooled melts (often occurring in dendritic growth); matter diffusion (crystal growth from solution or from vapor in presence of an inert gas or from melt with impurity or "solvent" present), are all regarded as Stefan problems. This is not to say that other factors and processes—crystallographic anisotropy, interface kinetics, or nucleation—do not play a role in crystallization mechanisms. It does say that the role of heat flow and/or matter diffusion must be understood in any particular case before the entire process can be regarded as understood.

Stefan problems may be categorized in the following (not necessarily mutually exclusive) ways: (a) whether the shape of the solid which is forming is known, or not known, (b) whether this shape can preserve itself during growth or dissolution, (c) whether the interface velocity and its time dependence are known, (d) whether the interface temperature and/or concentration are known (e) whether an analytical exact treatment or an analytical approximate treatment, or a numerical treatment is performed, (f) whether the shape is stable or not to slight shape perturbations, (g) whether the liquid is supercooled or not, (h) whether or not complicating factors are considered, such as interfacial free energy, interface kinetics (departure of interface temperature or concentration from local equilibrium

[46] J. Stefan, *Ann. Phys. Chem.* (n.F.) **42**, 269 (1891) [*Ann. Phys.* **278**].

as a function of interface velocity), fluid flow such as stirring or natural convection, presence of impurity, and effects of electric or magnetic fields or mechanical vibration.

Horvay[47] categorizes Stefan problems similarly according to which of the three: interface shape, interface temperature or concentration, and interface velocity, are known. The conventional problem assumes shape and temperature, and calculates velocity. The modified (sometimes called inverse) Stefan problem assumes shape and velocity, and calculates interface temperature. The alternate problem is to compute shape, knowing the velocity at some definite point of the shape, i.e. at the tip of a dendrite, and knowing the temperature.

Stefan problems are among the more difficult boundary value problems in mathematical physics, as we shall see in more detail just below; for two reasons. First, in general the position of the boundary on which one wishes to apply boundary conditions, is not known; finding its position is part of the problem. Second, the system of equations describing the problem is nonlinear so that solutions may not be built up by superposition. Friedman's[49] book discusses some of the existence and uniqueness questions involved. A review of the Stefan problem is given by Bankoff[50] with references to the purely mathematical as well as the numerous physical papers. Carslaw and Jaeger[51] and Horvay[52] also have written valuable reviews.

9. Stefan Problem without Complications: Various Shapes

a. The Plane

(i) *Liquid at Temperature above the Melting Point.* Stefan[46] first published the solution to planar freezing, although Neumann (unpublished lectures, see Riemann-Weber[53]) is said to have first solved it. We suppose a large mass of fluid at a uniform temperature $T_\infty \geq T_M$ where T_M is the melting point. At time $t = 0$, the planar surface of this mass is suddenly lowered to the temperature $T = 0$. The equations for conduction of heat in

---

[47] See discussion appended to Wilcox and Duty.[48]
[48] W. R. Wilcox and R. L. Duty, *J. Heat Transfer* **88c**, 45 (1966).
[49] A. Friedman, "Partial Differential Equations of Parabolic Type." Prentice-Hall, Englewood Cliffs, New Jersey, 1964.
[50] S. G. Bankoff, *Advan. Chem. Eng.* **5**, 75 (1964).
[51] H. S. Carslaw and J. C. Jaeger, "Conduction of Heat in Solids," 2nd ed. Oxford Univ. Press (Clarendon), London and New York, 1959.
[52] G. Horvay, *J. Eng. Phys. (Russian)* **8**, 779 (1965) [English transl.: Rept. 64-RL-3733M, General Electric Res. Lab., Schenectady, New York].
[53] H. Weber, "Die Partiellen Differential-Gleichungen der Mathematischen Physik," Vol. 2. Vieweg, Braunschweig, 1901.

an (isotropic) solid and in liquid are (L = liquid, S = solid), where x is distance measured from a fixed coordinate system, along a normal to the planar surface,

$$\nabla^2 T_L = \partial^2 T_L/\partial x^2 = \kappa_L^{-1}(\partial T_L/\partial t) \tag{9.1}$$

$$\nabla^2 T_S = \partial^2 T_S/\partial x^2 = \kappa_S^{-1}(\partial T_S/\partial t) \tag{9.2}$$

where the κ are thermal diffusivities (cm²/s) given by $\kappa_L = K_L/\rho c_L$, $\kappa_S = K_S/\rho c_S$ where K_S, K_L are thermal conductivities, c_L, c_S are specific heats, and ρ is the density of each phase. Fluid flow effects due to the actual density difference are here ignored and discussed below. At the solid–liquid interface $x = X(t)$ it is assumed that the temperature is equal to the melting point T_M:

$$T_L(X) = T_S(X) = T_M. \tag{9.3}$$

The continuity condition for heat flow at the interface may be written

$$K_S \, \partial T_S/\partial x - K_L \, \partial T_L/\partial x = L\rho \, dX/dt \tag{9.4}$$

where L is the latent heat of fusion. [It is Eq. (9.4) which introduces the nonlinearity of the system although the differential equations (9.1) and (9.2) are linear. It can be shown[51] that

$$dX/dt = -(\partial T_S/\partial t)/(\partial T_S/\partial x), \tag{9.5}$$

and hence the nonlinearity of Eq. (9.4) is clear.] Finally, we have the particular boundary conditions for this problem,

$$T_L \to T_\infty \quad \text{as} \quad x \to \infty \tag{9.6}$$

and

$$T_S \to 0 \quad \text{as} \quad x \to 0, \tag{9.7}$$

and the initial condition

$$X = 0, \quad \text{and} \quad T_L = T_\infty \quad \text{when} \quad t = 0. \tag{9.8, 9.8'}$$

To solve this system, we first note that, for dimensional reasons,

$$T_L(x, t) = T_L(S_L) \tag{9.9}$$

where

$$S_L = x\kappa_L^{-1/2}t^{-1/2}. \tag{9.10}$$

Following Frank,[54] differentiation of (9.9) and (9.10) converts (9.1) to

$$\partial^2 T_L/\partial S_L^2 = -(S_L/2)(\partial T_L/\partial S_L). \tag{9.11}$$

[54] F. C. Frank, *Proc. Roy. Soc.* **A201**, 586 (1950).

Letting $\eta = \partial T_L/\partial S_L$ and integrating, one gets

$$T_L(S_L) - T_L(\infty) = -\int_{S_L}^{\infty} \eta(z)\, dz = -A_1 \operatorname{erfc}(S_L/2) \tag{9.12}$$

where A_1 is a constant and

$$1 - \operatorname{erf}(x) \equiv \operatorname{erfc}(x) \equiv \int_x^{\infty} (2/\pi^{1/2}) \exp(-\xi^2)\, d\xi$$

(Carslaw and Jaeger[51]). Therefore

$$T_L[x/(\kappa_L t)^{1/2}] = T_\infty - A_1 \operatorname{erfc}[x/2(\kappa_L t)^{1/2}]. \tag{9.13}$$

Similarly one finds

$$T_S = B_1 \operatorname{erf}[x/2(\kappa_S t)^{1/2}]. \tag{9.13'}$$

Then Eq. (9.3) requires that, on $x = X$,

$$B_1 \operatorname{erf}[X/2(\kappa_S t)^{1/2}] = T_\infty - A_1 \operatorname{erfc}[X/2(\kappa_L t)^{1/2}] = T_M = \text{const.} \tag{9.14}$$

It follows then that X must be proportional to $t^{1/2}$, i.e.

$$X = 2\lambda_1(\kappa_S t)^{1/2} \tag{9.15}$$

where λ_1 is the "growth constant." Substituting (9.15), (9.13), and (9.13') into (9.4) and then using (9.14), we get

$$\frac{\exp(-\lambda_1^2)}{(\operatorname{erf} \lambda_1)} - \frac{K_L \kappa_S^{1/2}(T_\infty - T_M) \exp(-\kappa_S \lambda_1^2/\kappa_L)}{K_S \kappa_L^{1/2} T_M \operatorname{erfc} \lambda_1(\kappa_S/\kappa_L)^{1/2}} = \frac{\lambda_1 L \pi^{1/2}}{c_S T_M}, \tag{9.16}$$

which gives λ_1 in terms of the physical properties of the material K_L, K_S, κ_L, κ_S, T_M, L, c_S, and of the superheating of the liquid $T_\infty - T_M$. Thus for a particular value of λ_1, the temperatures are

$$T_S = \frac{T_M}{\operatorname{erf} \lambda_1} \operatorname{erf}\left(\frac{x}{2(\kappa_S t)^{1/2}}\right)$$

and

$$T_L = T_\infty - \frac{(T_\infty - T_M)\operatorname{erfc}[x/2(\kappa_L t)^{1/2}]}{\operatorname{erfc}[\lambda_1(\kappa_S/\kappa_L)^{1/2}]}. \tag{9.17}$$

We may verify that (9.17) and (9.17') satisfy (9.6), (9.7), and (9.8); we know already that they satisfy (9.1)–(9.4). We may consider the freezing of water for an appropriate numerical example: let the zero of the temperature scale be so defined that $T_M = 5$ Celsius degrees, then $x = 0$ will be at 0, or 5 degrees below the melting point of ice. Let the initial water mass be at 2 degrees above the melting point, i.e. $T_\infty - T_M = 2$. Then[51] $K_L = 0.00144$, $\kappa_L = 0.00144$ cm²/s, $K_S = 0.0053$, $\kappa_S = 0.0115$ cm²/s, $L\rho = 307$

J/cm^3, and for the temperatures given $\lambda_1 = 0.121$ from Eq. (9.16). Thus the ice sheet will be 1 cm thick in 1.48×10^3 s and only 10 cm thick in 1.48×10^5 s or about 42 hr. (We may observe that the corresponding case of melting of a semi-infinite mass is very similar to the case of freezing—see Carslaw and Jaeger[51].)

(*ii*) *Planar Freezing: Liquid Undercooled.* In this case[51] the liquid is presumed to be initially all at temperature $T_\infty < T_M$ and at $t = 0$ a solid skin begins to form, the temperature at all points of which remains at T_M, so that no heat is withdrawn through the solid; it all goes toward warming up the liquid. In place of Eq. (9.7) we now have

$$T_S = T_M, \quad 0 \leq x \leq X(t), \tag{9.18}$$

otherwise the system of equations is identical to (9.3)–(9.8), noting of course that now $T_\infty < T_M$. The solutions are

$$X = 2\lambda_2(\kappa_L t)^{1/2} \tag{9.19}$$

$$T_S = T_M \tag{9.20}$$

$$T_L = T_\infty + [\lambda_2 L \pi^{1/2}/c_L \exp(-\lambda_2^2)] \operatorname{erfc}[x/2(\kappa_L t)^{1/2}] \tag{9.21}$$

where the growth constant λ_2 is given by

$$\lambda_2 \exp(\lambda_2^2) \operatorname{erfc} \lambda_2 = (T_M - T_\infty)c_L/L\pi^{1/2}. \tag{9.22}$$

For $T_M - T_\infty = 1.3°C$ of undercooling, λ_2 for water is found to be 0.12, giving the same freezing rates as in the previous example.

(*iii*) *Planar Freezing: Constant Rate.*[46,51,55] One may ask what would be necessary for the freezing to take place at a constant rate $dX/dt =$ const instead of a steadily decreasing rate as above $(dX/dt = \lambda_2(\kappa_L)^{1/2}/t^{1/2})$. In the case of liquid at the melting point T_M [special case of case (i)] this may be done by continually reducing the temperature at $x = 0$ instead of keeping it at a constant value $< T_M$. The solution is, for

$$dX/dt = \kappa_S m_1, \tag{9.23}$$

$$T_S = T_M + (L/c_S)[1 - \exp(\kappa_S m_1^2 t - m_1 x)] \tag{9.24}$$

and

$$T_L = T_M, \tag{9.25}$$

where m_1 is a constant.

This solution requires the undercooling at $x = 0$ to increase exponentially with time. It may be noted that this is, in one sense, an example of an inverse or modified Stefan problem, in that the interface boundary motion is here prescribed instead of being unknown, but the temperature

[55] D. Langford, *Quart. Appl. Math.* **24**, 315 (1967).

(at $x = 0$) is unknown. Of course, at the boundary, both temperature and velocity (and shape) are prescribed.

(iv) *Planar Freezing: Boundary Condition at External Surface a Function of Time.*[56-58] The problem is identical to case (i) except for Eq. (9.7) which is now replaced by one of the following conditions:

$$T_S = f_1(t) \quad \text{at} \quad x = 0 \quad (9.26a)$$

$$K_S(\partial T_S/\partial x)_{x=0} = f_2(t) \quad (9.26b)$$

$$(\partial T_S/\partial x)_{x=0} = h[T_S(0, t) - f_3(t)] \quad (9.26c)$$

where f_1, f_2, and f_3 are arbitrary functions of the time and h is a heat transfer coefficient. In example (9.26a) the use of a series expansion method leads to the result that

$$\sum_{n=1}^{\infty} \frac{d^n[X^{2n}(t)]}{(2n)! \kappa_S^n \, dt^n} = \frac{T_M - f_1(t)}{\kappa_S \cdot L\rho_S/K_S}, \quad (9.27)$$

expressing the unknown interface position $X(t)$ in the form of an infinite series of its derivatives. It appears that this form is more useful for deriving the inverse or modified problem, i.e. given the boundary motion, to find the external temperature. Similar results are obtained for cases (9.26b) and (9.26c).[57]

b. The Sphere

(i) *Liquid Undercooled.* Rieck[59] first obtained, in an unpublished thesis, the solution to the spherical freezing problem. Other treatments of the sphere have been those of Ivantsov,[61] Zener,[62] and Frank.[54] We consider the growth of an initially vanishingly small solid sphere from supercooled liquid of temperature $T_\infty \leq T_M$. The flow of latent heat is now out into the liquid. The equation of conduction of heat in the liquid is now, where r is the coordinate radius,

$$\kappa_L \left(\frac{\partial^2 T_L}{\partial r^2} + \frac{2}{r} \frac{\partial T_L}{\partial r} \right) = \frac{\partial T_L}{\partial t}. \quad (9.28)$$

---

[56] B. Ya. Lyubov, *Dokl. Akad. Nauk SSSR* **68**, 847 (1949).
[57] B. Ya. Lyubov, A. L. Roitburd, and D. E. Temkin, *in* "Growth of Crystals" (A. V. Shubnikov and N. N. Sheftal, eds.), Vol. III, p. 47. Consultants Bureau, New York, 1962.
[58] V. T. Borisov, B. Ya. Lyubov, and D. E. Temkin, *Dokl. Akad. Nauk SSSR* **104**, 223 (1955).
[59] Rieck, 1924 (unpubl.); see Huber.[60]
[60] A. Huber, *Z. angew. Math. Mech.* **19**, 1 (1939).
[61] G. P. Ivantsov, *Dokl. Akad. Nauk SSSR* **58**, 567 (1947) [English transl.: G. Horvay, Rept. 60-RL-2511M, General Electric Res. Lab., Schenectady, New York].
[62] C. Zener, *J. Appl. Phys.* **20**, 950 (1949).

The boundary conditions are the same as in the planar case except that

$$T_S = T_M \qquad (0 \le r \le R(t)). \tag{9.29}$$

Making the same substitutions: $T_L(r, t) = T_L(S_L)$ as before, where $S_L = r\kappa_L^{-1/2}t^{-1/2}$, one arrives, following the same procedure, at the following results:

$$R = 2\lambda_3(\kappa_L t)^{1/2}, \tag{9.30}$$

$$T_L = T_\infty + \frac{2\lambda_3(T_M - T_\infty)}{\exp(-\lambda_3^2) - \lambda_3 \pi^{1/2} \operatorname{erfc} \lambda_3}$$

$$\times \left\{ \frac{(\kappa_L t)^{1/2}}{r} \exp\left(-\frac{r^2}{4\kappa_L t}\right) - \frac{\pi^{1/2}}{2} \operatorname{erfc}\left(\frac{r}{2(\kappa_L t)^{1/2}}\right) \right\} \tag{9.31}$$

where λ_3 is given by

$$\lambda_3^2 \exp(\lambda_3^2)[\exp(-\lambda_3^2) - \lambda_3 \pi^{1/2} \operatorname{erfc} \lambda_3] = \tfrac{1}{2} c_L (T_M - T_\infty)/L. \tag{9.32}$$

(*ii*) *Sphere Growing at Constant Rate, from Supercooled Melt.* For the sphere to grow at constant rate,[63] the latent heat must be taken up both by the liquid and by the sphere. For this to be true the sphere must rise in temperature and hence it (and the interface) must be below the melting point, contrary to the previous assumptions. Given this somewhat artificial set of conditions (note that interface kinetics, which could at least initially make this a feasible boundary condition, are not included in this treatment) it is found that the crystal boundary temperature rises linearly in time in the initial stages of growth.

(*iii*) *Sphere Crystallizing Inwards.*[57] A sphere of radius R of liquid, initially at the melting point, has its surface quickly cooled to a temperature below the melting point. Both the inverse problem (boundary motion specified) and direct (external temperature specified) problems have been studied. In particular, for two cases of the inverse problem, namely, when the skin thickness is linear in time, giving the interface position $y(t)$ as

$$y(t) = R - \beta_1 t, \tag{9.33}$$

and when

$$y(t) = R - \beta_2 t^{1/2}, \qquad \beta_1 \text{ and } \beta_2 \text{ being constants}, \tag{9.34}$$

the external surface temperatures have been found[57] as analytic functions of the time t, by a series method similar to that used by Lyubov in planar freezing (see above). The direct problem has been treated only approximately, for the case where the external surface temperature $T_{su} < T_M$ is

---

[63] G. P. Ivantsov, *in* "Growth of Crystals" (A. V. Shubnikov and N. N. Sheftal, eds.), Vol. I, p. 76. Consultants Bureau, New York, 1958.

constant in time. Then, approximately,

$$\tfrac{1}{6}R^3 + \tfrac{1}{3}y^3 - \tfrac{1}{2}(Ry^2) = [(T_M - T_{su})/(L\rho_S/K_S)]Rt. \qquad (9.35)$$

Langford[55] has discussed constant velocity melting of spheres with results similar to those of Lyubov.

c. The Cylinder (Infinite, Right Circular)

(i) *Solution Supersaturated (or Liquid Supercooled).* We discuss here the growth of a cylinder from supersaturated solution; we introduce the notation of the diffusion of matter problem, which is similar mathematically to the heat flow problem. The time-dependent diffusion equation for matter diffusing in solution in cylindrical symmetry (coordinates r radius, t time) is (see e.g. Crank[64]), corresponding to Eq. (9.2),

$$D \, \nabla^2 C = (\partial C/\partial t)(r, t) = D[(\partial^2 C/\partial r^2) + r^{-1}(\partial C/\partial r)] \qquad (9.36)$$

where $C(r, t)$ is the concentration of solute and D (cm^2/s) is the diffusion coefficient. Analogous to Eq. (9.3) we again make the assumption of local equilibrium, i.e. we assume that the concentration at the surface of the cylinder is the equilibrium value C_0:

$$C(R, t) = C_0 \qquad (9.37)$$

whereas at $r \to \infty$ the concentration is the supersaturated value C_∞,

$$C(\infty, t) = C_\infty \qquad (9.37')$$

as for Eq. (9.6). The continuity equation for the incorporation of matter at the moving boundary $R(t)$ is, corresponding to Eq. (9.4),

$$D \, \partial C/\partial r \, |_{r=R} = (\mathbf{C} - C_0) \, dR/dt \qquad (9.38)$$

where $\mathbf{C}$ is the concentration (g/cm^3) of solute in the solid. We again suppose the radius of the cylinder to be vanishingly small at $t = 0$.

Following a procedure very similar to that of Eqs. (9.9)–(9.17) we find for the concentration in the liquid

$$C - C_\infty = \frac{(C_0 - C_\infty)\{\mathrm{Ei}[-(\lambda_4 r/R)^2]\}}{\mathrm{Ei}[-\lambda_4^2]} \qquad (9.39)$$

(Rieck,[59] Zener,[62] Frank,[54] Ivantsov[61]; also Carslaw and Jaeger[51]) where

$$-\mathrm{Ei}(-x) = \int_x^\infty u^{-1} e^{-u} \, du$$

[64] J. Crank, "The Mathematics of Diffusion." Oxford Univ. Press (Clarendon), London and New York, 1956.

and where
$$\lambda_4^2 \exp(\lambda_4^2) \, \text{Ei}(-\lambda_4^2) + S = 0 \tag{9.40}$$
where
$$S = (C_\infty - C_0)/(\mathbf{C} - C_0) \tag{9.41}$$
and where
$$R = 2\lambda_4 (Dt)^{1/2}. \tag{9.42}$$

As a numerical example we may consider the growth of sodium chlorate from aqueous solution.[65,66] For a value of $C_\infty - C_0 = 0.02$ g/cm³ and for $\mathbf{C} = 2.5$ g/cm³ and $C_0 \approx 0.50$ g/cm³, then $S \approx 0.01$. For $S \ll 1$, Eq. (9.40) may be approximated by $\lambda_4^2 (0.577 + \ln \lambda_4^2) = -S$, from which we get $\lambda_4 \approx 0.05$. For $D \approx 2 \times 10^{-5}$ cm²/s one gets, for the time required to grow a cylinder 1 cm in radius, about 5×10^6 s or about 2 months (assuming no interface kinetic effects, no stirring or convection, of course). We may compare this with the much faster growth from the melt (ice example above) and note that the low diffusivity of matter compared to the much higher diffusivity of heat is a major contributing factor to the slow growth from solution (but not, of course, the only factor since some interface control is necessary from the existence of flat faces in the growth of sodium chlorate).

(*ii*) *Cylinder Crystallizing Inwards.*[57,67] This case is similar to that for the sphere crystallizing inwards (see above) in that both the inverse and the direct Stefan problems are studied by a series method. For the inverse problem for which the boundary position $y(t)$ is assumed to be of the form $\beta_3(t_0 - t)^{1/2}$, the temperature within the cylinder is found to be of the form $\text{Ei}[r^2/a_3(t_0 - t)]$ for $t < t_0$, where $t_0 = R^3/\beta_3^2$ and β_3 and a_3 are constants; R is the outside radius.

d. *The Ellipsoid*

Supersaturated Solution; Shape Preservation. Prior to the papers by Ham,[68–70] it was generally believed (see e.g. Doremus *et al.*[71]) that a non-spherical particle—say an ellipsoid—growing in a diffusion field would not be shape-preserving i.e. would, as it got bigger, also become more eccentric in shape, the ratio of long to short dimensions increasing, because of the point effect of diffusion. Ham showed, however, that an ellipsoid could be a

---

[65] C. W. Bunn, *Discussions Faraday Soc.* **5**, 132 (1949).
[66] S. P. F. Humphreys-Owen, *Discussions Faraday Soc.* **5**, 144 (1949).
[67] D. E. Temkin, *J. Eng. Phys. (Russian)* **5** (4), 89 (1962).
[68] F. S. Ham, *J. Phys. Chem. Solids* **6**, 335 (1958).
[69] F. S. Ham, *J. Appl. Phys.* **30**, 1518 (1959).
[70] F. S. Ham, *Quart. Appl. Math.* **17**, 137 (1959).
[71] R. H. Doremus, B. W. Roberts and D. Turnbull, eds., "Growth and Perfection of Crystals." Wiley, New York, 1958.

shape-preserving solution to the time-dependent diffusion equation. [He did not test the stability of this shape to small shape perturbations; this is discussed in Part VI].

The general procedure[70] is to introduce reduced variables for the same (dimensional) reasons as in Eq. (9.9). Then the ellipsoidal coordinate system is introduced so that the diffusion equation (9.36) can be written in this system. Finally, a solution to the equation is found which is a function of that variable which if constant specifies an ellipsoid. This solution must satisfy the two boundary conditions on the ellipsoid (flux and concentration) and the one at infinity.

The reduced variables are

$$u = xt^{-1/2}, \quad v = yt^{-1/2}, \quad w = zt^{-1/2}. \quad (9.43)$$

The ellipsoidal coordinate system is ξ_1, ξ_2, ξ_3 where

$$u = [(\xi_1^2 - a^2)(\xi_2^2 - a^2)(\xi_3^2 - a^2)/a^2(a^2 - b^2)]^{1/2}$$
$$v = [(\xi_1^2 - b^2)(\xi_2^2 - b^2)(\xi_3^2 - b^2)/b^2(b^2 - a^2)]^{1/2} \quad (9.44)$$
$$w = \xi_1 \xi_2 \xi_3 / ab$$

where $\xi_1 > a > \xi_2 > b > \xi_3 > 0$; a and b are given constants. Following the procedure outlined by Morse and Feshbach,[72] Eq. (9.36) becomes

$$-\frac{1}{2}\sum_i^3 \left(\frac{\xi_i}{h_i^2}\frac{\partial C}{\partial \xi_i}\right) = \frac{D}{h_1 h_2 h_3}\sum_{i=1}^3 \frac{\partial}{\partial \xi_i}\left(\frac{h_1 h_2 h_3}{h_i^2}\frac{\partial C}{\partial \xi_i}\right) \quad (9.45)$$

where the h_i (arc length scale factors) are known functions of the ξ_i. The coordinate $\xi_1 = \text{const} = \xi_0$ specifies a particular ellipsoid having semimajor axes of length $(\xi_0^2 - a^2)^{1/2} t^{1/2}$, $(\xi_0^2 - b^2)^{1/2} t^{1/2}$, and $\xi_0 t^{1/2}$. A solution, which is a function of ξ_1 only, of (9.45) and which equals C_0 on the above surface and C_∞ at infinity is

$$C(\xi_1) = C_0 + (C_\infty - C_0)\{1 - [f(\xi_1)/f(\xi_0)]\} \quad (9.46)$$

where

$$f(\xi_1) = \int_{\xi_1}^\infty \{\exp[-t^2/4D]\}[(t^2 - a^2)(t^2 - b^2)]^{-1/2} dt.$$

The flux boundary condition is used to determine ξ_0, which is then given by

$$\tfrac{1}{2}(\mathbf{C} - C_0)\xi_0 f(\xi_0) = -D(C_\infty - C_0)(df(\xi)/d\xi)_{\xi=\xi_0} \quad (9.47)$$

where $\mathbf{C}$ is as before the density of diffusing matter in the precipitate phase. The ellipsoid dimensions are seen to be proportional to $t^{1/2}$, just as for the

[72] P. M. Morse and H. Feshbach, "Methods of Theoretical Physics." McGraw-Hill, New York, 1953.

sphere, cylinder, and plane, the coordinate systems of which are special cases of the ellipsoidal coordinates, as discussed by Ham.[70] It may also be noted that all ellipsoids having any aspect ratio satisfy the given problem (i.e. parameters a, b are not determined). This says that a variety of particles of any ellipsoidal shape could all grow, in shape-preserving ways, under the same concentration, within the limits of the diffusion-limited treatment. This point will be returned to in discussing dendritic growth. A special case of the above, ellipsoids of rotation (spheroids), has also been discussed by Horvay and Cahn[73] and by Ivantsov.[74]

e. Paraboloidal Dendrites

The term "dendrite" referring to crystals means tree-shaped or having some attributes of the shape of a tree—particularly the branching. Paraboloidal dendrites refer to the tips of the trunk or the tips of the branches, the shapes of which, to some degree of approximation, may be represented by a paraboloid.

(*i*) *Circular Paraboloid from Supercooled Melt.* Papapetrou[75] in a classical study of dendritic growth, suggested that the tip of a dendrite had the form of a paraboloid of rotation. Ivantsov[61,63] first obtained this solution to the moving boundary problem diffusion equation. He showed that, if the surface of the dendrite was isothermal, and was moving forward at constant velocity, then the shape of a paraboloid of revolution satisfied the differential equation and boundary conditions; he also obtained the temperature field in the liquid.

Ivantsov's method, which is fully discussed by Horvay and Cahn,[73] is to find solutions to the nonlinear differential equation representing the continuity of heat or matter at the moving boundary [Eq. (9.4)] and which also satisfy the diffusion equation (9.1) and the boundary conditions (9.3), (9.6), and (9.7). The general solution to (9.4) contains five arbitrary constants and one arbitrary function, which are reduced by the boundary conditions and by Eq. (9.1), and by prescribing relations among the constants. For dendrite travel along the positive z axis, the result for the temperature in the liquid is

$$\frac{T_L(g) - T_\infty}{T_M - T_\infty} = \frac{\text{Ei}[-v\rho_t g/2\kappa_l]}{\text{Ei}[-v\rho_t/2\kappa_l]} \quad (9.48)$$

where T_∞ is the temperature of the liquid at infinite z, ρ_t is the radius of curvature of the paraboloid at its tip, v is the constant growth velocity, κ_L

---

[73] G. Horvay and J. W. Cahn, *Acta Met.* **9**, 695 (1961).
[74] G. P. Ivantsov, *in* "Growth of Crystals" (A. V. Shubnikov and N. N. Sheftal, eds.), Vol. III, p. 53. Consultants Bureau, New York, 1962.
[75] A. Papapetrou, *Z. Krist.* **92**, 89 (1935).

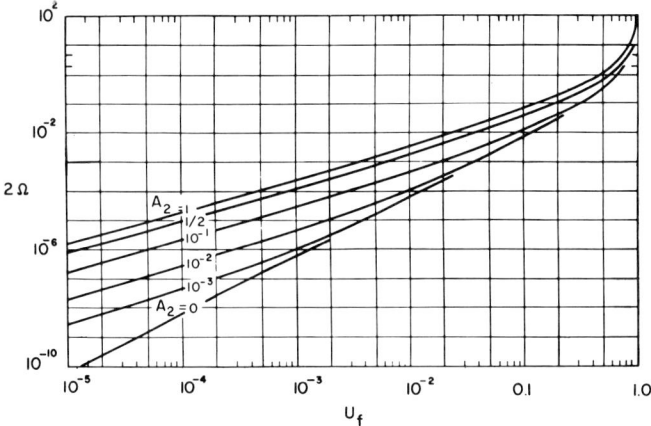

FIG. 8. Growth of elliptical paraboloids from supercooled melt showing reduced growth velocity 2Ω vs reduced undercooling U_f for several values of the ellipse aspect ratio A_2 [G. Horvay and J. W. Cahn, *Acta Met.* **9**, 695 (1961)].

the thermal diffusivity of the liquid. Here

$$g = \frac{z - vt}{\rho_t} + \left[\left(\frac{r}{\rho_t}\right)^2 + \left(\frac{z - vt}{\rho_t}\right)^2\right]^{1/2}.$$

Tables of Ei are given by Jahnke *et al.*[76]; for small x, $\text{Ei}(-x) \cong 0.577 + \ln x$. The quantity $v\rho_t/2\kappa_l$ is determined by

$$\frac{c_L(T_M - T_\infty)}{L} = \frac{-v\rho_t}{2\kappa_l} \exp\left(\frac{v\rho_t}{2\kappa_l}\right) \cdot \text{Ei}\left(\frac{-v\rho_t}{2\kappa_l}\right). \quad (9.49)$$

As in the case with ellipsoids (noted above) the analysis gives only the product of $v\rho_t$—not each separately. Thus, under the same thermal conditions, either a blunt needle could grow slowly, or a thin one rapidly; both are at the melting temperature, on their surface and in the solid interior. Ivantsov[77] also gave (9.49) in the form required for growth from solution.

(*ii*) *Elliptical Paraboloid—Supercooled Melt.* Horvay and Cahn,[73] using Ivantsov's[61] method, generalized the treatment to paraboloids having an elliptical cross section. Figure 8 taken from their paper gives the variation of dimensionless growth velocity $2\Omega = v\rho_t/\kappa_l$ with aspect ratio (ratio of axes) of the elliptical cross section, A_2 as a function of dimensionless undercooling $U_f = c_L(T_M - T_\infty)/L$. Ω is found to vary, at small undercoolings,

---

[76] E. Jahnke, F. Emde, and F. Lösch, "Tables of Higher Functions." McGraw-Hill, New York, 1960.

[77] G. P. Ivantsov, *Dokl. Akad. Nauk SSSR* **83**, 573 (1952) (see Ivantsov[63]).

approximately as U_f^2 for the parabolic cylinder ($A_2 = 0$) (see below) to about $U_f^{1.2}$ for circular cross section. Again, only the product $v\rho_t$ is determined.

f. Other Shapes

In this section we briefly summarize a number of Stefan problems on shapes other than those discussed above.

(*i*) Horvay and Cahn[73] and Ivantsov[74] both discussed the problem of a cylinder of paraboloidal cross section growing at a constant velocity parallel to the axis of the parabola from undercooled melt. It may be regarded as a two-dimensional form of the (three-dimensional) dendrite problem and is a special (degenerate) case of the ellipsoid. The interface is assumed to be at the melting point. Only the product $v\rho_t$ is found (product of velocity times radius of curvature of tip), as in the three-dimensional case:

$$(c_L/L)(T_M - T_\infty) = (\pi p')^{1/2} \exp(p') \operatorname{erfc}(p')^{1/2} \qquad (9.50)$$

where $p' = v\rho_t/2\kappa_L$; this is called the Peclet number, Ω.

(*ii*) *Polyhedra and Polygons.* Seeger[78] and Ivantsov[63] have studied the growth, from supersaturated solution and undercooled melt, respectively, of crystals having such shapes as regular polygons, cubes, and octahedra. Seeger, in particular, restricts his treatment to very small concentrations and constant growth rate, and deals with the non-time-dependent diffusion equation

$$\nabla^2 C = 0. \qquad (9.51)$$

Since the growth rate of a face is assumed constant over its surface, as well as constant in time, and hence the flux condition Eq. (9.4) or (9.38) is constant over the surface, the surface concentration must vary over the surface, in contrast to treatments for other shapes, above. For a square, the difference in concentration between a corner (where it is highest) and the center of a side (where it is least), expressed as a fraction of the integrated normal concentration gradient, amounts to 3.5%. A rough treatment for a cube gave about 2.5 to 3%, between the vertex and face center. Ivantsov[63] also restricted his treatment to very small (linear) growth rates, finding for the case of melt growth that the face centers are hottest and the corners coolest, in qualitative agreement with Seeger; however, he assumed the crystal surface temperature to be a function of time, which makes a quantitative comparison with Seeger's work difficult.

(*iii*) *Various Needle Shapes (Other Than Paraboloids).* Fisher[79] studied

---

[78] A. Seeger, *Phil. Mag.* **44**, 1 (1953).
[79] J. C. Fisher (unpublished); see Chalmers[80] or Horvay.[52]
[80] B. Chalmers, "Principles of Solidification." Wiley, New York, 1964.

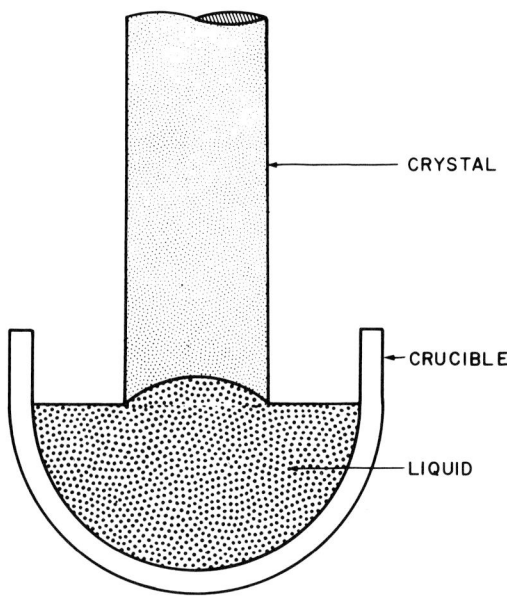

Fig. 9. Czochralski arrangement for crystal growth by pulling from the melt (schematic).

the lengthwise growth at speed v of a cylinder with a spherical cap of radius ρ_t, from the undercooled melt. Using an approximate form for the temperature distribution, he obtained (by setting the latent heat production equal to the approximate heat conduction) (see e.g. Horvay[52])

$$(c_L/L)(T_{su} - T_\infty) = v\rho_t/2\kappa_L \qquad (9.52)$$

where T_{su} is the (asumed unknown) interface temperature at the needle tip. Ivantsov[63] studied the growth at constant velocity from undercooled melt of pencil-shaped and of square-ended needles, finding the (assumed unknown) tip temperatures as a function of growth rate: in the latter case the result is quite similar to that of Fisher for a hemispherical cap.

Horvay,[52] in a very extensive discussion of the modified Stefan problem (shape, velocity specified; surface temperature assumed unknown), studies the cylinder with a spheroidal cap, including Fisher's needle as a special case, and more exactly, finding $0.94v\rho_t/2\kappa_L$ in place of 1.0 as coefficient of $v\rho_t/2\kappa_L$ in Eq. (9.52).

(iv) *Miscellaneous Other Shapes.* Wilcox[48] discusses what we will call the "Czochralski shape," i.e. the interface shape in a Czochralski melt growth apparatus (Fig. 9). Uniform velocity, melt above the melting point and interface at the melting point, are assumed. Numerical computation gives the result that the interface curves upward into the crystal in an approximately paraboloidal shape, whose curvature depends on such thermal parameters as the heat transfer coefficient between crystal and surroundings.

Horvay[52] discusses the two-dimensional slab with an elliptical cap, with extensive numerical calculations.

10. The Stefan Problem with Complications; Various Shapes

There are several types of complications which may be imposed on the simple or "bare" Stefan problem discussed above. Of course, the complications are introduced in order to make the model discussed conform more precisely to reality. Included among these are the effects of interface kinetics on the moving boundary temperature or concentration and shape; the effects of surface energy on boundary temperature or concentration and shape; whether the fluid (liquid, solution, or vapor) medium has hydrodynamic flows in it, affecting matter or heat transport through stirring or natural convection (density differences); the effects of impurities on interface temperature or concentration and shape; the combined roles of heat flow and of solute flow in growth from concentrated solutions; and the effects of applied force fields. Perhaps interface kinetics, impurities, and surface energy, together with their dependence on crystallographic orientation, are the most important of these complications.

As noted above, even the "bare" Stefan problem is a quite difficult boundary value problem in mathematical physics; the addition of just one complicating factor can render an exact treatment impossible to perform. Much more detailed descriptions of interface kinetics, impurity effects, and fluid flow effects are the subjects of subsequent parts of this article. Here we discuss these effects in the spirit of the Stefan problem, i.e. their effects on the boundary conditions or on the transport equation itself.

a. *The Plane, Sphere, and Cylinder*

(i) *Effects of Impurities.* As we have seen above, the diffusion of heat and the diffusion of matter obey identical differential equations (9.1) and (9.36); and similar boundary conditions (9.3), (9.4), (9.37), and (9.38) are appropriate. Until now we have discussed problems in which only one of the two effects is considered. When we dealt with planar freezing, e.g., we assumed that we had a pure material. When we studied the cylinder growing from supersaturated solution, we ignored any thermal effect that might be present. As Frank[54] has pointed out, however, growth may be controlled by both effects acting together. In particular, for growth from the melt, there may be enough impurity present to accumulate at the surface of the freezing solid and lower the melting point there. Or in growth from solution, the latent heat released at the boundary may cause enough temperature rise to change the equilibrium solubility there. That is, the values of temperature and concentration in the boundary conditions are altered, but not the form of the boundary conditions themselves. The transport equations themselves are not affected as long as the diffusivities D and κ are assumed to be independent of temperature and composition.

Frank[54] has studied the $t^{1/2}$ growth of a sphere from the undercooled melt in the presence of impurities which are assumed to lower the equilibrium melting point T_M at the interface by an amount $K_i C_s$ where C_s is concentration of impurity in the liquid at the interface and K_i is a constant. Thus it is required to solve simultaneously four equations; the growth constants λ_T, λ_C are related by their description of the same sphere:

$$\lambda_T \kappa_L^{1/2} = \lambda_C D^{1/2}; \tag{10.1}$$

the melting point depression, for the temperature at the surface T_{su}

$$T_{su} = T_M - K_i C_s; \tag{10.2}$$

and finally, Eq. (9.32) and its counterpart for solution growth, expressing the growth constants in terms of the temperature and concentration values at the surface of the sphere. This system of equations may be solved graphically by an instrument with calibrated scales as shown by Frank. A numerical example shows that with even a small amount of impurity in the bulk of the liquid, the buildup of impurity at the interface, owing to its low diffusivity, can be sufficient to substantially reduce the equilibrium freezing temperature at the interface and hence to reduce the effective value of the thermal undercooling, in turn substantially reducing the growth rate. This supposes little or no impurity to be trapped in the solid phase. It is implied in this treatment that the amount of impurity or "solvent" is small. If it is not, as in growth from solution, a similar treatment can be worked out; but in this problem the thermal correction is found to be generally small compared to the impurity correction in the freezing problem.

Ivantsov[63,77,81] studied problems similar to those of Frank, particularly the planar crystallization of a binary alloy from superheated melt.[81] He also made a similar treatment of a spherical alloy crystal from supercooled alloy melt.[77] Both, as expected, grow as $t^{1/2}$. In the case of the plane, which is the analog of Stefan's classical problem of pure freezing, Ivantsov concluded (without, it appears, quantitatively solving the coupled problem) as had Frank for the sphere, that owing to its low diffusivity, matter could build up in concentration at the interface, and so lower the equilibrium freezing

---

[81] G. P. Ivantsov, *Dokl. Akad. Nauk SSSR* **81**, 179 (1951).

temperature there. In an example he chose the preferential inclusion in the solid of the high-melting constituent (equivalent to Frank's buildup in the liquid by rejection of the low melting constituent.) Consequently a layer of supersaturated melt exists ahead of the crystallization front, even though the melt is above the melting temperature at a large distance from the front, and even though the interface itself is at equilibrium. This phenomenon was independently discovered by Rutter and Chalmers[82] and called by them constitutional supercooling, to distinguish it from pure thermal supercooling. Its consequences, for the stability of shape of interfaces, will be discussed below.

Lyubov and Temkin[83] have also studied the problem of planar alloy crystallization as formulated by Ivantsov[81] but neglect the thermal problem, discussing the distribution of solute only. It is assumed that the growth law is known, including the growth constant. The impurity distributions in the solid for $t^{1/2}$ and t^1 growth laws are calculated.

(ii) *Effects of Solid–Liquid Density Differences.* Fluid flow in the melt or in the supersaturated solution may arise from several sources, viz. thermal convection due to density differences in the fluid due in turn to temperature gradients in the fluid; composition differences leading to density differences; stirring or agitation of the fluid by external means; and differences in density between the fluid and the solid. The first two sources will be discussed in Part VIII. The third source has been treated explicitly in the spirit of the Stefan problem and we consider it here, discussing it only briefly in Part VIII.

It is clear that in the case of freezing, where in a typical case a given volume of crystal requires several per cent more than the same volume of liquid to form it, the liquid must flow toward the solid. In solving this Stefan problem, the temperature distribution may be affected by this mass flow.

Carslaw and Jaeger[51] and Horvay[84] have studied this effect for planar freezing; Horvay[84] has also considered this effect on the freezing of a cylinder and a sphere, from undercooled liquid. Equation (9.1) for conduction of heat in the liquid is now modified[84]:

$$\frac{\partial T_L}{\partial t} + u \frac{\partial T_L}{\partial r} = \kappa_L \left[\frac{\partial^2 T_L}{\partial r^2} + \frac{n}{r} \frac{\partial T_L}{\partial r} \right] \quad (10.3)$$

where $n = 0$ for a plane, 1 for a cylinder, and 2 for a sphere. Here u is the radial velocity of flow of the fluid. For the motion of the fluid, which is

---

[82] J. W. Rutter and B. Chalmers, *Can. J. Phys.* **31**, 15 (1953).
[83] B. Ya. Lyubov and D. E. Temkin, *in* "Growth of Crystals" (A. V. Shubnikov and N. N. Sheftal, eds.), Vol. III, p. 40. Consultants Bureau, New York, 1962.
[84] G. Horvay, *Proc. Natl. Cong. Appl. Mech. ASME, 4th, 1962* p. 1315.

assumed incompressible, two additional relations are used: the equation of continuity

$$r^{-n}(\partial/\partial r)(r^n \cdot u) = 0 \qquad (10.4)$$

and the equation of fluid motion (Navier–Stokes equation)

$$\frac{\partial u}{\partial t} + u\frac{\partial u}{\partial r} = \nu\left[\frac{\partial^2 u}{\partial r^2} + n\left(\frac{\partial u/\partial r}{r} - \frac{u}{r^2}\right)\right] - \frac{\partial p/\partial r}{\rho_L} \qquad (10.5)$$

where the fluid has pressure p, density ρ_L, and kinematic viscosity ν.

Horvay found that the bracket in (10.5) is zero and hence the flow is independent of viscosity ν; it is also irrotational. The condition (9.4) must be supplemented by the mass balance at the freezing interface R,

$$\rho_S \, dR/dt = \rho_L[dR/dt - u(R)] \qquad (10.6)$$

where ρ_S is the density of the solid; we define ϵ also as $\rho_S = \rho_L(1 + \epsilon)$. For all three shapes the growth law $R = 2\beta(\kappa_L t)^{1/2}$ is found to hold. For the case of plane front freezing, the growth constant β is given by $\beta' = (1 + \epsilon)\beta$ where $c_L(T_M - T_\infty) = L\pi^{1/2}\beta' e^{\beta'^2} \text{erfc}(\beta')$ which is of exactly the same form as Eq. (9.22). Horvay shows that β is also relatively insensitive to small ϵ for the case of sphere and cylinder.

However, the result shows that the pressure of the fluid is extremely high for plane front freezing, because the initial ($t = 0$) velocity of the liquid is infinite, as is the velocity of the interface as $t \to 0$. From this infinite velocity, the flow then decelerates, giving a high (positive) pressure—indicating that the assumption of incompressibility is not valid. Rather different results obtain for the pressure around the cylinder and sphere—negative pressures can exist which can presumably cause cavitation or nucleation. Glicksman[85] has also considered these dynamic effects, particularly for high freezing velocities. (See also Part VIII.)

(iii) Other Effects. The effects of interface kinetics in the Stefan problem have been considered by Lyubov[86] who has discussed the role of the undercooling, below the equilibrium melting point, of the solid-liquid interface of a pure substance in relation to heat flow and the one-dimensional problem. Chernov and Lyubov[87] have also studied the sphere taking into account interface kinetics, as have Glicksman and Schaefer.[88]

Cosgrove[89] has studied the effect of an electric field on the diffusion-

---

[85] M. E. Glicksman, *Acta Met.* **13**, 1231 (1965).
[86] B. Ya. Lyubov *in* "Growth of Crystals" (N. N. Sheftal, ed.), Vol. 5a, p. 80. Consultants Bureau, New York, 1968.
[87] A. A. Chernov and B. Ya. Lyubov *in* "Growth of Crystals" (N. N. Sheftal, ed.), Vol. 5a, p. 7. Consultants Bureau, New York, 1968.
[88] R. J. Schaefer and M. E. Glicksman, *J. Cryst. Growth* **5**, 44 (1969).
[89] G. J. Cosgrove, *Acta Met.* **13**, 1209 (1965).

controlled growth of a sphere growing from charged species in solution. The net result of the field is effectively to translate the sphere while it continues its radial growth, by causing extra deposition on one side and less than normal deposition on the opposite side of the sphere.

b. Dendrite Shapes—Paraboloidal and Others

(*i*) *Effects of Surface Energy and of Interface Kinetics.* As noted above, in the work of Ivantsov[61] and of Horvay and Cahn[73] the problem of the paraboloidal dendrite has a certain indeterminacy to it, in that only the product of dendrite velocity v and radius of curvature of the tip ρ_t is given, not v and ρ_t separately. Hence, very shary needles could grow very fast, or blunt ones slowly, from the same undercooled liquid. Partly in order to introduce a measure of length, or of velocity, as well as to make the problem treated more representative of the real situation, the effects of surface energy and of interface kinetics have been incorporated, to some degree of approximation.

Temkin[90] approached the problem by starting with Ivantsov's[61] "unadorned" treatment of the paraboloid of revolution moving at constant velocity v [Eq. (9.49)]. In that treatment, the dendrite surface was assumed to be isothermal and at the equilibrium melting point. If surface energy is now taken into account, its effect on the interface temperature as predicted by the Gibbs–Thomson equation, will be that those portions of the surface having a sharp curvature will have a lower equilibrium melting temperature. In practice, only for radii of curvature of the order of 10^{-2} mm or less is this effect appreciable, but dendrite tips in practice may well have such radii. Hence,

$$T_M(\rho) = T_M(\infty) - [2\gamma_{SL}T_M(\infty)/L\rho_S \cdot \rho] \qquad (10.7)$$

where $T_M(\infty)$ is the equilibrium melting temperature of a plane surface, $T_M(\rho)$ the equilibrium melting temperature of a surface having radius of curvature ρ, γ_{SL} is the crystal–liquid interfacial tension (interfacial free energy), and ρ_S the density of the solid. Temkin also introduced linear interface kinetics into the problem. "Interface kinetics" (see Part V) refers to the molecular process controlling how much the interface temperature must depart from the equilibrium temperature to permit a given net rate of deposition of molecules from the liquid onto the solid. The effect of linear kinetics can be expressed as

$$v_N = K(-T_{su} + T_M(\rho)) \qquad (10.8)$$

[90] D. E. Temkin, *Dokl. Akad. Nauk SSSR* **132,** 1307 (1960) [English transl.: *Soviet Phys.* **5,** 609 (1960)].

where v_N is the growth rate normal to the element of surface in question, K is a constant called the linear kinetic coefficient, and T_{su} is the actual interface temperature at that point. For a paraboloid of revolution, the surface $z(x, y)$ is given by $z = (\rho_t/2) - (x^2 + y^2)/2\rho_t$ where ρ_t is the radius of curvature of the tip which has forward velocity v. Hence the new interface temperature boundary condition of Temkin is

$$T_{su}(z) = T_M(\infty) - \frac{\gamma_{SL} T_M(\infty)}{\sqrt{2}} \frac{[\frac{3}{2} - z/\rho_t]}{(\rho_S L \rho_t)[1 - z/\rho_t]^{3/2}} - \frac{v}{K}[2(1 - z/\rho_t)]^{-1/2}. \tag{10.9}$$

In principle, one would now solve this Stefan problem for shape and temperature field, inquiring if there were a shape-preserving shape and if so, what it is. If the paraboloid is, as appears to be so, the unique result of the isothermal boundary condition, the shape from (10.9) could not be a paraboloid. For mathematical simplicity, and as a first approximation, Temkin assumes that the shape is still a paraboloid of revolution. Temkin also assumes that Laplace's equation is sufficient for the problem for small bath undercoolings $T_M(\infty) - T_\infty$. The result is, where κ_L is the thermal diffusivity of the liquid,

$$-\frac{v\rho_t}{2\kappa_L} \exp\left(\frac{v\rho_t}{2\kappa_L}\right) \text{Ei}\left(\frac{-v\rho_t}{2\kappa_L}\right) = \frac{(T_M(\infty) - T_\infty)(c_L/L)(\rho_L/\rho_S)}{[1 + L_1/K\rho_t + L_2\gamma_{SL}/v\rho_t^2]} \tag{10.10}$$

where L_1 and L_2 are constants; Eq. (10.10) differs from Eq. (9.49) primarily in that the denominator on the right-hand side is no longer unity, giving, compared to Eq. (9.49), a reduced growth velocity v for the same ρ_t and bath undercooling $(T_M(\infty) - T_\infty)$. This may be viewed as an effective reduction of the interface temperature, and hence a reduction of the driving force for growth, partly by curvature which lowers it as in Eq. (10.7); and partly by interface kinetics which "takes up" part of the supercooling for driving the interface kinetic reaction, leaving less for driving the heat flow. Clearly Eq. (10.10) reduces to (9.49) if interfacial energy is neglected by setting $\gamma_{SL} = 0$ and if kinetics are assumed fast, i.e. $K \to \infty$ (no interface supercooling for finite growth rate). For $v\rho_t/2\kappa_L \leq 0.003$, the left-hand side of (10.10) can be approximated by $\text{const}(v\rho_t/2\kappa_L)^{0.83}$ and it is found that setting $dv/d\rho_t = 0$ gives a maximum growth velocity $v_{\max}$. Temkin[90] and Bolling and Tiller[91] in a similar, independent treatment assume $v_{\max}$ is the one found in nature, but there appears to be no clear justification for this assumption.

(ii) *Effects of Impurity on Dendrite Growth.* Temkin[92] and Bolling and

[91] G. F. Bolling and W. A. Tiller, *J. Appl. Phys.* **32**, 2587 (1961).
[92] D. E. Temkin, *Soviet Phys.—Cryst. (English Transl.)* **7**, 354 (1962).

Tiller[91] have studied the growth of paraboloid and paraboloid-like shapes from an impure melt. Before considering their studies, it is worthwhile to note that both Ivantsov's[61] solution and Horvay and Cahn's[73] of the unadorned paraboloid growing from the melt, as well as Temkin's solution of the paraboloid taking account of interface kinetics and curvature, although given in terms of temperatures and melt growth, can equally well be transcribed into dilute solution growth, as was pointed out above in discussing growth of a cylinder. The only difference will be that, for equal driving forces [reduced undercooling $(c_L/L)(T_M(\infty) - T_\infty)$ or reduced supersaturation $S = (C_\infty - C_0)/(\mathbf{C} - C_0)$], the growth velocity will, for equal tip radii, be reduced by the same ratio for growth from dilute solution, since as noted above the matter diffusivity is typically several hundred times less than the thermal diffusivity. Furthermore it is expected that, as Frank pointed out for growth of spheres, the thermal effects on dilute solution growth will be small and can generally be neglected. However, one can expect, as for spheres, that the impurity effects on melt growth will not be small and cannot be ignored since the impurity buildup effect will lower the equilibrium freezing point and hence reduce the effective undercooling available for driving heat flow.

Temkin[92] studied the problem of simultaneous thermal and diffusion processes for the growth of a paraboloid of revolution from a binary undercooled alloy. He solved simultaneously the growth from solution form of Eq. (9.49) (Ivantsov[77]) for the unadorned paraboloid, together with a form of Eq. (10.10) for the thermal problem (Temkin[90]); this, however, included both linear kinetics and surface energy, i.e. the adorned paraboloid. Also assumed were values for the freezing point depression for dilute alloys and values for the segregation coefficient. He also assumed that the "maximum velocity" dendrite of the thermal problem was the one observed. Temkin found that, for given bath undercooling, the effect of even dilute impurity could reduce tip growth velocities by several orders of magnitude, for dilute alloys of Pb in Sn. It would appear that this treatment, of a hybrid nature, and with the unverified maximum velocity postulate, can be regarded as only a first approximation to the solution of the problem.

Bolling and Tiller[91] in their study of the growth of paraboloidal shapes also assumed the maximum velocity criterion. They made an extensive comparison of the theory with data on KCl solutions and on acetic acid solutions.

(iii) Effects of Density Difference on Dendrite Growth. Horvay[84] has studied the effects of solid–liquid density differences on the growth of the paraboloidal dendrite and has shown these to be small. The reason is that in the equal-density problem[73] the dendrite is regarded as moving with a velocity v through the fluid. Density differences simply cause the fluid to

flow toward the dendrite with the rate ϵv and hence in the original problem v is replaced by $(1 + \epsilon)v$ [i.e. in Eq. (3.49)].

c. *Various Other Shapes*

In this section we discuss Stefan problems for some shapes more complicated than those discussed above.

(*i*) *Whisker Shape.* The whisker shape resembles a face whisker in that it is long, thin, and often straight. Whisker crystals can grow by many different mechanisms.[93] One model of growth, believed applicable to growth of whiskers from the pure vapor phase without the presence of an inert or other diffusion-limiting gas, is that formulated by Sears[94] and treated, in various approximations, by Dittmar and Neumann,[95] Blakely and Jackson,[96] Gomer,[97,98] Ruth and Hirth,[99] and Simmons *et al.*[100] The first exact treatment was given by Simmons *et al.*[101] In the Sears model, the whisker sits on a substrate. Vapor atoms strike the whisker sides, adhere there, diffuse as "adatoms" by surface diffusion to the tip where they come to rest at a trap for adatoms, such as the emergent step of a screw dislocation. It is also assumed that atoms do not diffuse to or from the substrate; they do not come to rest on the whisker sides; and the tip area is small compared to that of the sides so that direct impingement onto the tip is insignificant in causing growth. The problem is to calculate the whisker length as a function of time, $l(t)$.

Although the problem is essentially one-dimensional, the diffusion equation, being that for surface diffusion on the sides of the whisker, is not the usual one-dimensional diffusion equation. It is

$$\partial n_S/\partial t = D_S(\partial^2 n_S/\partial x^2) - (n_S/\tau_S) + J' \qquad (10.11)$$

where $n_S = n_S(x, t)$ is the adatom concentration, x is the distance from the substrate, τ_S the adatom lifetime, D_S the adatom surface diffusion coefficient, and J' the rate of deposition from the vapor. The n_S/τ_S term represents the rate of reevaporation. The boundary conditions are: flux condition at tip, equilibrium concentration at tip, and no flux at base. The initial

---

[93] F. R. N. Nabarro and P. J. Jackson, *in* "Growth and Perfection of Crystals" (R. H. Doremus, B. W. Roberts, and D. Turnbull, eds.), p. 13. Wiley, New York, 1958.
[94] G. W. Sears, *Acta Met.* **3**, 361 (1955).
[95] W. Dittmar and K. Neumann, *Z. Elektrochem.* **64**, 297 (1960).
[96] J. M. Blakely and K. A. Jackson, *J. Chem. Phys.* **37**, 428 (1962).
[97] R. Gomer, *J. Chem. Phys.* **28**, 457 (1958).
[98] R. Gomer, *J. Chem. Phys.* **38**, 273 (1963).
[99] V. Ruth and J. P. Hirth, *J. Chem. Phys.* **41**, 3139 (1964).
[100] J. A. Simmons, R. L. Parker, and R. E. Howard, *J. Appl. Phys.* **35**, 2271 (1964).
[101] J. A. Simmons, H. Oser, and S. R. Coriell, *in* "Crystal Growth" (H. S. Peiser, ed.), Suppl. *J. Phys. Chem. Solids*, p. 255. Pergamon Press, Oxford, 1967.

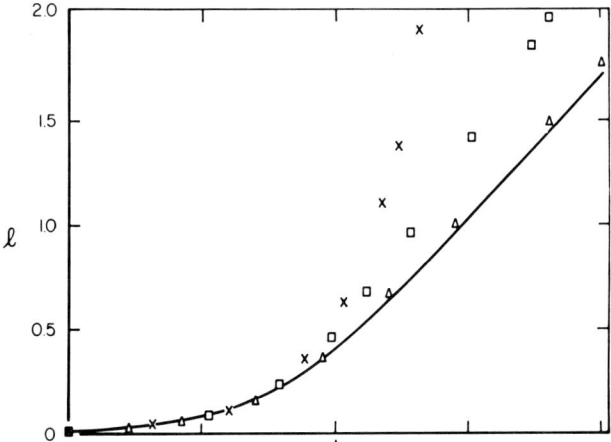

FIG. 10. Growth of whiskers from the vapor phase, showing exact numerical result and earlier approximations. Reduced length l vs reduced time t, for particular values of vapor saturation and initial length. ×, exponential approximation; □, Dittmar-Neumann approximation; △, Curtis approximation; —, exact numerical solution [J. A. Simmons, H. Oser, and S. R. Coriell, in "Crystal Growth" (H. S. Peiser, ed.), Suppl. *J. Phys. Chem. Solids*, p. 255. Pergamon Press, Oxford, 1967].

condition was a constant adatom surface density at $t = 0$, at which time the vapor is "turned on." The solution to (10.11) and the boundary and initial conditions was obtained numerically, using an iterative numerical method,[101] which gave the entire growth curve and the associated concentration profiles. Figure 10 compares the exact result (solid line) with several earlier approximations.

(ii) *Tammann Tube Shape.* The interface shape and velocity in the Tammann tube have been studied by Lyubov,[86] by Berlad,[102-104] and by many others.[105-107] The Tammann tube has been used primarily in attempts to obtain interface kinetics data—as will be seen below. From the point of view of a Stefan problem it is a complex system. The Tammann tube con-

[102] A. L. Berlad, Gen. Dynamics/Astronaut. Rept. (Jan. 1963).
[103] A. L. Berlad in *Proc. Heat Trans. Fluid Mech. Inst. 1964* p. 100. Stanford Univ. Press, Stanford, California, 1964.
[104] A. L. Berlad, *J. Chem. Phys.* **40**, 282 (1964).
[105] W. B. Hillig, in "Growth and Perfection of Crystals" (R. H. Doremus, B. W. Roberts, and D. Turnbull, eds.), p. 350. Wiley, New York, 1958.
[106] D. Kirtisinghe and R. F. Strickland-Constable, *Nature* **204**, 465 (1964) [See also: P. J. Morris, D. Kirtisinghe, and R. F. Strickland-Constable, *J. Cryst. Growth* **2**, 97 (1968)].
[107] G. Tammann, "The States of Aggregation" (translated by R. F. Mehl). Van Nostrand, Princeton, New Jersey, 1925.

sists of a long tube, often of glass, closed at one or both ends and containing the liquid to be supercooled. The tube is immersed in a constant-temperature bath of some fluid. Crystallization is started, by one of several methods, at one end of the supercooled liquid in the tube and proceeds down (or up) the tube. The velocities of crystallization for various bath temperatures are obtained by the measurement, often directly by optical means, of the solid–liquid interface position as a function of the time. Whether the data thus obtained are fundamental or are disturbed in an unknown way by the presence of the tube will be discussed below.

The classical one-dimensional Stefan problem, i.e. Eqs. (9.19) and (9.22), of growth from a supercooled liquid may be applied to this system if the heat flow is truly one-dimensional along the tube axis, and if complications such as interface kinetics, surface energy, and impurity; catalytic and similar effects of the tube walls are not significant. Under these assumptions, the interface shape is planar, heat flow is entirely into the supercooled liquid, and from Eq. (9.19) the velocity dX/dt is proportional to $t^{-1/2}$ and hence there is no unique or steady state velocity of crystallization in the Tammann tube. In order to achieve planar, one-dimensional freezing at a constant rate, as noted for the case of liquid initially at the melting point in connection with Eq. (9.23), it is necessary to extract more heat than can be done with a fixed temperature at $x = 0$. This was accomplished there by continually lowering the boundary temperature as an exponential in time. This procedure does not correspond to the boundary conditions in the Tammann tube experiment. Hence it may be concluded, if constant velocities are observed, that either the heat flow is not one-dimensional or some of the complications above are significant, or both. Hence, it would seem desirable to solve quantitatively the three-dimensional (or two-dimensional if cylindrical symmetry is assumed) Stefan problem in the Tammann tube. Without such a study it seems unlikely that interface shapes or velocities can be satisfactorily predicted. It is possible however, to derive a lower limit for the kinetic coefficient from such measurements, based on the fact that the interface undercooling $\delta T \leq \Delta T$, the bulk liquid undercooling. Hillig[105] has made some progress by studying a flat interface moving at constant velocity and calculating temperature gradients for various tube wall thicknesses. Michaels et al.[108] have made similar calculations. Lyubov[86] has examined the one-dimensional problem taking into account interface kinetics in the interface temperature boundary condition. It is possible, of course, that the interface kinetics can be so sluggish that the entire bath undercooling is used to drive the interface kinetics, and none for heat flow; i.e. that the amount of heat per unit time produced by crystallization is so small that negligible thermal gradients are required to carry it away. But

[108] A. S. Michaels, P. L. T. Brian, and P. R. Sperry, *J. Appl. Phys.* **37**, 4649 (1966).

this situation is no longer a Stefan problem—it will be discussed under interface kinetics.

11. Some Approximate Solutions to the Stefan Problem

We give only a very brief discussion of approximate methods for solving Stefan problems, for our main interest has been in exact solutions or attempts to get exact solutions. Bankoff[50] has given a quite extensive review of a number of analytic approximation methods. Here we consider the Laplace's equation approximations:

a. Planar Freezing; Liquid at Melting Point

In the case of planar freezing where the heat is withdrawn through the solid whose thickness is $X(t) = 2\lambda_1(\kappa_S t)^{1/2}$, and when the liquid is at the melting point, Eq. (9.16) becomes

$$\lambda_1 \exp(\lambda_1^2)\, \text{erf}\, \lambda_1 = (c_S/L\pi^{1/2})\, T_M \tag{11.1}$$

which is exact; we recall that the melting point T_M is here defined on a scale where the solid surface at $x = 0$ is at $0°$. Should the "reduced temperature" $c_S T_M/L$, which is the driving force for freezing, be small compared to one, then Eq. (11.1) becomes, approximately (Carslaw and Jaeger[51])

$$\lambda_1^2 \cong c_S T_M/2L. \tag{11.2}$$

(For example, $c_S T_M/L = 0.012$ for ice when the surface at $x = 0$ is held at $-2°C$.) Now the approximation (11.2) may also be obtained by assuming that Laplace's equation

$$\nabla^2 T_S = 0 \tag{11.3}$$

holds in the solid, rather than the exact equation (9.2). The solution of (11.3) fitting the temperatures $T_S = T_M$ at $x = X$ and $T_S = 0$ at $x = 0$ is

$$T_S = xT_M/X. \tag{11.4}$$

If now we apply (9.4) to (11.4) relating the boundary motion to the heat flow, ignoring the fact that the solution (11.4) to (11.3) does not permit time dependence, we get, on integrating,

$$X^2 = 4\lambda_1^2 \kappa_S t, \tag{11.5}$$

with λ_1^2 given by (11.2). Had the liquid been superheated or undercooled, and thus contained a thermal gradient in it, as discussed earlier, Laplace's equation for the liquid $\nabla^2 T_L = 0$ could not satisfy the boundary condition $T_L = $ finite at $x = \infty$ since the only planar solution permitting heat flow in the liquid at the interface would be a linear function of x. In the particular case for $T_L = T_M$, Laplace's equation is satisfied for the liquid.

b. Growth of a Sphere from Supersaturated Solution

In the case of a sphere it proves to be possible to use Laplace's equation, for small driving forces, and to have the boundary condition at ∞ satisfied, unlike the case for planar growth from supersaturated solution (or supercooled liquid). For a sphere, the solution–growth counterparts of (9.31) and (9.32) are[109]

$$C - C_\infty = \frac{\{C_0 - C_\infty\}\{(R/r)\exp[-(\lambda_5 r/R)^2] - \lambda_5 \pi^{1/2} \mathrm{erfc}[\lambda_5 r/R]\}}{\exp[-\lambda_5^2] - \lambda_5 \pi^{1/2}\mathrm{erfc}(\lambda_5)}$$

(11.6)

and

$$2\lambda_5^2\{1 - \pi^{1/2}\lambda_5 \exp[\lambda_5^2]\mathrm{erfc}[\lambda_5]\} = S \tag{11.7}$$

where $S = (C_\infty - C_0)/(\mathbf{C} - C_0)$ is the "reduced supersaturation," and where C_∞, C_0, and $\mathbf{C}$ are the concentrations of solute at infinity, sphere surface (assumed equilibrium concentration), and in the sphere, respectively. If $S \ll 1$, then $\lambda_5^2 \ll 1$ and (11.7) becomes

$$\lambda_5^2 \cong S/2. \tag{11.8}$$

The approximation (11.8) may also be derived by assuming that Laplace's equation (11.9) holds in the solution, rather than the exact diffusion equation:

$$\nabla^2 C = \left(\frac{\partial^2}{\partial r^2} + \frac{2}{r}\frac{\partial}{\partial r}\right) C = 0. \tag{11.9}$$

The solution of (11.9) fitting the boundary conditions at $r = R$ and at $r = \infty$ is

$$C - C_\infty = (C_0 - C_\infty)(R/r). \tag{11.10}$$

Applying the flux equation

$$D\, \partial C/\partial r \,|_{r=R} = (\mathbf{C} - C_0)\, dR/dt \tag{11.11}$$

as before, one gets $R^2 = 4\lambda_5^2 Dt$ with λ_5^2 given by the approximate equation (11.8). One can also easily show that (11.6) reduces to (11.10) if $\lambda_5 \ll 1$ and if $(\lambda_5 r/R) \ll 1$. The latter condition does not hold for $r \gg R$, in which case (11.10) is not a good approximation for (11.6).

c. Growth of a Cylinder from Supersaturated Solution

Laplace's equation can be used here[109] but only if the boundary condition at infinity $C(\infty, t) = C_\infty$ is not satisfied, since the solution to Laplace's

---

[109] S. R. Coriell and R. L. Parker, *J. Appl. Phys.* **36**, 632 (1965).

equation is of the form $\ln(r)$ which is infinite at infinity. Instead the boundary condition at infinity must be replaced with one of the form $C_\infty = C(R_\lambda)$ where R_λ is $R/\nu\lambda$, λ is the growth constant, ν a numerical constant, and R the cylinder radius. The same approximations as for the sphere are then made.

We thus see from the above examples that the exact solution to the Stefan problem may reduce to that given by Laplace's equation if the dimensionless driving force is small compared to unity. However, this Laplace's equation solution may have to be derived from a boundary condition at large distances which is not the same as the original given boundary condition, and which may not be known beforehand. Near the surface of the growing solid the two solutions agree well enough to give the same growth rate.

12. Some Questions Regarding Stefan Problems

Horvay[52] has posed a number of problems on the subject. These may be stated (using the heat flow language) as:

(a) State all shape-preserving isothermal surfaces growing as $t^{1/2}$.

(b) State all shape-preserving isothermal surfaces growing as t^1.

(c) State any shape-preserving isothermal surfaces growing with laws other than $t^{1/2}$ or t^1.

(d) Discuss the growth of dumbell shapes and other shapes (e.g. a torus), with regard to their time dependence and their stability against splitting or developing into a topologically new shape (as contrasted to stability against slight perturbations, which is very important and is discussed in Part VI).

Some further problems include

(e) Shape spectrum problems (the study of smoothly changing but not shape-preserving shapes), e.g., let the ratio of major to minor axes of an ellipsoid of revolution be a given, simple function of time or of size, and discuss the growth.

(f) The dendrite problem: the v–ρ_t indeterminacy is still outstanding. And why there are two temperatures at infinity (in liquid and in solid).

(g) Effects of kinetics, surface energy, and anisotropy on the above.

IV. Nucleation

13. Introduction

As we have noted earlier, crystallization (i.e. freezing, condensation of vapor) does not occur uniformly throughout the body of the metastable medium, in general. Instead, here and there in the matrix or mother phase, small regions of the new phase will originate and then grow.[9] Nucleation is the act of the origination of the very small particles of new phase from the mother medium. Growth is the process whereby these particles reach macroscopic size. Since our primary interest is the problem of growth, we give only a rather sketchy survey of classical nucleation, referring primarily to recent reviews, and of some particularly interesting modern aspects of nucleation.

As Cahn and Hilliard[110] have pointed out, Gibbs[111] distinguished two ways in which phase changes may originate; the assumed metastable phase tends to resist such changes. The first is a change infinitesimal in degree but large in extent; the second, called nucleation, is a change large in degree but small in extent, such as the formation of a liquid drop from supersaturated vapor. The first had been little studied until Cahn and Hilliard (see below) whereas nucleation has been extensively studied. Recent reviews of nucleation include those by Turnbull,[9] Hirth and Pound,[112] Christian,[19] Uhlmann and Chalmers,[113] Feder et al.,[114] Nielsen,[115] Walton,[116] and Pashley.[117]

14. CLASSICAL NUCLEATION

a. *Homogeneous Nucleation of Liquid Drops in Vapor; Thermodynamics*

Although our main interest is in crystal growth, we discuss first the nucleation of liquid drops from supersaturated vapor since this problem is in some ways the simplest, and illustrates the primary features well. We follow Christians'[19] treatment. It is assumed that thermodynamics may be applied to the (very) small droplets and that their surfaces are sharp. We first consider only homogeneous nucleation, i.e. no foreign particles or surfaces present to catalyze (heterogeneous nucleation) the nucleation event. (It should be noted that the nomenclature is somewhat confusing, in that homogeneous nucleation is a phase transformation occurring *heterogeneously*, e.g. at points here and there in the mother medium: nucleation catalysis seems to be a better word than heterogeneous nucleation.) We begin by

---

[110] J. W. Cahn and J. E. Hilliard, *J. Chem. Phys.* **31**, 688 (1959).
[111] J. W. Gibbs, "Collected Works," Vol. I. Yale Univ. Press, New Haven, Connecticut, 1948.
[112] J. P. Hirth and G. M. Pound, *Progr. Mater. Sci.* **11**, 1 (1963).
[113] D. R. Uhlmann and B. Chalmers, *Ind. Eng. Chem.* **57** (9), 19 (1965).
[114] J. Feder, K. C. Russell, J. Lothe, and G. M. Pound, *Advan. Phys.* **15**, 111 (1966).
[115] A. E. Nielsen, "Kinetics of Precipitation." Pergamon Press, Oxford, 1964.
[116] A. G. Walton, "The Formation and Properties of Precipitates." Wiley (Interscience), New York, 1967.
[117] D. W. Pashley, *Advan. Phys.* **14**, 327 (1965).

giving the thermodynamics of such droplets; the kinetics of their formation by "heterophase" fluctuations comes next.

The change in Gibbs free energy of a system, ΔG, when a small liquid droplet of radius r is formed from vapor of pressure p and equilibrium pressure p_0 (at temperature T) is composed of two parts:

$$\Delta G = \Delta G_{\text{bulk}} + \Delta G_{\text{surface}}. \tag{14.1}$$

ΔG_{bulk} is the change in free energy for condensation which for an ideal vapor is $-kT \ln(p/p_0)$ per atom (derived from the free energy change in isothermally compressing an ideal gas) or $(-\tfrac{4}{3}\pi r^3/v^l)kT \ln(p/p_0)$ to form a spherical droplet of radius r; v^l is the atomic volume of the liquid (assumed constant). $\Delta G_{\text{surface}}$ is just $4\pi r^2 \gamma_{LV}$ where γ_{LV} is the liquid–vapor interfacial free energy. Hence

$$\Delta G = 4\pi r^2 \gamma_{LV} - (\tfrac{4}{3}\pi r^3/v^l)kT \ln(p/p_0). \tag{14.2}$$

For sufficiently small r the first term dominates so that ΔG is positive; for sufficiently large r the second term dominates and then ΔG is negative. ΔG is a maximum at ΔG_c which then constitutes an activation barrier to nucleation; now

$$r_c = 2v^l \gamma_{LV}/kT \ln(p/p_0), \tag{14.3}$$

so that

$$\Delta G_c = 16(v^l)^2 \gamma_{LV}{}^3/3[kT \ln(p/p_0)]^2. \tag{14.4}$$

A droplet or embryo of size $r < r_c$ will tend to evaporate and those of $r > r_c$ will tend to grow, because the free energy ΔG will be reduced in either case. Therefore r_c is a "critical" radius for nucleation. If $p/p_0 \leq 1$, Eq. (14.2) always increases with r, and there is no critical size droplet—all are unstable to evaporation.

b. Fluctuations

It is assumed that fluctuations ("heterophase fluctuations") produce dimers, trimers, and higher species from the molecules in the vapor (whether supersaturated or not),[19,118] although these species will have short lifetimes and increasingly dilute concentration in the larger sizes. These fluctuations are analogous to the more familiar ones of density fluctuation in a gas or of concentration in a solution. Volmer assumed, as an approximation, that the embryo distribution was not significantly influenced by any subsequent growth of clusters of size r_c or greater. The primary process is assumed to be a long series of bimolecular reactions, i.e. one molecule adding to another to form a dimer, then another molecule adding to the dimer, etc., plus the reverse reactions; two n-mers ($n > 1$) are not assumed to interact.

[118] M. Volmer and A. Weber, Z. Physik. Chem. **119**, 277 (1926).

To calculate the distribution of embryos (which may also be assumed to be a Boltzmann distribution), one may treat embryos of n atoms, of which there are N_n, as present in "dilute solution" in the monomolecular vapor. Then the entropy of mixing S_m is that of N^V vapor molecules in N' total particles and of $\sum_n N_n$ embryos in N' total particles where $N' = N^V + \sum_n N_n$:

$$S_m = -k[N^V \ln(N^V/N') + \sum_n \{N_n \ln(N_n/N')\}]. \tag{14.5}$$

The change δG in the free energy of the system when the embryo with $n - 1$ molecules is converted to one with n molecules by the addition of one vapor molecule includes terms involving δN^V, $\delta N'$, and δN_n in (14.5) as well as terms in the volume and interfacial free energies. If, as Volmer assumed, the distribution of embryos is that of the equilibrium state then $\delta G = 0$ and the result is obtained that

$$N_n = N'(N^V/N')^n \exp(-\Delta G_n/kT) \tag{14.6}$$

where ΔG_n is given by the appropriate form of Eq. (14.2), namely $\Delta G_n = n^{2/3}\gamma_{LV}\eta - nkT \ln(p/p_0)$ where the shape factor $\eta = (36\pi)^{1/3}v^{l2/3}$ for a sphere. Since by assumption the total number of embryos is very small, (14.6) becomes, approximately,

$$N_n \cong N \exp(-\Delta G_n/kT) \tag{14.7}$$

where N is the total number of molecules. As mentioned, Volmer assumed this distribution of embryos to hold only for all $n < n_c$, and for $n > n_c$ to drop to zero, where n_c is the number of molecules in the critical embryo of radius r_c.

The nucleation rate I is now given by the rate $q_0 O_c$ at which single molecules strike (and adhere) to critical embryos of area O_c and is proportional to the incoming flux q_0, the area O_c and to N_n; also, ΔG_n becomes ΔG_c:

$$I = q_0 O_c N_n = N q_0 O_c \exp(-\Delta G_c/kT) \tag{14.8}$$

or

$$\frac{\text{nucl. rate}}{\text{unit volume}} = \frac{I}{Nv^l} = \frac{q_0 O_c}{v^l} \exp\left\{\frac{-16\pi v^{l2}\gamma_{LV}^3}{3(kT)^3[\ln(p/p_0)]^2}\right\} \tag{14.9}$$

where q_0 is given by the kinetic theory of gases as $p(2\pi mkT)^{1/2}$. Equation (14.9) is an extremely sensitive function of p/p_0. Hence the notion of a critical supersaturation occurs, i.e. one for which (14.9) becomes measurable.

c. *Becker and Döring*[119] *Correction*

Becker and Döring[119] corrected Volmer's assumption that the distribution function was (14.7) up to n_c and zero thereafter; in fact, critical

---

[119] R. Becker and W. Döring, *Ann. Phys.* **24**, 719 (1935).

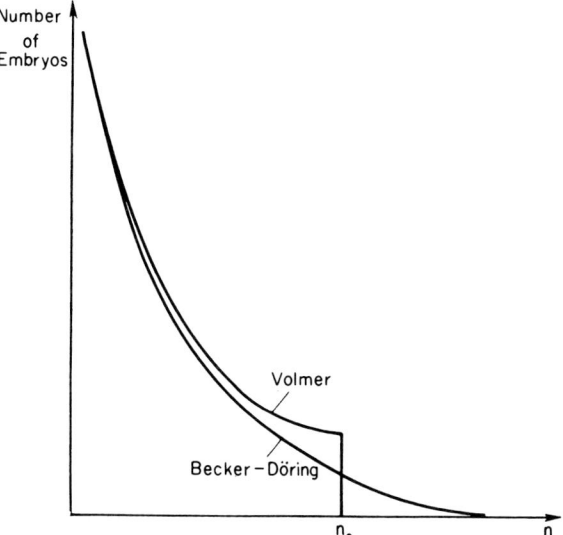

Fig. 11. Size distribution of embryos in Volmer and in Becker–Döring theories of nucleation. n is number of molecules in embryo [J. W. Christian, "The Theory of Transformations in Metal and Alloys." Pergamon Press, Oxford, 1965].

size embryos can either grow or evaporate with equal probability, and nuclei of even greater than critical size have some chance of shrinking. The method of Becker and Döring involves a set of difference equations describing the steady state distribution. The resulting distribution is shown in Fig. 11 and the resulting rate equation is Eq. (14.9) times $(1/n_c)(\Delta G_c/3\pi kT)^{1/2}$, i.e. is

$$\frac{I}{Nv^l} = \frac{\text{nucl. rate}}{\text{unit vol.}} = \frac{q_0 O_c}{v^l n^c}\left(\frac{\Delta G_c}{3\pi kT}\right)^{1/2} \exp\left\{\frac{-16\pi v^{l2}\gamma_{LV}^3}{3(kT)^3[\ln(p/p_0)]^2}\right\}. \quad (14.9a)$$

Compared to the exponential factor, the correction factor is generally not a substantial one, i.e. is within one or two orders of magnitude of unity.

Cloud chamber experiments[9] give for water and benzene droplets a surface free energy within about 13% of the bulk values, thought to be relatively good agreement.

Vapor–Solid. The nucleation rate equation for nucleation of solid crystalline nuclei from supersaturated vapor differs only slightly from (14.9) in the preexponential term, assuming that the appropriate shape factor η has been used in ΔG_c. The embryo crystal of minimum surface free energy will not in general be a sphere but will generally be a body, given by Wulff's theorem, which may be a polygonal figure.

d. Nucleation of Solid from Liquid

The nucleation of solid from liquid differs from the above treatment primarily by replacing the gaseous collision frequency q_0 with an absolute rate theory frequency ν_1 with which an atom surmounts the activation energy barrier $\Delta_\alpha g_n^\pm$ at the interface between liquid and solid embryo containing n atoms. The ν_1 is given by

$$\nu_1 = \nu_0 \exp(-\Delta_\alpha g_n^\pm / kT) \tag{14.10}$$

where ν_0 is the fundamental atomic vibration frequency. The result obtained is[9,19]

$$I = N \frac{kT}{h} \frac{n_c'}{n_c} \left(\frac{\Delta G_c}{3\pi kT}\right)^{1/2} \exp\left\{\frac{-(\Delta G_c + \Delta_\alpha g^\pm)}{kT}\right\} \tag{14.11}$$

where now $\Delta_\alpha g^\pm$ is independent of n, n_c' is the number of surface atoms in the nucleus, and ν_0 has been set approximately equal to kT/h.

An approximate version of (14.11) for metals is (see e.g. Christian[19])

$$I \text{(nuclei/cm}^3 \text{ sec)} \cong 10^{33} \exp(-\Delta G_c/kT) \tag{14.12}$$

where $\Delta_\alpha g^\pm$ has been estimated as approximately equal to the activation energy for viscous flow.

Here

$$\Delta G_c = \frac{4\eta^3 \gamma_{SL}^3}{27(g^l - g^s)^2}$$

where γ_{SL} is the solid–liquid interfacial free energy and

$$g^l - g^s \cong \Delta S(T_M - T) = (L/T_M)\,\Delta T$$

where T_M is the melting point, T the liquid temperature, L the latent heat of fusion, and ΔS the entropy of fusion.

Experiments on the freezing of Hg droplets[9] are believed to be in fair agreement with Eq. (14.11), although the absolute value of γ_{SL} required for a complete check of Eq. (14.11) is not in general known.

e. Time-dependent Nucleation

The question arises as to how long will it take to establish the steady state distribution of embryos assumed above, when the system is suddenly changed from one pressure or temperature to another. Zeldovitch and Frenkel (see Christian[19]) discussed this question by finding an analytical approximation to the set of difference equations used in the Becker–Döring theory. The theory is complex and the result, called the Fokker–Planck equation, is a type of one-dimensional diffusion equation with an imposed force field. Zeldovitch's solution to this equation has the form

$$I_t \cong I \exp(-\tau'/t), \qquad (14.13)$$

describing the approach of the time-dependent nucleation rate I_t to the steady state nucleation rate I, where $\tau' \approx n_c^2/D_c$ and D_c is a kind of diffusion coefficient proportional to the number of molecules in the surface of the nucleus and to an impingement frequency. For vapor–liquid nucleation τ' may be only a few microseconds but for liquid–solid transformations, where the activation energy barrier $\Delta_a g^{\pm}$ may be appreciable, τ' may be much greater.

f. Heterogeneous Nucleation

Heterogeneous nucleation or nucleation catalysis may take place when a solid surface or an impurity particle is present in the vapor or in the liquid; then the embryo may be formed on this surface. Hence some of the catalyst-medium interfacial free energy is removed, and the free energy so obtained assists in forming the embryo, making the formation easier (i.e. to occur at a lower critical supersaturation or undercooling than for homogeneous nucleation). Due to the common presence of foreign particles in most liquids this mode of nucleation occurs frequently. For liquids nucleating from the vapor onto flat solid surfaces it is usually assumed that the embryo shape is that of a portion of a sphere, called a spherical cap, and making a characteristic contact angle θ with the substrate surface. Then (see e.g. Turnbull[9]) by geometry we find for the critical free energy

$$\Delta G_c{}^s = 16\pi(\gamma_{LV})^3 f(\theta)/3(\Delta G_v)^2 \qquad (14.14)$$

where $\Delta G_v = (-kT/v^l) \ln(p/p_0)$ and where $f(\theta) = (2 + \cos\theta)(1 - \cos\theta)^2/4$. It is clear that as $\theta \to 0$ (solid is perfectly "wet" by the liquid), $f(\theta)$ and $\Delta G_c{}^s \to 0$ and the nucleation barrier is removed.

For other substrate shapes, e.g. holes or cavities, nucleation, for a given value of the contact angle, may be even easier than above. That is, more catalyst–mother medium surface energy can be removed by a given volume of embryo.[9]

The steady state nucleation rate is then, for vapor phase nucleation, given by Eq. (14.9a) with (14.14) for $\Delta G_c{}^s$ replacing ΔG_c. In addition, the quantity N is no longer the total number of molecules in the system, but is N^s, the number of molecules in contact with the impurity surface. A similar modification of Eq. (14.11) obtains for nucleation of solid onto foreign surfaces in a liquid.

15. Some Modern Developments in Nucleation

a. Heterogeneous Nucleation from Vapor and Surface Diffusion

Pound et al.[120] (see Hirth and Pound[112]) have discussed the surface diffusion process in connection with the heterogeneous nucleation of cap-shaped nuclei from the vapor onto substrates. This will control the factor q_0 in the surface counterpart to Eq. (14.9a), since direct condensation from the vapor onto the embryos is found to be relatively small. The rate equation is found to be

$$I_{SD} = [a \sin \theta \ln(p/p_0)/8\pi\bar{\omega}m][3/f(\theta)\gamma_{SV}kT]^{1/2}p^2$$
$$\times \exp[2\,\Delta G_{\text{des}}^* - \Delta G_{SD}^* - \Delta G_c{}^s)/kT] \qquad (15.1)$$

where ΔG_{SD}^* is the free energy of activation for surface diffusion, ΔG_{des}^* that for desorption of an adatom, γ_{SV} is the vapor-solid interfacial free energy, and $\Delta G_c{}^s$ is given by Eq. (14.14) with γ_{LV} replaced by γ_{SV}. Also, $\bar{\omega}$ is an attempt frequency and a is a jump distance, both for an adatom to join the nucleus.

Other examples of catalysis by various substrate shapes include the step or ledge on a substrate. In this case ΔG, as given by (14.14) is reduced because the step introduces more catalyst–medium surface area while not substantially altering the contact angle θ or nucleus–vapor surface area. Hirth and Pound[112] also consider the case where the nucleus forms at the juncture of two planes meeting at an arbitrary angle.

Another possible nucleus shape is that of a monatomic disk on the substrate. According to Hirth and Pound,[112] such a shape may be preferred to a cap-shape when the interfacial free energy between vapor and nucleus is sufficiently anisotropic (see Part V, γ plots). The free energy of formation of such an embryo of radius r and height h is

$$\Delta G = 2\pi r \gamma_{e,SV} + \pi r^2 (\gamma_{SV} + \gamma_{SX} - \gamma_{XV}) + \pi r^2 h\,\Delta G_v \qquad (15.2)$$

where ΔG_v is given just below Eq. (14.14), where γ_{SX} and γ_{XV} are the solid–substrate and vapor–substrate surface energies, and where $\gamma_{e,SV}$ is a new parameter, the edge free energy per unit length. As with other nuclei, ΔG_c is obtained by maximizing (15.2) with respect to r.

It may be noted here that recent field emission microscope studies[121-123]

---

[120] G. M. Pound, M. T. Simnad, and L. Yang, *J. Chem. Phys.* **22**, 1215 (1954).

[121] S. C. Hardy, in "Crystal Growth" (H. S. Peiser, ed.), Suppl. to *J. Phys. Chem. Solids*, p. 287. Pergamon Press, Oxford, 1967.

[122] R. D. Gretz and G. M. Pound, in "Condensation and Evaporation of Solids" (E. Rutner, P. Goldfinger, and J. P. Hirth, eds.), p. 575. Gordon and Breach, New York, 1964.

[123] K. L. Moazed and G. M. Pound, *Trans. AIME* **230**, 234 (1964).

of the heterogeneous nucleation of metal vapors upon tungsten point cathodes indicate that an adsorbed layer of as much as a monolayer or more equivalent thickness is necessary before nucleation will take place—an observation not in agreement with the dilute surface medium envisaged in the theory.

b. Surface Nuclei of One to Ten Atoms

Walton[124] pointed out that, in many cases where the substrate temperature is low, the heterogeneous nucleation of vapors must take place with very small critical nuclei (i.e. the embryo which, when one atom is added to it, has a greater chance of growing than of decaying) of perhaps less than ten atoms. The assumption that such nuclei can be treated as having surface energy by the liquid drop model then becomes less tenable. For critical nuclei of 2, 3, or 4 atoms, the possible number of arrangements of the atoms on the substrate is limited. In the limit of high supersaturation, the critical nucleus can be a single atom, i.e. a pair will be more likely to grow than to decay. At lower supersaturations, the pair will be more likely to decay and then it will require an additional atom to be more stable; hence, a pair will be the critical nucleus, and a triangle cluster having two bonds per atom will form. This triangle cluster may also be energetically favored on a (111) plane substrate crystal, for epitaxial reasons. For each range of substrate temperature corresponding to a particular critical nucleus size, a nucleation rate equation for the critical nucleus of n_c atoms can be written down. It should be noted that n_c must be determined by a trial and error process together with experimental observation of the crystallographic orientation of the deposit.[125]

c. Dynamic Nucleation

Chalmers[126] and Hunt and Jackson[127] have discussed the nucleation of solid from supercooled liquid by dynamic means. This means the introduction of relative motion of different parts of the liquid itself or of the liquid with respect to the container. Methods for this include nucleation by ultrasonics, by single sound pulses, by friction (scratching) applied at the liquid–container interface, and by cavitation. The process is distinct from effects that produce small crystals in the liquid by the tearing off of parts of fine dendrite tips by flowing liquid.

[124] D. Walton, *Phil. Mag.* **7**, 1671 (1962).
[125] T. N. Rhodin and D. Walton, *in* "Single-Crystal Films" (M. H. Francombe and H. Sato, eds.), p. 31. Macmillan, New York, 1964.
[126] B. Chalmers, *in* "Liquids: Structure, Properties, Solid Interactions" (T. J. Hughel, ed.), p. 308. Elsevier, Amsterdam, 1965.
[127] J. D. Hunt and K. A. Jackson, *J. Appl. Phys.* **37**, 254 (1966).

Hunt and Jackson[127] have studied the role of cavitation in nucleation of solid in undercooled water and in benzene. It was found that cavities, produced in the liquid, do not nucleate solid when they are formed, but do nucleate solid when they collapse. On collapsing, very high positive and negative pressures, of the order of 100 kbar, can result which can alter the freezing point sufficiently for homogeneous nucleation to take place. Nucleation by scratching is believed to occur by this cavitation mechanism.

d. Time Effects

Courtney[128] has performed computer calculations of the time lags or time dependence of embryo concentration in supersaturated water vapor. Instead of using the single Fokker–Planck equation analytic approximation to the series of bimolecular reactions, Courtney solved approximately 100 interdependent differential equations describing these bimolecular reactions. A typical result for water vapor, at a saturation ratio p/p_0 of 10 and temperature $-40°C$, was of the order of 1 μs to reach 95% of the steady state concentrations, for a critical nucleus of 43 molecules.

Buckle and Ubbelohde[129] have studied the freezing of aerosols of supercooled liquid droplets of the alkali halides and suggest that the observation time for certain freezing experiments may be comparable to the nucleation time lag rather than to the reciprocal of the steady state nucleation frequency. This estimate depended sensitively, however, on the figure chosen for the solid–liquid interfacial free energy.

e. Diffuse Interfaces in Nucleation: Spinodal Decomposition

(i) *Diffuse Interfaces.* Cahn and Hilliard[16,110] (see also Burke,[130] Uhlmann and Chalmers[113]) have discussed a modification to the classical nucleation assumption that the interface between the nucleus or embryo and the medium is sharp. In particular, they studied the case of nucleation in systems of two components with a phase diagram having a miscibility gap. The free energy of such an incompressible fluid having fluctuations in composition C is composed of two terms: one is the free energy per unit volume of a homogeneous fluid of composition C, and the other is a gradient energy arising due to composition gradients; the latter can be reduced by making the interface more diffuse [see Eq. (16.9)]. It is found by an analysis of the Euler equation involving ∇C, that for low supersaturations the nucleus has a uniform composition and a sharp interface, as in the

---

[128] W. G. Courtney, *J. Chem. Phys.* **36**, 2009 (1962).
[129] E. R. Buckle and A. R. Ubbelohde, *Proc. Roy. Soc.* **A261**, 197 (1961).
[130] J. Burke, "The Kinetics of Phase Transformations in Metals." Pergamon Press, Oxford, 1965.

classical model. As the supersaturation increases the work of forming a critical nucleus becomes less than in the classical theory, and the interface becomes increasingly diffuse. At the inflection point on the free energy versus composition curve, for this system, known as the spinodal, the work required for forming the critical nucleus goes to zero.

(ii) *Spinodal Decomposition.* Within the spinodal, i.e. for compositions within the two inflection points on the free energy vs composition diagram, it was shown by Cahn[131] that a second and very different class of phase change exists, as in fact Gibbs had much earlier distinguished. This is the class of continuous phase changes small in degree but large in extent, and having no thermodynamic barrier to its initiation. The theoretical analysis treats the stability of the spacewise Fourier components of a composition fluctuation; the solution is found to be unstable to composition fluctuations of wavelengths greater than a critical value λ_c, for the increase in them of gradient (surface) energy is smaller than the decrease of volume free energy.

f. Statistical Mechanics and Nucleation

Physical cluster theory, as an approach to the nucleation problem through statistical mechanics, was first employed by Frenkel and by Bijl (see Frenkel[132]). Hill[133] and Stillinger[134] discuss modern developments.

Kikuchi[135] has devised a statistical mechanical method for dealing with irreversible cooperative phenomena called the path probability method, and has applied this method to study the boundary motion of a domain wall in the Ising model,[136] and to study the nucleation and growth of a thin film by vapor deposition.[137] In this method, which is an extension of an earlier equilibrium theory,[31,138] a change of the state of the system is called a path and the variables which described the path are called the path variables. An example of a path variable in the case of a thin film would be the probability, over a given time interval, for a change of a neighbor-neighbor bond to a neighbor-vacancy bond. The probability that a certain change of state of a "cluster" occurs is called the path probability and is

---

[131] J. W. Cahn, *Acta Met.* **9**, 795 (1961).
[132] J. Frenkel, "Kinetic Theory of Liquids." Oxford Univ. Press, London and New York, 1946.
[133] T. L. Hill, "Statistical Mechanics." McGraw-Hill, New York, 1956.
[134] F. H. Stillinger, Jr., *J. Chem. Phys.* **47**, 2513 (1967).
[135] R. Kikuchi, *Ann. Phys.* **10**, 127 (1960).
[136] R. Kikuchi, *Ann. Phys.* **11**, 328 (1960).
[137] R. Kikuchi, in "Crystal Growth" (H. S. Peiser, ed.), Suppl. *J. Phys. Chem. Solids*, p. 605. Pergamon Press, Oxford, 1967.
[138] R. Kikuchi, *Phys. Rev.* **81**, 988 (1951).

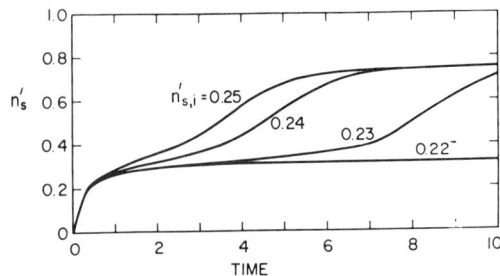

Fig. 12. Thin film nucleation and growth according to Kikuchi's path probability method. Reduced adatom surface density n_s' vs time, for various nucleating center densities $n_{s,i}'$ [R. Kikuchi, in "Crystal Growth" (H. S. Peiser, ed.), Suppl. *J. Phys. Chem. Solids*, p. 605. Pergamon Press, Oxford, 1967].

written in terms of the path variables. The path probability method asserts that the macroscopic change of state corresponds to the most probable path, i.e. maximizes the path probability with respect to the path variables. It is shown that the path probability method is consistent with irreversible thermodynamics, including the Onsager reciprocal relations. The maximizing leads to a set of nonlinear differential equations, and these are integrated. Figure 12 illustrates the film results, showing normalized adatom surface density n_s' versus time, assuming several different densities of foreign nucleating centers to be present. The jump to the second plateau represents the nucleation event.

V. Interface Structures and Interface Kinetics

16. Interface Structures in Equilibrium

Introduction

As Herring[139] has pointed out, the theory of surfaces or interfaces can be separated into three parts depending primarily on what branch of physics one wishes to use in examining surface structures of solids. These parts are macroscopic theory, atomic or atomistic theory, and electronic theory.

Macroscopic theory is the application of equilibrium and nonequilibrium thermodynamics to surfaces. This phenomenological approach, as for thermodynamics of bulk systems, postulates no model of, and few hypotheses regarding the detailed nature of the interface.

[139] C. Herring, *discussion in* "Structure and Properties of Thin Films" (C. A. Neugebauer, J. B. Newkirk, and D. A. Vermilyea, eds.), p. 527. Wiley, New York, 1959.

Atomic or atomistic theory constructs atom model-type pictures of surface structures and theories involving atoms treated as spheres attracting neighbors in the crystal lattice through central forces of short range; the atom movements are discussed as diffusion processes. Classical mechanics and kinetic theory are used.

Electronic theory is the application of quantum mechanics to the many body system of electrons and ions in the crystal with a view toward calculation of surface effects on the bulk system.

Another way of dividing up the subject, as pointed out by Cabrera and Coleman[13] is a classification of the types of surfaces using the $\gamma(\theta)$ plot of surface free energy versus orientation of the solid, to be discussed in more detail below. Three types of surfaces include singular surfaces, vicinal surfaces, and nonsingular and diffuse surfaces.

Singular surfaces are those having an orientation corresponding to a cusped minimum in the $\gamma(\theta)$ plot.

Vicinal surfaces are those having orientations very near that of a singular surface.

Nonsingular surfaces are all other orientations.

We will be concerned with all three types of surfaces, but only with macroscopic and atomistic theories. We say nothing about electronic theory (see Gatos[140] and also remarks by Herring[139] and by Juretschke[141]).

We concern ourselves with equilibrium structures in this section, and with the effects of kinetics, i.e. departures from equilibrium, in Sections 17–20. The study of equilibrium interface structures provides a starting point for studies of kinetics, and also remains valid for sufficiently small departures from equilibrium.

a. Thermodynamics of Surfaces—Phenomenological Treatment

(*i*) *Surface Tension* γ. We discuss here the thermodynamic definition of the surface tension of solids, sometimes called the surface free energy. The surface tension γ is[11,12] defined for plane surfaces as the reversible work dW required to create, by separation, unit surface area dA at constant temperature T, volume V, and chemical potentials, μ_i, i.e.

$$\gamma = dW/dA. \tag{16.1}$$

The concept of surface stress is a quantity which, although numerically equal to γ in the case of liquid surfaces, is not in general equal to γ in the

---

[140] H. C. Gatos, ed., "Solid Surfaces." North-Holland Publ., Amsterdam, 1964.

[141] H. J. Juretschke, *discussion in* "Structure and Properties of Thin Films" (C. A. Neugebauer, J. B. Newkirk, and D. A. Vermilyea, eds.), p. 538. Wiley, New York, 1959.

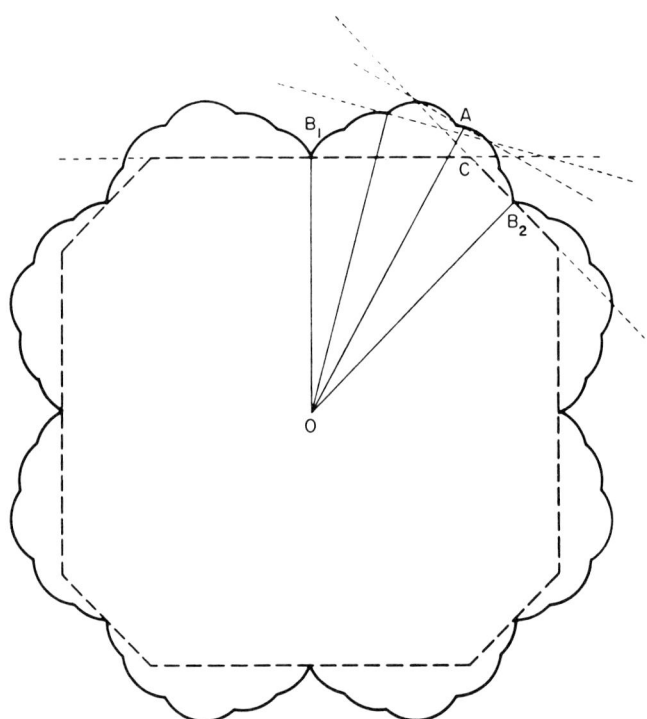

Fig. 13. Schematic plot of surface free energy γ of crystal, and Wulff construction. —, polar plot of surface free energy; - - -, samples of planes normal to radius vectors of this plot; – – –, equilibrium polyhedron [C. Herring, *in* "Structures and Properties of Solid Surfaces" (R. Gomer and C. S. Smith, eds.), p. 5. Univ. of Chicago Press, Chicago, Illinois, 1953].

case of solids. It may be positive, negative, or zero for a solid[12] as noted in Part I.

(ii) The Polar γ Plot, Wulff's Theorem. The surface tension or surface free energy γ when plotted on a polar diagram as a function of the crystallographic orientation of the surface considered should in general vary with orientation and reflect the symmetry of the crystal. A schematic γ plot is shown, in two-dimensional section, in Fig. 13. A number of cusps (first derivative discontinuous) are shown in this example.

For a crystal at $0°K$, there will be cusps at all rational orientations (Miller indices rational). For $T > 0°K$, many cusps disappear due to thermal fluctuations, and there are only a finite number of cusps.[11]

Wulff's theorem concerns the question of what is the equilibrium shape of a crystal having a given $\gamma(\bar{n})$ plot where $\bar{n}$ is the unit normal to the plane

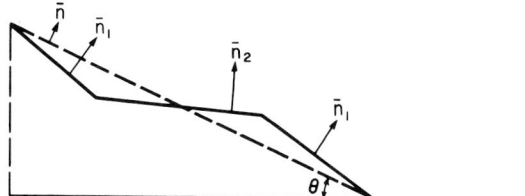

FIG. 14. Formation of hill-and-valley structure. Breakup of surface of orientation $\bar{n}$ into new surfaces of orientations $\bar{n}_1$, $\bar{n}_2$ [N. Cabrera and R. V. Coleman, *in* "The Art and Science of Growing Crystals" (J. Gilman, ed.), p. 3. Wiley, New York, 1963].

in question. The equilibrium shape will be that shape which minimizes the quantity γA taken over the entire crystal, i.e. at constant volume:

$$\int \gamma(\bar{n}) \, dA \quad \text{is a minimum.} \tag{16.2}$$

In a one-component system, the quantity to be minimized is just the total surface free energy of the crystal. Clearly for a liquid, where $\gamma(\bar{n}) = \gamma =$ constant, the equilibrium shape is a sphere, which has minimum surface for a given volume. A crystal will try to be bounded as much as possible by faces $\bar{n}$ having low $\gamma(\bar{n})$. Wulff's theorem states that, given the $\gamma(\bar{n})$ surface, if a plane is drawn through the end of each radius vector to this surface, and perpendicular to the vector, then the inner envelope, or body formed by all points reachable from the origin without crossing any of these planes, is the body of minimum surface free energy. Wulff's construction is illustrated also in Fig. 13 which in the particular case shown gives a polyhedron. (Discussions of the proofs of Wulff's theorem are given by Herring.[11] See also Johnson and Chakerian.[142]) The body need not be a polyhedron however; it may have rounded corners with neighboring flats, or some smoothly curved regions and some flats, depending on the γ plot. Some cusps in the plot are necessary for the plane regions.

It should not be expected that large crystals in practice will achieve their equilibrium form; as Frank[3] pointed out, for crystals much larger than a micron in size, the supersaturation or other driving force necessary to grow the crystal at a finite rate will be large compared to, and therefore swamp, any change of free energy due to a departure from equilibrium shape. The latter size dependence comes from the crystal counterpart of the Gibbs–Thomson equation. Experiments on the equilibrium form of small crystals will be mentioned below.

(iii) Faceting of Plane Surfaces and β Plots. Given a large crystal surface of an orientation $\bar{n}$ such that $\gamma(\bar{n})$ is sufficiently high, it is possible that the surface energy $\int \gamma \, dA$ may be reduced by this plane breaking up into

[142] C. A. Johnson and G. D. Chakerian, *J. Math. Phys.* **6**, 1403 (1965).

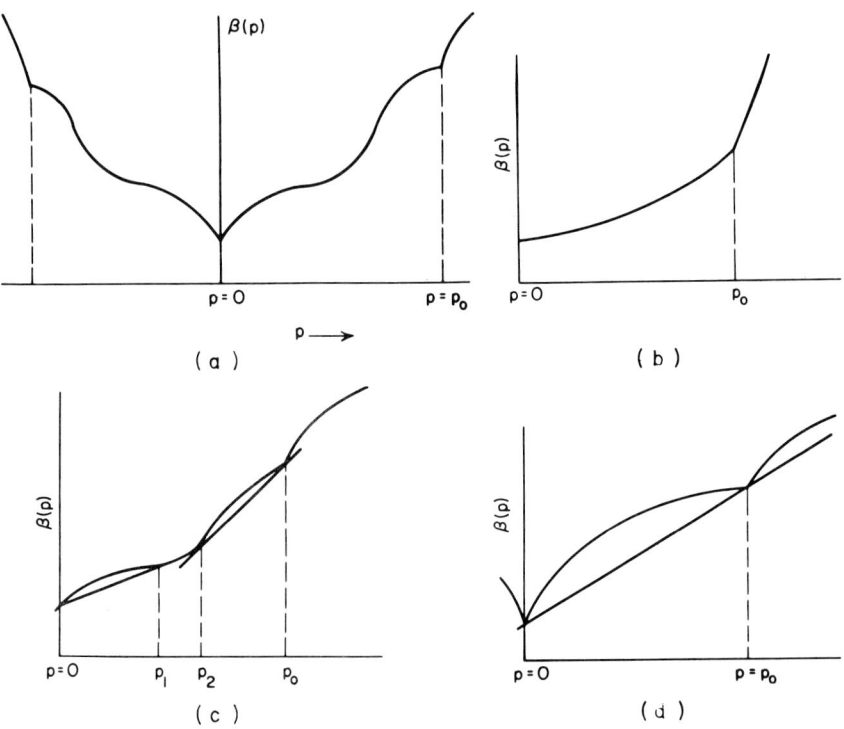

Fig. 15. The surface free energy per unit area of reference surface $\beta(p)$. (a–d) Several possible schematic plots and their stability. $p = 0$ and $p = p_0$ are singular (cusped) orientations [N. Cabrera and R. V. Coleman, in "The Art and Science of Growing Crystals" (J. Gilman, ed.), p. 3. Wiley, New York, 1963].

a hill and valley structure (Fig. 14). The new surfaces of orientation say $\bar{n}_1$ and $\bar{n}_2$ have sufficiently lower γ values to overcompensate the increase in surface area. Experiments relating to such faceting or thermal etching have been reviewed in detail by Moore.[143] Cabrera and Coleman[13] have given a discussion of how to predict relative stability, in the case of an infinite plane surface, against break up. The function

$$\beta(p) = (1 + p^2)^{1/2}\gamma(p), \qquad (16.3)$$

which is the surface free energy per unit area of reference surface, is plotted versus orientation p in Fig. 15a. The p is defined as the slope

$$p = -\partial z/\partial x \qquad (16.4)$$

[143] A. J. W. Moore, in "Metal Surfaces: Structure, Energetics, and Kinetics," p. 155. Am. Soc. Metals, Cleveland, Ohio, 1963.

of the crystal surface

$$z = z(x, y). \tag{16.5}$$

The cusps at $p = 0$ and $p = p_0$ correspond to singular surfaces. Cabrera and Coleman[13] show that by drawing tangents to the $\beta(p)$ curve, the stable orientations can be deduced. In Fig. 15b all orientations between $p = 0$ and p_0 are stable. In Fig. 15c all orientations between 0 and p_1, and p_2 and p_0, are unstable; but all orientations between p_1 and p_2 are stable in addition to p_0 and $p = 0$. In Fig. 15d only $p = 0$ and p_0 are stable. (Hirth[144] has shown that other forms of $\beta(p)$ plots can give different breakup effects.)

(*iv*) *Chemical Potential and the* $\gamma(\theta)$. The classical Gibbs–Thomson equation gives the chemical potential increase of small droplets of liquid. We give it here in the form

$$\Delta\mu = \Omega\gamma(1/R_1 + 1/R_2). \tag{16.6}$$

For the case of a crystal, where $\gamma = \gamma(\theta)$, Herring[11] (see Mullins[12]) has generalized this formula to give the chemical potential at a point P:

$$\Delta\mu(P) = \Omega\left[\left(\gamma + \frac{d^2\gamma}{d\theta_1^2}\right)K_1' + \left(\gamma + \frac{d^2\gamma}{d\theta_2^2}\right)K_2'\right] \tag{16.7}$$

where

$$K_1' = 1/R_1 \quad \text{and} \quad K_2' = 1/R_2. \tag{16.8}$$

In Eqs. (16.6)–(16.8) K_1' and K_2' are the principal curvatures at P, R_1 and R_2 the corresponding radii of curvature, Ω is the atomic volume, and θ_1, θ_2 are the appropriate angles with respect to the point P. The new terms are the second derivatives of γ. For the equilibrium form, Eq. (16.7) must of course take on a constant value for all points of the surface, otherwise the free energy could be lowered by redistribution. [At cusps Eq. (16.7) is not defined because the second derivatives are not defined.] Chernov[17] has discussed the relationship between (16.7) and Wulff's theorem.

(*v*) *Nonuniform Systems—Diffuse Interface.* As noted in Part IV on nucleation, Cahn and Hilliard[16] (see also Widom[145]) have studied the thermodynamics of nonuniform systems, and in particular have shown that the interfacial energy may be reduced by permitting, instead of an abrupt change in concentration or density at the interface, a gradual change. They showed this by taking the case of composition C as a variable, and expanding the local free energy per molecule of a nonuniform system $G(C)$ in a Taylor series about the point G_0, the free energy per molecule of a

[144] J. P. Hirth, *in* "Energetics in Metallurgical Phenomena" (W. M. Mueller, ed.), Vol. II, p. 1. Gordon and Breach, New York, 1965.
[145] B. Widom, *J. Chem. Phys.* **43**, 3892 (1965).

solution of uniform composition C. From this they found that the free energy of a small volume of nonuniform solution, G' is

$$G' = N_V \int_V [G_0 + \kappa(\nabla C)^2 + \cdots] \, dV \tag{16.9}$$

where N_V is the number of molecules per unit volume, V is the volume, and

$$\kappa = -[\partial^2 G/\partial C \, \partial \, \nabla^2 C]_0 + [\partial^2 G/\partial \mid \nabla C \mid^2]_0/2. \tag{16.10}$$

Equation (16.9) says that the free energy of a small volume of nonuniform solution can be expressed as the sum of two contributions, the free energy in a uniform solution and the gradient energy term varying with local composition. Applying this theory to a flat interface between two coexisting isotropic phases, α and β, of composition C_α and C_β of the components A and B, the result is obtained that the specific interfacial free energy γ is

$$\gamma = N_V \int_{-\infty}^{\infty} [\Delta G(C) + \kappa (dC/dx)^2] \, dx \tag{16.11}$$

where $\Delta G(C)$ is the free energy, referred to a standard state, of an equilibrium mixture of α and β, given by $\Delta G(C) G_0(C) - [C \mu_B(C) + (1 - C) \mu_A(C)]$ where the μ's are chemical potentials, per molecule, of A and B. In Eq. (16.11), the wider or more diffuse the interface the smaller is the gradient energy term $\kappa (dC/dx)^2$, but this decrease is only achieved by introducing more nonequilibrium composition material near the interface and hence increasing the first term. The minimum condition gives

$$\gamma = 2N_V \int_{C_\beta}^{C_\alpha} [\kappa \, \Delta G(C)]^{1/2} \, dC. \tag{16.12}$$

Application of this thermodynamic relationship to specific systems requires a specific model for evaluation of κ and of ΔG.

b. *Structure and Energetics of Atomistic Models*

Introduction. As noted above, atomistics is the construction of various atomic models of surface or interface structures, with the assumption of certain types of force laws between the atoms, and calculation by classical mechanics, statistical mechanics, kinetic theory or transport theory of the energetics and kinetics; we discuss mostly in this section structures and energetics.

A model of an interface structure is both a statement about the geometry of that structure, plus some physical calculations which are consistent with the structure and which predict some measurable properties of the structure.

The ideal model would be none at all, i.e. it would be best not to have to postulate a model but instead to calculate all properties *ab initio* from the partition function knowing the interatomic forces. This, of course, cannot be done by statistical mechanics treatment of such complex, many-body systems as crystals or crystal/fluid interfaces. The extensive use of models, as in crystal growth interface mechanisms, is some measure of the large amount of ignorance and the lack of sophistication of the subject. Given an interface observing microscope of 10^{-9} cm resolution that does not disturb the interfacial process, the proliferation of models would greatly decrease.

One should require that model-type calculations give numerical predictions of measurable quantities, and furthermore, given the arbitrariness and crudity of most models, that the prediction must be stable with respect to slight changes in the model. That is, one must try to separate those predictions of the model which have general validity from those peculiar to the specific model used (this is seldom done). Also, it may be hard to prove the uniqueness of a particular model to fit the observations. Finally, the old question also arises of which is better: an exactly solvable, simple but unreal model versus an approximately solvable, complex realistic model.

(*i*) *Surface Roughness Models.* In these models the primary question raised is how many surface vacancies or how many surface adsorbed atoms (adatoms) form as the temperature is raised. (The reason that this is of interest is that an atomically smooth surface is generally believed, at least on the Kossel model, to require that steps be formed upon it (see below) in order to grow, and hence the origin and number and motion of these steps becomes the primary matter of interest in calculating crystal growth rates. Whereas, if the surface is atomically rough, then no steps are needed (or there are steps everywhere) and the growth mechanism can be much simpler.) The various models differ in the number of surface layers examined in which the vacancies are assumed to form, and they differ as to what the external medium is.

(a) *Onsager*[29] *and Burton and Cabrera.*[40] We have already discussed this calculation in some detail in Part II. The model is of a two-dimensional square lattice, with two levels, i.e. with the reference surface and adsorbed molecules. The calculation is exact. For the square lattice, with four nearest neighbors of interaction energy φ,

$$kT_c^I/\varphi \approx 0.57 \qquad (16.13)$$

where T_c^I is the critical temperature for surface melting, believed usually greater than the bulk melting point for close packed (singular surfaces) of such solids as metals. The surface roughness analysis showed that for $T \ll T_c^I$, there are only single adatoms, not clusters of them, or steps, on the surface.

This calculation ignores any interactions of the surface with the external phase (liquid or vapor). Any adatoms would have to come, presumably, from the layer itself.

(b) *Jackson Theory.*[42,43] This theory was also discussed in Part II. It is a two-level, approximate Bragg–Williams treatment of a surface layer in contact with a second phase in equilibrium—in particular, with the melt. Roughness (surface half populated with adatoms) or smoothness (few adatoms) is not studied as a function of temperature, for the system is assumed to be at the equilibrium temperature (melting point) of the solid in contact with the liquid, or with the vapor at a temperature presumably near the melting point. Roughness is studied as a function of the latent heat of transformation L and of the relative amount of binding energy (first neighbor bonds) for an atom in the plane in question.

Jackson found for the vapor phase that close-packed orientations should generally be smooth, but whether or not these surfaces are smooth when in contact with the melt depends strongly on the class of material as discussed in detail in Section 19 on melt growth interface kinetics. Jackson noted that the macroscopically observed interface morphologies of many liquid–solid one-component systems are in accord with the model. As an approximate treatment, however, it may significantly overestimate the critical temperature compared to the exact (Onsager) approach.[146,147]

(c) *Mullins'*[148] *Theory.* Mullins[148] has given a theory of surface melting, following the Burton–Cabrera picture but using the simpler Bragg–Williams approach, and dealing with a three-level problem (i.e. both adatoms and vacancies). The (100) plane of a simple cubic crystal is discussed, with the assignment of $\varphi/2$ for the first neighbor energy per atom per bond. The resulting curve for surface roughness (number of bonds per site parallel to the surface) is given in Fig. 16 and is quite similar to the Burton–Cabrera result. A similar conclusion is made as to the absence of surface melting. Again interactions between the surface layer and the external phase are considered negligible.

(d) *Theory of Temkin.*[147] Temkin studied the crystal–liquid equilibrium interface of a (100) plane of a simple cubic crystal, assuming an atom in either to have 6 nearest neighbors, 4 in the plane parallel to (100), and using the Bragg–Williams method. An arbitrary number n of levels is considered, and a numerical solution is obtained. It is found that the

---

[146] N. Cabrera, *discussion in* "Growth and Perfection of Crystals" (R. H. Doremus, B. W. Roberts, and D. Turnbull, eds.), p. 324. Wiley, New York, 1958.

[147] D. E. Temkin *in* "Crystallization Processes" (N. N. Sirota, F. K. Gorskii, and V. M. Varikash, eds.), p. 15. Consultants Bureau, New York, 1966.

[148] W. W. Mullins, *Acta Met.* **7,** 746 (1959).

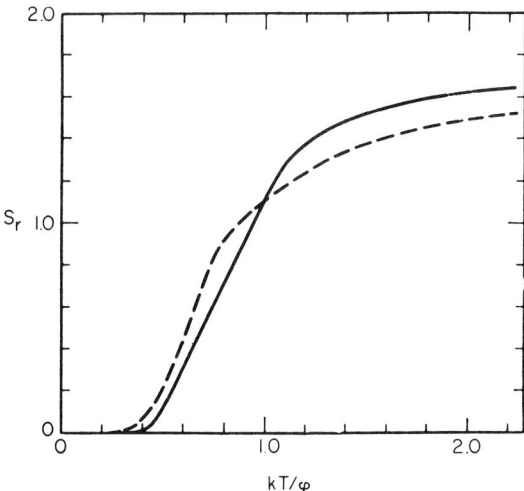

FIG. 16. Surface roughness S_r vs reduced temperature kT/φ. —, Theory of Mullins; – – –, theory of Burton and Cabrera [W. W. Mullins, *Acta Met.* **7**, 746 (1959)].

effective width of the interface depends smoothly on the value of the quantity

$$\gamma'' = 4w'/kT_M \tag{16.14}$$

where T_M is the melting temperature (see Fig. 17). Here w' (called by Temkin the surface energy per atom) may, as in Jackson's theory, be

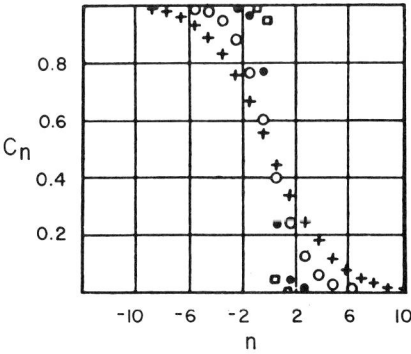

FIG. 17. Interface width in Temkin theory. Concentration C_n of "solid" atoms versus layer position n for various values of γ'' (defined in text.) (+) $\gamma'' = 0.446$, (○) 0.769, (●) 1.889, (□) 3.310 [D. E. Temkin, *in* "Crystallization Processes" (N. N. Sirota, F. K. Gorskii, and V. M. Varikash, eds.), p. 15. Consultants Bureau, New York, 1966].

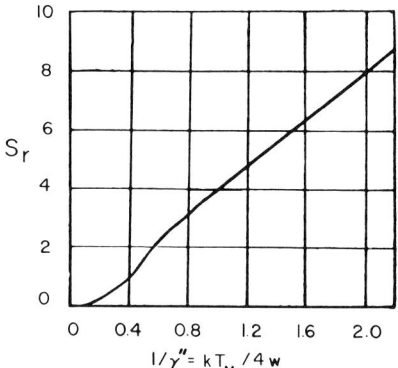

FIG. 18. Surface roughness S_r vs $1/\gamma''$ in Temkin theory [D. E. Temkin, *in* "Crystallization Processes" (N. N. Sirota, F. K. Gorskii, and V. M. Varikash, eds.), p. 15. Consultants Bureau, New York, 1966].

thought of as the bond energy found by dividing the latent heat of fusion by the number of first neighbors, i.e. $w' = \frac{1}{6}L$ where L is the latent heat of fusion, in the simple cubic case. (Note that the Jackson criterion, $\alpha = 2$, becomes for a (100) plane where $f_k = \frac{2}{3}$, $kT_M/\varphi_j = 2$ where $6\varphi_j = L$. The Temkin theory gives $kT_M/w' = 1.6$. As noted, w' may also be identified with $\frac{1}{6}L$.) There is an inflection but no sudden drop at

$$kT_M/2w' \approx 0.82 = 2/\gamma'' \qquad (16.15)$$

(in the Jackson theory this value is unity), hence $1/\gamma'' = 0.41$; see Fig. 18. At the inflection point $\gamma'' \approx 2.4$ giving (see Fig. 17) an interface width less than an atomic diameter. This is in qualitative agreement with Jackson's theory. Of course, at larger γ'' values, very large interface widths are predicted from the theory of Temkin.

(e) *Other Theories.* Voronkov and Chernov[149] studied the atomic roughness of a crystal–solution interface, in the case of a one-component simple cubic crystal and an ideal binary solution. They predict a critical transition temperature which depends on solution concentration.

Kerr and Winegard[150] also studied a binary system, following in their approach the two-level method of Jackson[42,43]; they applied the results to the interface structure of Bi–Ag and Bi–Sn eutectic alloys.

---

[149] V. V. Voronkov and A. A. Chernov, *in* "Crystal Growth" (H. S. Peiser, ed.), Suppl. *J. Phys. Chem. Solids*, p. 593. Pergamon Press, Oxford, 1967.

[150] H. W. Kerr and W. C. Winegard, *in* "Crystal Growth" (H. S. Peiser, ed.), Suppl. *J. Phys. Chem. Solids*, p. 179. Pergamon Press, Oxford, 1967.

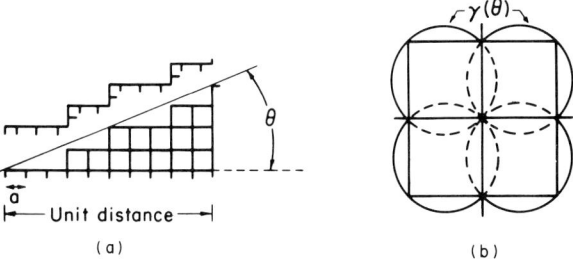

Fig. 19. Anisotropy of surface free energy $\gamma(\theta)$ in two-dimensional nearest-neighbor broken bond model. (a) Surface, and (b) resulting γ plot [W. W. Mullins, *in* "Metal Surfaces: Structure, Energetics, and Kinetics," p. 17. Am. Soc. Metals, Cleveland, Ohio, 1963].

Mutaftschiev[151] studied surface roughness of crystal–vapor and crystal–liquid interfaces by a method of the Bragg–Williams type. However, he explicitly considered the degree of surface roughness ("autoadsorption") as a function of the departure from equilibrium or supersaturation.

(*ii*) *Surface Free Energy Calculation from Atomic Models.* In these calculations the question of primary interest is the magnitude of the surface or interfacial free energy and its variation with orientation, i.e. anisotropy.

(a) *Mullins.* Mullins[12] has studied the surface energy versus orientation (γ plot) at 0°K of a two-dimensional simple cubic structure having nearest-neighbor bonding. Figure 19 shows the model and the resulting (two-dimensional) γ plot. The atoms are represented by squares. To create the surface shown, of total "area" $2/\cos \theta$ (two sides), requires the bond energy φ times the number of bonds broken. The latter is, for unit length and for atom of side a, $1/a$ for the vertical free bonds and $(\tan \theta)/a$ for the horizontal free bonds. The ratio is γ:

$$\gamma(\theta) = \frac{\varphi(a^{-1} + a^{-1}\tan \theta)}{2/\cos \theta} = \frac{\varphi}{2a}(\cos \theta + \sin \theta) \qquad (16.16)$$

where θ is the angle between the normal to the new surface, and the positive y axis. Equation (16.16) is a circle passing through the origin and occupying the upper left quadrant of Fig. 19b. From the symmetry of the atom model there are four such circles as shown and the $\gamma(\theta)$ curve is their outside bounding curve. Thus there are four cusps in the $\gamma(\theta)$ curve, where the derivative $d\gamma/d\theta$ is discontinuous. The Wulff construction giving the equilibrium body is also shown in Fig. 19b—this is simply the square,

[151] B. Mutaftschiev, *in* "Adsorption et Croissance Cristalline," p. 231. *Colloq. Intern. Centre Natl. Rech. Sci. (Paris)* No. 152 (1965).

corresponding to these four cusps. All other planes perpendicular to other $\gamma(\theta)$ values do not alter the square inner envelope, as they pass through the corners of it. Mullins also discussed other consequences and limitations of this very simple model: more cusps would be obtained if the interaction between atoms were longer-ranged; the three-dimensional case gives 8 spheres analogous to the 4 circles. These conclusions follow from a theorem due to Herring, according to which any bond theory of finite range attractions between pairs of atoms on a constant lattice spacing must have a γ plot composed of portions of spheres.

(b) *Cabrera and Coleman,*[13] *following Chernov.*[17] As we saw earlier, the $\beta(p) = \gamma(p)(1 + p^2)^{1/2}$ plot of Cabrera could be used to predict the stability of a particular orientation to breakup into a hill and valley structure. Here if $p = -\partial z/\partial x$, the slope of the crystal surface, then

$$p/h = k \qquad (16.17)$$

is the density of steps on the crystal surface, where h is the step height. If $\beta(p)$ is expanded as

$$\beta(p) = b_0 + b_1 p + (b_2/2)p^2, \qquad (16.18)$$

then the first term is just the surface energy of the low index surface, the b_1 term is just the energy of formation of the steps and the b_2 term the energy of interaction between the steps. Both b_0 and b_1 are positive but b_2 could be either positive or negative corresponding to repulsion or attraction of steps. The first two terms correspond to a γ plot composed of portions of spheres as we noted above, and a $\beta(p)$ plot that is a straight line. Reasons for step interaction have been discussed by Mullins[12] and by Gruber and Mullins.[152] One is that steps tend to roughen (develop kinks) as we will see below, and this reduces the free energy; if they are close together, their freedom to roughen without overlapping is restricted, which increases the free energy as a function of step separation, i.e. gives repulsion. Mullins has described this repulsion as analogous to the force with which a high-polymer molecule resists the straightening accompanying extension of it—the step resists the reduction in configurational entropy resulting from lateral compression.

(c) *Gruber and Mullins.*[152] They have calculated the anisotropy of surface energy at temperatures greater than 0°K by explicitly calculating the configurational effect of step roughness, i.e. of introducing thermally produced "kinks" in the steps. Steps are assumed not to interact energetically (i.e. no attraction or repulsion at 0°K), but are constrained to occupy only a finite width. The results, obtained by a matrix method, for an $[01\bar{1}]$ step in a (100) face of an fcc crystal are shown in Fig. 20. Here, θ is the average orientation of the surface with respect to a terrace, so that the larger

[152] E. E. Gruber and W. W. Mullins, *J. Phys. Chem. Solids* **28**, 875 (1967).

θ the closer are the steps together, the straighter they are and hence the closer their free energy will approach their straight (0°K) step energy (i.e. the ordinate Φ in Fig. 20 approaches unity). For β'', a quantity inversely proportional to temperature, a high temperature (small β'') leads to a large difference between 0°K energy and T°K free energy (i.e. rough steps). When the data are plotted (as $\Phi \tan \theta$ vs $\tan \theta$) to study the step stability, it is found that the curve is convex (i.e. step repulsion) so that on this model no faceting would occur, but rounded areas could occur on the equilibrium form.

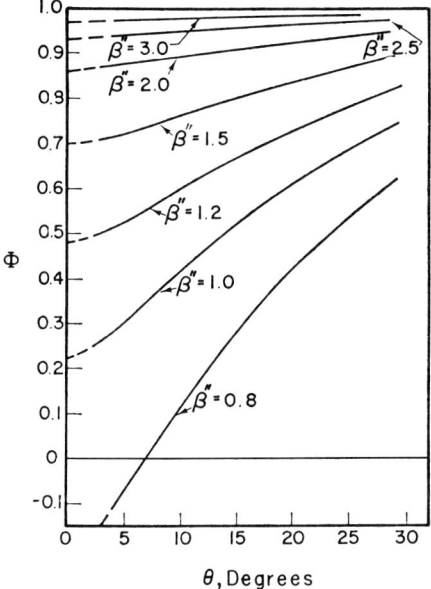

Fig. 20. Anisotropy of surface energy in step-roughness model. Φ (ratio of free energy of single (kinked) step to that of single straight step) versus surface orientation θ for various temperatures T (proportional to $1/\beta''$) for an fcc (100) [01$\bar{1}$] step system [E. E. Gruber and W. W. Mullins, *J. Phys. Chem. Solids* **28**, 875 (1967)].

(d) *Mackenzie et al.*[153] These authors have developed a detailed geometric theory of the number of atoms per unit area having broken bonds of first and second neighbors in cubic crystals when an atomically flat surface is formed by separation of a crystal. As they point out, the resulting γ plot must consist of portions of spheres passing through the origin. For nearest-neighbor bonds in an fcc crystal, the γ plot obtained is

[153] J. K. Mackenzie, A. J. W. Moore, and J. F. Nicholas, *J. Phys. Chem. Solids* **23**, 185 (1962).

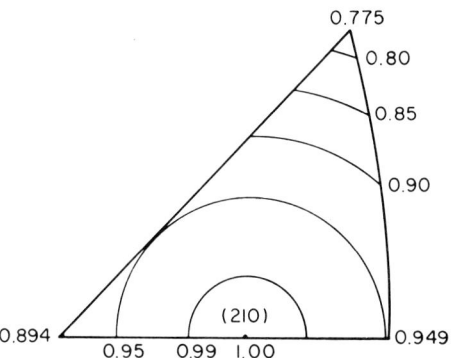

Fig. 21. Anisotropy of surface energy in three-dimensional nearest-neighbor broken bond model, fcc crystal. The value of $\gamma(210)$ is taken as unity [J. K. MacKenzie, A. J. W. Moore and J. F. Nicholas, *J. Phys. Chem. Solids* **23**, 185 (1962)].

shown as Fig. 21 in the unit stereographic triangle, where $\gamma(210)$ is taken as unity.

(e) *Benson and Yun*.[154] They reviewed calculations of the surface energy of atomically flat surfaces of rare gas crystals and simple ionic crystals, based on the model of pairwise interactions between atoms or ions, but taking into account possible changes in position of the surface atoms, from their bulk values, by relaxation. Various types of force law were assumed, and interactions with all ions or atoms in the lattice are taken into account. Thus, lattice sums enter into the theory, which is highly complex. The temperature variation of surface energy is calculated on the basis of model of atomic vibrations. For example, the surface energy at $0°K$ for argon was calculated to be 19.70 for $\{111\}$, 20.34 for $\{100\}$, and 21.34 for $\{110\}$ in units of 10^{-7} J/cm².

(f) *Sundquist*.[155] He experimentally studied the equilibrium shapes of small ($\sim 10^{-3}$ mm) particles of a number of fcc metals, obtained by heating thin films on an inert substrate. These shapes had planar faces connected by rounded regions. The γ values, obtained as a function of orientation, varied about 16% from the mean value. There appear to be some doubts about the interpretation of these results, however.[152,156] Shewmon and Robertson[157] also discuss experiments on $\gamma(\theta)$.

[154] G. C. Benson and K. S. Yun, *in* "The Solid-Gas Interface" (E. A. Flood, ed.), p. 203. Dekker, New York, 1967.
[155] B. E. Sundquist, *Acta Met.* **12**, 67 (1964).
[156] W. L. Winterbottom and N. A. Gjostein, *Acta Met.* **14**, 1033, 1041 (1966).
[157] P. G. Shewmon and W. M. Robertson, *in* "Metal Surfaces: Structure, Energetics and Kinetics," p. 67. Am. Soc. Metals, Cleveland, Ohio, 1963.

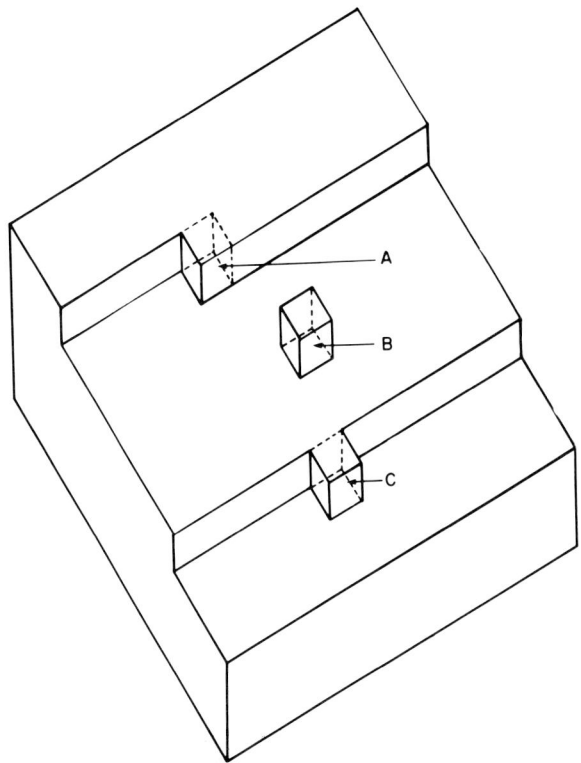

Fig. 22. Model of crystal surface showing steps or ledges and adatoms. The repeatable growth process or growth step of Kossel occurs with atom A.

(*iii*) *The Kossel Model for Crystal Growth.* Kossel[158] proposed that crystal growth takes place by means of a repeatable step or repeatable process of adding a molecule at point A in Fig. 22. This point has a much stronger bonding to the crystal, based on the number of first and second neighbors, than such a point as B or even C. Atoms at B will be much more likely to be desorbed.

The structures on the singular surface shown are called steps (or ledges). Two important questions about these steps are: how do they originate, and what is their structure or form? The question of the origin we defer to a later discussion of step sources, mentioning only here that the Burton–Cabrera[40] calculation showing the presence of only single adatoms and the absence of large clumps of adatoms prohibited the formation of steps from a surface-melting process. Of course, the equilibrium form may consist, as

[158] W. Kossel, *Nachr. Gesell. Wiss. Gottingen, Math.-Phys. Kl.* 135 (1927).

we saw in surface energy calculations, of flat regions and steps, i.e. vicinal surfaces. But according to crystal growth theory these will quickly grow out under growth conditions, leaving the flat singular surface and its need for steps for further growth. Thus, sources of steps on singular surfaces are required.

Burton et al.[41] showed that, unlike the case for the plane, the step though straight at 0°K could indeed "melt" or develop roughness at higher temperatures, but well below the melting point; the "melting" consists in the formation of many kinks or corners in the step. Frenkel[159] had also obtained this result. A step following, on the average, a close-packed direction, will contain equal numbers of positive n_+ and negative n_- kinks. If $2n_k$ and q are the probabilities for having a kink or no kink, respectively, at a given site, then as shown by detailed balancing,

$$n_k/q = \exp(-w/kT) \qquad (16.19)$$

and

$$2n_k + q = 1 \qquad (16.20)$$

where w is the energy required to form a kink. Hence $X_0 = a/2n_k$, the mean distance between kinks, is then

$$X_0 = \tfrac{1}{2}a\{\exp(w/kT) + 2\} \qquad (16.21)$$

where a is the interatomic spacing. It may also be shown that, under equilibrium conditions, the step must have a constant mean direction, i.e. be straight, on the average; and conversely, that it must be curved under nonequilibrium conditions. If the step is inclined at the small angle θ to a close-packed direction, then for geometrical reasons the number of kinks increases and now

$$X_0(\theta) = X_0\{1 - \tfrac{1}{2}(X_0/a)^2\theta^2\}. \qquad (16.22)$$

Taking as an example a (11) step on a (111) fcc crystal, for first neighbors only, $w = \tfrac{1}{2}\varphi = \tfrac{1}{12}W$, where φ is the nearest-neighbor interaction energy and W the heat of sublimation. For typical substances and temperatures $\varphi \sim 4kT$; hence, $X_0 \sim 4a$ or one kink every four lattice spacings. It will be shown later that this is a sufficiently high density of kinks for the step to serve as an infinite sink for adatoms during crystal growth.

Another feature of the Kossel model is the presence of surface adsorbed atoms, which were also shown to exist in the Burton–Cabrera[40] surface roughness calculation. The density of adatoms can be given as[41]

$$n_s = n_0 \exp(-w_s/kT) \qquad (16.23)$$

where n_0 is approximately the density of sites on the surface, and w_s the

[159] J. Frenkel, *J. Phys. USSR* **9**, 392 (1945).

kink to surface evaporation energy (energy required to take an atom from a kink to a surface site). The lifetime τ_s of an adatom is given by

$$1/\tau_s = \nu_2 \exp -(w_s'/kT) \qquad (16.24)$$

where the frequency of attempts to escape ν_2 is of the order of an atomic vibration frequency and where the surface to vapor evaporation energy (from point B to vapor in Fig. 22) is w_s'. Note that $W = w_s' + w_s$, i.e. the total evaporation energy is the sum of the kink–surface and surface–vapor energies. Similar relations[13] hold for the concentration and lifetime of surface vacancies, and such vacancies may also take part in the transport processes in growth to be described below, if their mobility is sufficient.

(iv) *The Screw Dislocation Step of Frank.*[160] Frank[160] (see also Burton et al.[41]) introduced the hypothesis that screw dislocations could be an important source of steps on crystal surfaces. A step is produced when a dislocation having a component of its Burgers' vector normal to the surface intersects the surface. This particular source of steps proves to be very important in crystal growth processes. The step is pinned at the point of emergence of the dislocation, but can otherwise move as adatoms lodge in its kinks to form a spiral as will be seen below. Under equilibrium conditions the step is straight. Other properties of the Kossel model are appropriate to the Frank model.

(v) *Model of Stranski and Kaischew.*[161] These authors, using a first- and second-neighbors bond picture, studied the equilibrium form of crystals by taking away all atoms less strongly bonded than those in the half-crystal position (point A in Fig. 22) or "repeatable step" position. Considering first neighbor bonds only, for the fcc structure, the {111} and {100} planes are found to bound the equilibrium form; {110} bound the equilibrium form for the bcc structure. If one includes second neighbors as well as first neighbors, one obtains as additional bounding planes, the {110} for fcc and {100} for bcc.

(vi) *Periodic Bond Chain (PBC) Theory.* This is a theory of crystal morphology, based on crystal chemistry. The theory was applied particularly to inorganic crystals, grown from dilute solution. The observation of such fibrous crystals as asbestos, where strongly bonded chains run parallel to the fiber axis, led Hartman and Perdok[162] (see also Hartman[163]) to assume that crystals should grow most rapidly in the direction in which the chemical bonding is strongest. A crystal will grow in a given direction only

---

[160] F. C. Frank, *Discussions Faraday Soc.* **5**, 48 (1949).
[161] I. N. Stranski and R. Kaischew, *Z. Krist.* **78**, 373 (1931).
[162] P. Hartman and W. G. Perdok, *Acta Cryst.* **8**, 49, 521, 525 (1955).
[163] P. Hartman *in* "Physics and Chemistry of the Organic Solid State" (D. Fox, M. M. Labes, and A. Weissberger, eds.), p. 369. Wiley (Interscience), New York, 1963.

when an uninterrupted chain of strong bonds exists in the structure. Three classes of crystal faces are defined: F or flat faces, which are parallel to at least two PBC's; S or stepped faces, parallel to one PBC; and K or kinked faces, parallel to no PBC. The PBC's are along cube directions. Hence, cube faces are F type, (110) are S type, and (111) are K type. The growth velocities increase as F–S–K. The theory appears to give qualitative agreement with many observations, but has no clear thermodynamic or kinetic basis. It is related to earlier theories by Bravais, Donnay and Harker and Niggli.

(vii) Step Sources. For growth on singular surfaces some source of steps is required. This source will often be a screw dislocation, but it may be also: the contact between the crystal and its container or mount; grain boundaries between two adjacent crystals of differing orientation; foreign macroscopic particles; twins; two-dimensional nuclei; and in evaporation or dissolution, crystal edges and edge dislocations.

17. Vapor Phase Growth Interface Kinetics: Theoretical and Experimental

a. Burton, Cabrera, and Frank Theory[41]

This theory occupies a central position in the field of vapor phase growth, and is also significant for growth from dilute solution, and to some extent, for growth from the melt.

(i) Mobility and Mean Path Length of Adatoms. We have seen [Eq. (16.23)] that there exists a concentration n_s of adsorbed atoms or molecules when the crystal is in contact with its vapor with a formation energy $w_s < W$, the heat of evaporation. Due to fluctuation of the thermal oscillations of the particles in their potential wells, there is a certain probability of escape to the vapor, which provides for the evaporation of the adatoms, and hence the lifetime τ_s is the inverse of this, given by Eq. (16.24). Similarly there will be potential barriers to lateral motion on the surface, as can be better visualized by examining a cork-ball model of a surface. The activation energy is now that required to surmount the saddle position of a row of surface atoms. Figure 23 shows the various potential energies schematically. Here the time τ required for a lateral jump is given by (see Fig. 23)

$$1/\tau = \nu' \exp(-U_s/kT) \qquad (17.1)$$

where U_s is the activation energy for surface diffusion (Fig. 23) and ν' is another frequency of the same order of magnitude as ν_2, i.e. 10^{12} or 10^{13} s^{-1}. Einstein's relation for random walk surface diffusion is, where D_s is

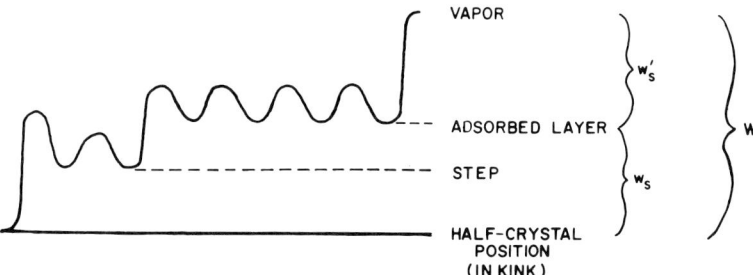

Fig. 23. Potential energy of an adatom at various positions on a crystal surface (schematic). The half-crystal position is at a kink (repeatable step site) [O. Knacke and I. N. Stranski, *in* "Progress in Metal Physics" (B. Chalmers and R. King, eds.), Vol. 6, p. 181. Pergamon Press, Oxford, 1956].

the surface diffusion coefficient,

$$D_s = a^2/\tau \tag{17.2}$$

where a is the separation between adjacent sites. (We omit a factor $\frac{1}{4}$ in Einstein's relation to preserve consistency with BCF's original treatment.) Thus

$$D_s = a^2 \nu' \exp(-U_s/kT). \tag{17.3}$$

The mean displacement before reevaporation, X_s, is given by

$$X_s^2 = D_s \tau_s \tag{17.3'}$$

and, invoking Eq. (16.24), is

$$X_s = a(\nu'/\nu_2)^{1/2} \exp[(w_s' - U_s)/2kT] \tag{17.4}$$

where the coefficient $(\nu'/\nu_2)^{1/2}$ may be assumed to be not far from unity. A rough estimate for X_s can be obtained by assuming nearest-neighbor interaction energies φ on a (111) surface—then $w_s' = 3\varphi = W/2$; assuming U_s small compared to w_s gives $X_s \sim ae^{3\varphi/2kT}$, which for $\varphi/kT = 4$ gives $\sim 400a$, i.e. perhaps 10^{-5} cm. We note that for this example X_s is about 100 times X_0, the mean separation between kinks in a step.

(*ii*) *Motion of a Single Straight Step.* The growth process, using the Kossel model, is visualized as

vapor ↔ adsorbed atom

adsorbed atom ↔ step

along step ↔ kink

The probability of direct condensation of an atom from the vapor into the kink or step is small, so long as the surface area occupied by steps is small (i.e. relatively small step density, vicinal surface). If the kink density is sufficiently high, as noted, the problem of diffusion along the step to the kink may be ignored since it would not be rate controlling. Fick's law of diffusion is (for one-dimensional flow along the y axis, perpendicular to the step)

$$j_s = -D_s \, \partial n_s/\partial y = D_s n_{s0} \, \partial \psi/\partial y \tag{17.5}$$

where j_s is the surface current and n_s is the actual and n_{s0} the equilibrium surface concentration of adatoms, and $\psi = \sigma - \sigma_s = \alpha - \alpha_s$, where $\alpha = p/p_0$ is the saturation ratio of the vapor assumed to have the uniform pressure p over the surface, p_0 is the equilibrium pressure, the supersaturation $\sigma = \alpha - 1$, α_s is the surface saturation ratio n_s/n_{s0}, $\sigma_s = \alpha_s - 1$. There will also be a net flow of atoms j_v from the vapor to the surface given by

$$j_v = (\alpha - \alpha_s) n_{s0}/\tau_s = n_{s0} \psi/\tau_s, \tag{17.6}$$

since the number of adatoms leaving the surface is $n_s/\tau_s = \alpha_s n_{s0}/\tau_s$; and the number arriving, $\alpha n_{s0}/\tau_s$, is just α times the number, n_{s0}/τ_s, arriving or evaporating at equilibrium. If it is assumed that the step motion can be ignored and that a steady state condition is established, then conservation of adatoms requires

$$\text{div } j_s = j_v \tag{17.7}$$

or

$$X_s^2 (d^2\psi/dy^2) = \psi. \tag{17.8}$$

If it is assumed that the exchange of adatoms with kinks is very rapid, the surface saturation ratio at the step may be assumed to be the equilibrium value ($\alpha_s = 1$ at $y = 0$); far from the step α_s has its maximum value ($\alpha_s = \alpha$ at $y \to \infty$). Then a solution fitting Eq. (17.8) and the boundary conditions is

$$\psi = \sigma \exp(-y/X_s) \tag{17.9}$$

and hence the flux into the step at $y = 0$ is

$$j_s = -D_s n_{s0} (\sigma/X_s). \tag{17.10}$$

The step velocity v_∞ is $-j_s/n_0$ where $1/n_0$ is the area per molecular position; thus

$$v_\infty = 2\sigma \nu_2 X_s \exp(-W/kT) \tag{17.11}$$

where $W = w_s + w_s'$ and where the factor of 2 arises because the step is fed from both sides. The step speed is seen to be independent of its crystallographic orientation.

Burton et al.[41] then introduce two multiplicative factors β and c_0, each less than or equal to unity, on the right-hand side of Eq. (17.11), to allow for the possibility that certain of the assumptions made in deriving (17.11) might not hold for particular materials—molecular substances, e.g. in contrast to monatomic substances. If there are not sufficient kinks in the steps, i.e. if $X_s \gg X_0$ is not satisfied, then $c_0 < 1$; if there is not sufficiently rapid exchange of adatoms or molecules between the adsorbed layer and the kinks (as might happen if there were steric hindrance to a molecule fitting into the kink) so as to maintain the equilibrium concentration at the step, then $\beta < 1$. Several limiting cases for the calculation of c_0 are treated: one is the extreme case where $X_s \ll X_0$ and where diffusion in (i.e. along) the step is slow and is neglected. The result of solving the diffusion problem for isolated kinks in the step is

$$c_0 = (\pi X_s)/[X_0 \ln(2X_s/\gamma' a)] \qquad (17.12)$$

where $\gamma' = 1.78$. Also treated is the case where the diffusion in the edge of the step, with a diffusion constant D_e, is so fast that direct diffusion from the surface to the kink may be ignored. The general case is very complicated and requires estimates of the relative magnitudes of the current diffusing in the edge of the step and the current diffusing on the surface to the step.

(*iii*) *Motion of Parallel Straight Steps.* Equation (17.8) may be solved for the case where there is a sequence of parallel steps separated by equal distances y_0, with the surface supersaturation $\sigma_s = 0$ at each step as assumed in deriving Eq. (17.9). The result is

$$v_\infty = 2\beta c_0 \sigma X_s \nu_2 \exp(-W/kT) \tanh(y_0/2X_s) \qquad (17.13)$$

which reduces to (17.11) when the steps are far apart, and where we have now inserted the factors β and c_0; the calculation of c_0 is again complicated.

(*iv*) *Motion of a Circular Step.* The solution of the counterpart of Eq. (17.8) for circular symmetry gives the result that for a step of radius ρ,

$$v(\rho) = v_\infty(1 - \rho_c/\rho), \qquad (17.14)$$

valid for small supersaturations, where

$$\rho_c = \gamma_e a/kT \ln \alpha, \qquad (17.15)$$

and where v_∞ is given by Eq. (17.11), where ρ_c is the critical radius for two-dimensional nucleation and γ_e the edge energy of the nucleus, expressed as energy per *molecule*.

It is also shown by BCF that for a series of concentric circles the expression is the same (17.14) where v_∞ is now given by Eq. (17.13).

(*v*) *Screw Dislocation, Growth Spiral, and Growth Rate.* There needs to be a source of steps on crystal faces composed of close-packed planes, for although a given crystal may initially be bounded by high index surfaces with steps present for geometrical reasons, these steps will disappear after a certain amount of growth, leaving the nonstepped surfaces as bounding planes. Frank[160] proposed that crystal dislocations, which had been studied earlier primarily in connection with mechanical properties of crystals, of the "screw" type (as noted earlier the dislocation need not be pure screw but must have a component of its Burger's vector normal to the surface) could provide a continuing source of steps for crystal growth on close-packed surfaces.

As atoms are added to the growth step, it begins to curve, for the point at the dislocation is fixed. Since the step velocity v_∞ is independent of orientation (if $X_0 \ll X_s$), all points on the step, except near the dislocation, move with the same linear velocity and a spiral is formed, which gives a flat cone for the shape of the crystal face. The spiral winds itself up until the curvature at the center becomes $1/\rho_c$ [see Eq. (17.14)], and then the whole spiral "rotates" with stationary shape.

An approximate form for the steady state spiral is the Archimedes spiral, in radial coordinates r, θ

$$r = 2\rho_c \theta \tag{17.16}$$

which has the proper curvature at the origin. Thus for each turn the step advances $4\pi\rho_c$ in a time $4\pi\rho_c/v_\infty = 2\pi/\omega$ where ω is the angular velocity of the spiral rotation. Then the normal growth rate R of the crystal, where the step height (Burgers' vector normal component) is a,

$$R = a\omega/2\pi = n_0\Omega\omega/2\pi = n_0\Omega v_\infty/4\pi\rho_c \tag{17.17}$$

where Ω is the volume per molecule. From (17.13), where $y_0 = 4\pi\rho_c$, (17.15) and assuming for small σ that $\ln(1 + \sigma) \cong \sigma$, one gets the BCF law

$$R = \beta c_0 \Omega n_0 \nu_2 \exp(-W/kT)(\sigma^2/\sigma_1)\tanh(\sigma_1/\sigma) \tag{17.18}$$

where

$$\sigma_1 = 2\pi\gamma_e a/kTX_s. \tag{17.19}$$

When $\sigma \ll \sigma_1$, then one has the parabolic law

$$R \propto \sigma^2 \tag{17.20}$$

and for $\sigma \gg \sigma_1$, the linear law

$$R \propto \sigma. \tag{17.21}$$

For the parabolic law case, the step separation $4\pi\rho_c$ is much greater than X_s and hence the diffusion fields of adjacent steps do not overlap (as seen from Eq. (17.9). Consequently adatoms landing on the crystal surface much further than X_s from the nearest step will tend to evaporate before reaching it, and the growth rate is reduced below that given by the linear law. The form $(\sigma/\sigma_1)\tanh(\sigma_1/\sigma)$ can be regarded as a type of "condensation coefficient" for the step diffusion problem: it is unity for large σ (steps close together) and approaches zero linearly with σ for small σ (steps far apart). The term $\sigma n_0 \nu_2 \exp(-W/kT)$ is (note $n_0 \nu_2 \exp(-W/kT) = p_0(2\pi mkT)^{-1/2}$ is the incoming or outgoing vapor flux under equilibrium conditions[41]) the net incoming vapor flux, which when multiplied by the given condensation coefficient gives the crystal growth rate (apart from the factors β and c_0).

For the more general situation, where the crystal face has more than one screw dislocation emerging from it, a number of cases can be distinguished. First we note that one dislocation is sufficient for growth of the entire crystal face. For two dislocations of opposite sign with equal magnitudes of Burgers' vectors normal components, closed loops will be emitted, and hence growth will proceed, but only if the dislocations are separated by a distance d, greater than $2\rho_c$. For two dislocations of the same sign with $d > 2\rho_c$, the growth rate will be the same as for one dislocation. But if $d < 2\rho_c$, two cases arise: (i) d much less than $2\rho_c$—gives twice the ω value of one dislocation (ii) d less than $2\rho_c$ apart: between one and two times the ω value of one dislocation. Finally, by suitable grouping of dislocations, the above results may be generalized to the case of many dislocations. A group of strength s (excess of screw dislocations of one sign) will have an activity up to $|s|$ times that due to one dislocation, so long as the steps remain sufficiently far apart, compared to X_s. However, the crystal growth rate R can never surpass the linear law for one dislocation nor can it be less than the square law for one dislocation, except for balanced groups $< 2\rho_c$ apart as noted above. The resultant activity will be that of the most active group.

b. Growth Rates in Classical Surface Nucleation Theory

The disk-shaped cluster of Part IV is assumed to form by statistical fluctuations of adatoms on a close-packed surface. We now give, in more detail, this calculation for a cluster–substrate system of the same material. The rate of two-dimensional nucleation is given by[112,144]

$$I_{SD} = Z\omega' n_i \qquad (17.22)$$

where the Zeldovitch nonequilibrium factor Z has a typical value of 10^{-2}, ω' is the frequency with which subcritical embryos (i.e. lacking one particle to be a "stable" nucleus) become nuclei, which is just the rate at which individual adatoms add on to the subcritical clusters, and n_i the concentration of just subcritical embryos. The frequency ω' may be estimated as

$$\omega' = 2\pi r^* a n_s \nu_3 \exp(-U_s/kT), \qquad (17.23)$$

i.e. the product of the number of adatoms $2\pi r^* a n_s$ just adjacent to the embryo of radius r^*, times the attempt frequency ν_3 at overcoming the diffusional barrier to joining the cluster, times the probability $\exp(-U_s/kT)$

of overcoming the said barrier. The concentration n_i is given by

$$n_i = n_0 \exp(-\Delta G_c/kT) \qquad (17.24)$$

where, from Eq. (15.2) setting the nucleus area term equal to zero,

$$\Delta G_c = -\pi \gamma_e^2/h \, \Delta G_v \qquad (17.25)$$

where γ_e is the edge energy per unit length and where

$$\Delta G_v = -(kT/\Omega) \ln(p/p_0). \qquad (17.26)$$

There are several uncertainties involved in estimating ω' from (17.23), including the problem of estimating surface mobilities (especially U_s). We take $\omega' \sim 10^8$. Then for $n_0 = 10^5$, (17.22) and (17.24) give

$$I_{SD} = 10^{21} \exp[-\pi \gamma_e^2 \Omega/h(kT)^2 \ln(p/p_0)]. \qquad (17.27)$$

The nucleation rate, and hence the crystal growth rate, is a very strong function of $p/p_0 = \alpha$. Below a certain supersaturation $\sigma_c = \alpha_c - 1$, the growth rate is essentially zero. As an example, let the lower limit of observable crystal growth rate be 1 atomic layer in 1000 sec on a crystal face of 1 cm² area. Thus, for one nucleus per monotomic layer, $I_{SD} \approx 1 \times 10^{-3}/s$. Letting $\gamma_e h \cong \varphi/2$ and $\varphi/kT \simeq 4$ one gets approximately $10^{-3} = 10^{21} \exp -[(\pi/4)(\varphi/kT)^2(1/\ln(p/p_0)]$ or $\exp[-4\pi/\ln(p/p_0)] = \exp(-55)$, i.e. $\ln(p/p_0) \cong 0.23$, $p/p_0 = 1.26$ or 26% supersaturation. Conversely at 1% supersaturation the growth rate would be quite negligible. The fact that crystals were known to grow at measurable rates even at very small departures from equilibrium led Frank[160] to conclude that such crystals are not perfect and that screw dislocations provide the steps required for their growth.

c. *Further Study of Vapor Phase Growth Theory. Whisker Growth*

(i) *Surface Vacancy Diffusion.* Cabrera and Coleman[13] have pointed out that surface vacancies, having a concentration at least as large as that of the surface adatoms, will exist in a crystal surface and that these vacancies may be able to take part in the transport process as a surface flow in addition to the adatom flow. Hirth[144] showed that for a bond model of a (111) fcc surface the ratio of adatom flux J_{ad} to surface vacancy flux J_{vac} is given by

$$J_{ad}/J_{vac} = \tfrac{1}{3} \exp[(2\varphi + E_{st} - 3\varphi_r)/kT] \qquad (17.28)$$

where φ is the first neighbor energy, E_{st} is the strain energy due to displacement of neighboring atoms as the vacancy exchanges places with an atom, and φ_r is the decrease in bond energy due to relaxation at a vacant site. On the assumption that $E_{st} > 3\varphi_r$, (17.28) would predict that for, e.g. $\varphi/kT = 4$, the surface vacancy flow is negligible compared to that of adatoms.

Hirth[144] has also discussed the rather large preexponential factors obtained in certain measurements[164] of surface diffusion. These may be explained[165] by assuming jump distances a in Eq. (17.3) much larger than a lattice spacing. This is the "hot atom" hypothesis which requires that the atom in the activated state have a long relaxation time for losing the excess energy. Or these large factors may result from entropy effects involving the various vibrational frequencies ν affecting the surface concentration and the surface diffusion coefficient. For a recent detailed discussion of surface diffusion see Gjostein and Winterbottom.[166]

(ii) *Motion of a Single Step—Further Considerations.* As discussed in Part III, crystal growth is in general a moving boundary problem and the differential equations used to describe the flow of matter or heat must take this into account. In the treatment by BCF, Eq. (17.8) does not include a term in the velocity, although the authors did propose a criterion involving the mean travel distance X_s before reevaporation. Hirth[144] includes a velocity term in the diffusion equation (however, still assuming a steady state condition). The result is that the BCF solution (17.9) is valid when

$$(n_s - n_{s0})/n_0 \ll 1, \qquad (17.29)$$

i.e. a condition on the surface concentration, which will certainly hold if $n_s/n_0 \ll 1$. In Eq. (16.23), where a typical value for w_s, the kink to surface evaporation energy is 3φ for the (111) plane, $\varphi/kT = 4$ gives $n_s/n_0 \sim \exp(-12) \ll 1$.

Although BCF explicitly included two "efficiency" factors β and c_0 [Eq. (17.13)], there is yet another factor implicitly assumed to be unity by them. This is the condensation coefficient,[167] α_k, for the process vapor → adsorbed layer, which only enters in (17.13) or (17.18) when the term $\sigma n_0 \nu_2 \exp(-W/kT)$ is replaced by the term $\alpha_k \sigma p_0 (2\pi mkT)^{-1/2}$. The α_k is less than unity if some of the incoming molecules are reflected during the collision with the surface; it can also be less than unity for complex molecules for reasons involving rotational degrees of freedom and steric effects.

---

[164] N. A. Gjostein *in* "Metal Surfaces: Structure, Energetics, and Kinetics," p. 99. Am. Soc. Metals, Cleveland, Ohio, 1963.
[165] J. Y. Choi and P. G. Shewmon, *Trans. AIME* **230**, 123 (1964).
[166] N. A. Gjostein and W. L. Winterbottom, *in* "Fundamentals of Gas-Surface Interactions" (H. Saltsburg, J. N. Smith, Jr., and M. Rogers, eds.), p. 42. Academic Press, New York, 1967.
[167] O. Knacke and I. N. Stranski, *Progr. Met. Phys.* **6**, 181 (1956).

Cabrera[168] and Zwanzig[169] have calculated α_k on the basis of a classical mechanics, one-dimensional spring and mass chain type model, with the result that for an atom impinging on like atoms, α_k is unity. More recent calculations on three-dimensional models have been made by F. O. Goodman and others. (A recent review is that by Trilling.[170])

(iii) Parallel Steps; Spiral Growth. Chernov[17] has emphasized that the BCF calculation for the shape and rotation rate of a spiral depends on the mechanism of growth only through the step velocity, v_∞, and hence should apply not only to vapor growth but also to solution growth and melt growth, assuming of course that the general model of growth by step motion over close-packed surfaces is still relevant. In any case v_∞ will in general be different from the BCF vapor phase model due to different transport effects near the steps.

Cabrera and Levine[171] have made a more accurate calculation of the spiral step spacing, which was $4\pi\rho_c$ in the original theory, and have found a value $y_0 = 19\rho_c$. This result increases σ_1 in Eq. (17.19) to

$$\sigma_1 = (19/2)\gamma_e a/kTX_s, \qquad (17.30)$$

which has the effect of reducing somewhat the condensation coefficient term $(\sigma/\sigma_1)\tanh(\sigma_1/\sigma)$ for a given value of σ, and hence reduces the growth rate for low $\sigma \lesssim \sigma_1$, as might be expected for the greater step separation permitting more reevaporation. For $\sigma \gg \sigma_1$ there is no change in R. In fact, BCF[41] had already discussed an improvement over the Archimedes spiral, obtaining $y_0 = 4\pi(1 + 3^{-1/2})\rho_c \approx 20\rho_c$. Cabrera and Levine also considered the effect of the elastic strain energy of the crystal around the dislocation on the behavior of the spiral, particularly in the case of evaporation, where an asymmetry with growth is introduced since the dislocation strain energy opposes growth and favors evaporation.

Hirth and Pound[112,172,173] and Hirth,[144] in studying the spacing of steps emitted from a step source (originally[172,173] a crystal edge in evaporation) concluded that when $\sigma \gtrsim \sigma_1$, the leading growth steps of a finite train (i.e. those relatively far from the dislocation origin) tend to accelerate their motion. This acceleration continued to the point where the step spacing y_0 becomes $y_0 \approx 6X_s$, consequently the condensation coefficient $\alpha = (\sigma/\sigma_1)\tanh(\sigma_1/\sigma)$ will approach $\frac{1}{3}$. The arguments leading to these con-

---

[168] N. Cabrera, *Discussions Faraday Soc.* **28**, 16 (1959).
[169] R. W. Zwanzig, *J. Chem. Phys.* **32**, 1173 (1960).
[170] L. Trilling, *in* "Fundamentals of Gas-Surface Interactions" (H. Saltsburg, J. N. Smith, Jr., and M. Rogers, eds.), p. 392. Academic Press, New York, 1967.
[171] N. Cabrera and M. M. Levine, *Phil. Mag.* **1**, 450 (1956).
[172] J. P. Hirth and G. M. Pound, *J. Chem. Phys.* **26**, 1216 (1957).
[173] J. P. Hirth and G. M. Pound, *Acta Met.* **5**, 649 (1957).

clusions on acceleration and step spacing do not appear entirely clear, however.

(iv) Back Stress Effect. Cabrera and Coleman[13] have pointed out that the surface supersaturation σ_s existing near the first turn of the spiral (which in the case of the simple Archimedes spiral influences the curvature [see Eqs. (17.15) and (17.16)] of the spiral at the origin and also the spiral arm spacing) may not be as high as the bulk vapor supersaturation σ, as originally assumed. This is so since there will be a surface diffusion flow to that portion of the spiral, reducing the value σ_s it will "see" to $\sigma_s < \sigma$. This "back stress" effect is approximately calculated by replacing $\ln \alpha \approx \sigma$ in Eq. (17.15) with σ_s, where σ_s is then found by solving the diffusion continuity equation (17.8) (generalized to two-dimensions) for a circular step of radius ρ_0. We obtain

$$\frac{kT \ln[1 + \sigma_s(\rho_0)]}{kT \ln[1 + \sigma]} \equiv \frac{\Delta\mu_0(\rho_0)}{\Delta\mu} = 1 - \frac{1}{I_0(\rho_0/X_s)}. \qquad (17.31)$$

Here the surface supersaturation at the origin has been assumed to equal σ, I_0 is the modified Bessel function of imaginary argument, and the $\Delta\mu$ are chemical potentials given by (17.31). Defining $\Delta\mu_1 = 19\Omega\gamma_e/hX_s$ where γ_e is edge energy per centimeter (the same argument holds for the simple Archimedes spiral where 19 is replaced by 4π), one finds from (17.31) and the definition of I_0, where $\Delta\mu_0 = 19\Omega\gamma_e/h\rho_0$, that

$$(\rho_0/2X_s)^3 = \Delta\mu_1/2\,\Delta\mu \qquad (17.32)$$

for large supersaturations, i.e. $\Delta\mu \gg \Delta\mu_1$, $\rho_0 \ll X_s$. Consequently, the rate of growth given by (17.18) as $R = R_m(2X_s/\rho_0) \tanh(\rho_0/2X_s)$, where R_m is the maximum growth rate, becomes, for large $\Delta\mu$,

$$R = R_m[1 - \tfrac{1}{2}(\Delta\mu_1/2\,\Delta\mu)^{2/3}]. \qquad (17.33)$$

As Cabrera notes, this expression approaches R_m very slowly as $\Delta\mu$ increases, compared to the behavior of the earlier treatment which had, for $\sigma \gg \sigma_1$ corresponding to large $\Delta\mu$, $(\sigma/\sigma_1) \tanh(\sigma_1/\sigma) \sim [1 - \tfrac{1}{2}(\sigma_1/\sigma)^2]$. Physically, as the supersaturation $\Delta\mu$ increases, the spiral would, according to the 1951 treatment, wind up more tightly in direct proportion to $1/\Delta\mu$, whereas in the present treatment, as $\Delta\mu$ increases, the effective value $\Delta\mu_0$ increases less rapidly. Thus the growth rate increases more slowly with $\Delta\mu$ than before.

(v) Whisker Growth. The surface diffusion treatment for vapor phase whisker growth has already been discussed in Part III as a Stefan problem. The basic model is of course similar to that of the BCF picture for surface diffusion to a step, the step in this case being at the end of the whisker, where the screw dislocation emerges. Simple theory (e.g. Sears,[94] Dittmar

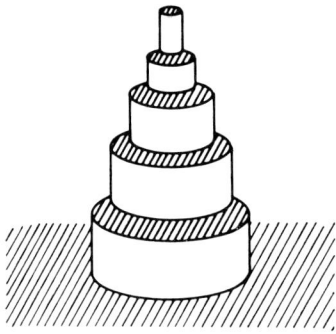

Fig. 24. Whisker growth model involving anisotropy in step capture probability for adatoms. Adatoms are captured on shaded (step) regions [R. L. Schwoebel, *J. Appl. Phys.* **38**, 1759 (1967)].

and Neumann[95]) predicts that the length will initially increase exponentially in time, since the collection area is the entire whisker. As the whisker length becomes large compared to X_s it then increases linearly with time, the collection area than accompanying the tip motion. The exact theory[101] shows that deviations from simple theory may permit D_s and τ_s, in principle, to be separately measured, although surface nucleation on the whisker sides may intervene first. Usually, it is only their product, $X_s = (D_s\tau_s)^{1/2}$, which is determined.

Schwoebel[174] has proposed a vapor phase whisker growth model involving surface diffusion but not requiring a screw dislocation. A conical structure of 20 concentric tiers receives atoms from the vapor phase at a uniform rate, and none reevaporate. All atoms striking the sides of the tiers move by surface diffusion to the steps and are incorporated only at steps. Unlike the BCF assumption, a step (horizontal portions of Fig. 24) is assumed to have a different capture probability for an adatom moving toward it from below than from above. This difference could result from adatom coordination differences near the step, although the sum of the capture probabilities is unity. It is this anisotropy which results in the step motion anisotropy. No diffusion fields or concentration gradients are introduced in the theory, which is essentially geometrical in nature. The mobility of adatoms on the sides of the tier is assumed to be higher than on the steps, but is not otherwise specified. Twenty coupled differential equations of continuity are then solved numerically in order to obtain the height of each tier as a function of time. It is found that, for sufficient difference of the capture coefficients, an initial conical shape can quickly become a sharp-pointed whisker, and the length quickly becomes an exponential function

[174] R. L. Schwoebel, *J. Appl. Phys.* **38**, 1759 (1967).

of time. Even when the starting system is a very flat cone with just one cylindrical nucleus, the structure is found to peak.

d. Macroscopic Steps: Vapor Phase (Excluding Kinematic Wave Studies)

The expression [Eq. (17.11)] for the (tangential) velocity of a single step does not contain a factor allowing for the step height, since it was implicitly assumed that the step is of unit height when writing $v_\infty = -j_s/n_0$. A very high step, however, will not have a high kink density, for such a step may well correspond to a close-packed crystal face; we have already noted how, in the equilibrium case, a hill-and-valley structure (due to step attraction) composed of close-packed surfaces could lower the total surface free energy. Thus such a step could serve as a sink for adatoms if secondary steps, such as a surface nucleus, were formed on it.[175]

In addition to the hill-and-valley effect, the origins of high steps include simply a dislocation of large Burgers' vector b. Since the elastic strain energy of a dislocation is proportional to b^2, such dislocations will tend, however, to decompose into a number of parallel dislocations of smaller Burgers' vector, unless a hollow core is formed to relieve this strain energy.[176] Other sources of high steps include kinematic wave effects and impurity bunching effects, to be discussed below.

Chernov[17] has studied the shape profile of a high step as it is supplied with adatoms at both its top and bottom. He assumed that the step is sufficiently rough not to require steps for its growth. The result is that an S-shaped profile develops, the step bulging at the top and at the bottom where it is fed by the adjacent surfaces.

e. Experimental Studies of Vapor Phase Growth

Quantitative vapor phase growth rate studies in which the rate R is measured as a function of supersaturation σ, for various crystal temperatures, in a clean system, are still relatively few; particularly those in which the surface topography was closely followed. There have been, however, numerous observations of the predicted growth spirals, together with the simple observation, often made, that growth occurs at measurable rates for real crystals at supersaturations much less than would be permitted by surface nucleation theory. The experimental measurements have been reviewed in detail by Hirth and Pound[112] and also by Heyer.[177] We mention

---

[175] N. Cabrera, *discussion in* "Structure and Properties of Solid Surfaces" (R. Gomer and C. S. Smith, eds.), p. 294. Univ. of Chicago Press, Chicago, Illinois, 1953.

[176] F. C. Frank, *Acta Cryst.* **4,** 497 (1951).

[177] H. Heyer, *Angew. Chem. Intern. Ed. Engl.* **5,** 67 (1966); or see *Angew. Chem.* **78,** 130 (1966).

here three quite recent careful experiments: Heyer,[178] Schwoebel,[179] and Bethge.[180]

Heyer[178] measured by optical means R vs σ on crystals of urotropine, arsenolite, and iodine, growing from the vapor in evacuated, sealed-off glass tubes. Two stages of growth were observed for all three substances: undisturbed and disturbed growth. In the latter, visible defects appear on the crystal surface as the crystals become larger than a certain size, and a linear growth law is obtained; for undisturbed (small) crystals a square law is obtained. The results are in agreement with the BCF[41] predictions if it is assumed that the appearance of visible defects corresponds to a substantial increase in dislocation density so as to make step separations less than or comparable to X_s. The condensation coefficients measured were generally less than unity even in the linear range, which may indicate foreign gas effects in the system. Schwoebel[179] studied the condensation of gold atoms upon gold single crystals by use of a microbalance in an ultrahigh vacuum system. The effective condensation coefficient varied from unity at $\sim 900°K$ to near zero at $1200°K$, and was interpreted to be in reasonable agreement with the diffusion-to-steps part of the BCF (1951) model. It was assumed, however, that the step spacings were constant as the crystal temperature was varied, which would not be true of screw dislocation step sources. Further interesting work by Schwoebel[181] on the detailed morphology of vapor-deposited gold crystals, in particular the observation of trigonal growth centers as seen by electron microscope replica techniques, led to the conclusion that the steps were not perfect sinks for the capture of gold atoms but that the probability of capture depends on step orientation. He further suggested that the activation energy for surface diffusion in such a case will involve not only the motional activation energy U_s but also the energy required to dissociate the atom from a step to the crystal surface.

Bethge[180] has studied the topography of the evaporation and growth from the vapor of cleavage faces of NaCl crystals by using the gold decoration technique which sensitively reveals steps of monatomic spirals, formed in evaporation around screw dislocations. It was also observed, for regions far from a step that sufficient undersaturation produces disk-shaped holes of monatomic depth by the surface nucleation process.

[178] H. Heyer, *in* "Crystal Growth" (H. S. Peiser, ed.), Suppl. *J. Phys. Chem. Solids*, p. 265. Pergamon Press, Oxford, 1967.
[179] R. L. Schwoebel, *in* "Solid Surfaces" (H. C. Gatos, ed.), p. 356. North-Holland Publ., Amsterdam, 1964.
[180] H. Bethge, *in* "Crystal Growth" (H. S. Peiser, ed.), Suppl. *J. Phys. Chem. Solids*, p. 623. Pergamon Press, Oxford, 1967.
[181] R. L. Schwoebel, *J. Appl. Phys.* **38**, 672 (1967).

18. Solution Growth Interface Kinetics

a. Burton, Cabrera, and Frank[41]

In the case of vapor phase growth, the calculated path for an atom was as follows: (i) vapor to surface, surface to step, and along step to kink. Other conceivable parallel routes include (ii) vapor to kink, (iii) medium to step to kink, and (iv) medium to surface to kink. Route (i) is appropriate for pure vapor growth, for it clearly results in the predominant mass transport in that case. However, if in the case of growth from solution, one or more of the components of (i) has a substantial resistance (small transport for given driving force) then (ii), (iii), or (iv) may be more appropriate. Clearly the relative magnitude of the surface D_s and volume D (i.e. in the medium) diffusion coefficients, together with the relevant concentrations, and the step and kink morphology, will determine the appropriate route for solution growth. Unfortunately, not all of these quantities are sufficiently well known.

BCF,[41] in their solution growth theory, chose to study path (ii), direct deposition from the medium to the kink, as would be the case if the surface mobility and the mobility in the step were very low. The geometry assumed is a crystal surface on which there is a set of parallel steps separated by a distance y_0 from each other, and having kinks a distance X_0 apart. This rather difficult diffusion problem is approached by assuming diffusion fields having the equilibrium concentration at the kink site, and extending out to a radial distance X_0 from the kinks. Half-cylindrical diffusion fields around each step match the hemispherical field at a radial distance $r = X_0$ and hold on out to $r = y_0$. Finally, a planar diffusion field (coordinate z) matches the half-cylinder at $r = z = y_0$ and holds out to $z = \delta_c$, which is the thickness of the unstirred portion of the solution near the crystal (more precisely, concentration boundary layer thickness). From a consideration of the flux through one hemisphere, the step velocity is given by

$$v_\infty = DC_0\Omega[2\pi\sigma(X_0)/X_0] \qquad (18.1)$$

where $\sigma(X_0)$ is the supersaturation at radial distance $r = X_0$ from the kink, and C_0 the equilibrium concentration in the solution (not to be confused with the BCF c_0 defined just before Eq. (17.12)). By equating the three series of fluxes—hemispherical, cylindrical, planar—the supersaturation σ at $z = \delta_c$, which equals the bulk supersaturation σ, can be expressed in terms of $\sigma(X_0)$ as

$$\sigma(X_0)/\sigma = \{1 + [2\pi a(\delta_c - y_0)/X_0 y_0] + (2a/X_0)\ln(y_0/X_0)\}^{-1}. \qquad (18.2)$$

Using (17.16) and (17.17) for the growth spiral, which is assumed to have the same properties as before,

$$R = DC_0\Omega a\sigma(X_0)/2X_0\rho_c = DC_0\Omega kT\sigma^2(X_0)/2X_0\gamma_e \qquad (18.3)$$

where $\rho_c = \gamma_e a/kT\sigma(X_0)$ [compare Eq. (17.15) for low supersaturation], and where $y_0 = 4\pi\rho_c$ as before. For sufficiently low supersaturations $\sigma(X_0)$, y_0 will be large and the logarithmic term in Eq. (18.2) will predominate, giving an approximately linear relation between $\sigma(X_0)$ and σ, hence giving a parabolic law for Eq. (18.3). (Note that it is parabolic for a somewhat different reason than in the vapor phase case: there it was due to reevaporation of surface diffusing adatoms between widely spaced steps; here there is no surface diffusion, but available kink sites per unit area are decreased as the step-separation increases.) For high supersaturations, the middle term of Eq. (18.2) will be large as y_0 becomes small compared to δ_c, and $\sigma^2(X_0)$ is proportional to σ, giving a linear law.

b. Chernov[17]

Chernov has also studied the growth of crystals from solution by analysis of the diffusion field associated with parallel sequence of steps. He also considers the route from the medium to step to kink, and thus ignores surface diffusion. Diffusion within the step to the kink is also considered, and the diffusion profiles around kinks are not individually studied as in the BCF treatment. Instead, their influence is accounted for in a linear kinetic law for flow of molecules into the steps by the formula

$$D\, \partial C/\partial r = \beta(C - C_0) \qquad (18.4)$$

where C is the concentration, C_0 the equilibrium concentration, r the (cylindrical) coordinate around the step, and where β will be large (fast kinetics) if the kink density is large. Figure 25 shows the diffusion field as studied by Chernov; at the boundary layer of thickness δ_c, C is assumed to be a constant C_δ, equal to the bulk concentration C_∞. Laplace's equation for this system is then solved by a conformal mapping technique. The result for the growth rate due to a screw dislocation is

$$R = \frac{\beta akTC_0}{4\gamma_e} \frac{\sigma^2}{1 + (\beta a/D)\ln[(\delta_c\sigma_c'/a\sigma)\sinh(\sigma/\sigma_c')]} \qquad (18.5)$$

where $\sigma_c' = 4\Omega\gamma_e/kT\delta_c$, and γ_e is the edge free energy per unit area. The product βa is regarded by Chernov as the diffusion coefficient for passage through the solution–crystal interface, and may be approximated roughly by the volume diffusion coefficient D. For very low supersaturations

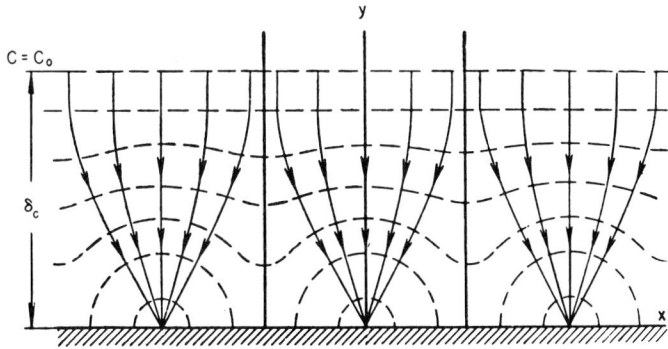

FIG. 25. Diffusion fields of parallel steps in solution growth, showing lines of flux and of equal concentration [A. A. Chernov, *Soviet Phys.-Usp. (English Transl.)* **4**, 116 (1961)].

$\sigma \ll \sigma_c$, a parabolic law $V \propto \sigma^2$ is again predicted. For large σ a nearly linear relation results.

It should be noted that although volume diffusion fields are being considered here, the scale of dimensions involved—say 10^{-7} cm for X_0, 10^{-5} cm for y_0, and 10^{-3}–10^{-5} cm for δ_c—is still generally quite small compared to crystals having observable faces. Therefore, the scale is quite small compared to macroscopic diffusion fields around crystals. The relatively microscopic fields dealt with here are thus quite properly regarded as aspects of the interface kinetics problem and may, as we see, provide substantial impedance to growth (σ^2 laws). Of course, in general, both macroscopic and microscopic diffusion fields will be involved.

c. Bennema[182]

Bennema has performed a careful experimental study of the growth of single crystals from aqueous solution at low supersaturations of potassium aluminum alum and of sodium chlorate, in order to test the above theories. The growth rate was determined by suspending the crystal from an analytical balance in a supersaturated solution. Figure 26 shows the data for sodium chlorate. For the conditions of these experiments, Chernov's σ_c was estimated to be $\sim 10^{-5}$, so that the linear law should hold. When the linear form of (18.5) is extrapolated to $R = 0$ it does not go through the origin but intersects the σ axis at a value of about $\sigma = 10^{-4}$ theoretically, and 2×10^{-4} experimentally as seen in Fig. 26. Also, the slope of the line,

[182] P. Bennema, *in* "Crystal Growth" (H. S. Peiser, ed.), Suppl. *J. Phys. Chem. Solids*, p. 413. Pergamon Press, Oxford, 1967.

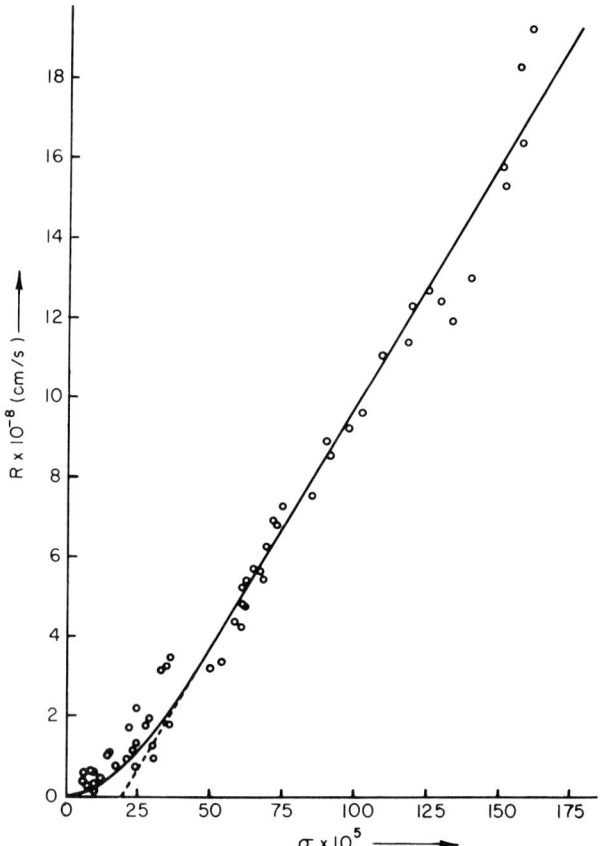

FIG. 26. Crystal growth rate versus solution supersaturation for {100} faces of sodium chlorate (○); BCF surface diffusion equation, —[P. Bennema, in "Crystal Growth" (H. S. Peiser, ed.), Suppl. *J. Phys. Chem. Solids*, p. 413. Pergamon Press, Oxford, 1967].

$D\Omega C_0/\delta_c$ agrees with the experimental slope. Here δ_c is estimated by a formula from hydrodynamics (see Part VIII). However, when δ_c is experimentally decreased by about 50% (increasing stirring rate) the growth rate does not change, although it would be expected to vary by the same factor in the linear case. (The same agreement with the experimental slope and the same criticism with regard to δ_c would apply to the BCF theory equation (18.3) for which the linear form gives the same slope $DC_0\Omega/\delta_c$.) Bennema also suggests that the approximation $\beta a = D$ made by Chernov for the solution–step kinetic coefficient may be unwarranted. The activation energy for an ion to enter a kink is composed of the dehydration activation energy of the kinks and the dehydration energy of the ions. Together these, according to Bennema, may total 6.27 to 12.5×10^4 J/mol (15 to 30 kcal/mol) and even if only 4.18×10^4 J/mol (10 kcal/mol) the diffusion coefficient for entry into kinks would be perhaps only 10^{-4} that of the volume diffusion coefficient, which would make $\beta a/D \sim 10^{-4}$ instead of the value of unity assumed. This situation in turn would predict a nonlinear law for the range of σ values studied. Also shown in Fig. 26 is a *surface* diffusion BCF plot $\text{const}(\sigma^2/\sigma_1)\tanh(\sigma_1/\sigma)$. Bennema feels that the good fit observed, together with the problems of the volume diffusion models mentioned above, makes the surface diffusion process the more likely one, although no theoretical estimates of σ_1 or of surface mobilities are given. The dehydration process for the ion would then consist of both a surface stage and a kink stage.

d. Growth Spirals

Among the numerous examples of growth spirals observed, both in as-growing crystals and after growth, we mention the work of Forty[183] and of Newkirk[184] on CdI_2 and of Kozlovskii[185] on beta methyl napthalene. Also, even before the screw dislocation was discovered in 1939 (by Burgers), spirals on paraffin crystals were observed by Heck in 1936 and 1937.[186,187] Also, the book by Verma[188] has many beautiful pictures of spirals on SiC.

e. Doremus[189]

This author studied the growth of ionic crystals from dilute supersaturated aqueous solution with special attention to the interface mechanisms. He interpreted precipitation experiments where many very small crystals of microns or less in size—e.g. in the precipitation of barium sulfate—are formed, and where only a kind of average growth rate, averaged over the size and orientation, is determined by such a method as the electrical conductivity of the remaining solute. The problem was to explain why the solute flux per unit area of particle surface J had the form, where b_3 is a constant,

$$J = b_3(C_s - C_0) \tag{18.6}$$

[183] A. J. Forty, *Phil. Mag.* **43**, 377 (1952).
[184] J. B. Newkirk, *Acta Met.* **3**, 121 (1955).
[185] M. I. Kozlovskii, *in* "Growth of Crystals" (A. V. Shubnikov and N. N. Sheftal, eds.), Vol. III, p. 101. Consultants Bureau, New York, 1962.
[186] C. M. Heck, *Phys. Rev.* **51**, 686 (1937).
[187] See Doremus *et al.*,[71] p. 6.
[188] A. R. Verma, "Crystal Growth and Dislocations." Academic Press, London, 1953.
[189] R. H. Doremus, *J. Phys. Chem.* **62**, 1068 (1958).

where b_3 is a constant, C_0 is the equilibrium concentration, C_s the concentration at the interface (assumed equal to the bulk original concentration C_∞ in the medium reduced only by the amount of precipitation that has occurred—i.e. it was assumed that the process for the particles considered was interface controlled, diffusion from the medium to the surface being believed sufficiently fast), and n was often an integer—2, 3, or 4 for example, depending on the experimental system studied. Two models were considered—both involving stoichiometric reactions on the surface. The first assumed that the two adsorbed ions of opposite sign combined on the crystal surface in the presence of a third ion, in analogy with gas phase reactions, to form a neutral molecule which could then diffuse to a kink in a step and that the rate of adding of such molecules to the step was rapid. The second model assumed that the ad-ions of both signs in the adsorbed layer on the crystal surface proceeded alternately, as opposite charges, to the kinks. For one–one electrolytes the first model was found to predict $n = 3$, the second $n = 2$. The rate coefficients are presumably quite complex and are not evaluated in this theory, involving such quantities as collision probabilities.

Summaries of other studies on precipitation of fine particles are given

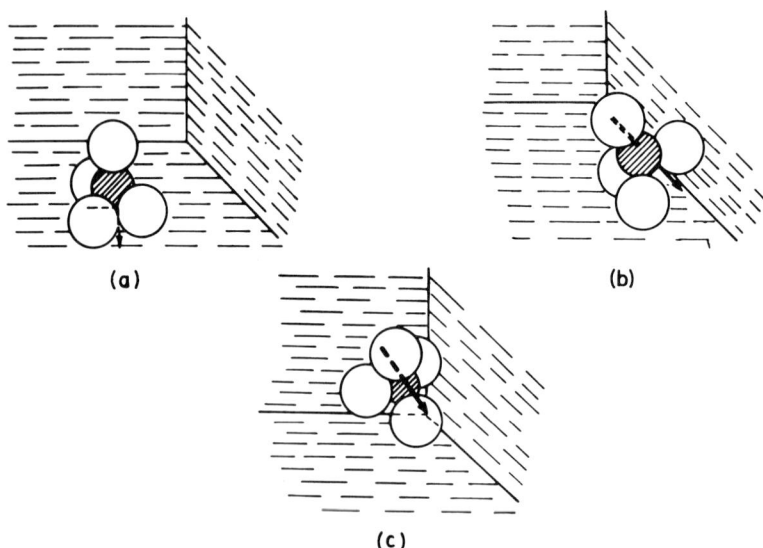

FIG. 27. Displacement of hydration sheath around (cross-hatched) ion in electrodeposition at (a) plane surface, (b) edge, and (c) corner [J. O'M. Bockris and G. A. Razumney, "Fundamental Aspects of Electrocrystallization." Plenum Press, New York, 1967].

in the review by Nancollas and Purdie,[190] and in the books by Nielsen[115] and Walton.[116]

f. Electrodeposition of Crystals

In electrodeposition of metals, which can be regarded as a low temperature type of crystal growth of metals from solution, a key step is the charge transfer from the diffusing charged ion to the metallic crystal. For the ion with a hydration sheath of water molecules, the sheath must be partially displaced for the transfer to take place. The activation energy for such displacement can be substantial and also can vary with the crystal position at which the displacement takes place (Fig. 27). Conway and Bockris (see Bockris and Razumney[191]) have calculated that the heat of activation for direct transfer to a surface site is much less than that for direct transfer to an edge or to a kink, so that the former is the likely first stage of the process. Then the partially hydrated adion must surface diffuse to a step and finally to a kink, losing molecules of hydration at each point. Whether the charge transfer or surface diffusion or kink incorporation is rate controlling is not known.

19. Melt Growth Interface Kinetics

a. Jackson Model

The model of Jackson[42,43] for the roughness of a crystal surface in contact with its melt was presented in Part II and further discussed in Section 16 on interface structure. The parameter α is defined as $(L/kT_0)f_k$ where T_0 is equilibrium temperature, i.e. melting temperature in melt growth, and L is the latent heat of fusion. $f_k \leq 1$ is the crystallographic factor, e.g. is the fraction of the total binding energy, on a nearest-neighbor model, binding the molecule in a layer parallel to the plane considered. If $\alpha < 2$, the interface is rough and hence steps and stepwise growth would not be needed for the growth of a crystal; for $\alpha > 2$ the interface is smooth (Fig. 7) and steps will be necessary.

Jackson et al.[192] and Jackson[193] distinguished three classes of materials according to their entropies of fusion L/kT_0, and found remarkable experimental verification for the surface model. The lowest entropy of fusion range, which includes most metals and certain compounds, includes CBr_4

[190] G. H. Nancollas and N. Purdie, *Quart. Rev.* **18**, 1 (1964).
[191] J. O'M. Bockris and G. A. Razumney, "Fundamental Aspects of Electrocrystallization." Plenum Press, New York, 1967.
[192] K. A. Jackson, D. R. Uhlmann, and J. D. Hunt, *J. Cryst. Growth* **1**, 1 (1967).
[193] K. A. Jackson, *Progr. Solid State Chem.* **4**, 53 (1967).

for which $L/kT_0 = 0.8$. Direct observation,[192] during melt growth, of the interface shape of CBr_4 shows that it follows the isotherm strictly, as would be expected for a rough interface. Similarly, metals growing from undercooled melts show dendritic morphology indicating a rough interface with heat flow control. The intermediate range, containing most inorganic and organic crystals, has L/kT_0 values ranging from about 2.0 to 4.0. Depending on the f value for the particular face in question crystals in this range can exhibit morphology corresponding to either a rough or a smooth interface; the semimetal bismuth has $L/kT_0 = 2.4$ and shows both dendritic growth and on a finer scale, well-defined crystal facets. Salol $(L/kT_0 = 7)$ is in the third (high entropy of fusion) class and shows highly anisotropic, faceted growth where temperature differences of several degrees over a given face are possible.

b. Jackson et al. Reviews

Jackson et al.[192,193] recently reviewed the status of current theories of the interface kinetics of melt growth and also carefully evaluated many of the significant measurements of interface kinetics. We summarize relevant parts of these reviews here.

(i) *Rough Interface or Normal Growth Theory.* This theory, due originally to Wilson[194] and further studied by Jackson and Chalmers,[195] assumes that the processes of melting or of freezing occur simultaneously and independently and that growth or melting can take place on any site on the crystal surface. A further assumption is that melting and freezing are activated processes, the difference in activation energies being just the latent heat of fusion. Thus the net growth rate $\bar{R}$ of an interface at temperature T is given by the difference in the freezing rate R_F and melting rate R_M as

$$\bar{R} = R_F - R_M = R_{F0}\exp(-Q_F/kT) - R_{M0}\exp(-Q_M/kT). \quad (19.1)$$

The difference in the corresponding activation energies Q_F and Q_M is

$$Q_M - Q_F = L, \quad (19.2)$$

the latent heat of fusion per molecule. At the equilibrium temperature T_E, $R_{M0}/R_{F0} = \exp(L/kT_E)$. Letting $\delta T = T_E - T$, and assuming $L\,\delta T/kTT_E \ll 1$,

$$\bar{R} = R_{F0}\exp(-Q_F/kT)(L\,\delta T/kTT_E) \quad (19.3)$$

which is the linear law. If we describe the atom movements across the interface in a way analogous to that involved in bulk diffusion, then Q_F will be

---
[194] H. A. Wilson, *Proc. Camb. Phil. Soc.* **10**, 25 (1898).
[195] K. A. Jackson and B. Chalmers, *Can. J. Phys.* **34**, 473 (1956).

the activation energy for this atom transport, and $R_{F0} \approx a\nu_4$ where a is a jump distance and ν_4 an atomic vibration frequency. Hence $a\nu_4 \exp(-Q_F/kT) = D_i/a$ where D_i is the diffusion coefficient for transport across the interface, and (19.3) becomes the form given by Hillig and Turnbull[196]

$$\bar{R} = (D_i/a)(\Delta S/kT)\,\delta T \qquad (19.4)$$

where $\Delta S = L/T_E$.

(ii) Growth by Screw Dislocation Mechanism. Hillig and Turnbull[196] pointed out that Eq. (19.4) should be multiplied by a factor $\alpha_{HT} < 1$ when not all sites on the crystal surface are available for growth. In screw dislocation growth α is the fraction of surface sites in the steps of the Archimedes spiral formed by a "screw" dislocation intersecting the crystal surface. Assuming that the step kink spacing is low, i.e. comparable to the lattice spacing a, and that only atoms within a of a step are added to it, $\alpha_{HT} = a/4\pi\rho_{c,m}$ where $4\pi\rho_{c,m}$ is the separation of steps. Since

$$\rho_{c,m} = \gamma_{e,SL}\Omega/\Delta S\,\delta Ta \qquad (19.5)$$

where $\gamma_{e,SL}$ is the edge free energy per unit area and Ω the volume per molecule, then

$$\alpha_{HT} = a\,\Delta S\,\delta T/4\pi\gamma_{e,SL}\Omega$$

and from (19.4)

$$\bar{R} = D_l(\Delta S)^2(\delta T)^2/4\pi\Omega kT\gamma_{e,SL}, \qquad (19.6)$$

the square law.[196]

(iii) Growth by Surface Nucleation. The general form for the crystal growth rate in the case of surface–nucleation controlled growth from the melt is[105,193,197,198]

$$\bar{R} = K_1 \exp(-K_2/T\,\delta T). \qquad (19.7)$$

Although there appear to be a number of different theoretical values for K_1 and K_2, there is general agreement that the $1/\delta T$ term in the exponential dominates all other temperature dependence.

(iv) Growth by Diffuse Interface Mechanism.[198,199] In this theory it was suggested that all liquid–solid interfaces would advance by a layer-passage or step motion mechanism at low driving forces where the driving force is $\Delta G_V = -L\,\delta T/\Omega T_E$ and by continuous growth at high driving forces; a transitional regime joins the two extreme cases. The values of the two driving forces separating the three regimes are $(\gamma g/a)$ and $\pi(\gamma g/a)$ as

---

[196] W. B. Hillig and D. Turnbull, *J. Chem. Phys.* **24**, 914 (1956).
[197] G. A. Alfintzev and D. E. Ovsienko, in "Crystal Growth" (H. S. Peiser, ed.), Suppl. *J. Phys. Chem. Solids*, p. 757. Pergamon Press, Oxford, 1967.
[198] J. W. Cahn, W. B. Hillig, and G. W. Sears, *Acta Met.* **12**, 1421 (1964).
[199] J. W. Cahn, *Acta Met.* **8**, 554 (1960).

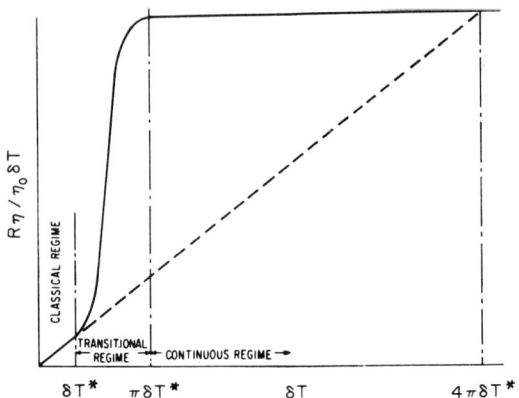

FIG. 28. Transition from layer growth (classical regime) to continuous growth in growth from the melt according to the diffuse interface theory of Cahn. $R\eta/\eta_0 \, \delta T$ involves growth velocity R divided by interface undercooling δT [J. W. Cahn, W. B. Hillig, and G. W. Sears, *Acta Met.* **12**, 1421 (1964)].

shown in Fig. 28 where $\delta T^* = (\Omega T_E/L)(\gamma g/a)$. Here g is a diffuseness parameter, approximately unity for a sharp interface, and smaller for a diffuse interface of larger n; n is the number of atomic layers in the transition zone from liquid to solid at the melting temperature; γ is the interfacial free energy. The value of g for a particular material was not predicted, however, but had to be obtained by curve-fitting with experiment.

(*v*) *Evaluation of Experimental Data and Comparison with Theory.* Jackson and his colleagues[192,193] have made a careful study of interface kinetic data in growth from the melt of about a dozen materials. They noted that there are very few meaningful and reliable data; and that heat flow effects can often interfere with a correct determination of the interface undercooling. Indeed, heat flow can give, when not recognized, a false impression of agreement of theory with experiment, as noted by Turnbull.[200] We emphasize that δT is the *interface* undercooling $T_E - T$, not the bulk liquid undercooling ΔT, which latter is either equal to or greater than δT, and is greater if heat flow plays a role. Preferred techniques for circumventing the heat flow problem included letting growth occur in small capillary tubes under the right conditions[105] (but see discussion of Tammann tube in Part III; also Berlad[103,104]; and also Kirtisinghe[106] who found that growth rates were greatly influenced by the tube), the thermal wave technique,[201] and the study of materials of low fluidity (low growth rate at large undercoolings).

[200] D. Turnbull, *J. Phys. Chem.* **66**, 609 (1962).
[201] J. J. Kramer and W. A. Tiller, *J. Chem. Phys.* **42**, 257 (1965).

Jackson et al. conclude that for metals there are no data that can be reliably interpreted, partly due to heat flow problems, partly to uncertainties about the thermal wave technique. For water, heat flow control may have been present in all studies of growth in directions parallel to the basal plane of ice whereas perpendicular to the basal plane (along the c axis) an exponential law is found for δT between 0.03 and 0.07°C; and a $(\delta T)^2$ law between 0.07 and 0.3°C. For organic compounds, uncertainties about the liquid viscosity rendered doubtful the interpretation of much of the data. For tri-α-napthyl benzene, an exponential law in $(1/T\ \delta T)$ was obtained. For salol and glycerine, both much studied, impurity and other effects appear to introduce significant scatter in the growth rates. For two inorganic materials, sodium disilicate and potassium disilicate, linear laws of "reduced" growth rate versus undercooling were obtained. ("Reduced" means corrected for temperature dependence of viscosity and taking the exact form Eq. (19.1) without assuming $L\ \delta T/kTT_E \ll 1$.)

Jackson et al. also noted the lack of evidence for a transition from the lateral growth to the continuous growth mechanism, of the type predicted by Cahn[199] or by a modification[192] of Cahn's theory. However, they also noted that the coefficient of $(\delta T)^2$ in Eq. (19.6) in those materials having square law behavior, does not agree with theory; similarly with the coefficients K_1 and K_2 in Eq. (19.7).

c. More Recent Studies

Alfintzev and Ovsienko[197] measured the growth rate versus interface undercooling of both Ga and Bi crystals. For undeformed Ga crystals of (001) and (111) orientations, kinetic laws of the type Eq. (19.7) were observed, i.e. $\bar{R} = 70 \cdot \exp(-3900/T\ \delta T)$ m/s (001) and $\bar{R} = 54{,}600\ \exp - (11{,}000/T\ \delta T)$ for (111). For deformed (111) Ga, a law of the type (19.6) was observed, i.e. $\bar{R} = 4.2 \times 10^{-3}(\delta T)^2$ m/s as would be expected since dislocations would be introduced by deformation. For Bi, a law of the form (19.6) was observed. In both cases the coefficients in the parabolic law were within about one order of magnitude of the value predicted by Eq. (19.6). Both Ga and Bi were observed to grow with strongly faceted interfaces. Comparison with the Cahn[199] theory was made and the thickness n of the interface on this theory was calculated to be about two atomic layers.

James[202] using the thermal wave technique, measured the interface kinetics of the growth of ice crystals normal to the basal plane. The kinetic law observed was $\bar{R} = (1 \pm 0.5) \times 10^{-1}\ \delta T^{1.3 \pm 0.2}$ in the interface under-

---

[202] D. W. James, in "Crystal Growth" (H. S. Peiser, ed.), Suppl. *J. Phys. Chem. Solids*, p. 767. Pergamon Press, Oxford, 1967.

cooling range from 0 to 2×10^{-3}°C, a substantially lower undercooling than in earlier studies; this law does not seem to agree with either the linear, screw dislocation or nucleation theories.

20. Kinematic Theory of Crystal Growth

a. *Kinematic Waves*

Frank[203] and Cabrera and Vermilyea[204] introduced a more general way to look at step motion on crystal surfaces following methods used in an earlier theoretical study by Lighthill and Whitham[205] on the problems of traffic flow on crowded roads and of floods in rivers. Let the crystal surface $y(x)$ have steps all of height h. At a particular point in the surface, let the number of steps per unit length in the x direction be k, the step density. Let q be the step flux, i.e. the number of steps per unit time passing the point. Then the slope $\partial y/\partial x$ is

$$\partial y/\partial x = -hk \tag{20.1}$$

and the growth rate in the y direction is

$$\partial y/\partial t = hq. \tag{20.2}$$

The basic assumption of the theory is that the step flux depends only on the step density, i.e.

$$q = q(k). \tag{20.3}$$

That the flux will depend, actually, on the step separation (as well as upon other quantities) or density is clear since steps which are very far apart will achieve their maximum velocity, whereas as they approach each other their diffusion fields overlap, and they will slow down, as in the growth spiral problem. Differentiating (20.1) with respect to t, and (20.2) with respect to x, and adding gives the continuity equation (law of conservation of steps)

$$\partial q/\partial x + \partial k/\partial t = 0. \tag{20.4}$$

Defining the kinematic wave velocity $c(k) = dq/dk$, (20.4) becomes

$$c(k)\, \partial k/\partial x + \partial k/\partial t = 0, \tag{20.5}$$

a nonlinear differential equation. In the xt plane, along a line having the slope $dx/dt = c(k) = dq/dk$, k (and hence q) is constant, from Eq. (20.5).

[203] F. C. Frank, *in* "Growth and Perfection of Crystals" (R. H. Doremus, B. W. Roberts, and D. Turnbull, eds.), p. 411. Wiley, New York, 1958.

[204] N. Cabrera and D. A. Vermilyea, *in* "Growth and Perfection of Crystals" (R. H. Doremus, B. W. Roberts, and D. Turnbull, eds.), p. 393. Wiley, New York, 1958.

[205] M. J. Lighthill and G. B. Whitham, *Proc. Roy. Soc.* **A229**, 281, 317 (1955).

Such a line is called a characteristic. A kinematic wave is a region of the crystal surface of constant step density k (and constant flux q) which moves at a velocity $c(k) = dx/dt$. These waves do not consist of the same steps in time, as the wave velocity $c = dq/dk$ need not be the same as the step velocity $v = q/k$. The word kinematic is used because no "dynamics" (i.e. laws of mechanics, driving force, diffusion fields) have entered into the fundamental equation (20.4).

b. Crystal Profiles

Frank[203] proved two theorems relating the kinematical theory to crystal profiles in growth or dissolution. The first theorem is that points of a given orientation on a crystal surface have a straight line trajectory. We have $dy/dx = (\partial y/\partial x) + (\partial y/\partial t)(dt/dx)$; from Eqs. (20.1) and (20.2) we then have

$$dy/dx = -h[k - (q/c)], \tag{20.6}$$

which, along a characteristic (dx/dt) is constant since k and q are constant. The relation $y = y(x, t)$ defines a surface, and the characteristics are lines

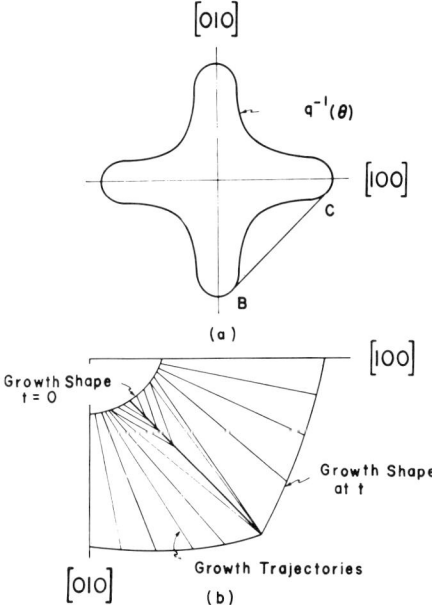

FIG. 29. Crystal profile in kinematic theory. (a) Reciprocal growth step flux q^{-1} vs orientation for a cubic crystal (schematic). (b) Corresponding growth profiles [J. P. Hirth, in "Energetics in Metallurgical Phenomena" (W. M. Mueller, ed.), Vol. II, p. 1. Gordon and Breach, New York, 1965].

on this surface which have straight projections in both xt and xy planes, hence the characteristics are straight lines on the $y(x, t)$ surface.

The second theorem says that if the polar diagram of the reciprocal of the growth velocity measured normal to the crystal surface is formed, that the straight line trajectory of a point of given orientation on the crystal surface will be parallel to the normal to the polar diagram at this orientation. Figure 29 from Hirth[144] illustrates the development of crystal shape as a function of time, showing the disappearance of some orientations in time. The dissolution of Ge was studied by Frank and Ives[206] and LiF by Ives[207] and the behavior was found to agree with both of Frank's theorems.

c. Shock Waves

Figure 30 from Frank[203] gives a typical $q(k)$ curve, and also illustrates what happens to a bunch of steps in the case of such a curve. Using the relation $c = (dx/dt) = (dq/dk)$, the characteristics for small k will have larger dq/dk and hence smaller slope in the $(t-x)$ diagram (Fig. 30), so that the characteristics at the left or trailing side of the bunch converge, making a discontinuity in density k as shown. This discontinuity, according

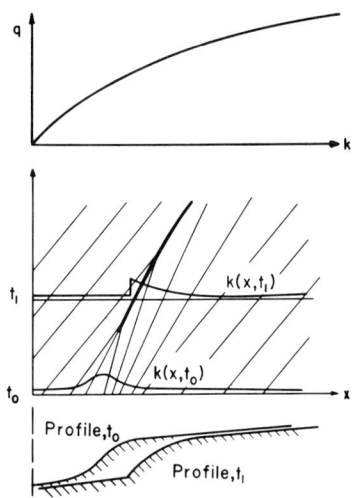

FIG. 30. Development of a bunch of steps in kinematic theory. Shown are a $q(k)$ curve, $X-t$ characteristics, step density distributions $k(X, t)$, and resulting crystal profiles with discontinuity in slope of trailing edge [after F. C. Frank, in "Growth and Perfection of Crystals" (R. H. Doremus, B. W. Roberts, and D. Turnbull, eds.), p. 411. Wiley, New York, 1958].

[206] F. C. Frank and M. B. Ives, J. Appl. Phys. **31,** 1996 (1960).
[207] M. B. Ives, J. Appl. Phys. **32,** 1534 (1961).

to Frank, is analogous to a shock wave in aerodynamics, and can be shown to travel with a velocity $c = (q_2 - q_1)/(k_2 - k_1)$ where the q's and k's are those just adjacent to the resulting sharp jump in slope of crystal profile, shown in Fig. 30. Frank also points out that to make the leading edge of the step sharp, as desired in an etch pit, d^2q/dk^2 must be positive, unlike Fig. 30; he suggests that impurities do this (see Part VII).

Chernov[17] has made a further study of shock waves, in particular of their variation with time; the time required for them to form and then to disappear is estimated.

d. *Some More Recent Developments in Kinematic Theory*

(i) Mullins and Hirth[208] have proposed a detailed microscopic kinematic theory, in which they explicitly write a system of differential equations, each one corresponding to the motion of one step. The theory is found to agree with the Frank "continuum" theory when q is a continuous function of k. Some new conclusions relate to the behavior of finite step trains, the trailing portion of which is found to break up into a train of double steps.

(ii) Bartini[209] made an experimental study, using an optical technique, of the surface profile of β-methyl napthalene crystals as they are growing from alcohol solution, with a claimed sensitivity of 38 Å or about two unit cell parameters. Kinematic wave profiles were observed directly. The profiles had a smoothly rounded leading edge and a sharp corner point at the trailing edge as indicated in Fig. 30, in qualitative agreement with the theory of Frank.

In addition, a new form of kinematic wave, just the reverse of that mentioned above, was observed, which would indicate an increase in individual step velocity as they approach each other (d^2q/dk^2 positive). Chernov and Budurov[210] have interpreted such waves as "diffusion-kinematic" waves in which the existence of diffusion in the surrounding medium is explicitly included.

VI. Morphological Stability

21. INTRODUCTION

The main problem concerning morphological stability in crystal growth is whether a particular shape of solid, growing from melt, solution, or vapor

---

[208] W. W. Mullins and J. P. Hirth, *J. Phys. Chem. Solids* **24,** 1391 (1963).
[209] G. R. Bartini, E. D. Dukova, I. P. Korshunov, and A. A. Chernov, *Soviet Phys.—Cryst. (English Transl.)* **8,** 605 (1964).
[210] A. A. Chernov and S. I. Budurov, *Soviet Phys.—Cryst. (English Transl.)* **9,** 388 (1965).

by diffusional processes, is a stable shape; that is, whether it is stable against small perturbations in shape. The stability is tested in the theoretical development, by assuming an arbitrary small perturbation in shape to be present, and then seeing whether this perturbation will grow or decay. This is different from the question of whether the shape and equations are shape-preserving. The concept of a shape-preserving system, as we have seen in Part III on Stefan problems, is simply that of a shape which can continue to satisfy the diffusion equations and boundary conditions as it grows larger in size but preserves its geometric form.

Mullins and Sekerka[211] (MS) in a classic paper, introduced into crystal growth studies the concept of quantitative testing for shape stability. Prior to this paper, the shape-preserving ellipsoids of Ham[68,69] were generally believed to imply that such shapes were stable whereas in fact they had not been tested for stability. An important motive for studying stability is that, in fact, crystals often grow with a dendritic form. Dendrites, or tree-shaped crystals, having a main stalk, primary branches, secondary branches, etc. are quite common in nature and in the laboratory. Their growth is a clear example of the instability of ellipsoidal or polyhedral shapes, under some conditions.

Analogous treatments of stability occur in hydrodynamics (see, e.g. Landau and Lifshitz,[212] Chandrasekhar,[213] Lin,[214] Donnelley et al.,[215] and Stuart[216]). The onset of the transition from laminar to turbulent flow is studied, by superposing on a known steady flow velocity a nonsteady small velocity perturbation. Then one determines whether, in the linearized differential equations for the perturbing velocity thus obtained, the perturbation in velocity tends to increase or decrease in time. An example of velocity instability involving both fluid flow and mass transfer by diffusion is treated by Ball[217] in the case of gas–liquid interphase mass transfer. The classical Benard problem (see e.g. Chandrasekhar[213]; see also Part VIII) of the motional instability of a fluid heated from below also involves both diffusion fields and convective motion.

[211] W. W. Mullins and R. F. Sekerka, *J. Appl. Phys.* **34**, 323 (1963).
[212] L. D. Landau and E. M. Lifshitz, "Fluid Mechanics." Addison-Wesley, Reading, Massachusetts, 1959.
[213] S. Chandrasekhar, "Hydrodynamic and Hydromagnetic Stability." Oxford Univ. Press (Clarendon), London and New York, 1961.
[214] C. C. Lin, "The Theory of Hydrodynamic Stability." Cambridge Univ. Press, London and New York, 1955.
[215] R. J. Donnelly, R. Herman, and I. Prigogine, eds., "Non-Equilibrium Thermodynamics, Variational Techniques, and Stability." Univ. of Chicago Press, Chicago, Illinois, 1966.
[216] J. T. Stuart, *Appl. Mech. Rev.* **18**, 523 (1965).
[217] J. G. Ball and D. M. Himmelblau, *Advan. Chem. Phys.* **13**, 267 (1967).

22. Morphological Stability Problem without Complications: Various Shapes

a. The Sphere

We repeat here for convenience the justification for the use of Laplace's equation instead of the exact, time-dependent diffusion equation, as given in Part III on Stefan problems. The latter is, for a sphere,

$$(\partial C/\partial t)(r, t) = D \nabla^2 C \qquad (22.1)$$

where $C(r, t)$ is the concentration, and D the diffusion coefficient. The boundary conditions are

$$C(\infty, t) = C_\infty \qquad (22.2)$$

$$C(R, t) = C_0 \qquad (22.3)$$

where the concentration at the surface of the sphere of radius R is assumed equal to the equilibrium concentration C_0. The flux boundary condition giving the growth rate is

$$D \, \partial C/\partial r \,|_{r=R} = (\mathbf{C} - C_0) \, dR/dt \qquad (22.4)$$

where $\mathbf{C}$ is the concentration of solute in the precipitate. As we saw in Part III, the solution satisfying (22.1)–(22.4) is, as may be verified by substitution,

$$C(r, t) - C_\infty = \frac{(C_0 - C_\infty)[(R/r)\exp(-(\lambda_5 r/R)^2) - \lambda_5 \pi^{1/2} \operatorname{erfc}(\lambda_5 r/R)]}{\exp(-\lambda_5^2) - \lambda_5 \pi^{1/2} \operatorname{erfc} \lambda_5} \qquad (22.5)$$

where

$$R = 2\lambda_5 (Dt)^{1/2} \qquad (22.6)$$

and where

$$2\lambda_5^2 [1 - \pi^{1/2} \lambda_5 \exp(\lambda_5^2) \operatorname{erfc}(\lambda_5)] = S \equiv (C_\infty - C_0)/(\mathbf{C} - C_0). \qquad (22.7)$$

(Note: $\operatorname{erfc} x = 1 - \operatorname{erf} x$; $\operatorname{erf}(x) = (2/\pi^{1/2}) \int_0^x \exp(-u^2) \, du$; $(d/dz)(\operatorname{erf} z) = +2\pi^{-1/2} \exp(-z^2)$.) The quantity S is a measure of the driving force for growth. For the case $S \ll 1$ (low driving force), Eq. (22.7) shows that $\lambda_5^2 \ll 1$ and hence (22.7) becomes

$$\lambda_5 \approx (S/2)^{1/2}. \qquad (22.8)$$

Hence (22.5) becomes

$$C - C_\infty = (C_0 - C_\infty)(R/r), \qquad (22.9)$$

if it is assumed that $(\lambda_5 r/R) \ll 1$. Now (22.9) is also the solution to Laplace's equation $\nabla^2 C = 0$ and also satisfies (22.2) and (22.3). Thus, so long as

$S \ll 1$, the concentration field and gradient at the surface of the growing sphere given by Laplace's equation are the same as given by the exact time dependent equation (22.1). The condition $(\lambda_5 r/R) \ll 1$ will also be valid near the sphere surface.

Then to study perturbations on the sphere, it must be assumed that, for small perturbations, the concentration field near the perturbed sphere is still correctly given by Laplace's equation but with a Gibbs–Thomson boundary condition on the surface of the perturbed sphere; and furthermore that $C = C_\infty$, at $r = \infty$, and $\lambda_5 = (S/2)^{1/2}$.

The equation of a distorted sphere of radius R and perturbation magnitude δ, using spherical coordinates r, θ, φ, is

$$r(\theta, \varphi) = R + \delta Y_{lm}(\theta, \varphi) \qquad (22.10)$$

where $(\delta/R) \ll 1$ and where the Y_{lm} are spherical harmonics, i.e.

$$\Lambda Y_{lm} \equiv \left[\frac{1}{\sin\theta}\frac{\partial}{\partial\theta}\left(\sin\theta\frac{\partial}{\partial\theta}\right) + \frac{1}{\sin^2\theta}\frac{\partial^2}{\partial\varphi^2}\right] Y_{lm} = -l(l+1) Y_{lm}. \qquad (22.10')$$

The equilibrium concentration on the distorted surface, $C_{0,s}$ as given by the Gibbs–Thomson relation is

$$C_{0,s}(\theta, \varphi) = C_0 + C_0 \Gamma_D K' \qquad (22.11)$$

(local equilibrium boundary condition), where C_0 is the equilibrium concentration at a flat interface, Γ_D is the capillary constant $\gamma\Omega'/R_g T$, K' is the curvature, γ the interfacial free energy, Ω' the precipitate volume per mole of added solute, and R_g the gas constant.

It can be shown[211] that for the slightly perturbed sphere

$$K'(\theta, \varphi) = (2/R)[1 - (\delta Y_{lm}/R)] - (\delta \Lambda Y_{lm}/R^2). \qquad (22.12)$$

Using (22.12), (22.11), and (22.10') we get

$$C_{0,s}(\theta, \varphi) = C_0[1 + (2\Gamma_D/R) + (l+2)(l-1)(\Gamma_D \delta Y_{lm}/R^2)] \qquad (22.13)$$

for the equilibrium concentration on the distorted sphere. The solution to Laplace's equation reducing to (22.13) on the surface given by (22.10) and having the value C_∞ at $r = \infty$ is now required. A suggested solution is

$$C(r, \theta, \varphi) = (A_3/r) + (B_3 \delta Y_{lm}/r^{l+1}) + C_\infty \qquad (22.14)$$

where A_3 and B_3 are constants. This is clearly C_∞ at $r \to \infty$; using (22.10') it satisfies Laplace's equation

$$\frac{1}{r^2}\left[\frac{\partial}{\partial r}\left(r^2\frac{\partial}{\partial r}\right) + \frac{1}{\sin\theta}\frac{\partial}{\partial\theta}\left(\sin\theta\frac{\partial}{\partial\theta}\right) + \frac{1}{\sin^2\theta}\frac{\partial^2}{\partial\varphi^2}\right] C = 0 \qquad (22.14')$$

and is equal to (22.13) on the surface (22.10) provided that

$$A_3 = R[C_0 + (2C_0\Gamma_D/R) - C_\infty] \tag{22.15a}$$

and

$$B_3 = R^{l+1}\{R^{-1}[C_0 + (2C_0\Gamma_D/R) - C_\infty] + C_0(l+2)(l-1)(\Gamma_D/R^2)\}. \tag{22.15b}$$

To obtain (22.15a) and (22.15b) the coefficients in (22.13) and (22.14) of Y_{lm} have been equated and so have those not involving Y_{lm}. Thus

$$C(r, \theta, \varphi) = C_\infty + \frac{(C_0 - C_\infty)R + 2C_0\Gamma_D}{r}$$

$$+ \frac{\{(C_0 - C_\infty)R^l + C_0\Gamma_D R^{l-1}l(l+1)\}\delta Y_{lm}}{r^{l+1}}. \tag{22.16}$$

To calculate the rate of growth of the perturbation, we form the time derivative of (22.10) and use the counterpart of (22.4):

$$v = \frac{dr}{dt} = \frac{dR}{dt} + \left(\frac{d\delta}{dt}\right) Y_{lm} = \left(\frac{D}{\mathbf{C} - C_{0,S}}\right) \frac{\partial C}{\partial r}\bigg|_{\text{pert surf}} \tag{22.17}$$

valid for small perturbations. Hence

$$v = \frac{D}{\mathbf{C} - C_{0,S}} \left\{ \frac{C_\infty - C_{0,R}}{R} \right.$$

$$\left. + \left[(l-1)\frac{(C_\infty - C_0)}{R^2} - \frac{C_0\Gamma_D}{R^3}(l(l+1)^2 - 4)\right]\delta Y_{lm} \right\} \tag{22.18}$$

where $C_{0,R} \equiv C_0[1 + (2\Gamma_D/R)]$ is the Gibbs–Thomson equilibrium concentration on the surface of the undistorted sphere. Equating coefficients of Y_{lm} in (22.18) and (22.17) gives[211]

$$\dot\delta(l) = \frac{D(l-1)}{(\mathbf{C} - C_{0,R})R}\left[G_{\text{cn}} - \frac{C_0\Gamma_D}{R^2}(l+1)(l+2)\right]\delta_l \tag{22.19}$$

for the rate of growth of the lth harmonic, where $G_{\text{cn}} = (C_\infty - C_{0,R})/R$ is the concentration gradient at the surface of the undistorted sphere. The positive term, proportional to the gradient, favors growth of the harmonic; the negative term, proportional to the surface energy, favors decay. All harmonics l for which the bracket [ ] is positive must grow: these are l for which

$$(l+1)(l+2) + 2 < (R/C_0\Gamma_D)(C_\infty - C_0). \tag{22.20}$$

All higher harmonics (shorter wavelengths on the surface) must decay. For a given harmonic l there is a radius $R_c(l)$ for which the harmonic is just stable (i.e. sphere unstable) given by (22.20) as

$$R_c(l) = [(l+1)(l+2)/2 + 1]R_{\text{sph}}{}^* \qquad (22.21)$$

where $R_{\text{sph}}{}^* = 2\Gamma_D/[(C_\infty - C_0)/C_0]$ is the sphere critical nucleation radius of nucleation theory [corresponding to r_c, in Eq. (14.3)]. The lowest radius below which no harmonic will grow is for $l = 2$. Above this radius the $l = 2$ perturbation will grow. This radius is $R_c(2) = 7R_{\text{sph}}{}^*$ and hence $R_c(2) \sim 10^{-5}$ cm at 10% supersaturation (for $\Gamma_D = 10^{-7}$ cm, a typical value). We note that the $l = 1$ perturbation only translates the sphere.

Thus the sphere growing from a diffusion field is not a stable shape, so long as it is larger than, in a typical case, microns in size, and so long as the only stabilizing effect considered is surface energy.

This very important result[211] shows that certainly the sphere and perhaps all previous "shape-preserving" shapes (Part III) are actually unstable shapes, and hence throws open the whole question of the morphology of crystals—how crystals can develop large flat faces for example. The result also applies to the growth of a sphere from supercooled melt—the case of solidification—as the correspondence between the two formally equivalent (heat and matter) diffusion problems is outlined by MS.[211] It is also interesting to note that in dissolution or melting spheres will be stable shapes since then G_{cn} changes sign and (22.19) is always negative.

b. *The Cylinder*

Coriell and Parker[109] extended the result of MS on the sphere to the case of a cylinder. This is perhaps the next simplest case. Also, certain crystals (dendrite branches, whiskers) are approximately cylindrical in form. Both perturbations in circular cross-section shape were considered as well as perturbations along the length of the cylinder. For perturbations of the form $r = R + \delta e^{ik\varphi}$ in the circular cross section, where k is a positive integer, the cylinder was found to be stable when its radius is less than a critical radius R_c, and unstable when greater than R_c, analogous to the case for a sphere. However, the ratio is not seven but is, for the $k = 2$ perturbation,

$$R_c(2) = [1 + 6A_\lambda]R_{\text{cyl}} \qquad (22.22)$$

where $A_\lambda = -\frac{1}{2}\ln(\nu^2\lambda_4{}^2)$, where λ_4 is the growth constant, and $\ln \nu^2 = 0.5772$ (Euler's constant).

Hence $R_c/R_{\text{cyl}}{}^*$, unlike the case for the sphere, is now a function of the concentration ratio S, since [discussion following Eq. (9.42)]

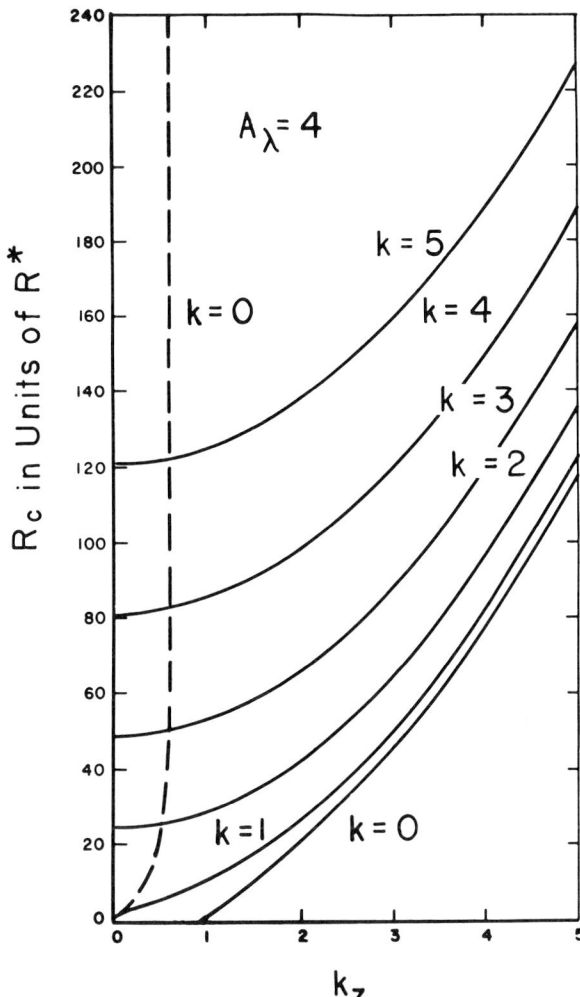

Fig. 31. Theory of morphological stability of a cylinder growing in a diffusion field. Radius R_c above which cylinder is unstable to perturbations of wave number k versus wave number k_z [S. R. Coriell and R. L. Parker, *J. Appl. Phys.* **36**, 632 (1965)].

$(0.577 + \ln \lambda_4^2) \cong -S$, for $S \ll 1$. [Note: R_{cyl}^* for a cylinder is defined as $\Gamma_D/[(C_\infty - C_0)/C_0]$.] The case of perturbations (hourglass type) in the z direction of the form $\exp[i(k_z/R)z]$ where k_z is real, positive but not necessarily an integer, together with the circular perturbations (fluted-column type) is somewhat more complicated. Figure 31 gives R_c/R_{cyl}^* as a function of k_z, for 6 values of k, all for $A_\lambda = 4$ ($S \approx 0.0015$). The cylinder is unstable to a particular perturbation (k, k_z) when its radius $R_c/R_{cyl}{}^*$ is greater than or equal to the value shown. For the special case of hourglass perturbations only, of long wavelength, for $0.6 \leq k_z \leq 1$, the cylinder is always unstable since these lower the surface to volume ratio of the cylinder. For $k_z < 0.6$, the diffusion effect itself reverses sign and becomes stabilizing, and for this unusual case the cylinder is stable *above* the R_c curve and unstable below (dotted curve in Fig. 31).

c. Planar Alloy

Mullins and Sekerka,[218] Sekerka,[219] and Voronkov[220] have studied the stability of the shape of a planar solid–liquid interface during the freezing of a dilute alloy, using a technique generally similar to that used by MS.[211] Steady state diffusion equations for both thermal and diffusion fields are studied in the moving coordinate system fixed at the interface which is assumed to be moving with a uniform velocity, v. There is a positive temperature gradient imposed externally, i.e. temperature increases as one moves into the liquid from the interface. Local equilibrium is again assumed at the interface, but now the interface temperature depends on the amount of solute present there. It is found that there are three factors influencing decay or growth of a perturbation. Surface energy, as before, is a stabilizing factor that favors decay of a perturbation. The thermal gradient also favors decay of a perturbation, unlike the previous cases of sphere and cylinder, for in the present case the temperature increases into the liquid. The solute gradient, formed by partial rejection of solute at the freezing interface with the distribution coefficient k_0 (see Part VIII) on the contrary favors growth of the perturbation. Ignoring for the moment the surface energy effect, which can be small compared to the thermal gradient term, the condition for stability of the interface is

$$-1/2(g' + g) + m'G_{cn} < 0 \quad \text{stable}$$
$$> 0 \quad \text{unstable} \quad (22.23)$$

where $g' = (K_S/\bar{K})G_{tS}$, $g = (K_L/\bar{K})G_{tL}$. Here G_{tS} is the temperature gradient in the solid, G_{tL} that for the liquid, K_S and K_L the corresponding thermal conductivities, $\bar{K} = \frac{1}{2}(K_S + K_L)$, G_{cn} is the concentration gradient in the liquid at the unperturbed interface [given by $G_{cn} = (C_\infty v/D)[(1-k_0)/k_0]$ where C_∞ is the composition of bulk liquid, k_0 the equilibrium distribution coefficient, and v the interface velocity], and m'

---

[218] W. W. Mullins and R. F. Sekerka, *J. Appl. Phys.* **35**, 444 (1964).
[219] R. F. Sekerka, *J. Appl. Phys.* **36**, 264 (1965).
[220] V. V. Voronkov, *Soviet Phys.—Solid State* **6**, 2378 (1965).

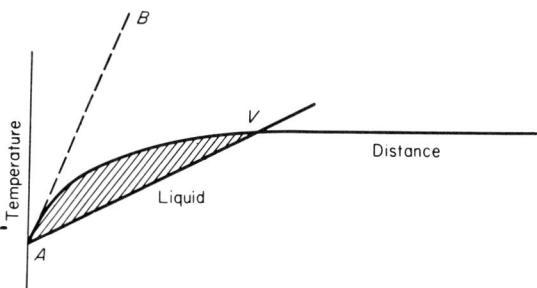

FIG. 32. Constitutional supercooling in advance of a planar interface (schematic). Hatched region denotes departure of actual temperature below liquidus temperature [B. Chalmers, "Principles of Solidification." Wiley, New York, 1964].

is the slope of the liquidus line on the phase diagram. Equation (22.23) is similar, but not identical, to the constitutional supercooling criterion (C.S.) of Tiller et al.,[221] which is

$$-G_{tL} + m'G_{cn} < 0 \quad \text{stable}$$
$$> 0 \quad \text{unstable.} \quad (22.24)$$

This result may be understood from examination of Fig. 32, where the actual temperature in the liquid (straight line) may be less than the liquidus temperature resulting from the solute distribution, giving a constitutionally supercooled (hatched) region of the liquid near the interface; Eq. (22.24) gives the condition for the two slopes to be equal at the interface. As discussed below, the C.S. criterion has been experimentally verified by Walton et al.[222] and by Tiller and Rutter[223] (see Chalmers[80]). Equation (22.23) differs from (22.24) in that it also involves the temperature gradient in the solid, which, if expressed in terms of that in the liquid, will then involve the interface velocity v and latent heat L. It is then possible to have instability without C.S. and C.S. without instability. When surface energy is included the stability criterion becomes much more complicated.[218,219] A limiting case occurs, wherein the stabilizing effect of surface energy dominates the unstabilizing solute effect, as can apparently result for velocities sufficiently high.

Voronkov[220] in an independent study of the same problem as MS[218] has reached an identical conclusion [Eq. (22.23)].

[221] W. A. Tiller, K. A. Jackson, J. W. Rutter, and B. Chalmers, *Acta Met.* **1**, 428 (1953).
[222] D. Walton, W. A. Tiller, J. W. Rutter, and W. C. Winegard, *Trans. AIME* **203**, 1023 (1955).
[223] W. A. Tiller and J. W. Rutter, *Can. J. Phys.* **34**, 96 (1956).

Temkin[224,225] studied the velocity of growth of a hemispherical protuberance on an otherwise plane face in the freezing of a dilute alloy. It appears[218] that the study of this rather special shape does not constitute a test of stability of the shape of the plane to arbitrary perturbations.

23. Stability Problem with Complications

a. Surface Diffusion

Coriell and Parker[226] have studied the effect of surface diffusion on the shape stability of a cylinder and of a sphere growing from supersaturated solution or supersaturated vapor with inert gas present. Nichols and Mullins[227] have also studied the sphere with surface diffusion, and Shewmon[228] has studied the plane. It is found that surface diffusion can enhance the stability of these shapes. The reason is that the Gibbs–Thomson effect on a region of high curvature sets up a higher equilibrium concentration there than on a flat region; when surface diffusion is permitted, this concentration gradient along the surface is the driving force for surface flow from a protuberance to a flat. This makes protuberances less stable, and hence makes the original shape more stable.

The result for perturbations of the form $\exp(ik\varphi)$ in the circular shape of the cylinder (fluted column) is obtained simply by adding algebraically the mass transport term due to surface diffusion to the already obtained[109] expression for the contribution of the volume diffusion term to the rate of change of amplitude of a perturbation. The result is

$$R_c(k)/R_{\text{cyl}}^* = \tfrac{1}{2}\Lambda'\{1 + [1 + (4P/R^*\Lambda'^2)]^{1/2}\} \qquad (23.1)$$

where $\Lambda' = 1 + A_\lambda k(k+1)$ and $P = (\mathbf{C} - C_{0,R})\Omega^{1/3}A_\lambda D_s k^2(k+1)/DC_0$; for a cylinder $C_{0,R} = C_0 + C_0\Gamma_D/R$, where C_0 is equilibrium concentration at a flat surface, and D_s is the surface diffusion coefficient; Ω volume per molecule; the other quantities were defined above. It is clear that large values of D_s/D contribute to larger critical radii. A numerical estimate[226] for vapor growth in the presence of an inert gas give a stability enhancement due to surface diffusion of about 40 times, i.e. $R_c(3)$ increased from 1.1×10^{-6} cm to $\sim 4.6 \times 10^{-5}$ cm.

The results for the sphere[226,227] are similar to those for a cylinder.

---

[224] D. E. Temkin, *Dokl. Akad. Nauk SSSR* **133**, 174 (1960).

[225] D. E. Temkin, *in* "Kristallizatsiya i Fasovye Perekhody," p. 249. Akad. Nauk Belorussk, Minsk, 1962.

[226] S. R. Coriell and R. L. Parker, *J. Appl. Phys.* **37**, 1548 (1966).

[227] F. A. Nichols and W. W. Mullins, *Trans. AIME* **233**, 1840 (1965).

[228] P. G. Shewmon, *Trans. AIME* **233**, 736 (1965).

Nichols and Mullins give a numerical example for which the enhancement is a factor of about 10^2, from 1×10^{-5} cm to 1×10^{-3} cm.

b. Interface Kinetics

(i) *The Sphere with Interface Kinetics.* Cahn[229] and Coriell and Parker[230] have studied the role of interface kinetics in the stability of the shape of a sphere growing in a diffusion field, thus removing the assumption of local equilibrium made in the earlier treatments. The other assumptions are retained, i.e. isotropic surface free energy and the use of Laplace's equation (low driving forces). Coriell and Parker considered both linear and quadratic interface kinetic laws. It was found that slow interface kinetics can greatly enhance the stability.

The procedure used differs from the nonkinetic case in the flux boundary condition which is now (for the case of melt growth), for linear kinetics,

$$v = \frac{K_S}{L}\left(\frac{\partial T_S}{\partial r}\right)_I - \frac{K_L}{L}\left(\frac{\partial T_L}{\partial r}\right)_I = K(T_0 - T_{\text{su}}) \qquad (23.2)$$

where v is the velocity of growth of the solid, T_{su} is the interface temperature, T_0 the equilibrium temperature [for a curved surface, see Eq. (10.7)], K the linear kinetic coefficient, K_S and K_L the thermal conductivities, and $T_S(r, \theta, \varphi)$ and $T_L(r, \theta, \varphi)$ are the temperatures in solid and liquid, respectively, which are complicated functions of K. Using the criterion[230] for relative stability

$$(\dot{\delta}/\delta)/(\dot{R}/R) = 1 \qquad (23.3)$$

[for absolute stability (23.3) is zero giving no growth of the perturbation] which gives the radius R_r at which the perturbation grows just as fast as does the sphere itself, and above which the perturbation grows faster, it is found that the minimum value of $R_r(l)$ occurs at $l = 3$ ($l = 1$ translates the sphere, $l = 2$ changes the sphere to an ellipsoid which is unstable in the absolute sense but not in the relative sense). This is so in the limit $K \to \infty$ (infinitely fast kinetics). In the limit $K \to 0$ (infinitely sluggish kinetics), the minimum value of $R_r(l)$ occurs at infinite l, and for the real case (finite K) at finite l. The R_r is found to be approximately a linear function of the reciprocal of the kinetic coefficient, when the coefficient is small. Generally similar results are found for the case of square law kinetics.

As a numerical example, salol, having sluggish kinetics (Cahn et al.[198]) is found, using the theory of Coriell and Parker,[230] to have a maximum

---

[229] J. W. Cahn, *in* "Crystal Growth" (H. S. Peiser, ed.), Suppl. *J. Phys. Chem. Solids*, p. 681. Pergamon Press, Oxford, 1967.

[230] S. R. Coriell and R. L. Parker, *in* "Crystal Growth" (H. S. Peiser, ed.), Suppl. *J. Phys. Chem. Solids*, p. 703. Pergamon Press, Oxford, 1967.

stable radius of about 0.5 cm in growth from the melt at 5°C bath undercooling, compared to about 10^{-4} cm as the maximum stable size if kinetics are ignored. The figure of 0.5 cm is consistent with the large ($\sim$centimeters) flat stable faces of salol observed in experiment.[198]

Another interesting result of the theory is that the sphere can be stable in the relative sense even though the surface energy is permitted to go to zero, so long as the kinetic coefficient is finite.

Cahn[229] has studied, in a qualitative way, the stability of a sphere growing from supersaturated solution with linear interface kinetics. Using the solution counterpart of Eqs. (23.2) and (22.17) and (22.16) and ignoring perturbations and curvature, Cahn showed that

$$(dR/dt) = K(C_\infty - C_0)/[1 + (KCR/D)] \qquad (23.4)$$

where K is the (linear) interface kinetic coefficient, and D the diffusion coefficient in the medium. Examining the term KCR/D, for sufficiently small radii R, Eq. (23.4) gives the interface control law, linear in time: $\dot{R} = K(C_\infty - C_0)$; for sufficiently large radii (23.4) gives $\dot{R} = -D(C_0 - C_\infty)/RC$, i.e. a diffusion control law, with radius proportional to $t^{1/2}$. The former case is stable since the surface concentration C_s is then approximately C_∞ and there is no gradient to drive the unstabilizing point effect of diffusion, whereas the stabilizing effect of surface energy would still be present. The latter case is unstable as already discussed above. An approximate value for the transition between the two cases is $KCR/D \sim 1$, and the sphere should be stable up to radii $R \sim D/KC$.

(ii) *The Planar Binary Alloy with Kinetics.* Tarshis and Tiller,[231] Seidensticker,[232] and Jackson[193] have studied the morphological stability of a planar alloy interface, in a positive temperature gradient, taking into account linear interface kinetics in addition to the solute and curvature (surface energy) aspects of this problem already dealt with by Mullins and Sekerka[218] and discussed above. The interface is presumed to be moving with a uniform velocity v; solved are the steady state diffusion equations for heat and solute, with mass continuity and interface temperature boundary conditions. Jackson and Seidensticker find that the region of the stability is not affected by kinetics although the Fourier spectrum of the instability is. Tarshis and Tiller find, using a computer analysis, that not only the Fourier spectrum of the instability but also the region and even the existence of the instability can be altered by the kinetic coefficient. An explanation for the differences found may lie in whether temperature

---

[231] L. A. Tarshis and W. A. Tiller, *in* "Crystal Growth" (H. S. Peiser, ed.), Suppl. *J. Phys. Chem. Solids*, p. 709. Pergamon Press, Oxford, 1967.

[232] R. G. Seidensticker, *in* "Crystal Growth" (H. S. Peiser, ed.), Suppl. *J. Phys. Chem. Solids*, p. 733. Pergamon Press, Oxford, 1967.

gradients are regarded as fixed in the problem, as in the last two papers mentioned,[232,193] or are permitted to vary if the kinetic coefficient varies, as in Tarshis and Tiller's paper.

(iii) Other Studies of Stability with Interface Kinetics. Kotler and Tiller[233] have studied the stability of a cylinder to both radial and longitudinal perturbations in the presence of both solute and (undercooled) temperature gradients, taking into account interface kinetics and curvature as well. The wavelengths of the most rapidly growing longitudinal (z) perturbations were calculated as a function of bath undercooling and (linear) kinetic coefficient.

Shewmon[228] studied the growth from supersaturated solution of a perturbed planar interface taking linear interface kinetics into account. For a given gradient of solute, stability or nonstability of a perturbation is determined by the opposing effects of the gradient and surface energy, and is not affected by the interface kinetic coefficient. Of course, if the gradient is permitted to vary, then an arbitrarily small kinetic coefficient corresponds to zero gradient and hence stability.

Chernov[234] using an entirely different approach has made a qualitative study of the shape of polygonal crystals growing from supersaturated solution. Noting that Seeger's[78] calculation showed that for such a crystal the supersaturation is greatest at the edges of the face and least at the face center, Chernov assumes that this inconstancy over the face is compensated for (to obtain a uniform face growth rate) by an interface kinetic coefficient (linear kinetics assumed) which varies significantly for slight deviations from close-packed directions in orientation of adjacent regions on the face. Chernov finds, without explicitly solving Laplace's equation for the perturbed shape, that for certain types of shape perturbation (i.e. depressions in a face) the anisotropy of the kinetic coefficient leads to stability of shape, up to a critical size which may be 1 mm in a particular example. However, in the absence of anisotropy of the kinetic coefficient no large-size stability is predicted, apparently contrary to the quantitative results of Coriell and Parker[230] for the sphere.

c. Anisotropic Surface Tension

Cahn[229] has studied quantitatively the role of slightly anisotropic surface tension in the morphological stability of a spherical crystal growing from supersaturated solution. The slight variations in surface energy from a spherical γ plot are expressed as a sum of spherical harmonics with coeffi-

---

[233] G. R. Kotler and W. A. Tiller, *in* "Crystal Growth" (H. S. Peiser, ed.), Suppl. *J. Phys. Chem. Solids*, p. 721. Pergamon Press, Oxford, 1967.
[234] A. A. Chernov, *Soviet Phys.—Cryst. (English Transl.)* **8**, 63 (1963).

cients ϵ_{lm}, all small except for ϵ_{00} which is related to the average surface energy and is normalized to 2. Following a procedure similar to that of MS,[211] the surface concentration (counterpart of Eq. (22.13)) for the slightly distorted sphere is obtained. Then the equilibrium form of the crystal, given by the condition that $C_{0,s}$, the Gibbs–Thomson equilibrium concentration on the distorted surface, be constant over the crystal surface, is found to be a slightly distorted sphere. With the notation of Eq. (22.10)

$$\delta_{lm}{}^{\mathrm{eq}} = -R\epsilon_{lm}/(l+2)(l-1). \tag{23.5}$$

From Eq. (23.5) the surface concentration may be expressed entirely in terms of the $\delta_{lm}{}^{\mathrm{eq}}$, replacing the ϵ_{lm}, and the resulting equation is very similar to (22.13). The time rate of change of the perturbation is obtained by a procedure identical to that outlined earlier, and is

$$\frac{d\delta_{lm}}{dt} = \frac{C_0 D(l-1)}{(\mathbf{C} - C_{0,s})R^2}\left[\left\{\frac{C_\infty - C_0}{C_0} - \frac{2\Gamma_D}{R}\right\}\delta_{lm}\right.$$

$$\left. - \frac{\Gamma_D}{R}(l+1)(l+2)(\delta_{lm} - \delta_{lm}{}^{\mathrm{eq}})\right], \tag{23.6}$$

which reduces to (22.19) when $\delta_{lm}{}^{\mathrm{eq}} = 0$ (no anisotropy). If Eq. (23.6) is divided by the growth rate of the average radius of the sphere, then the

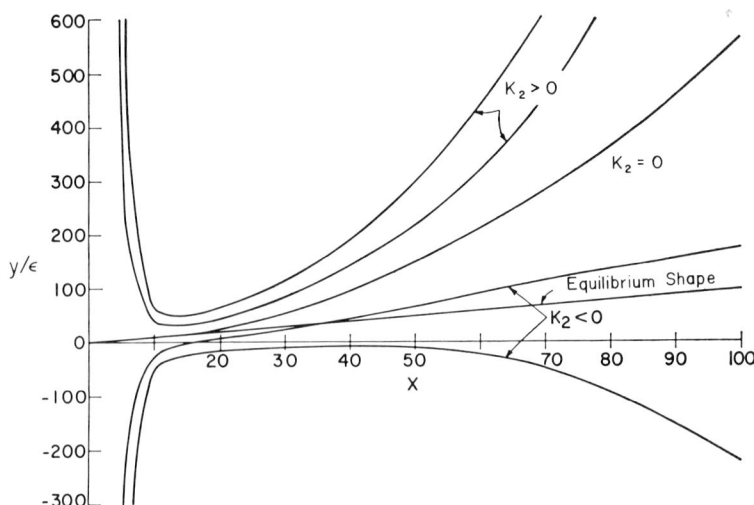

FIG. 33. Theory of shape distortion of a sphere with anisotropic surface free energy. Plot of normalized distortion amplitude versus normalized sphere radius, for various initial distortions (see text) [J. W. Cahn, in "Crystal Growth" (H. S. Peiser, ed.), Suppl. J. Phys. Chem. Solids, p. 681. Pergamon Press, Oxford, 1967].

time dependence of δ is converted to a geometrical dependence of δ on the average size of the sphere. Figure 33 gives a plot of the resulting equation for an $l = 4$ distortion of the spherical shape, and an $l = 4$ variation of the surface energy. Here $y = \delta_4/R^*$ and $x = R/R^*$ where R^* is that for the sphere, defined following Eq. (22.21). Following the top two curves, which are for initial distortions of the sphere that are in phase with the equilibrium shape, the distortions are unstable, i.e. grow less rapidly than the average radius, for $R \lesssim 10R^*$. As the shape grows larger, though, these distortions begin to grow at a faster rate than the average radius, indicating instability. However, even the unperturbed sphere is not stable (curve for $K_2 = 0$), as it tends to develop perturbations corresponding to the equilibrium shape, and these then are enhanced, for larger r, by the effects of the diffusion field. The straight line through the origin represents the growth of the equilibrium shape. The $K_2 < 0$ curves represent distortions initially out of phase with the equilibrium shape.

d. Time-Dependence Studies of Stability

Sekerka[235,236] has formulated a theory of stability of a planar interface which does not assume, as did Mullins and Sekerka,[218] that the solute and temperature fields obey steady state diffusion equations, but instead uses the general time-dependent diffusion equations. The mathematical procedure, which is quite complex, involves linearizing the nonlinear flux boundary condition by writing the concentration and temperature fields as the sum of a small time varying part and a large part satisfying the steady state equations. Then both Laplace transform and Fourier transform techniques are applied. For the dilute alloy problem,[235] involving both thermal and solute fields, it was necessary to make the assumption that the thermal fields were steady state in order to arrive at a stability condition. The stability condition is found to be in agreement with the criterion given above in the MS[218] steady state analysis.

The case of a planar interface with solute field only (isothermal) is solved exactly in a fully time-dependent treatment[236] and compared with steady state and with Laplace equation approximations. The stability condition, for a perturbation of wavelength λ, is

$$\lambda < \lambda_0 = 2\pi[C_0\Gamma_D D/(\mathbf{C} - C_0)V]^{1/2}, \qquad (23.7)$$

and is found to be the same in all three cases. However, the behavior of a particular wavelength as a function of time is quite different in the three

---

[235] R. F. Sekerka, *in* "Crystal Growth" (H. S. Peiser, ed.), Suppl. *J. Phys. Chem. Solids*, p. 691. Pergamon Press, Oxford, 1967.
[236] R. F. Sekerka, *J. Phys. Chem. Solids* **28**, 983 (1967).

methods; in particular, the wavelength and velocity of the fastest growing perturbation differ in the three cases, although for sufficiently large D/V they approach each other. Finally, the time evolution of an actual interface perturbation [a large central peak plus small side-bands of the form $(\sin x)/x$] is numerically computed; the result indicates that a periodic structure may spread over the surface from this perturbation.

Delves[237,238] has also given a time-dependent theory of the dilute planar alloy and finds the same stability conditions as did MS.[218]

24. Experimental Studies of Morphological Stability

a. Quantitative Comparisons with Theory

Hardy and Coriell[239,240] have made measurement of the growth (and decay) of perturbations on single crystal cylinders of ice growing from slightly supercooled distilled water and compared the observed growth rates with the morphological stability theory for a cylinder.[109] The ice crystal cylinder axis was perpendicular to the basal plane of the ice unit cell. Hence growth was occurring parallel to this plane; the kinetic coefficient for such growth is known to be sufficiently large so that the kinetic contribution to the undercooling is negligible in these experiments. Due to the low undercoolings and small growth velocities, the steady state approximation, used in the theory, was sufficient. The equation for the growth of the cylinder then follows from Coriell and Parker[109] with modifications necessary for the heat flow case, and for the fact that the liquid bath has a finite radius. The result is

$$\frac{(\dot{\delta}/\delta)}{(\dot{R}/R)} = (H_K - 1) - \frac{[k^2 + k_z^2 - 1]\ln(R_b/R)(H_K + H_I K_S/K_L)}{(R/R_{\text{cyl}}^* - 1)} \quad (24.1)$$

where

$$H_K = k_z[K_{k-1}(k_z) + K_{k+1}(k_z)]/[2K_k(k_z)]$$

and

$$H_I = k_z[I_{k-1}(k_z) + I_{k+1}(k_z)]/[2I_k(k_z)]$$

where the K_k and I_k are modified Bessel functions and where R_b is the radius of the (cylindrical) bath; the other quantities have been previously defined (note K_S and K_L are thermal conductivities).

[237] R. T. Delves, Phys. Status Solidi **16**, 621 (1966).
[238] R. T. Delves, Phys. Status Solidi **17**, 119 (1966).
[239] S. C. Hardy and S. R. Coriell, J. Appl. Phys. **39**, 3505 (1968).
[240] S. C. Hardy and S. R. Coriell, J. Cryst. Growth **3/4**, 569 (1968).

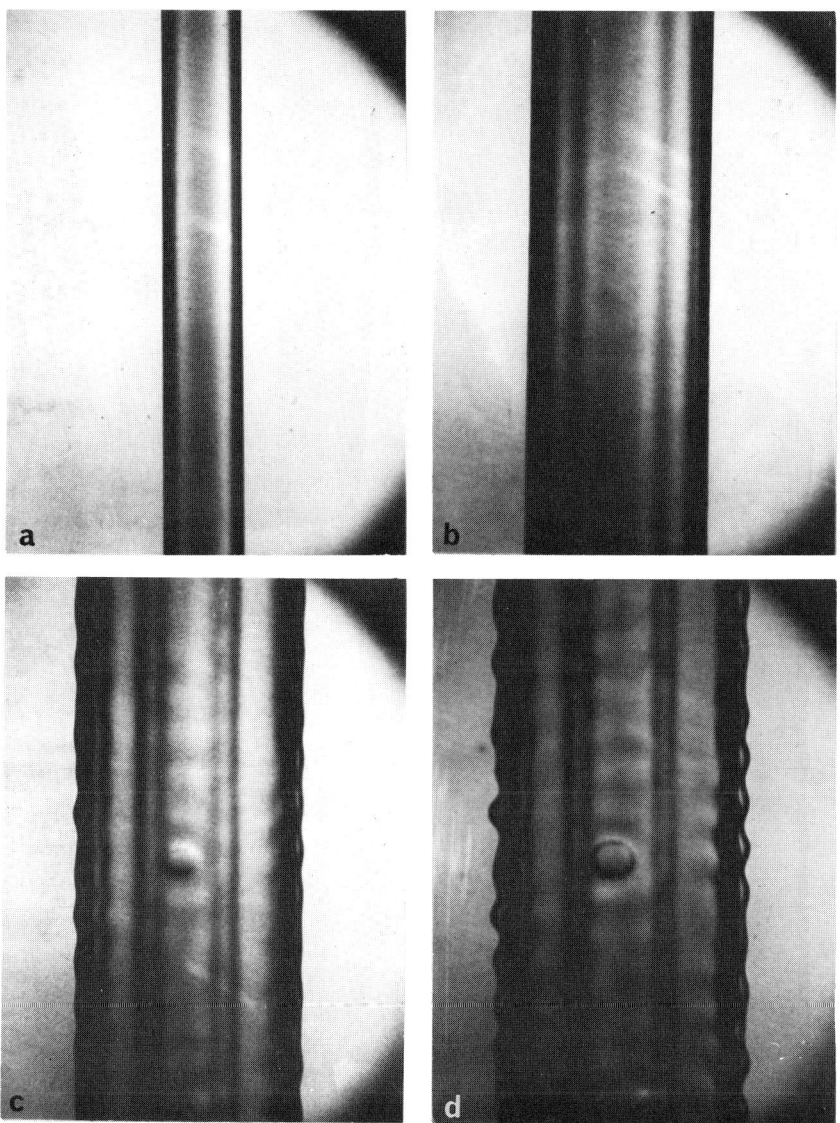

Fig. 34. Morphological instability of an initially smooth ice cylinder growing radially from supercooled water. Bath undercooling 0.095°C. (a) Radius 0.053 cm, $t = 0$; (b) $t = 8.52 \times 10^3$ sec; (c) 13.02×10^3 sec; (d) 14.82×10^3 sec [S. C. Hardy and S. R. Coriell, *J. Appl. Phys.* **39**, 3505 (1968)].

TABLE I. GROWTH

δ (cm)	λ_z (cm)	R (cm)	$\dot{R} \times 10^6$ (cm/s)	$H_K - 1$	$(R/\delta)(\dot{\delta}/\dot{R})$	$\gamma \times 10^3$ (J/m²)
0.0032	0.052	0.182	5.49	22.2	16.4	11.6
0.0030	0.057	0.168	6.00	19.0	12.3	19.3
0.0095	0.055	0.154	5.90	18.0	12.2	16.1
0.0072	0.052	0.135	6.25	16.8	10.6	17.1
0.0031	0.062	0.185	5.92	19.2	14.0	17.0

The quantities on the left of Eq. (24.1) are experimentally measured; those on the right can be calculated if the interfacial free energy γ_{SL} of ice–water and hence $R_{\text{cyl}}^* = T_M \gamma_{SL}/L_V(T_M - T_b)$ are known where $L_V =$ (latent heat)/(unit volume), and T_b is the bath temperature. Conversely, the equation predicts a value of γ_{SL} which is compared to other values of γ_{SL}. Figure 34 shows photographs[239] of the growth of an ice cylinder and perturbations at $\Delta T = -0.095°C$. Table I[240] gives several measured values of $\delta, \lambda_z, R, \dot{R}$, and the resulting computed values of interfacial energy which average to 16×10^{-3} J/m² (16 ergs/cm²). From nucleation measurements,[241] values of 20×10^{-3} J/m² at $-40°C$ and 24×10^{-3} J/m² at $0°C$ are obtained. The agreement is quite satisfactory particularly considering that the morphological instability is an average over planes parallel to the c axis, while the nucleation value involves other planes as well.

b. *Region of Instability Experiments—"Go, No-Go"*

We have already mentioned the constitutional supercooling criterion of Tiller et al.[221] [Eq. (22.24)], and how the morphological stability equation (22.23) of Mullins and Sekerka reduces to it in the rough approximation that the temperature gradient in the solid equals that in the liquid, and that surface energy effects are excluded. The experiments of Walton et al.[222] and others which have verified the constitutional supercooling criterion will also therefore, within this degree of approximation, provide evidence for the morphological instability theory of the dilute planar alloy of Mullins and Sekerka. Figure 35 shows that the demarcation separating stability (no cells) from instability (cells) is fairly sharp; its slope, from Eq. (22.24) is given by

$$G_{tL}/v = m'(C_\infty/D)[(1 - k_0)/k_0] \qquad (24.2)$$

where k_0 is equilibrium distribution coefficient. For Pb in Sn, Eq. (24.2)

[241] L. Dufour and R. Defay, "Thermodynamics of Clouds." Academic Press, New York, 1963.

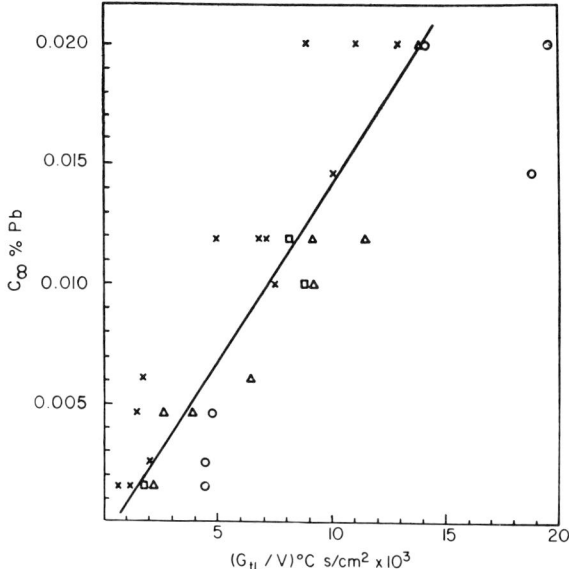

Fig. 35. Morphological instability of the dilute planar alloy. Bulk concentration C_∞ of Pb in Sn above which cells are formed versus ratio of temperature gradient to growth velocity. ×, cells; □, cells disappearing; △, pox; ○, no cells [D. Walton, W. A. Tiller, J. W. Rutter, and W. C. Winegard, *Trans. AIME* **203**, 1023 (1955)].

gives a diffusion coefficient D (in the liquid Sn) of 2.0×10^{-5} cm²/s which compares well, considering the approximations made and the possible difficulties involved in the decanting process,[242] with the value of 2.0×10^{-5} cm²/s, which is a value obtained from diffusion measurements[243] with a temperature correction by Walton et al.[222]

c. *Qualitative Observations of Morphological Instability*

Arakawa[244] photographed the growth and the development of instability of small disk crystals of ice growing parallel to the basal plane on the surface of supercooled water. The thin disks are circular in shape until they attain a size of perhaps 1 or 2 mm diam. Then numerous (i.e. a hundred or so) small perturbations develop around the edge of the disk. Initially, these show no evidence of the expected sixfold symmetry, but in later stages of the growth, certain perturbations grow faster than others, giving the sixfold

---

[242] J. A. Spittle, M. D. Hunt, and R. W. Smith, *J. Inst. Met.* **93**, 234 (1965).
[243] W. Jost, "Diffusion in Solids, Liquids, Gases." Academic Press, New York, 1952.
[244] K. Arakawa, *J. Glaciol.* **2**, 463 (1955).

symmetry. Williamson and Chalmers[245] have also photographed this behavior of ice disk crystals, attached at the edge to a substrate.

Dendritic growth itself, the onset of which is of course qualitative evidence of morphological instability, is described in many references; we mention the classical article of Papapetrou,[75] the beautiful photographs of snow crystals by Nakaya,[5] the work of Mason[246,247] also on ice, a review by Tiller,[248] and the books by Chalmers[80] and by Saratovkin.[249]

VII. Effects of Impurities on Crystal Growth Mechanisms

25. INTRODUCTION

It appears that impurities can affect the growth of crystals in very many ways. The effects will be different for small than for large concentrations. There can be equilibrium effects, nonequilibrium steady state effects, and time-dependent effects. There are effects which increase and effects which decrease growth rates, and effects on dissolution, melting, and evaporation. Other effects include crystallographic effects, of a static, structural type; or kinetic or dynamic effects; adsorption effects at the interface with various adsorption isotherms; impurities which are immobile at the interface and those which move; atomic, molecular, or ionic impurity particles in a system; and larger but still submicron impurity phases. There are effects on kinks, on steps, on step bunches, and on kinematic waves; electrical effects of impurities at an interface; and effects on surface and interfacial energy and therefore on nucleation probabilities.

Although there is overlapping in most categorizations of these effects, we have decided to choose the following breakdown:

(a) the effects of impurities on the growth process itself, (b) the amount and distribution of impurity in the crystal as a result of the growth process.

In category (b) the crystal growth rate and the equilibrium phase diagram are generally assumed as given, and calculations are made of the effects of solute rejection, solute diffusion, stirring, mixing, and instability on the actual distribution coefficient in the crystal. (See, e.g., the discussion

[245] R. B. Williamson and B. Chalmers, in "Crystal Growth" (H. S. Peiser, ed.), Suppl. J. Phys. Chem. Solids, p. 739. Pergamon Press, Oxford, 1967.
[246] B. J. Mason, "The Physics of Clouds." Oxford Univ. Press (Clarendon), London and New York, 1957.
[247] B. J. Mason, in "The Art and Science of Growing Crystals" (J. J. Gilman, ed.), p. 119. Wiley, New York, 1963.
[248] W. A. Tiller, Science 146, 871 (1964).
[249] D. D. Saratovkin, "Dendritic Crystallization." Consultants Bureau, New York, 1959.

of constitutional supercooling in Part VI on morphological stability.) The calculations generally avoid a molecular description of the process and are couched in macroscopic terms. This part of the subject has been extensively reviewed in recent years. We mention Chalmers,[80] Pfann,[250,251] Zief and Wilcox,[252] and Hurle.[253] We will instead concentrate on category (a), in which, generally, microscopic molecular scale models are put forward to describe the effects of impurity on kinks, step motion, step distribution, and growth rates. We do discuss an aspect of category (b) in connection with fluid flow, see Part VIII (Burton et al.[254]). We will not discuss the enormous amount of largely qualitative observation of impurity effect, e.g. on sectorial structure, hour-glasses;[255] or on habit change (Buckley,[255] van Hook,[256] Hartman,[257] plus other papers in the same volume[258]). The object of this section will be to describe a representative selection of the studies rather than to review them completely.

26. Growth of Binary Chains

Perhaps the most fundamental problem in growth in the presence of impurities is the growth of a binary chain, a one-dimensional crystal. [It is interesting that only recently, despite the long history of impurity studies, was this problem, including the backward reaction, solved[259,260]; this delay is similar to that in solving the equally fundamental but quite different classical mechanics problem of the collision of an atom with the end of a (one-component) one-dimensional chain.[168,169]] In this impurity problem a chain composed of two types of molecules is considered. Its end is supposed immersed in a matrix which also contains the two types of molecules described as one and two. The chain may grow either by adding a molecule one or a molecule two from the medium; the frequency of doing this will

[250] W. G. Pfann, *Solid State Phys.* **4**, 423 (1957).
[251] W. G. Pfann, "Zone Melting," 2nd ed. Wiley, New York, 1966.
[252] M. Zief and W. R. Wilcox, eds., "Fractional Solidification," Vol. I. Dekker, New York, 1967.
[253] D. T. J. Hurle, *Progr. Mat. Sci.* **10**, 79 (1963).
[254] J. A. Burton, R. C. Prim, and W. P. Slichter, *J. Chem. Phys.* **21**, 1987 (1953).
[255] H. E. Buckley, "Crystal Growth." Wiley, New York, 1951.
[256] A. Van Hook, "Crystallization Theory and Practice." Reinhold, New York, 1961.
[257] P. Hartman, *in* "Adsorption et Croissance Cristalline," p. 477. *Colloq. Intern. Centre Natl. Rech. Sci. (Paris)* No. 152 (1965).
[258] "Adsorption et Croissance Cristalline." *Colloq. Intern. Centre Natl. Rech. Sci. (Paris)* No. 152 (1965).
[259] A. A. Chernov, *in* "Adsorption et Croissance Cristalline," p. 283. *Colloq. Intern. Centre Natl. Rech. Sci. (Paris)* No. 152 (1965).
[260] J. I. Lauritzen, Jr., E. Passaglia, and E. A. DiMarzio, *J. Res. Natl. Bur. Std.* **71A**, 245 (1967).

depend on whether the added particle is 1 or 2, and on whether the already present end particle is a 1 or 2. Thus, there are four growth frequencies or rate constants α^{ij} ($i = 1$ or $2; j = 1$ or 2), where the right-hand superscript refers to the particle being added. The frequency of chain shortening by loss of the end particle will depend on whether the particle is 1 or 2, and on whether the next-to-the-end particle, is 1 or 2. There are thus four dissolution frequencies or rate constants β^{ij}. Therefore the chain end will perform a kind of random walk. The average displacement is zero if the chain is in equilibrium with the mother medium, but the chain end moves at a finite (positive or negative) average rate if the medium is (super- or under-)saturated.

Following Lauritzen et al.[260,261] we define $N_{\nu+1}{}^j$ and $P_{\nu+1}{}^{ij}$, called occupation numbers, as the number of chains in the system, of $(\nu + 1)$ units in length with species j in the end position, and species i and j in the next to the end and the end position. Differential equations are then formulated for the rates of change $dN_{\nu+1}{}^j/dt$ and $dP_{\nu+1}{}^{ij}/dt$ in the following way:

$$dN_{\nu+1}{}^j/dt = \sum_i \alpha^{ij} N_\nu{}^i - \sum_i \beta^{ij} P_{\nu+1}{}^{ij} - N_{\nu+1}{}^j \sum_k \alpha^{jk} + \sum_k \beta^{jk} P_{\nu+2}{}^{jk}. \quad (26.1)$$

Equation (26.1) gives the net rate of change of the number of chains $\nu + 1$ units long having the species j in the end position. The term $\sum_i \alpha^{ij} N_\nu{}^i$ represents the formation of such chains by the addition of a particle j to all chains of length ν, including both those previously ended in $i = 1$ and $i = 2$. The term $-\sum_i \beta^{ij} P_{\nu+1}{}^{ij}$ represents the loss of such chains by the removal of the end particle j from all chains of length $\nu + 1$ which have the particle i in the next to the last position, also summed over i. The term $-N_{\nu+1}{}^j \sum_k \alpha^{jk}$ represents the loss of such chains by their growth, i.e. by adding an additional particle of either type. The term $\sum_k \beta^{jk} P_{\nu+2}{}^{jk}$ represents the formation of such chains by removing the species k (both 1 and 2) from all chains of length $\nu + 2$ which have the particle j in the next to the last position. A similar differential equation is written for $P_{\nu+1}{}^{ij}$.

These (coupled equations) are solved for the steady state case in order to obtain the occupation numbers. Then the partial currents $S_\nu{}^i$ and $S_\nu{}^{ij}$ are defined in terms of the α, β, and the occupation numbers. Finally, a total current or growth rate $S_T = \sum_{i,j} S_\nu{}^{ij}$ is defined and the chain composition f_j, the fraction of the jth species in the chain, is defined.

For application to a specific system the rate constants α^{ij} and β^{ij} must be obtained. Lauritzen et al.[260] did this for an ideal solution of the two paraffin species $C_{24}H_{50}$ and $C_{26}H_{54}$ for which there is a free energy term involving the surface energy γ of the solid due to the different lengths of the molecules forming the strip or chain of material which is considered to

[261] J. I. Lauritzen, Jr., E. A. DiMarzio, and E. Passaglia, *J. Chem. Phys.* **45**, 4444 (1966).

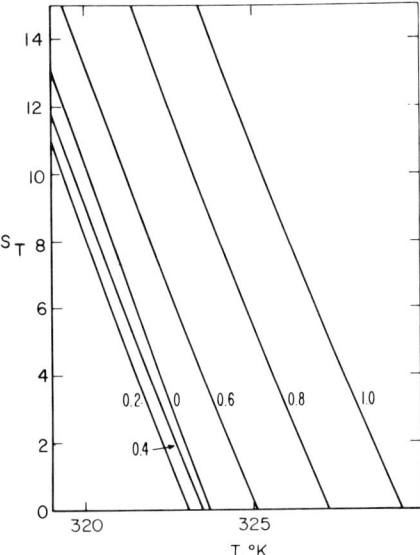

FIG. 36. Application of binary chain theory to crystallization of paraffin mixture. Total flux of crystallization S_T vs temperature for six values of concentration of $C_{26}H_{54}$ in $C_{24}H_{50}$. Assumed surface energy, $\gamma = 8 \times 10^{-3}$ J/m²; $\phi = 1$ [J. I. Lauritzen, Jr., E. Passaglia, and E. A. DiMarzio, *J. Res. Natl. Bur. Std.* **71A**, 245 (1967)].

be crystallizing onto a substrate of intermediate thickness. Figure 36 presents computer-obtained curves of S_T vs temperature for a value of 8×10^{-3} J/m² (8 ergs/cm²) of surface energy, of various compositions of the liquid (0 to 1.0), and shows that the total flux is nearly linear with temperature. The slope is largely independent of composition, i.e., the growth rate is proportional to and only dependent upon the degree of undercooling appropriate to that composition. Also calculated was the chain composition f, for all liquid compositions, and both for $S_T = 0$ (equilibrium phase diagram) and $S_T \neq 0$ (kinetic "phase diagram"). For the conditions and substances chosen, these were found to be very similar.

The theory of multicomponent chains has been studied by Lauritzen et al.[261] and by Chernov.[262,263] In addition, Chernov and Lewis[264] have performed very interesting computer modeling or simulation of the growth of binary one-, two-, and three-dimensional crystals. The one-dimensional calculation served to verify Chernov's[259] one-dimensional (analytical)

[262] A. A. Chernov *in* "Crystal Growth" (H. S. Peiser, ed.), Suppl. *J. Phys. Chem. Solids*, p. 25. Pergamon Press, Oxford, 1967.
[263] A. A. Chernov, *Soviet. Phys.—Doklady* **11**, 751 (1967).
[264] A. A. Chernov and J. Lewis, *J. Phys. Chem. Solids* **28**, 2185 (1967).

solution of the dilute binary system. In the two-dimensional case the degree of order was studied as a function of the supersaturation for a constant (50–50) crystal composition.

27. Poisoning of Kinks during Layer Growth by Step Motion around Screw Dislocations

The theory of Burton et al.,[41] as discussed in Part V, required the presence of sufficient kinks in the step to serve as final deposition sites for surface diffusing adatoms. If these kinks were rendered ineffective as growth sites, such as, e.g., by adsorbing impurity molecules into the kinks, then the net rate of step motion and hence of crystal growth could be reduced. Chernov[17,265,266] has made semiquantitative estimates of this effect by calculating a Langmuir-type adsorption isotherm for the impurities to be adsorbed at the kinks and then using the BCF formulas for the "efficiency" factor c_0 [see discussion of Eqs. (17.12) and (17.13)]. This gives the reduction in step velocity when the mean separation between (active) kinks X_0 is greater than the mean diffusion distance X_s.

Let the mean separation between kinks in a step in the pure system be X_0 (typically of the order of 5 to 10 lattice parameters). Then in the case of vapor phase growth in the presence of an impurity gas at partial pressure p_i, if the mean separation between (active, unpoisoned) kinks is increased to X_c, the Langmuir adsorption isotherm for one-dimension gives

$$X_c = X_0 + \xi p_i. \tag{27.1}$$

Here ξ is a constant. Equation (27.1) can be derived by noting that the one-dimensional "coverage" θ_{1D} is $(1 - X_0/X_c)$ and hence is $\xi p_i/(\xi p_i + X_0)$. Chernov derives by statistical mechanics an expression for ξ in terms of the adsorption energy w'' in a kink and other parameters. A rough numerical estimate is[266]

$$\xi_{\text{vapor}} = \frac{a(kT)^{1/2} \exp(w/kT)}{2\pi m^{3/2} \nu_i^3}, \tag{27.2}$$

m and ν_i being the impurity molecule mass and vibrational frequency in the kink. For $w'' = 4.18 \times 10^4$ J/mol and $\nu_i \approx 10^{12}$/s, a rough estimate for ξ is

$$\xi_v = 0.075 a/(\text{N/m}^2) \tag{27.3}$$

so that significant poisoning of kinks could occur at 133 N/m² (1 Torr)

---

[265] A. A. Chernov, in "Growth of Crystals" (A. V. Shubnikov and N. N. Sheftal, eds.), Vol. III, p. 31. Consultants Bureau, New York, 1962.

[266] A. A. Chernov in "Adsorption et Croissance Cristalline," p. 265. Colloq. Intern. Centre Natl. Rech. Sci. (Paris) No. 152 (1965).

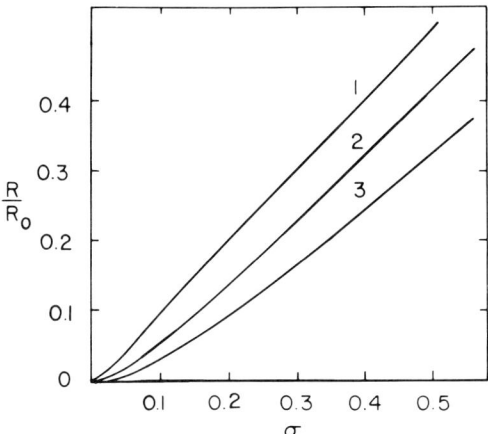

Fig. 37. Theory of impurity poisoning of kinks in vapor growth at screw dislocations. Reduced growth rate versus supersaturation for $X_s = 400a$, $\sigma_1 = 5 \times 10^{-2}$. Curve 1, $X_c \ll X_s$ (pure case); curve 2, $X_c = 100a$; curve 3, $X_c = 400a$ [A. A. Chernov, in "Growth of Crystals" (A. V. Shubnikov and N. N. Sheftal, eds.), Vol. III, p. 31. Consultants Bureau, New York, 1962].

pressure of foreign gas. A similar expression for an ideal solution gives $\xi_{\text{solu}} \approx 10^5 a$ in the expression $X_c = X_0 + \xi_{\text{solu}} C$. Thus a fractional concentration C of 10^{-4} impurity can cause significant poisoning. Clearly, sufficient impurity pressure or concentration could reduce the growth rate to zero; with other adsorption isotherms this might not be so.

The ideal normal growth velocity round a screw dislocation is reduced by a factor c_0 ($0 \leq c_0 \leq 1$), when there are insufficient kinks in the step; c_0 is given by BCF[41] as

$$1/c_0 = 1 + 2b_4 \tanh(y_0/X_s)[\ln\{(4b_4 X_s/a)/(1 + (1 + b_4^2)^{1/2})\} + (2X_s/y_0) \tan^{-1} b_4] \quad (27.4)$$

where $b_4 = X_0/2\pi X_s$ and where y_0 is the distance $4\pi\rho_c$ between turns in the steady state growth spiral. Replacing X_0 now by $X_0 + \xi p$ or $X_0 + \xi C$ and recalling that $\rho_c \cong \gamma_e a/kT\sigma$, and $\sigma_1 = 2\pi\gamma_e a/kTX_s$, one gets the desired expression for the growth velocity using Eq. (17.18). We do not reproduce it here but instead in Fig. 37 give a plot[265] showing the main effect: it is to reduce the velocity, and greatly increase the region of nonlinearity, which in the pure case lay near the origin between 0 and σ_1.

A related type of poisoning at steps was proposed by Sears.[267] He suggested that for poisoning to be effective in greatly reducing the growth

[267] G. W. Sears, *J. Chem. Phys.* **29**, 1045 (1958).

rate, the entire step had to be covered by impurity for otherwise fresh kinks would be produced, by statistical fluctuations, in the unpoisoned parts and these fresh kinks would be more likely to grow than to adsorb impurity molecules, there being far fewer of the latter than of the growing species. Sears's calculations, though not quantitative, suggest that the Chernov calculation above could be improved by considering the dynamic effects of competition for kinks by the two species, but that Chernov's calculation should be valid in the limit of slow growth velocities where the adsorption equilibrium could take place.

Bliznakov[268] has also used the Langmuir isotherm to interpret impurity-reduced growth rates in aqueous solutions. Only the normal, not the step growth rate is considered. The normal rate is supposed decomposed into two parallel parts, one being the normal growth velocity for those parts of the entire crystal surface free of impurity, times the free surface $(1 - \theta)$. The other is the "impurity" growth velocity times the coverage, θ. Then θ is assumed to be of the Langmuir type. Agreement with data on normal velocity of crystal faces in several systems is found. As Chernov[17] noted, however, this approach ignores the fact of layer growth on faces at steps.

28. Sucrose Growth from Solution: Step Pinning

Albon and Dunning[269] have studied the effect of small amounts of raffinose impurity molecules upon the step velocity in the growth of sucrose crystals from aqueous solution. Raffinose at a concentration of a few parts per thousand reduced step velocities on the (100) face by 50% for low sucrose supersaturations ($\sim 1\%$), but the effect tailed off and the growth velocity as a function of impurity concentration appeared to approach zero only asymptotically. The authors proposed that raffinose molecules adsorbed onto the steps (presumably at kinks, although the kink density of the pure crystal does not appear in the theory), and when a step length between two adsorbed raffinose molecules was less than d_c, the diameter ($=2\rho_c$) of the critical nucleus, that portion of step was pinned and could not grow (see also below, Cabrera and Vermilyea[204]). The net or average step growth rate was then assumed to be proportional to the fraction of unpinned step length, i.e., the sum of all portions of step lengths greater than or equal to d_c, times their probability. This gave

$$(R/R_0) = (d_c - pd_c + p)p^{d_c} \tag{28.1}$$

where R is the growth velocity in presence of impurities, R_0 is that without

[268] G. Bliznakov, in "Adsorption et Croissance Cristalline," p. 291. *Colloq. Intern. Centre Natl. Rech. Sci. (Paris)* No. 152 (1965).
[269] N. Albon and W. J. Dunning, *Acta Cryst.* **15**, 474 (1962).

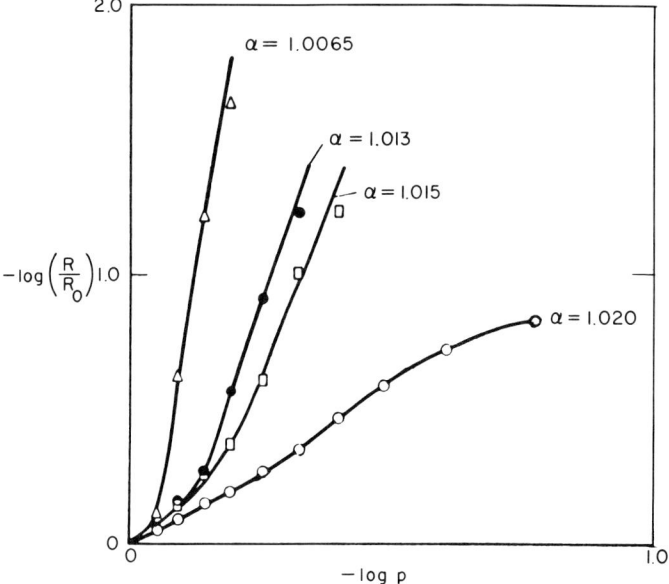

FIG. 38. Effect of raffinose impurity on sucrose growth from aqueous solution. Reduced growth rate versus $-\log p$ (see text) (pure case is $-\log p = 0$), for various saturation ratios α [N. Albon and W. J. Dunning, *Acta Cryst.* **15**, 474 (1962)].

impurities, p is the probability per lattice distance of finding no raffinose molecule adsorbed, and $C_a = 1 - p$ the concentration, per site length along the step, of such adsorbed molecules. The adsorbed amount, C_a, and hence p was related to the amount C_0 of raffinose in solution by the Freundlich adsorption isotherm

$$C_a = AC_0^{1/n}. \tag{28.2}$$

Equation (28.2) was found to hold experimentally if C_a is assumed to be proportional to the measured reduction $\Delta R = R_0 - R$ in growth velocity. From Eq. (28.1) a plot of $\log(R/R_0)$ vs $\log p$ should be approximately linear and have the slope d_c; this is seen in Fig. 38. The values of d_c so obtained run from 1.6 to 11.8 molecular diameters depending on α or σ. Step edge energies can be calculated from the relation $\rho_c = \frac{1}{2}d_c = \gamma_e a/kT\sigma$.

Dunning et al.[270] have related the highly differing potencies of different additives for step rate reduction in sucrose growth to the molecular structure of the additive. It appears that the most potent ones have a molecular structure such that they may be expected to be more strongly adsorbed.

[270] W. J. Dunning, R. W. Jackson, and D. G. Mead, *in* "Adsorption et Croissance Cristalline," p. 303. *Colloq. Intern. Centre Natl. Rech. Sci. (Paris)* No. 152 (1965).

29. STEP PINNING (CONTINUED)

Cabrera and Vermilyea[204] have also studied the pinning of steps by impurities. In their model the large, immobile impurity particles are assumed to exist on the crystal surface and to pin or stop the step where it touches the particle. The rest of the step bows out and can squeeze through, i.e. reform on the other side only if its radius of curvature, ρ, equals or exceeds the critical nucleation radius ρ_c for the system. If, just ahead of the step, the particle separation is assumed to be uniform at z, and if $z > 2\rho_c$, then the step can squeeze through. But if $z \leq 2\rho_c$ it will be stopped. The reduced step velocity of a curved step had been calculated by BCF[41] as

$$v/v_\infty = (1 - \rho_c/\rho) \tag{29.1}$$

[see Eq. (17.14)], so that when the step forms a semicircle of radius $z/2$, $v/v_\infty = (1 - 2\rho_c/z)$. The exact average velocity would require an integration over all step radii ρ as the step squeezes through. Cabrera and Vermilyea, without proof, give the approximate relation

$$v/v_\infty = (1 - 2\rho_c/z)^{1/2}, \tag{29.2}$$

which has the proper behavior at $z = 2\rho_c$ and at $z = \infty$. If, just before the step, the impurities are arranged on a two-dimensional lattice of spacing z and density d, then $z = d^{-1/2}$ and

$$v/v_\infty = (1 - 2\rho_c d^{1/2})^{1/2}. \tag{29.2'}$$

Further assume that the impurities are buried or otherwise rendered ineffective, if the step passes by them, and that fresh steps are flowing from the surroundings to the surface at a rate J_i. Then just ahead of a step moving at velocity v and with step density k:

$$d = J_i/kv. \tag{29.3}$$

Combining (29.3) and (29.2') gives a fifth degree equation for v/v_∞, which has a solution (i.e. steps will move) only if $2\rho_c(J_i/kv_\infty)^{1/2} < 0.54$ and hence only if the pure system step flow

$$kv_\infty \geq 14\rho_c^2 J_i. \tag{29.4}$$

Since for small σ, ρ_c is proportional to $1/\sigma$ and since kv_∞ is given by a linear or parabolic law, there will be a minimum supersaturation for growth given by

$$\sigma^3/J_i \quad \text{or} \quad \sigma^4/J_i = \text{const.} \tag{29.5}$$

Price et al.[271] have studied the effects of large molecule impurities on the

[271] P. B. Price, D. A. Vermilyea, and M. B. Webb, *Acta Met.* **6**, 524 (1958).

electrolytic growth of silver whiskers. The whiskers grow from the tip. It is found that for a given concentration C_i of added gelatin there is a critical driving force below which whiskers are not formed. Under the conditions of the experiments, the critical driving force is proportional to the current density I_c. The flux of impurities into the whisker tip was, using simple diffusion theory, believed to be proportional to the bulk solution concentration of impurities C_i. The experimentally obtained relation was $I_c =$ const $(C_i)^{1/3}$ as in Eq. (29.5), where I_c takes the place of σ and C_i takes the place of J_i. This gives experimental support to the theory of Cabrera and Vermilyea.

Van Damme[272] studied the effect of ferrocyanide ion on the dissolution of NaCl and interpreted the results in terms of a somewhat modified Cabrera–Vermilyea theory. A critical undersaturation σ_c at which no further dissolution of NaCl occurred was measured as a function of impurity concentration C_i. The equilibrium concentration in the presence of dilute ferrocyanide was assumed to be the same as with no impurity added. It was further assumed that an equilibrium adsorption isotherm held, namely Langmuir's, in contrast to Eq. (29.3) which gives the flux of impurities as the relevant factor. Thus for zero velocity, Eq. (29.2') gives $\sigma_c^2/d =$ const or

$$\sigma_c^2/\sigma_{c\infty}^2 = d_c/d_{c\infty} = \theta \tag{29.6}$$

where $\sigma_{c\infty}$ is the critical undersaturation for maximum coverage $\theta = 1$. With the Langmuir isotherm $\theta = C_i/(1/K_a + C_i)$, where K_a is a constant, one obtains

$$1/\sigma_c^2 = 1/C_i K_a \sigma_{c\infty}^2 + 1/\sigma_{c\infty}^2, \tag{29.6'}$$

a form which agrees with the experiments. Also, the adsorption of ferrocyanide on very small NaCl crystals was measured directly. The measured coverages were within a factor of ~ 10 of that obtained from Eq. (29.6') and the dissolution data.

30. Kinematic Waves and Impurities in Dissolution: Etch Pits

Hulett and Young[273] found that their observations on the formation of etch pits in copper crystals can be interpreted satisfactorily with the kinematic theory.[13,203,204,274] The surfaces of highly perfect copper crystals (200–1000 dislocations/cm^2) near a (111) orientation were etched anodically in HCl solutions with current densities of 5 to 30 mA/cm^2 in the

[272] M. A. van Damme, *in* "Adsorption et Croissance Cristalline," p. 433. *Colloq. Intern. Centre Natl. Rech. Sci. (Paris)* No. 152 (1965).
[273] L. D. Hulett, Jr. and F. W. Young, Jr., *J. Phys. Chem. Solids* **26**, 1287 (1965).
[274] N. Cabrera, *in* "Reactivity of Solids" (J. H. deBoer, ed.), p. 345. Elsevier, Amsterdam, 1961.

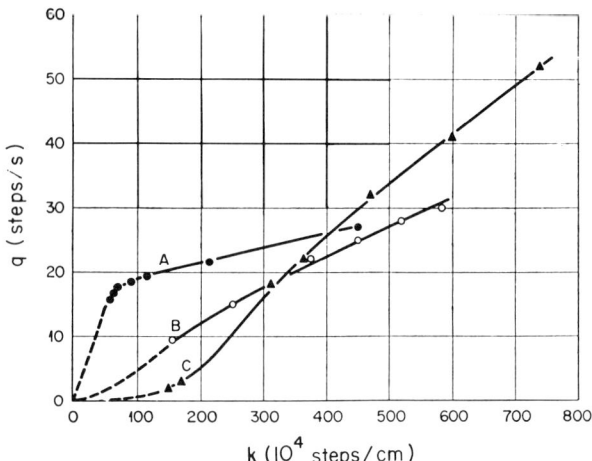

Fig. 39. Effect of HBr impurity on step flux q vs step density k in anodic etch-pitting of Cu crystal in 6 M HCl. Current density 5 mA/cm². (A) 0.03 M HBr, (B) 0.25 M, (C) 1.0 M [L. D. Hulett, Jr. and F. W. Young, Jr., *J. Phys. Chem. Solids* **26**, 1287 (1965)].

presence of 0.03 M to 1.0 M HBr added impurity. Etch pits were formed at the emergence of dislocations as was already well known.[275] The dissolution etch pit profiles were recorded as a function of time by interference microscopy. Step densities $k(x) = (+1/h)\, dy/dx$, where h is the elementary step height, were obtained from the plotted step profiles $y(x)$. A family of such $k(x)$ curves, obtained by following $y(x)$ as a function of time, permitted Hulett and Young to obtain, for a constant value of k, plots of x vs time, or trajectories of a point of constant step density. These trajectories were experimentally found to be linear, which as seen in Part V, is in accord (Frank's theorem) with kinematic theory.

Now from kinematical theory [Eq. (20.5)] the slope dx/dt of the trajectory of a point of step density k is just

$$\partial q/\partial k\,|_k = dx/dt\,|_k = c(k), \qquad (30.1)$$

the kinematic wave velocity, where $q = q(k)$ is the step flux in steps per second passing a point at which the step density is k steps per centimeter. From a plot of the measured slopes vs k, the integral $q = \int_{k_0}^{k} c\, dk$ was obtained, giving the basic flow-concentration curve $q = q(k)$. This was done for three values of the Br⁻ impurity as shown in Fig. 39. It is seen that the nature of the flow-concentration curves is drastically changed by the

---

[275] F. W. Young, Jr. and L. D. Hulett, *in* "Metal Surfaces: Structure, Energetics and Kinetics," p. 375. Am. Soc. Metals, Cleveland, Ohio, 1963.

impurity level. The form of these curves is in good agreement with the predictions of Frank[203] and of Cabrera.[274] Cabrera[274] (see also Cabrera and Coleman[13]) called A a type I curve and pointed out that it is in fact of the form predicted by the screw dislocation theory in the high purity limit. Curve C, called a type II curve, had been predicted by Frank[203] and Cabrera.[274] It is distinguished by its positive curvature ($\partial^2 q/\partial k^2 > 0$) at small k. The reason for this shape is that the time-dependent amount of adsorbed particles just next to a step, and hence the impedance to the motion of that step, depends on the period since the passage of the previous step, if it is assumed that the particles are rendered ineffective by step passage (burial in growth, removal in dissolution). Thus, if steps are far apart (small k) the velocity of steps q/k will be less than for steps closer together (large k) so that q/k will increase with k and $\partial^2 q/\partial k^2$ is positive.

Another consequence of the type of $q(k)$ curve studied by Hulett and Young is the formation of leading-edge and trailing-edge discontinuities. Frank had shown that the profile of a bunch of steps, in the case when $\partial^2 q/\partial k^2$ is negative, will in dissolution develop into a form shallow at the leading edge of the profile, with a sharp break at the trailing edge; and the opposite is expected for the case when $\partial^2 q/\partial k^2$ is positive. The latter case, where the leading edge develops the break, is the desired type of etch pit for visibility, since the break is then the sharply defined periphery of the pit. Hulett and Young found, indeed, that with type I curves their etch pits were rounded at the periphery, and with impurities added (type II curves, C in Fig. 39) the pits developed sharp edges.

31. Macroscopic Particles as the Impurity

The conditions under which small foreign particles of greater than molecular size are incorporated or rejected at the interface, seem to be much less well understood than for molecular size particles. Yet the subject is important, for as Turnbull's[9] droplet nucleation experiments indicate, most fluids may be expected to have many macroscopic foreign particles in them. Jackson[276] has suggested that these particles may be the main source of dislocations in melt-grown crystals.

Uhlmann et al.[277] performed experiments on the freezing of several organic materials, including salol, in which foreign particles of numerous types including Zn, MgO, diamond, and others, in the size range of 1 to several hundred $\times 10^{-4}$ cm, had been placed. It was found from microscope observations that for nearly all the materials studied, the particles in the liquid were rejected by the interface, at sufficiently small interface velocities.

[276] K. A. Jackson, *Phil. Mag.* **7**, 1615 (1962).
[277] D. R. Uhlmann, B. Chalmers, and K. A. Jackson, *J. Appl. Phys.* **35**, 2986 (1964).

As the interface velocity was increased beyond a critical value typically a few microns per second, the particles became trapped in the solid. This critical velocity depended on the composition of the impurity particle, and on that of the freezing matrix, but was independent of particle size, when $d < 15 \times 10^{-4}$ cm. According to the theory proposed by Uhlmann et al., a force keeping the particle out of the solid arises when the particle–solid interfacial energy is greater than the sum of the particle–liquid and liquid–solid interfacial energies. However, at a sufficiently high growth velocity the liquid cannot diffuse or flow rapidly enough in the restrictive narrow region between the particle and interface to maintain the interface velocity there; consequently, the particle becomes buried. The absence of size dependence was attributed to the presence on all solid particles of very small—$\sim 3 \times 10^{-6}$ cm—irregularities which determined the scale of the processes occuring at the interface.

Chernov and Melnikova[278,279] (independently of Uhlmann) studied, from a theoretical standpoint, a very similar problem for both solution and for melt growth. They calculated, for solution growth, the diffusional field for the case of an impurity sphere near an interface. The reduced supersaturation in the restricted area between sphere and interface would cause a depression there, that could be compensated partly by a particular form of crystallographic anisotropy of the kinetic coefficients. In general, however, there appears to be a critical value of a_4/R beneath which a deep pit or inclusion will form (or entrapment of sphere), where a_4 is the distance of center of sphere from surface, and R the radius of sphere. However, this value of a_4/R did not depend, for linear kinetics, on the rate of growth.

Chernov and Melnikova[279] also studied the melt growth case with emphasis on heat flow calculations. Results similar to those for solution growth were obtained, in the case where the thermal conductivity of the sphere was greater than that of the melt. Thus, their predictions are in disagreement with the observations of Uhlmann, and with other observations, as Chernov and Melnikova[278] note.

VIII. Fluid Flow Effects

32. INTRODUCTION

We have discussed, particularly in Part III, heat transport and mass transport effects in crystallization, assuming that the only mode of such transport to or from the interface is by diffusion, either of heat or matter or both. Another quite different mode of heat or mass transport is by con-

---

[278] A. A. Chernov and A. M. Melnikova, *Soviet Phys.—Cryst.* **10,** 666 (1966).
[279] A. A. Chernov and A. M. Melnikova, *Soviet Phys.—Cryst.* **10,** 672 (1966).

vection, in which the heat or mass is carried from one point to another by the hydrodynamic flow of a fluid containing the heat or mass. One may expect that growth from fluid phases, i.e. solution, vapor, and melt would, in general, be influenced by such flow. The fluid flow in turn may be motivated in several ways, such as (a) density differences between the fluid and crystal which requires a convective flow in the fluid as the interface moves into it; (b) density differences within the fluid itself caused by temperature gradients or concentration gradients in it, in turn acted on by gravity to cause natural convection flow; (c) forced convective flow, in which the fluid is caused, by external action-stirring or mixing, e.g., to flow by the crystal.

A common effect of such convective flow of heat or mass would be to break up long-range diffusion fields in the mother medium by translating them away. These are replaced by relatively short-range diffusion fields in which the fluid of bulk supersaturation is now quite close to the interface, being separated from it by a relatively thin hydrodynamic (momentum) boundary layer of thickness δ. Then diffusion would take place through this boundary layer, and concentration or temperature gradients would be correspondingly increased over their no-fluid-flow values. Growth rates would therefore be increased in cases where interface kinetics is not entirely the limiting process. As another example, the segregation of impurity at a freezing interface can cause a building up of impurity in the liquid near the interface. The next solid freezing from such impure liquid will have a higher concentration of impurity. Diffusion of the impurity in the liquid away from the interface can help somewhat to reduce the concentration near the interface, and thereby to reduce the trapping of impurities in the solid, but mixing by fluid flow can facilitate it much more, again by introducing a much thinner diffusion distance as a hydrodynamic boundary layer (see e.g. Zief and Wilcox,[252] and Pfann[251]).

Although fluid flow effects seem to be, in many cases, as important as purely diffusive transfer in crystallization, there does not seem to be a nearly comparable body of quantitative knowledge regarding them as there is, e.g., in the diffusive Stefan and related problems. (However, in the chemical industry the effects of stirring and mixing in certain crystallization processes have been studied frequently, see e.g. Bamforth[280] and Mullin.[281]) There seem to be several reasons for this neglect. One is that in experiments where no external mixing or stirring was applied, the potentiality of natural convection was often overlooked. Another is that even when the occurrence of fluid flow during crystal growth is recognized, the very formidable differential equations of hydrodynamics have discouraged analytical or even semiquantitative studies of the flow patterns.

[280] A. W. Bamforth, "Industrial Crystallization." Leonard Hill, London, 1965.
[281] J. W. Mullin, "Crystallization." Butterworth, London and Washington, D.C., 1961.

A third reason is that techniques for observing fluid flow patterns in crystal growth experiments, particularly in opaque fluids, are often lacking.

In this part we discuss studies of the effect of fluid flow on heat and mass transfer in crystallization, with an emphasis on quantitative considerations. In analogy with our breakdown of the Stefan problem, we categorize the studies made in terms of the shape or geometry of the system; this is the same categorization often made in books on hydrodynamics and heat and mass transfer (general references: Landau and Lifshitz,[212] Goldstein,[282] Schlichting,[283] Eckert and Drake,[284] also journal[285]).

33. Flow Past a Flat Plate

As an approximation to a crystal with flat faces growing from a stirred or flowing supersaturated solution, or being agitated in such a solution, the fluid flow problem of the boundary layer in parallel flow past a semi-infinite flat plate has been selected by a number of authors in crystal growth studies, including Carlson,[286] Bennema,[182] and Brice.[287]

The starting point, in principle, for analyzing fluid flow problems is the Navier–Stokes equations (see e.g. Landau and Lifshitz,[212] Schlichting[283]). These complex equations of motion are nonlinear and in only a few cases (i.e. flow in a pipe, flow near a rotating disk, creeping (very slow) flow around a sphere) are exact solutions known. It was shown by Prandtl that for the flow of fluids near solid bodies a layer next to the body, called the boundary layer, exists in which there are substantial velocity gradients and in which friction (viscosity) must be considered inside this layer but it may be neglected outside of it. Simple but still nonlinear boundary layer equations result from this approximation and are, for two-dimensional steady flow of an incompressible fluid with constant property values,

$$u \frac{\partial u}{\partial x} + v \frac{\partial u}{\partial y} = -\frac{1}{\rho_L} \frac{dp}{dx} + \nu \frac{\partial^2 u}{\partial y^2} \tag{33.1}$$

$$\frac{\partial u}{\partial x} + \frac{\partial v}{\partial y} = 0 \tag{33.2}$$

[282] S. Goldstein, ed., "Modern Developments in Fluid Dynamics," Vol. I, II. Oxford Univ. Press, London and New York, 1938.
[283] H. Schlichting, "Boundary Layer Theory." McGraw-Hill, New York, 1955.
[284] E. R. G. Eckert and R. M. Drake, Jr. "Heat and Mass Transfer." McGraw-Hill, New York, 1959.
[285] *Intern. J. Heat Mass Transfer* 1960 to present.
[286] A. Carlson, *in* "Growth and Perfection of Crystals" (R. H. Doremus, B. W. Roberts, and D. Turnbull, eds.), p. 421. Wiley, New York, 1958.
[287] J. C. Brice, *J. Cryst. Growth* **1**, 161 (1967).

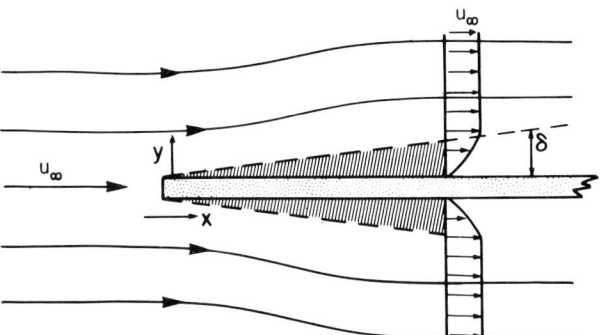

FIG. 40. Boundary layer in fluid flow past a flat plate [H. Schlichting, "Boundary Layer Theory." McGraw-Hill, New York, 1955].

where u, v are the x, y components of the flow velocity, ν is the kinematic viscosity η/ρ_L, η being the viscosity, ρ_L the density, and the pressure p is assumed known from the potential-type (frictionless) flow solution to the problem. For the case of fluid flow past a plate (see Fig. 40), $dp/dx = 0$. The boundary conditions are $u = v = 0$ at $y = 0$ and $u = u_\infty$ at $y = \infty$. The exact numerical solution for this problem is still quite difficult and was performed first by Blasius (see Schlichting[283]) using a series method. We give here instead only the results obtained[284] by a simpler method, using the momentum equation of the boundary layer. In this approximation the velocity profile is

$$u/u_\infty = \tfrac{3}{2}(y/\delta) - \tfrac{1}{2}(y/\delta)^3 \qquad (33.3)$$

where the boundary layer thickness δ is

$$\delta/x = 4.64/(\mathrm{Re}_x)^{1/2} \qquad (33.4)$$

and the Reynolds number $\mathrm{Re}_x = u_\infty x/\nu$. In this approximation δ is the thickness at which $u = u_\infty$, but in the exact treatment u only asymptotically approaches u_∞. In the exact treatment for δ, for which $u = 0.99u_\infty$, a factor ~ 5.0 appears instead of the 4.64 above. Thus from (33.4) the boundary layer at the leading edge has zero thickness and increases as $x^{1/2}$.

In order to treat heat transfer to or from the flat plate an energy equation, expressing the energy balance for a fluid element, is first derived. In principle this equation must be solved along with the Navier–Stokes equations. A simpler version of it, in the case of the boundary layer approximation, is

$$c_{LP}\left(u\frac{\partial T}{\partial x} + v\frac{\partial T}{\partial y}\right) = \frac{K_L}{\rho}\frac{\partial^2 T}{\partial y^2} + \nu\left(\frac{\partial u}{\partial y}\right)^2 \qquad (33.5)$$

where K_L is the thermal conductivity of the fluid, c_{LP} the specific heat of the fluid at constant pressure. Equation (33.5), along with (33.1) and (33.2) and their boundary conditions, has the boundary conditions $T = T_{su}$ at $y = 0$, $T = T_\infty$ at $y = \infty$. The velocity field is, for fluid properties independent of temperature, independent of the temperature field, so that Eqs. (33.1) and (33.2) may be solved first and then the linear equation (33.5) in $T(x, y)$ solved. The exact solution to (33.5) using (33.1) and (33.2) was carried out by Pohlhausen (see Schlichting[283]). An approximate solution[284] is, assuming the thermal boundary layer δ_t to be less than the momentum boundary layer δ,

$$(T - T_{su})/(T_\infty - T_{su}) = \tfrac{3}{2}(y/\delta_t) - \tfrac{1}{2}(y/\delta_t)^3 \qquad (33.6)$$

where

$$\delta_t/\delta = 1/1.026(\text{Pr})^{1/3} \qquad (33.7)$$

where the Prandtl number $\text{Pr} = \nu/\kappa_L$, κ_L being the thermal diffusivity $K_L/\rho_L c_{LP}$. The heat flow Q per unit area from a heated plate is then

$$Q = \tfrac{3}{2}(K_L/\delta_t)(T_\infty - T_{su}), \qquad (33.8)$$

and using (33.7) and (33.4)

$$Q = 0.332 K_L (T_\infty - T_{su})(\text{Pr})^{1/3}(u_\infty/\nu x)^{1/2}, \qquad (33.9)$$

thus Q is inversely proportional to the square root of the distance from the leading edge. We note from (33.7) that for viscous liquids, i.e. $\text{Pr} \sim 1000$ or more, the thickness of the thermal boundary layer is less than $\tfrac{1}{10}$ that of the momentum boundary layer. For water at room temperature, $\text{Pr} \approx 7$ and Pr is nearly unity for many gases. For liquid metals $\text{Pr} \ll 1$ (e.g. Hg, $\text{Pr} \sim 0.02$ at room temperature) but formula (33.9) does not then hold.

When the diffusing quantity is not heat but solute, as in dilute solution growth or dissolution, then[282] the corresponding equation for the matter flux, Q_{matter}, is[286]

$$Q_{\text{matter}} = 0.332 D (C_\infty - C_s)(\text{Sc})^{1/3}(u_\infty/\nu x)^{1/2} \qquad (33.10)$$

where the Schmidt number Sc is ν/D, D being the solute diffusion constant, and C_s the concentration at the surface, C_∞ that at infinity. Sc is near unity for simple molecules diffusing in water. The corresponding concentration boundary layer δ_c is then, from (33.7) and (33.4),

$$\delta_c = (4.64/1.026)(\text{Sc})^{-1/3}(u_\infty/\nu x)^{-1/2} \qquad (33.11)$$

or $\delta_c = 4.5 D^{1/3} \nu^{1/6} u_\infty^{-1/2} x^{1/2}$. These formulas for the flat plate have been used by Brice,[287] Carlson,[286] and Bennema[182] to interpret the results of measurements on the growth rate of crystals from stirred solutions where the flow velocity u_∞ is known.

Brice[287] has discussed the role of the concentration boundary layer in solution growth of crystals and in particular how it must be taken into account in deducing interface kinetics laws from experiment. The interface kinetic law may be written, at least for linear and square law kinetics, as

$$Q = A(C_s - C_0)^n \tag{33.12}$$

where C_s is the concentration at the interface, C_0 the equilibrium concentration, Q is the growth rate, and A and n are constants at a fixed temperature. But also

$$Q = D(C_\infty - C_s)/\delta_c \tag{33.13}$$

(the numerical factor in Eq. (33.11) will be 3.0 with this definition of δ_c instead of 4.5), and so

$$(Q\delta_c/D) + (Q/A)^{1/n} = C_\infty - C_0. \tag{33.14}$$

The two limits of negligible boundary layer effects ($\delta_c \to 0$) giving pure interface kinetic control, and controlling boundary layer effects ($\delta_c \to \infty$) are seen to be special cases of (33.14). It is also apparent that as δ_c is made to approach zero, i.e. by faster stirring, the observed growth rate Q will increase to an asymptotic value (see e.g. Mullin[281]). In the case when both terms on the left side of (33.14) are important, then, since δ_c is proportional to $u_\infty^{-1/2}$ [Eq. (33.11)], the growth rate at constant $(C_\infty - C_0)$ expressed as $Qu_\infty^{-1/2}$ should be linear with $Q^{1/n}$ for some value of n corresponding to the exponent in the interface kinetic law (33.12). (In this treatment it is assumed that δ_c is a constant over the surface, whereas of course in the hydrodynamic treatment it is a variable, starting out at zero at the leading edge.) The data of McCabe and Stevens[288] on growth of $CuSO_4 \cdot 5H_2O$ from aqueous solution were found to fit Eq. (33.14) with $n = 2$.

Bennema[182] has also used Eq. (33.11) to arrive at δ_c values of $\sim 10^{-2}$ cm in interpreting his data on solution growth rates of potassium aluminum alum and sodium chlorate crystals. However, altering the solution stirring rate at constant supersaturation made no difference to the growth rate, and Bennema concluded that growth was entirely controlled by an interface process, as in the BCF[41] surface diffusion model, rather than by volume diffusion as in the model of Chernov[17] which involves δ_c in the growth rate.

Carlson[286] pointed out that the variable thickness boundary layer δ_c proportional to $x^{1/2}$ predicts an infinite rate of mass transfer at the leading edge of the crystal (assuming of course that interface kinetics is exerting negligible control) and that a more satisfactory model would be a plate having variable temperature (or surface concentration) such that the rate of heat (or mass) transfer is a constant. This heat transfer problem had

[288] W. L. McCabe and R. P. Stevens, *Chem. Eng. Progr.* **47**, 168 (1951).

been solved by Fage and Falkner[282] with the result that the plate temperature excess over the bulk fluid temperature varied as $x^{1/2}$. For the solute problem the surface concentration departure from bulk solute concentration would also be proportional to $x^{1/2}$. The effect of this is that for a certain downstream distance x, the surface concentration would be reduced to the equilibrium concentration. Carlson suggested that this critical distance led to the flaws observed in large ADP crystals ("veils") when the solution flow rates were too low. There remains the difficulty that a variable (with x) surface concentration C_s introduces in affecting the interface kinetic law, Eq. (33.12), which would then require a variable growth rate over the surface. As mentioned earlier, a nonlocal boundary condition is involved.

A discussion of semiempirical correlations [of the general form of Eq. (33.10)] of crystallization, dissolution and melting, relating the Reynolds and Prandtl or Schmidt numbers to the Nusselt or Sherwood numbers (the latter two proportional to heat and mass transfer rates, respectively) is given by Mullin[281] and references therein.

34. THE ROTATING DISK AND THE CZOCHRALSKI APPARATUS

Burton et al.[254] studied the effect of fluid flow on the solute distribution in crystals grown from an impure or doped melt using the Czochralski process (see also Burton and Slichter[289]). The rotating crystal cylinder whose end is immersed in the fluid provides a hydrodynamics problem which was regarded[254] to be similar to that of the rotating disk immersed in a fluid. The latter problem had been treated by von Karman and by Cochran (see Schlichting[283]), and provides one of the few examples in which an exact solution of the Navier–Stokes equations is obtained. The disk problem in turn resembles that of the action of a centrifugal pump, in that the layer of fluid near the disk is carried tangentially by friction and hence also thrown outward by centrifugal force. To replace this fluid, other fluid is drawn up along the axis of the system toward the rotating disk. The flow is clearly three-dimensional: tangential, radial, and axial. The axial flow velocity toward the disk falls off rapidly near it, as the fluid then moves radially and tangentially. A momentum boundary layer for this problem, for the thickness of fluid "carried" by the disk rotating at angular frequency ω, is approximately[283]

$$\delta \sim (\nu/\omega)^{1/2}. \qquad (34.1)$$

Within this region the axial flow velocity u, perpendicular to the disk, and

[289] J. A. Burton and W. P. Slichter, in "Transistor Technology" (H. E. Bridgers, J. H. Scaff, and J. N. Shive, eds.), Vol. I, p. 71. Van Nostrand, Princeton, New Jersey, 1958.

at a distance x from it, is approximately given by

$$u = 0.51\omega^{3/2}\nu^{-1/2}x^2. \tag{34.2}$$

To derive the solute boundary layer thickness, δ_c, in the case of dilute solutions of concentration C, conservation of solute requires, for a coordinate system fixed to the interface and for steady state, that

$$D\, d^2C/dx^2 - (u + v_g)\, dC/dx = 0 \tag{34.3}$$

where v_g is the growth velocity. The boundary conditions are

$$C \to C_\infty \text{(bulk liquid conc.)} \quad \text{at} \quad x \to \infty \tag{34.4}$$

and

$$(C_s - \mathbf{C})v_g + D\, dc/dx = 0 \quad \text{at} \quad x = 0, \tag{34.5}$$

the latter expressing conservation of solute at the interface where C_s is the concentration of solute in the liquid at the interface; $\mathbf{C}$ is that in the solid. The solution to this system for u given by Eq. (34.2) gives the concentration $C = C_s$ at $x = 0$,

$$(C_s - C_\infty)/(C_s - \mathbf{C}) = \int_0^\infty \exp -[X + BX^3]\, dX, \tag{34.6}$$

$B = (0.51/3)v_g^{-3}\omega^{3/2}D^2\nu^{-1/2}$, where $X = v_g x/D$. To define and obtain δ_c, it is assumed that within δ_c the total axial flow velocity is just the growth velocity v_g, and that $C = C_\infty$ at $x = \delta_c$. This gives a diffusion problem similar to that problem using Eq. (34.3) for $u = 0$; the solution is

$$(C - \mathbf{C})/(C_\infty - \mathbf{C}) = \exp[(v_g/D)(\delta_c - x)]. \tag{34.7}$$

Evaluating (34.7) at $x = 0$ and using (34.6) gives, after numerical computation of the integral in (34.6),

$$\delta_c \cong 1.6\nu^{1/6}D^{1/3}/\omega^{1/2} \tag{34.8}$$

for the concentration boundary layer thickness.

If local equilibrium is assumed to hold at the interface, then the equilibrium distribution coefficient is given by $k_0 = \mathbf{C}/C_s$ and the effective distribution coefficient by $k_e = \mathbf{C}/C_\infty$. Then from Eq. (34.7) for $x = 0$,

$$k_e = k_0/\{k_0 + [\exp - (v_g\delta_c/D)](1 - k_0)\} \tag{34.9}$$

and δ_c is given by Eq. (34.8). The effective distribution coefficient may have any value between k_0 and unity. For perfect mixing ($\delta_c \to 0$), k_e approaches k_0; for no mixing—pure diffusion control—($\delta_c \to \infty$) and k_e goes to unity. Burton et al.[290] performed experiments on Czochralski growth of Sb-doped

---

[290] J. A. Burton, E. D. Kolb, W. P. Slichter, and J. D. Struthers, J. Chem. Phys. **21**, 1991 (1953).

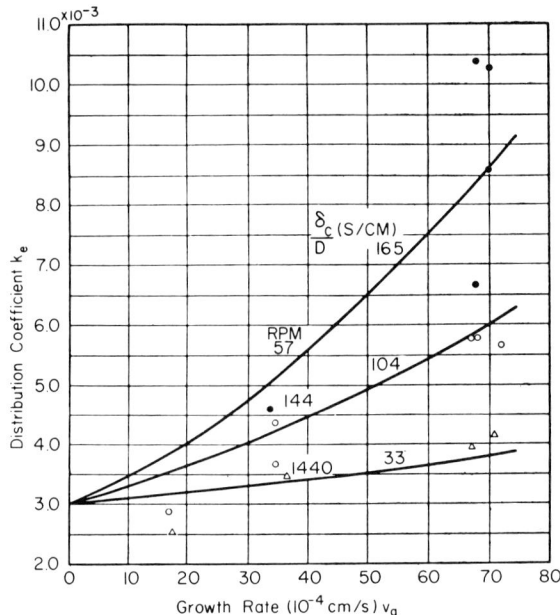

Fig. 41. Effective distribution coefficient k_e vs growth rate v_g. Czochralski growth of Sb-doped Ge, various rotation rates: ●, 57 rpm; ○, 144 rpm; △, 1440 rpm [J. A. Burton, E. D. Kolb, W. P. Slichter, and J. D. Struthers, *J. Chem. Phys.* **21**, 1991 (1953)].

Ge. Figure 41 shows their results and the prediction of Eq. (34.9) (solid lines). Reasonable diffusion coefficients D were obtained from the data. Zief and Wilcox[252] have deduced δ_c values from measured k_e values and known D values from experimental studies of numerous types, using Eq. (34.9). The values obtained for δ_c range from about 1 to 10^{-3} cm, depending on the type and vigor of mixing.

Improvements in the theory of the heat and mass transfer from the rotating disk were made by Sparrow and Gregg[291] and used by Emanuel and Olander[292] (see also Zief and Wilcox[252]). In particular, Eq. (34.8) was extended to include systems of any Schmidt number $Sc = \nu/D$, although (34.8) is a fairly good approximation for $Sc >$ unity. Emanuel and Olander[292] studied dissolution of benzoic acid and other substances in water, having high Schmidt numbers, and found agreement with the theory.

Experimental (and largely qualitative) studies of the fluid flow in Czochralski systems have been made by a number of authors, most often by modeling the system using water, glycerin, etc. Generally these studies

[291] E. M. Sparrow and J. L. Gregg, *J. Heat Transfer* **81**, 249 (1959).
[292] A. S. Emanuel and D. R. Olander, *Intern. J. Heat Mass Transfer* **7**, 539 (1964).

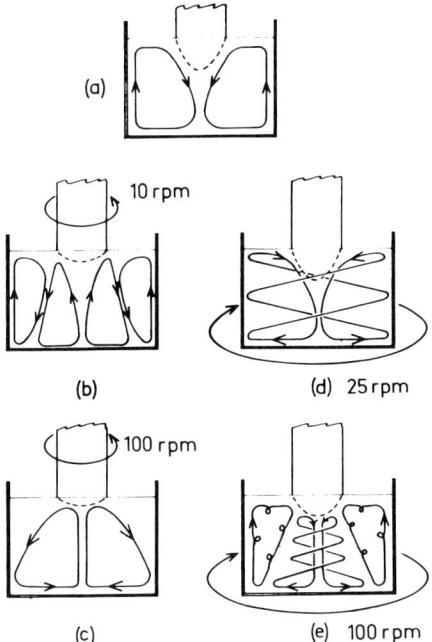

FIG. 42. Czochralski flow patterns (schematic) from transparent fluid model of process: (a) thermal convection only, no rotation; (b, c) crystal rotation; (d, e) crucible rotation [J. R. Carruthers, *J. Electrochem. Soc.* **114,** 959 (1967)].

confirm, in a qualitative way, the upward flow of fluid toward the rotating crystal surface. However, other more complex flow patterns result when the crucible and crystal are both rotated, as they often are, and still more complex patterns result when thermal convection (which we discuss in more detail below in connection with the horizontal boat) due to temperature gradients occurs. These studies include those by Goss and Adlington,[293] Turovskii and Mil'vidskii,[294] Robertson[295] (who finds that to establish the upward flow requires a minimum crystal rotation rate), and Carruthers[296] who studied thermal convection in connection with radial solute segregation (see also Zief and Wilcox[252]). The complexity of certain patterns with thermal convection and rotation is illustrated in the schematic Fig. 42 taken from Carruthers.[296] Wilcox and Fullmer[297] observed fluid turbulence (in

[293] A. J. Goss and R. E. Adlington, *Marconi Rev.* **22,** 18 (1959).
[294] B. M. Turovskii and M. G. Mil'vidskii, *Soviet Phys.—Cryst.* **6,** 606 (1962).
[295] D. S. Robertson, *Brit. J. Appl. Phys.* **17,** 1047 (1966).
[296] J. R. Carruthers, *J. Electrochem. Soc.* **114,** 959 (1967).
[297] W. R. Wilcox and L. D. Fullmer, *J. Appl. Phys.* **36,** 2201 (1965).

contrast to the laminar flow studies referred to above) caused by the large vertical temperature gradient, in Czochralski growth of CaF_2 near 1400–1500°C.

An application of the Burton et al.[254] fluid flow theory to constitutional supercooling in Czochralski growth was made by Hurle[298] (see also Bardsley et al.[299]). Brice[287] also has used the theory to interpret growth rates of crystals from solution at fixed supersaturation, in an analogous way to that discussed above for a flat plate. Additional discussion of the rotating disk is given by Levich[300] who notes that the assumption that the flow is laminar is correct up to Reynolds numbers approaching 10^4; he also considers the disk in turbulent flow.

35. The Sphere in Fluid Flow

As indicated above, the hydrodynamic problem of fluid flowing at velocity v past a sphere of radius R at very low Reynolds number $2vR/\nu$ (unity or less), called the Stokes problem, had been solved exactly by Stokes. The corresponding problem of mass transport to or from the sphere was solved by Levich.[300] Although there is no hydrodynamic boundary layer in this problem, the perturbed fluid flow region being of the order of the sphere diameter itself, there is a diffusional boundary layer, thin compared to the sphere diameter, in the case where the diffusion coefficient is small (large Schmidt number compared to unity). Hence a boundary layer approach to solving the convective diffusion equation [the mass-transport counterpart to the energy equation (33.5)], using Stokes's solution for the fluid flow velocity, could be used. The result for the flux of matter J to the surface of the sphere is[300]

$$J = \frac{D(C_\infty - C_s)}{1.15} \left(\frac{3v}{16DR^2}\right)^{1/3} \frac{\sin \theta}{[\theta - \frac{1}{2}(\sin 2\theta)]^{1/3}} \quad (35.1)$$

where C_s is the concentration on the sphere surface (assumed constant) and θ is the angle around the sphere beginning at the upstream point. The flux is highest at the front ($\theta = 0$) and decreases to zero at $\theta = \pi$ (downstream end). Sih and Newman[301] have studied the problem in more detail and point out that (35.1) is not accurate near the downstream end since there the thickness of the diffusion layer is comparable to that of the sphere. They delineated five regions having different mass-transfer mechanisms as

---

[298] D. T. J. Hurle, *Solid-State Electron.* **3**, 37 (1961).
[299] W. Bardsley, J. M. Callan, H. A. Chedzey, and D. T. J. Hurle, *Solid-State Electron.* **3**, 142 (1961).
[300] V. G. Levich, "Physicochemical Hydrodynamics." Prentice Hall, Englewood Cliffs, New Jersey, 1962.
[301] P. H. Sih and J. Newman, *Intern. J. Heat Mass Transfer* **10**, 1749 (1967).

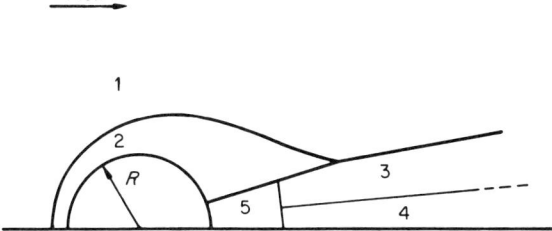

FIG. 43. Regions of different mass transfer mechanisms in very slow flow (Stokes flow) around a sphere (see text) [P. H. Sih and J. Newman, *Intern. J. Heat Mass Transfer* **10**, 1749 (1967)].

illustrated in Fig. 43. Region 1 is of uniform concentration; region 2 delineates the diffusion boundary layer. In region 3 the ratio of convective mass transport to diffusion is larger than assumed in the boundary layer treatment and a modified treatment is performed. Regions 4 and 5 are also discussed. The result still holds that mass transport to the rear of the sphere is very small. (We note that Re = 1 corresponds to a 1-mm-diam sphere moving at $\sim$1 mm/s in water.) Nielsen[115] has studied the growth of precipitates under convection and diffusion control of this type.

A review of mass transport to or from the sphere at larger Reynolds numbers, up to 10^6, including the phenomenon of separation of the streamlines from the sphere, and the turbulent region, has been made by Galloway and Sage.[302,303] Semiempirical relations were developed relating transport coefficient to angle, Reynolds number, and other variables.

36. Thermal or Natural Convection and Crystal Growth in Horizontal Boats

Utech and colleagues,[304–307] Cole and Winegard,[308] Hurle,[309,310] and Hurle *et al.*[311] have studied the fluid flow in long horizontal containers or

[302] T. R. Galloway and B. H. Sage, *Intern. J. Heat Mass Transfer* **10**, 1195 (1967).
[303] T. R. Galloway and B. H. Sage, *Intern. J. Heat Mass Transfer* **11**, 539 (1968).
[304] H. P. Utech, Sc.D. Thesis, Mass. Inst. Technol., Cambridge, Massachusetts (April, 1965).
[305] H. P. Utech and M. C. Flemings, *J. Appl. Phys.* **37**, 2021 (1966).
[306] H. P. Utech, W. S. Brower, and J. G. Early, *in* "Crystal Growth" (H. S. Peiser, ed.), Suppl. *J. Phys. Chem. Solids*, p. 201. Pergamon Press, Oxford, 1967.
[307] H. P. Utech and M. C. Flemings, *in* "Crystal Growth" (H. S. Peiser, ed.), Suppl. *J. Phys. Chem. Solids*, p. 651. Pergamon Press, Oxford, 1967.
[308] G. S. Cole and W. C. Winegard, *J. Inst. Metals* **93**, 153 (1965).
[309] D. T. J. Hurle, *Phil. Mag.* **13**, 305 (1966).
[310] D. T. J. Hurle, *in* "Crystal Growth" (H. S. Peiser, ed.), Suppl. *J. Phys. Chem. Solids*, p. 659. Pergamon Press, Oxford, 1967.
[311] D. T. J. Hurle, J. Gillman, and J. Harp, *Phil. Mag.* **14**, 205 (1966).

FIG. 44. Flow patterns in horizontal boat (schematic) with applied (horizontal) temperature gradient, from transparent fluid model of process [H. P. Utech, W. S. Brower, and J. G. Early, in "Crystal Growth" (H. S. Peiser, ed.), Suppl. *J. Phys. Chem. Solids*, p. 201. Pergamon Press, Oxford, 1967].

boats during crystal growth from the melt. Such flow, in the absence of stirring or other applied motion, is called natural or thermal convection, caused by density differences acted on by the gravitational field. Although there are theoretical treatments of allied problems, including heat transport in convective flow from a vertical heated plate,[284] heat transport between two close vertical plates,[312] and heat transport between two horizontal plates when heated from below[213] (the latter is the classical Rayleigh–Benard problem), the particular problem of heat and mass transfer in a long horizontal container of fluid having a temperature difference at the ends does not appear to have been treated theoretically. Some idea of the pattern of natural convective flow in such geometries is given by the modeling experiment of Rossby[313] in which a transparent container of clear fluid rested on a horizontal aluminum bar, which latter served as the bottom of the container and had a temperature gradient along its length. The flow pattern, made visible by suspended aluminum particles, consisted of flow along the bottom from the cold toward the warm end of the tank, rising flow at the warm end, then flow along the top from the hot to the cold end with descent there and with flow also descending from the top stream over its entire length. The fluid layer on the bottom was cooler than at the top of the tank. Figure 44[306] illustrates schematically the quite similar flow patterns with velocities ~ 2.5 cm/s observed visually in a horizontal boat of (transparent) molten sodium chloride, which in addition shows a pattern of cells with upward and downward flow throughout the boat length. Much of the motion was observed to be turbulent, at an applied longitudinal temperature gradient of about 30°C/cm, as could be seen by following small foreign particles visually in a motion picture[306]; the turbulence was also apparent from the erratic temperature fluctuations of a thermocouple immersed in the fluid. These had been observed earlier by Cole and Winegard[308] and attributed by them to turbulent thermal convection. In the case of metal crystal growth, the flow pattern could not be seen visually, but the

---

[312] G. K. Batchelor, *Quart. Appl. Math.* **12**, 209 (1954).
[313] H. T. Rossby, *Deep Sea Res.* **12**, 9 (1965).

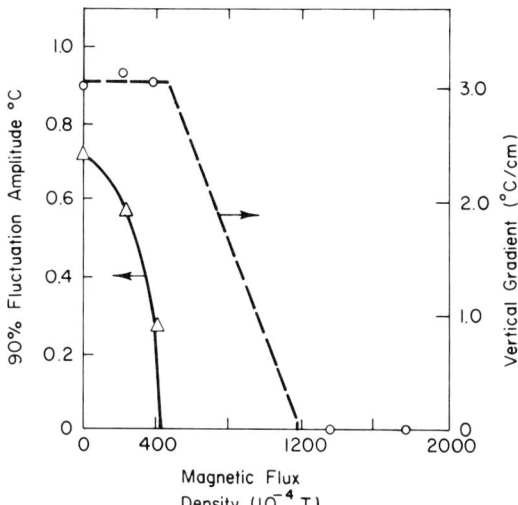

FIG. 45. Effect of applied magnetic field on temperature fluctuations and on vertical gradient. Horizontal boat of liquid tin with applied horizontal temperature gradient. (Note 1 tesla (T) = 10^4 G) [H. P. Utech and M. C. Flemings, *in* "Crystal Growth" (H. S. Peiser, ed.), Suppl. *J. Phys. Chem. Solids*, p. 651. Pergamon Press, Oxford, 1967]

same type of erratic temperature fluctuations was observed[304,307] for temperature gradients greater than 1.1°C/cm in Sn. The temperature fluctuations in the liquid metal could be made to disappear by applying a magnetic field of 4 to 6 $\times$ 10^{-2} T (400–600 G) which has the effect[213] of eddy current damping of the motion of the electrically conducting melt. The damping effect may be regarded[307] as an increase in viscosity or a "magnetic viscosity." Such viscosity dominates the intrinsic fluid viscosity if the Hartmann number M is large compared to unity, where

$$M^2 = (\sigma/\rho_L \nu)(\mu H l)^2 \qquad (36.1)$$

where μ is the magnetic permeability, H the magnetic field strength, σ the electrical conductivity, ρ_L the density, and l a characteristic length. When the turbulence was suppressed, still larger fields were required to suppress the remaining laminar motion. This could be shown by the disappearance at about 1.2×10^{-1} T (1200 G) of the vertical temperature gradient (hottest at top) which results from the total rotational flow in the boat (Fig. 45).

It was also shown[304,305] that the formation of solute concentration bands perpendicular to the growth direction in growth of Te-doped InSb could be prevented by eliminating the temperature fluctuations in the InSb melt by the magnetic field. The temperature fluctuations[314] introduced fluctuations in the growth rate and hence concentration.

Although it is known (see e.g. Landau and Lifshitz[212]) that the pertinent dimensionless numbers describing thermal convection are the Prandtl number $\mathrm{Pr} = \nu/\kappa_L$ and the Grashof number

$$\mathrm{Gr} = \beta_t g l^3 \, \Delta T / \nu^2 \qquad (36.2)$$

where l is a characteristic length, β_t the thermal coefficient of expansion, g the acceleration due to gravity, and ΔT a characteristic temperature difference, there does not seem to be a general rule giving the Grashof number for the onset of turbulence (there being no characteristic velocity, there is no Reynolds number defined). For the classical Benard case Gr is around 50,000, but there the turbulence results from the breakup of a Benard cell pattern, while in the horizontal boat the flow pattern is quite different as given above (see Dropkin and Somerscales[315] and Landis[316]).

Hurle[309,310] and Hurle et al.[311] independently of Utech, observed the temperature fluctuations in horizontal boats, and also that a magnetic field could stop the fluctuations and simultaneously stop the solute banding in doped conducting melts. However, certain of the fluctuations observed by Hurle are sinusoidal, i.e. quite regular in contrast to those observed by Utech. The period of the oscillations was a linear function of the length of the boat. The reason for the difference in the observations is not clear. Hurle has related his sinusoidal fluctuations to the concept of overstability as expressed by Chandrasekhar[215]; they had been earlier observed in a rotating Benard apparatus with liquid Hg.

37. Fluid Flow Resulting from Crystal/Fluid Density Differences

This was discussed in Part III from the point of view of a Stefan problem. Horvay[84,317] has made a theoretical study of the problem, following earlier theoretical work by Chambré[318] and experimental work in the freezing of undercooled nickel by Walker.[319] For the case of zero initial nucleus size[84] the solid–liquid surface grows as $t^{1/2}$. Horvay[317] points out that the inrush of fluid at $t = 0$ creates an infinite negative pressure for a

---

[314] A. Müller and M. Wilhelm, *Z. für Naturforsch.* **19a**, 254 (1964).
[315] D. Dropkin and E. Somerscales, *J. Heat Transfer* **87**, 77 (1965).
[316] See discussion appended to Dropkin and Somerscales.[315]
[317] G. Horvay, *Intern. J. Heat Mass Transfer* **8**, 195 (1965).
[318] P. L. Chambré, *Quart. J. Mech. Appl. Math.* **9**, 224 (1956).
[319] J. L. Walker, *in* "Physical Chemistry of Process Metallurgy" (G. R. St. Pierre, ed.), Pt. 2, p. 845. Wiley (Interscience), New York, 1961.

spherical nucleus of initial radius zero. He therefore considers the more realistic problem of finite nucleus size, together with a pressure dependence of the melting point. Very large negative pressures are still predicted; it is suggested that these cause cavitation and subsequent nucleation in the experiments of Walker.

ACKNOWLEDGMENTS

The author wishes to express his thanks to his many present and former colleagues in the National Bureau of Standards for numerous discussions, over the years, on the subject of crystallization. In particular he wishes to thank Dr. S. R. Coriell, Dr. C. M. Herzfeld, Dr. L. M. Kushner, and Mr. H. S. Peiser. Thanks are also due to Dr. E. Passaglia of the National Bureau of Standards and Professor D. Turnbull of Harvard University for helpful criticisms of the manuscript.

Thermoelastic Theory of Stressed Crystals and Higher-Order Elastic Constants*

DUANE C. WALLACE

Sandia Laboratories, Albuquerque, New Mexico

I. Introduction	302
II. Thermoelasticity of Stressed Materials	305
1. Thermodynamic Functions	305
2. Derivation of the Basic Equations	309
3. Small Amplitude Elastic Wave Propagation	314
4. Harmonic Generation	318
5. Attenuation of Elastic Waves	324
6. Stress–Strain Relations and Compressibility	328
7. Case of Initial Isotropic Pressure	334
III. Expansions for Small Initial Stress	336
8. Expansion of Elastic Constants	336
9. Expansion of Strains and Propagation Matrices	339
10. Stress Derivatives of the Transit Time	343
IV. Crystal Symmetry Properties	347
11. Schemes of Independent Elastic Constants	347
12. Cubic I Crystal under Isotropic Pressure	351
13. Tetragonal I Crystal under Uniaxial Stress	357
14. Isotropic Material	360
V. Measurement of Third-Order Elastic Constants	363
15. Wave Propagation in Stressed Crystals—Tables of Results	363
16. Wave Propagation in Stressed Crystals—Interpretation of Experiments	368
17. Harmonic Generation—Tables of Results	372
18. Harmonic Generation—Interpretation of Experiments	376
VI. Fourth-Order Elastic Constants of Cubic Crystals	380
19. Application of Isotropic Pressure or Uniaxial Stress	380
20. Application of Isotropic Pressure plus Uniaxial Stress	385
21. Harmonic Generation in Stressed Crystals	389
VII. Interpretation of Measurements in Terms of Atomic Theories	391
22. Application of Lattice Dynamics	391

* This work was supported by the U.S. Atomic Energy Commission.

I. Introduction

The purpose of the present article is to summarize, and in some cases extend, the general theory of thermoelasticity of stressed solids, to apply the theory to the interpretation of experiments which measure nonlinear thermoelastic properties of solids, and to illustrate this application by working out some examples. Our work has been motivated, in fact, by the development in recent years of the measurement of ultrasonic wave velocities as functions of stresses applied to the sample, and the measurement of the amplitude of the first harmonic generated by the passage of an ultrasonic fundamental wave through the sample. These experiments, and others such as the generation of sum- and difference-frequency waves by the interaction of two ultrasonic waves, are all interpreted from the same theoretical basis, namely nonlinear thermoelasticity of anisotropic solids. Since thermoelasticity is characterized by a rather cumbersome notation, and indeed throughout the years by many different sets of such notation, corresponding to different theoretical and experimental situations, it is worthwhile to set out the general theory in a unified notation, suitable for application to all experiments on nonlinear thermoelasticity.

In order to characterize processes as adiabatic or isothermal, the theory must be based on the thermodynamic potentials, e.g. the internal energy and the free energy, rather than on a mechanical elastic energy density which is independent of the entropy or temperature. In Chapter VII of his book, Voigt[1] discussed the theory of thermoelasticity, based on thermodynamic potentials, for the case of infinitesimal strains from a configuration of zero stress. However, to provide a complete discussion of the stress dependence of thermoelastic properties of solids, it is necessary to develop the theory for the case of arbitrary initial stress; the reason for this is as follows. If one does a stress–strain experiment for example, or observes the properties of ultrasonic waves in a solid, the situation is described in terms of the motion of the material particles about their initial equilibrium configuration, i.e. in terms of the strains measured from the initial configuration. Note these strains can vary in space and time, as in the case of wave propagation. If these experiments are done on a sample which is under an initial stress, one is then observing properties associated with strains from the initial stressed configuration, and it is both awkward and difficult to describe these properties in terms of strains measured from a zero-stress configuration. The convenience of measuring strains from an arbitrary initial state, which may correspond to an applied stress, was

---

[1] W. Voigt, "Lehrbuch der Kristallphysik." Teubner, Berlin, 1928. For previous papers on thermoelasticity, Voigt references W. Thomson (1857) *et seq.*

noted by Brillouin.[2] In addition, in order to include all nonlinear effects properly, it is convenient to use finite strain theory, such as that developed by Murnaghan,[3] instead of using infinitesimal strain parameters.

The general theory of finite-strain thermoelasticity of initially stressed solids is presented in Part II. There are three different sets of second-order thermoelastic coefficients which play a fundamental role in this theory, namely the elastic constants, which are finite-strain derivatives of the thermodynamic potentials, the stress–strain coefficients, which measure the variation of stress with strain away from the initial configuration, and the wave propagation coefficients, which govern the properties of small-amplitude thermoelastic waves propagating in the material. An interesting result is that components of the initial stress appear explicitly in the stress–strain coefficients and the wave propagation coefficients. This means a stress–strain experiment or a wave propagation experiment on an initially stressed solid measures not simply the strain derivatives of the thermodynamic potential of the material in that configuration, in the form of the elastic constants, but measures those constants plus certain combinations of the initially applied stress components. This reflects the physical picture that, in producing a further deformation of the solid by application of a further steady or time-dependent stress, the initial stresses work with (or against) the added stress and hence will be directly observed in the experiment. For the case of a stress–strain measurement this idea was recognized by Brillouin.[4] The appearance of the applied stresses in the wave propagation coefficients is of importance for understanding the stress derivatives of ultrasonic velocities, even at zero stress; it is also important for understanding wave propagation in materials under very high stresses, when the stresses become comparable to the elastic constants. An example of the latter is thermoelastic properties of the interior of the earth.

In order to prepare for the interpretation of experiments which measure third- and fourth-order elastic constants evaluated at zero stress, the general theory of thermoelastic motion about the stressed initial configuration is expanded for small initial stress in Part III. Each of the Laue groups has a common array of second-, third-, and higher-order elastic constants; in Part IV it is shown how the symmetries of these arrays are used to reduce the general problems of thermoelastic motion for a given crystal

---

[2] L. Brillouin, "Les Tenseurs en Méchanique et en Élasticité." Masson, Paris, 1938. (First American Ed.: Dover, New York, 1946; English Ed., "Tensors in Mechanics and Elasticity." Academic Press, New York, 1964.)

[3] F. D. Murnaghan, *Am. J. Math.* **59,** 235 (1937). For previous papers on finite strains, Murnaghan references G. Kirchhoff (1852), J. Boussinesq (1869), E. and F. Cosserat (1896), P. Duhem (1904–1906), and others.

[4] Brillouin,[2] Chapter 10, Section 8.

symmetry. Illustrative examples of the application of the theory to the determination of third-order elastic constants, by measuring the first stress derivatives of small-amplitude wave velocities, and by measuring the fundamental and harmonic amplitudes in harmonic generation experiments, are presented in Part V. As a further illustration of the generality of the theory, experiments designed to measure fourth-order elastic constants from the second stress derivatives of small-amplitude wave velocities, and from harmonic generation experiments on stressed crystals, are discussed in Part VI.

We wish to emphasize that it is not the purpose of the present article to provide tables for the interpretation of experiments on a large number of different crystals, but rather to provide a theory of wide applicability, with a few tables worked out mainly for the purpose of illustration. Certain results for many crystal symmetries have been published, but these have not been reproduced here on account of the high incidence of errors found in those parts we have checked. Nearly everything presented in this article has been derived or otherwise checked by the author.

It is appropriate to consider for a moment the value of measuring higher-order elastic constants. There are obviously two general areas of usefulness of such measurements, namely to provide equation of state data, and to assist in our understanding of the atomic nature of solids. According to their definition, higher-order elastic constants correct stress–strain relations for nonlinear effects, and hence give better extrapolation of the equation of state to higher stresses. In addition, higher-order elastic constants characterize the stress or strain dependence of elastic wave velocities, and the nonlinear interactions (scattering) between elastic waves. From the atomic point of view, in the language of lattice dynamics, the higher-order elastic constants represent anharmonic effects; they can be interpreted to a good approximation as higher-order strain derivatives of the total lattice potential. Alternatively, since the elastic waves are in leading (harmonic) approximation the same as the long-wavelength acoustic phonons, the higher-order elastic constants may be used to calculate the strain derivatives of these phonon frequencies, i.e. the generalized Grüneisen parameters, and the scattering cross sections among these phonons. In addition to providing tests of atomic theories of solids, these applications are important in understanding phonon lifetimes, thermal expansion, thermal conductivity, and many other properties. For example, Seeger and Mann[5] emphasized the use of higher-order elastic constants in calculating the strain field around a lattice defect. Further detailed applications will be discussed in a review article being prepared by Suzuki, Holder, and Granato.

[5] A. Seeger and E. Mann, *Z. Naturforsch.* **14a,** 154 (1959).

For all crystals except those composed of the very light atoms, and excepting certain other anomalous cases such as ferroelectrics near the transition temperature, the theory of lattice dynamics is developed as a rapidly converging perturbation expansion in the displacements of the ions from equilibrium. The interpretation of thermoelastic properties of crystals in terms of lattice dynamics is discussed in Part VII; it is not our purpose there to derive the detailed lattice dynamics equations, but rather to point out the order of approximation involved in comparing calculations with measurements.

Finally, a few comments on our notation are in order. We have tried to use the most commonplace notation to describe the thermoelastic properties of solids in the presence of arbitrary initial stress. In our derivations it is necessary to distinguish these same properties for a solid at zero initial stress, and this is done by using overbars to denote evaluation at zero stress. In many applications of the theory this distinction is not required, and the bars should be deleted. Another point is that the theory is characterized by partial derivatives, and we usually omit denoting which variables are held constant with the understanding that in any derivative with respect to a given variable, all other independent variables are held fixed. Evaluation of a function at constant entropy or constant temperature is indicated by a superscript S or T, respectively, and in many cases this superscript is omitted to show the equations are valid for either case.

II. Thermoelasticity of Stressed Materials

1. Thermodynamic Functions

We consider a material whose equilibrium states depend only on the configuration, denoted by $\mathbf{x}$, and the temperature T or the entropy per unit mass S. The state functions are the Helmholtz free energy per unit mass F, and the internal energy per unit mass U. It is convenient to work with functions per unit mass in elasticity theory because mass is conserved, while volume is not generally conserved, during elastic strains. The functional dependence of U and F is

$$U = U(\mathbf{x}, S), \qquad F = F(\mathbf{x}, T). \tag{1.1}$$

The internal energy is convenient for describing processes in which the independent variables are $\mathbf{x}$ and S; in this case the dependent variables are the stresses and T. On the other hand, the Helmholtz free energy is convenient when the independent variables are $\mathbf{x}$ and T; then the dependent variables are the stresses and S.

A theory of thermoelasticity for infinitesimal strains from configurations of zero stress was described by Voigt.[1] The present theory will be developed for finite strains from an arbitrary initial configuration **X**. The finite deformation theory has been developed by Murnaghan.[3,6] Allowing the initial configuration to be arbitrary, corresponding to arbitrary initial stress, gives no essential complications and greatly extends the usefulness of the results. Much of the theory of the present Part II was published recently.[7]

It is not necessary in general for the initial state to be obtained by elastic deformation from a state of zero stress; the theory could describe, e.g., a crystal under pressure which has undergone a phase transition during the application of the pressure. However, when the initial-state properties are expressed as expansions from a zero-stress state, as in Part III, then it is assumed that the deformation to the initial state is elastic so the state functions are analytic and the expansions are valid.

The position of a material particle in the initial configuration is denoted by **X**, and the position of the same material particle in the final configuration is **x**. There should be no confusion from using the same symbol to denote the position of a material particle, say **x**, and the configuration of the whole system, which is a function of the set $\{\mathbf{x}\}$. If the final configuration is obtained by applying a small additional uniform stress to the material, then the strain from **X** to **x** is homogeneous, i.e. uniform throughout the material. If the final configuration corresponds to a wave (or several waves) propagating in the material, the strain from **X** to **x** is nonuniform in space and in the time t, and the equations derived below are to be considered as local in **X** and t.

The deformation from **X** to **x** can be measured by either of two different sets of nine independent deformation parameters.

(a) The transformation coefficients α_{ij}, used mostly by classical elasticity theories:

$$\alpha_{ij} = \partial x_i/\partial X_j, \quad \text{where} \quad x_i = x_i(\mathbf{X}, t). \tag{1.2}$$

(b) The displacement gradients u_{ij}, used mostly by physicists:

$$u_{ij} = \partial u_i/\partial X_j, \quad \text{where} \quad x_i - X_i = u_i(\mathbf{X}, t). \tag{1.3}$$

The subscripts $i, j, \ldots$ indicate Cartesian components and each goes over three values. We always use the Einstein summation convention, implying a sum over repeated indices. Although the independent position variable for the deformation parameters has been denoted as **X** in their definitions, we can transform to a description in which **x** is the independent position

---
[6] F. D. Murnaghan, "Finite Deformation of an Elastic Solid." Wiley, New York, 1951.
[7] D. C. Wallace, *Phys. Rev.* **162**, 776 (1967).

variable whenever it may be convenient. The two sets of deformation parameters are obviously related according to

$$\alpha_{ij} = \delta_{ij} + u_{ij}, \tag{1.4}$$

where δ_{ij} is the Kronecker delta.

The symmetric finite strain parameters of Murnaghan,[3,6] also called Lagrangian strain parameters, are

$$\eta_{ij} = \tfrac{1}{2}(\alpha_{ki}\alpha_{kj} - \delta_{ij}) = \tfrac{1}{2}(u_{ij} + u_{ji} + u_{ki}u_{kj}). \tag{1.5}$$

We treat the η_{ij} as nine independent variables, but always make sure the calculations are consistent with $\eta_{ij} = \eta_{ji}$. If $\Delta\mathbf{x}$ and $\Delta\mathbf{X}$ are the vectors between two material particles in the final and initial configurations, respectively, and if η_{ij} is constant over the region of $\Delta\mathbf{X}$, then it is easy to show[6]

$$|\Delta\mathbf{x}|^2 - |\Delta\mathbf{X}|^2 = 2\eta_{ij}\,\Delta X_i\,\Delta X_j. \tag{1.6}$$

This is the important property of the symmetric finite strain parameters η_{ij}: the distances between all pairs of material particles, and hence the relative positions of all the material particles, are completely specified by the initial configuration $\mathbf{X}$ and the η_{ij}, for constant η_{ij}.

Now the internal energy and the Helmholtz free energy must depend only on the *relative* positions of all the material particles; more particularly this requirement applies over distances of the order of the range of interactions within the material. Thus if η_{ij} is essentially constant over such distances, then U and F depend on $\mathbf{x}$ only through $\mathbf{X}$ and the η_{ij}:

$$U(\mathbf{x}, S) = U(\mathbf{X}, \eta_{ij}, S), \qquad F(\mathbf{x}, T) = F(\mathbf{X}, \eta_{ij}, T). \tag{1.7}$$

The theory is strongly based on these relations, and we must remember that they are valid when the deformation varies significantly only over distances large compared to the range of effective interactions in the material. For elastic wave propagation, e.g., the wavelength must be large compared to the range of interactions; if the wavelength is too short, dispersion occurs, and a theory such as lattice dynamics (for crystals) applies. For a rigid rotation, $\eta_{ij} = 0$ throughout the material, and the conditions (1.7) are just the necessary rotational invariance of the state functions.

The density of the material is denoted by $\rho(\mathbf{x})$ or $\rho(\mathbf{X})$, with the abbreviations

$$\rho = \rho(\mathbf{x}); \qquad \rho_1 = \rho(\mathbf{X}). \tag{1.8}$$

Since the deformation from $\mathbf{X}$ to $\mathbf{x}$ is considered small (but not infinitesimal), it is convenient to expand the state functions about **X**:

$$\rho_1 U(\mathbf{X}, \eta_{ij}, S) = \rho_1 U(\mathbf{X}, 0, S) + C^S{}_{ij}\eta_{ij} + \tfrac{1}{2}C^S{}_{ijkl}\eta_{ij}\eta_{kl}$$
$$+ (1/3!)C^S{}_{ijklmn}\eta_{ij}\eta_{kl}\eta_{mn}$$
$$+ (1/4!)C^S{}_{ijklmnpq}\eta_{ij}\eta_{kl}\eta_{mn}\eta_{pq} + \cdots, \quad (1.9)$$

$$\rho_1 F(\mathbf{X}, \eta_{ij}, T) = \rho_1 F(\mathbf{X}, 0, T) + C^T{}_{ij}\eta_{ij} + \tfrac{1}{2}C^T{}_{ijkl}\eta_{ij}\eta_{kl}$$
$$+ (1/3!)C^T{}_{ijklmn}\eta_{ij}\eta_{kl}\eta_{mn}$$
$$+ (1/4!)C^T{}_{ijklmnpq}\eta_{ij}\eta_{kl}\eta_{mn}\eta_{pq} + \cdots. \quad (1.10)$$

By definition, the C coefficients are derivatives of $\rho_1 U$ or $\rho_1 F$, evaluated at **X**. For example, the adiabatic (constant S) coefficients are[8]

$$C^S{}_{ij} = \rho_1(\partial U/\partial \eta_{ij}), \qquad C^S{}_{ijklmn} = \rho_1(\partial^3 U/\partial\eta_{ij}\,\partial\eta_{kl}\,\partial\eta_{mn}),$$
$$C^S{}_{ijkl} = \rho_1(\partial^2 U/\partial\eta_{ij}\,\partial\eta_{kl}), \qquad C^S{}_{ijklmnpq} = \rho_1(\partial^4 U/\partial\eta_{ij}\,\partial\eta_{kl}\,\partial\eta_{mn}\,\partial\eta_{pq}).$$
$$(1.11)$$

Similar definitions hold for the isothermal (constant T) coefficients, with F in place of U. In writing partial derivatives as in (1.11), we will usually omit explicit statement of which variables are held constant, with the understanding that all *other* variables are fixed. For example, $(\partial U/\partial \eta_{ij})$ implies **X**, S, and all other η_{kl} are held constant.

Now we can interpret the C coefficients. As shown in Section 2, the symmetric Cauchy stress tensor $T_{ij} = T_{ji}$, evaluated at the initial configuration, is just

$$T_{ij}(\mathbf{X}) = \rho_1(\partial U/\partial \eta_{ij}) = \rho_1(\partial F/\partial \eta_{ij}). \quad (1.12)$$

In fact, this is just the thermodynamic equilibrium condition. It follows that

$$C^S{}_{ij} = C^T{}_{ij} = T_{ij}(\mathbf{X}). \quad (1.13)$$

Further, the $C^S{}_{ijkl}$ and $C^T{}_{ijkl}$ are the usual second-order adiabatic and isothermal elastic constants, respectively, evaluated at **X**. Finally, the coefficients $C^S{}_{ijklmn}$, $C^S{}_{ijklmnpq}$ are the third- and fourth-order adiabatic elastic constants, while $C^T{}_{ijklmn}$, $C^T{}_{ijklmnpq}$ are the third- and fourth-order isothermal elastic constants, evaluated at the arbitrary initial configuration **X**, in the notation proposed by Brugger[9] for the zero-stress case.

---

[8] Note we do *not* define higher-order elastic constants as strain derivatives of second-order elastic constants, i.e. $C_{ijklmn} \neq (\partial C_{ijkl}/\partial \eta_{mn})$; see e.g. Eq. (8.20) for illustration of this point.

[9] K. Brugger, *Phys. Rev.* **133**, A1611 (1964).

From the symmetry of the η_{ij}, and from the definition of the C coefficients, it follows that these coefficients have complete Voigt symmetry; for constant S or constant T coefficients this is

$$C_{ij} = C_{ji}, \qquad C_{ijkl} = C_{jikl} = C_{klij} = \cdots, \qquad \text{etc.} \qquad (1.14)$$

It will often be convenient to use Voigt notation, in which one Greek letter index replaces a pair ij, according to the following scheme:

$$\begin{array}{llllll} ij = 11 & 22 & 33 & 32 \text{ or } 23 & 31 \text{ or } 13 & 21 \text{ or } 12 \\ \alpha = 1 & 2 & 3 & 4 & 5 & 6 \end{array} \qquad (1.15)$$

In Voigt notation the elastic constants are completely symmetric in their indices:

$$C_{\alpha\beta} = C_{\beta\alpha}, \qquad C_{\alpha\beta\gamma} = C_{\beta\alpha\gamma} = C_{\gamma\beta\alpha} = \cdots. \qquad (1.16)$$

2. Derivation of the Basic Equations

The theory of thermoelastic motion of a material about the initial configuration is based on three equations, which are as follows.

(a) Equations for the dependent variables are obtained from the first and second laws of thermodynamics. From the internal energy,

$$T = (\partial U/\partial S), \qquad T_{ij} = \rho \alpha_{ik} \alpha_{jl} (\partial U/\partial \eta_{kl}). \qquad (2.1)$$

From the Helmholtz free energy,

$$S = -(\partial F/\partial T), \qquad T_{ij} = \rho \alpha_{ik} \alpha_{jl} (\partial F/\partial \eta_{kl}). \qquad (2.2)$$

Here T_{ij} are stress components; both sides of (2.1) and (2.2) are evaluated at **x**, with the deformation parameters measured from the arbitrary initial configuration **X**.

(b) The equation of continuity, or conservation of mass, is based on the geometry of strain and is well known[6,10]:

$$\rho_1/\rho = J = \det[\alpha_{ij}]. \qquad (2.3)$$

(c) The equation of motion in the absence of body forces is also well known, and is an expression of Newton's second law of motion[6,10]:

$$\rho \ddot{x}_i = \partial T_{ji}/\partial x_j. \qquad (2.4)$$

Here $\ddot{x}_i$ is the second time derivative of x_i, and if the motion is adiabatic or isothermal the right-hand side derivative is evaluated at constant S or T, respectively.

[10] C. Truesdell and W. Noll, in "Handbuch der Physik" (S. Flügge, ed.), Vol. III/3, p. 1. Springer-Verlag, Berlin, 1965.

To derive the equations for the dependent variables, we generalize a derivation of Leibfried and Ludwig[11] to the case of arbitrary initial configuration. The initial configuration is **X**, the final configuration is **x**, and in order to calculate *derivatives* of the stress evaluated at **x**, we need to calculate the virtual work done by the stresses $T_{ij}(\mathbf{x})$ when the material undergoes a virtual homogeneous deformation $\Delta \mathbf{x}$ from **x**. We can write

$$(x + \Delta x)_i = (\delta_{ij} + \Delta u_{ij}) x_j = (\delta_{ij} + \Delta u_{ij}) \alpha_{jk} X_k. \tag{2.5}$$

Since $\alpha_{ij} = \partial x_i / \partial X_j$, the $\Delta \mathbf{x}$ corresponds to a $\Delta \boldsymbol{\alpha}$ given by

$$\Delta \alpha_{ik} = \Delta u_{ij}\, \alpha_{jk}. \tag{2.6}$$

This in turn can be transformed to a $\Delta \boldsymbol{\eta}$ given by

$$\begin{aligned}
\Delta \eta_{mn} &= \tfrac{1}{2}(\alpha_{pm}\, \Delta \alpha_{pn} + \alpha_{pn}\, \Delta \alpha_{pm}), \\
&= \tfrac{1}{2}(\alpha_{pm}\, \Delta u_{pq}\, \alpha_{qn} + \alpha_{pn}\, \Delta u_{pq}\, \alpha_{qm}), \\
&= \tfrac{1}{2}\alpha_{pm}\alpha_{qn}(\Delta u_{pq} + \Delta u_{qp}).
\end{aligned} \tag{2.7}$$

Define $\boldsymbol{\gamma}$ as the matrix inverse to $\boldsymbol{\alpha}$, so that

$$\gamma_{ij}\alpha_{jk} = \delta_{ik}; \tag{2.8}$$

then (2.7) can be written

$$\Delta \eta_{mn}\, \gamma_{mi}\gamma_{nj} = \tfrac{1}{2}(\Delta u_{ij} + \Delta u_{ji}). \tag{2.9}$$

Let the surface of the material in configuration **x** be **s**, with surface element d**s**. The i component of force on d**s**, due to the stresses applied in configuration **x**, is $f_i = T_{ij}\, ds_j$. The virtual displacement of d**s** in the i direction is $\Delta x_i = \Delta u_{ik}\, x_k$, evaluated at d**s**. The virtual work done by the stress acting on d**s** is

$$f_i\, \Delta x_i = T_{ij}\, ds_j\, \Delta u_{ik}\, x_k. \tag{2.10}$$

The total virtual work done on the material is ΔW and is just the integral of (2.10) over the surface in configuration **x**; this integral is transformed by Gauss's theorem and evaluated as follows (since the deformation is homogeneous):

$$\Delta W = \int_s \Delta u_{ik}\, x_k T_{ij}\, ds_j = \int_\tau T_{ij}\, \Delta u_{ij}\, d\tau = T_{ij}\, \Delta u_{ij}\, V(\mathbf{x}), \tag{2.11}$$

where $V(\mathbf{x})$ is the total volume of the material in configuration **x**. Since $T_{ij} = T_{ji}$, only the symmetric part of Δu_{ij} survives the sum on ij in (2.11);

---

[11] G. Leibfried and W. Ludwig, *Solid State Phys.* **12**, 275 (1961).

this symmetric part is given by (2.9) and we have

$$\Delta W = V(\mathbf{x}) T_{ij} \Delta \eta_{mn} \gamma_{mi} \gamma_{nj}. \qquad (2.12)$$

Let us now write the combined first and second laws of the thermodynamics in differential form per unit mass:

$$dU = dW + T\, dS. \qquad (2.13)$$

Multiplying (2.13) by the density ρ in configuration $\mathbf{x}$, and identifying $\rho\, dW$ with $V^{-1} \Delta W$ of (2.12), we obtain

$$\rho\, dU = T_{ij}\gamma_{mi}\gamma_{nj}\, d\eta_{mn} + \rho T\, dS. \qquad (2.14)$$

Also, since $F = U - TS$ per unit mass, (2.14) leads to

$$\rho\, dF = T_{ij}\gamma_{mi}\gamma_{nj}\, d\eta_{mn} - \rho S\, dT. \qquad (2.15)$$

For a process at constant S, (2.14) gives

$$T_{ij}\gamma_{mi}\gamma_{nj} = \rho(\partial U/\partial \eta_{mn}), \qquad (2.16)$$

or in view of (2.8)

$$T_{kl} = \rho \alpha_{km}\alpha_{ln}(\partial U/\partial \eta_{mn}). \qquad (2.17)$$

Similarly, for a process at constant T, (2.15) gives

$$T_{kl} = \rho \alpha_{km}\alpha_{ln}(\partial F/\partial \eta_{mn}). \qquad (2.18)$$

For a process at constant η_{ij}, (2.14) gives $T = (\partial U/\partial S)$, while (2.15) gives $S = -(\partial F/\partial T)$. This completes the derivation of Eqs. (2.1) and (2.2) for the dependent variables. The derivation is valid on a local level by considering homogeneous deformations over small volume elements instead of over the entire material. Note when the equations for the stress are evaluated at the initial configuration $\mathbf{X}$, we have $\rho = \rho_1$ and $\alpha_{ij} = \delta_{ij}$ and the equilibrium condition quoted in (1.12) is obtained.

The equation of conservation of mass has been derived by Murnaghan,[6] and we need not repeat the derivation here. It is useful to derive two other identities which depend on the geometry of strain. Differentiating the Jacobian (2.3) gives

$$\partial J/\partial \alpha_{ij} = \alpha^{ij}, \qquad (2.19)$$

where α^{ij} is the cofactor of α_{ij} in the determinant. A well-known property of cofactors is

$$\sum_j \alpha_{kj}\alpha^{ij} = J\, \delta_{ik}; \qquad (2.20)$$

this is so since if $k = i$ it is just the expansion of J in cofactors, and if

$k \neq i$ it is the expansion of a determinant whose k, j rows are the same and hence is zero. But the set of equations (2.20) is sufficient to determine α^{ij} uniquely (e.g., in the present 3×3 case there are 9 equations). The solution is

$$\alpha^{ij} = \gamma_{ji} J, \qquad (2.21)$$

as can be seen with the aid of (2.8). Therefore, from (2.19), we have Jacobi's identity (see, e.g., Truesdell and Toupin[12])

$$\partial J / \partial \alpha_{ij} = \gamma_{ji} J. \qquad (2.22)$$

Further, since $\boldsymbol{\gamma}$ is inverse to $\boldsymbol{\alpha}$, and from the definition (1.2) of α_{ij},

$$\gamma_{ij} = \partial X_i / \partial x_j. \qquad (2.23)$$

Letting **X** be the independent configuration variable for any function, we can write

$$\partial/\partial x_i = (\partial \alpha_{jk}/\partial x_i)(\partial/\partial \alpha_{jk}) = (\partial \alpha_{jk}/\partial X_l)\gamma_{li}(\partial/\partial \alpha_{jk}). \qquad (2.24)$$

Applying this relation to the identity (2.22) gives

$$(\partial/\partial x_i)(\alpha_{im}/J) = (\partial \alpha_{jk}/\partial X_l)\gamma_{li}[\delta_{ij}\,\delta_{mk}/J - (\alpha_{im}/J^2)\gamma_{kj}J]$$
$$= (1/J)(\partial \alpha_{jk}/\partial X_l)[\gamma_{lj}\,\delta_{mk} - \gamma_{kj}\,\delta_{ml}] = 0, \qquad (2.25)$$

since $(\partial \alpha_{jk}/\partial X_l) = (\partial \alpha_{jl}/\partial X_k)$. Therefore we have the identity of Euler Piola, and Jacobi:

$$\partial(\alpha_{ij}/J)/\partial x_i = 0. \qquad (2.26)$$

It is convenient to derive the equation of motion with the aid of Lagrange's equations. We choose the independent variables as **X**, t, and the generalized coordinates as $\mathbf{x}(\mathbf{X}, t)$. Then in the absence of body forces Lagrange's equations of motion are[13]

$$(d/dt)(\partial \mathcal{L}/\partial \dot{x}_i) + (\partial/\partial X_k)(\partial \mathcal{L}/\partial \alpha_{ik}) = 0, \qquad (2.27)$$

where the Lagrangian density $\mathcal{L}$ is the density in **X** space, i.e. the density per unit initial volume. For adiabatic or isothermal motion the thermodynamic potential is the internal energy at constant S or the Helmholtz free energy at constant T, respectively. For adiabatic motion, e.g.,

$$\mathcal{L} = \tfrac{1}{2}\rho_1 \dot{x}_i \dot{x}_i - \rho_1 U(\mathbf{x}, S). \qquad (2.28)$$

[12] C. Truesdell and R. A. Toupin, *in* "Handbuch der Physik" (S. Flügge, ed.), Vol. III/1, p. 226. Springer-Verlag, Berlin, 1960.
[13] H. Goldstein, "Classical Mechanics." Addison-Wesley, Cambridge, Massachusetts, 1950.

Evaluation of (2.27) then proceeds as follows:

$$\rho_1 \ddot{x}_i = \rho_1(\partial/\partial X_k)(\partial U/\partial \alpha_{ik}) = \rho_1(\partial/\partial X_k)(\partial U/\partial \eta_{lm})(\partial \eta_{lm}/\partial \alpha_{ik})$$
$$= \rho_1(\partial/\partial X_k)\alpha_{il}(\partial U/\partial \eta_{kl}), \qquad (2.29)$$

where we calculated $(\partial \eta_{lm}/\partial \alpha_{ik})$ from the relation (1.5) and used the symmetry of η_{lm}. Equation (2.29) is a convenient form of the equation of motion, valid at arbitrary configuration. The customary Newton's law form (2.4) of the equation of motion is obtained by multiplying (2.29) by $\rho/\rho_1 = 1/J$ and using the identity (2.26) to write

$$\rho \ddot{x}_i = (\rho_1/J)(\partial/\partial X_k)\alpha_{il}(\partial U/\partial \eta_{kl})$$
$$= (\rho_1/J)\alpha_{jk}(\partial/\partial x_j)\alpha_{il}(\partial U/\partial \eta_{kl})$$
$$= (\partial/\partial x_j)\rho_1(\alpha_{jk}/J)\alpha_{il}(\partial U/\partial \eta_{kl}) = (\partial/\partial x_j)T_{ji}.$$

The procedure is similar for isothermal motion; in either case the derivation shows that the stress derivative in the equation of motion is to be evaluated for thermodynamic conditions appropriate to the type of motion (e.g. constant S or T).

There has been no assumption of small strains in the above derivations, so our resulting basic equations are not restricted in this sense and can be used to treat large deformations, nonlinear wave propagation, etc. However, the "rotational invariance" conditions (1.7) were relied upon in deriving the equations for the dependent variables and the equation of motion. Hence in the treatment of inhomogeneous deformation problems we will always assume the η_{ij} vary significantly only over distances large compared to the range of forces in the material, so that our equations may be considered to be local in **X** and t.

It is also useful to derive an expression for the energy flux **E** in the limit of small strains from the initial configuration. This will be illustrated for adiabatic motion, following the procedure of Love.[14] The internal energy per unit initial volume is $\rho_1 U$, and the internal energy contained in a small volume τ is $\int \rho_1 U \, d\tau$. The time rate of change of this volume integral is just minus the integral of the energy flux across the surface **s** of the volume:

$$(\partial/\partial t)\int \rho_1 U \, d\tau = -\int E^s{}_j \, ds_j. \qquad (2.30)$$

We evaluate the left-hand side of (2.30) in the limit of small strains, where

[14] A. E. H. Love, "Mathematical Theory of Elasticity," 4th ed., p. 177. Cambridge Univ. Press, London and New York, 1927.

η_{ij} may be replaced by the symmetric part of u_{ij}, as follows:

$$(\partial/\partial t)\int \rho_1 U \, d\tau = \int (\partial\rho_1 U/\partial\eta_{ij})(\partial\eta_{ij}/\partial t) \, d\tau = \int (\partial\rho_1 U/\partial\eta_{ij})(\partial u_{ij}/\partial t) \, d\tau$$

$$= \int (\partial\rho_1 U/\partial\eta_{ij})(\partial \dot{u}_i/\partial X_j) \, d\tau = \int (\partial\rho_1 U/\partial\eta_{ij}) \dot{u}_i \, ds_j, \quad (2.31)$$

where to lowest order in the strains $(\partial\rho_1 U/\partial\eta_{ij})$ is taken as constant over the volume τ, and the last equality follows from Gauss's theorem. Comparing (2.30) and (2.31) gives the energy flux for adiabatic motion

$$E^S{}_j = -\dot{u}_i \rho_1 (\partial U/\partial\eta_{ij}). \quad (2.32)$$

3. Small Amplitude Elastic Wave Propagation

We consider small motions about the initial configuration **X**, and linearize the equation of motion by evaluating it at **X**. The calculation is adequately illustrated by taking the case of adiabatic motion. Since $\rho = \rho_1/J$, the stress according to (2.1) may be written

$$T_{ji} = (\rho_1/J)\alpha_{jk}\alpha_{il}(\partial U/\partial\eta_{kl}). \quad (3.1)$$

With the aid of the identity (2.26) the stress derivative is

$$\partial T_{ji}/\partial x_j = \rho_1(\alpha_{jk}/J)(\partial/\partial x_j)[\alpha_{il}(\partial U/\partial\eta_{kl})]$$

$$= (\rho_1/J)(\partial x_j/\partial X_k)(\partial/\partial x_j)[\alpha_{il}(\partial U/\partial\eta_{kl})]$$

$$= (1/J)(\partial/\partial X_k)[\alpha_{il}\rho_1(\partial U/\partial\eta_{kl})], \quad (3.2)$$

since $\rho_1 = $ constant. Also from the internal energy expansion (1.9) we have

$$\rho_1(\partial U/\partial\eta_{kl}) = C^S{}_{kl} + C^S{}_{klmn}\eta_{mn} + \cdots,$$

$$= C^S{}_{kl} + \tfrac{1}{2}C^S{}_{klmn}(\alpha_{pm}\alpha_{pn} - \delta_{mn}) + \cdots. \quad (3.3)$$

The linearized stress derivative is obtained by carrying out the differentiation $(\partial/\partial X_k)$ in (3.2), with the aid of (3.3), and evaluating at **X**. Since at **X** we have $J = 1$, $\eta_{ij} = 0$, $\alpha_{ij} = \delta_{ij}$, the constant entropy stress derivative at **X** is

$$(\partial T_{ji}/\partial x_j) = [C^S{}_{jl}\delta_{ik} + C^S{}_{ijkl}](\partial^2 x_k/\partial X_j \, \partial X_l). \quad (3.4)$$

For the isothermal stress derivative, the C coefficients in (3.4) are replaced by constant T coefficients. Let us introduce the wave propagation coefficients A_{ijkl} for the bracket in (3.4) and write the equation of motion for *either* adiabatic or isothermal motion, with all coefficients evaluated at **X**:

$$\rho_1 \ddot{x}_i = A_{ijkl}(\partial^2 x_k/\partial X_j \, \partial X_l), \quad (3.5)$$

where for adiabatic processes one uses

$$A^S_{ijkl} = T_{jl}\delta_{ik} + C^S_{ijkl}, \tag{3.6}$$

and for isothermal processes one uses

$$A^T_{ijkl} = T_{jl}\delta_{ik} + C^T_{ijkl}. \tag{3.7}$$

The stresses at **X** have been introduced by means of (1.13).

The linear equation of motion (3.5) for a stressed material was derived from a mechanical energy density, instead of a thermodynamic potential, by Huang[15] and by Toupin and Bernstein.[16] Thermoelastic derivation of (3.5) was made by Green,[17] and lattice dynamical derivation was given by Wallace.[18] Note the propagation coefficients A_{ijkl} do not have Voigt symmetry and hence cannot be written as a 6 × 6 matrix, except of course at zero initial stress, where the A_{ijkl} and C_{ijkl} are equal. The adiabatic–isothermal differences, however, are the same as for the elastic constants at any initial stress:

$$A^S_{ijkl} - A^T_{ijkl} = C^S_{ijkl} - C^T_{ijkl}. \tag{3.8}$$

Solutions of the equation of motion are plane waves. Let **k** be the wave vector, ω the (circular) frequency, and **w** the magnitude and direction of the displacement; then the space- and time-dependent displacement for a plane wave may be written

$$\mathbf{x} - \mathbf{X} = \mathbf{w}\sin(\mathbf{k}\cdot\mathbf{X} - \omega t). \tag{3.9}$$

With this displacement, the equation of motion becomes

$$\rho_1 \omega^2 w_i = A_{ijkl}k_j k_l w_k, \tag{3.10}$$

for either adiabatic or isothermal motion. Or, with $\boldsymbol{\kappa}$ a unit vector along **k** and v the wave velocity,

$$\boldsymbol{\kappa} = \mathbf{k}/|\mathbf{k}|; \quad v = \omega/|\mathbf{k}|; \tag{3.11}$$

then the equation of motion is

$$\rho_1 v^2 w_i = A_{ijkl}\kappa_j \kappa_l w_k. \tag{3.12}$$

Equations (3.10) or (3.12) are the eigenvalue–eigenvector equations for the elastic wave solutions, in which the eigenvalues are $\rho_1\omega^2$ or $\rho_1 v^2$, respectively, and the eigenvectors are **w**. We define the propagation matrices

---

[15] K. Huang, *Proc. Roy. Soc. (London)* **A203**, 178 (1950).
[16] R. A. Toupin and B. Bernstein, *J. Acoust. Soc. Am.* **33**, 216 (1961).
[17] A. E. Green, *Proc. Roy. Soc. (London)* **A266**, 1 (1962).
[18] D. C. Wallace, *Rev. Mod. Phys.* **37**, 57 (1965).

L (either adiabatic or isothermal), whose elements L_{ik} are

$$L_{ik} = A_{ijkl}\kappa_j\kappa_l. \tag{3.13}$$

The matrix **L** depends on the direction, but not on the magnitude, of **k**. Because of the sum on j, l in (3.13) and from the symmetry of the A_{ijkl} according to (3.6) and (3.7), it is seen that **L** is a real-symmetric 3×3 matrix for any direction of **k**. Therefore the eigenvalues of **L** are real, and the eigenvectors form an orthogonal matrix. Hereafter, for computational convenience, we will take the eigenvectors to be normalized:

$$w_i w_i = 1. \tag{3.14}$$

The diagonal form of the eigenvalue–eigenvector problem is then written

$$\rho_1 v^2 = w_i L_{ik} w_k. \tag{3.15}$$

Actually for each propagation matrix **L** there are three eigenvectors and three corresponding eigenvalues; these solutions are normal modes of the problem and represent elastic waves which propagate independently of one another. To be quite proper we should add an index 1, 2, or 3 to each eigenvector and eigenvalue to denote the mode, but this notation will not be necessary in our work here. We have already mentioned the (macroscopic) thermodynamic equilibrium condition (1.12). The condition for stability of the material against all thermoelastic motion is that all eigenvalues for all propagation directions κ be positive.[19] Since for a given crystal structure there is a relatively small number of second-order elastic constants, the thermoelastic stability conditions can always be reduced to a small number of conditions involving the elastic constants and stress components. We note that there is no dispersion of the elastic normal modes, i.e., for a given direction of **k**, phase and group velocities are equal to $\omega/|\mathbf{k}|$ for any $|\mathbf{k}|$; this is correct of course only for long-wavelength plane waves, which is the region of validity of our theory. Also, there is no attenuation in the theory at this point; approximate inclusion of attenuation will be discussed in Section 5.

The starting point for working out wave propagation problems for a given crystal in the presence of a given stress system is to construct the propagation matrices; we give some examples in Part IV. For certain propagation directions, usually along directions of high crystalline symmetry, the three normal modes are found to be one purely longitudinal (**w** $\parallel$ κ) and two purely transverse (**w** $\perp$ κ); these are called pure modes. For zero-stress conditions, Borgnis[20] found those directions for which pure modes propagate for several crystal symmetries (trigonal, hexagonal,

[19] D. C. Wallace, *Advan. Mater. Res.* **3**, 331 (1968).
[20] F. E. Borgnis, *Phys. Rev.* **98**, 1000 (1955).

tetragonal, and cubic). Brugger[21] extended these calculations to include orthorhombic crystals. Except for pure mode directions, the eigenvectors depend explicitly on the values of the second-order elastic constants; for the example of nickel, and again at zero-stress conditions, Fein and Smith[22] calculated the eigenvectors for all propagation directions and found them to be always nearly longitudinal and transverse.

Another important aspect of elastic wave propagation in crystals is that the direction of energy flux may not coincide with the propagation direction κ. Note κ is the direction normal to planes of constant phase. The energy flux for small amplitude motion is given by (2.32). For a normal mode of amplitude a, the displacement is

$$\mathbf{u} = a\mathbf{w} \sin(\mathbf{k}\cdot\mathbf{X} - \omega t),$$

and the adiabatic energy flux averaged over one cycle is

$$E^S{}_i = (a^2\omega^2/2v) C^S{}_{ijkl} w_j w_k \kappa_l. \tag{3.16}$$

According to Waterman,[23] for cubic, hexagonal, tetragonal, and trigonal crystal classes, the energy flux is parallel to the propagation direction for all pure longitudinal modes, and all pure transverse modes except for cubic along $\langle 111 \rangle$ and trigonal along $\langle 001 \rangle$. Holt[24] has shown for the example of copper that the energy flux direction may deviate as much as 15° from the propagation direction for nonpure mode directions in the crystal.

It is instructive to relate the present treatment of the linear equation of motion to that of Huang.[15] He expanded the mechanical energy density in powers of the displacement gradients u_{ij}, for the case of initial homogeneous stress. We generalize the theory of Huang by expanding the state functions U and F, and thus differentiate between adiabatic and isothermal coefficients:

$$\rho_1 U(\mathbf{X}, u_{ij}, S) = \rho_1 U(\mathbf{X}, 0, S) + A^S{}_{ij} u_{ij} + \tfrac{1}{2} A^S{}_{ijkl} u_{ij} u_{kl} + \cdots, \tag{3.17}$$

with a similar expansion for F in terms of isothermal A coefficients. These expansions serve as the fundamental definitions of the A coefficients. From this energy expansion, Huang derived the equation of motion (3.5). He also showed that only when the stresses vanish do the A_{ijkl} have complete Voigt symmetry. This lack of Voigt symmetry led Huang to say that elastic constants can be defined only at zero stress.[15] We can clear up this difficulty by relating the Huang coefficients[25] to the elastic constants; this may be

---

[21] K. Brugger, *J. Appl. Phys.* **36**, 759 (1965).
[22] A. E. Fein and C. S. Smith, *J. Appl. Phys.* **23**, 1212 (1952).
[23] P. C. Waterman, *Phys. Rev.* **113**, 1240 (1959).
[24] A. C. Holt, *J. Appl. Phys.* **39**, 3046 (1968).
[25] Huang used the notation S_{ij}, S_{ijkl}, in place of our A_{ij}, A_{ijkl}.

done by eliminating η_{ij} in favor of u_{ij} in our original expansions (1.9) and (1.10) of the state functions. The internal energy, e.g., becomes

$$\rho_1 U(\mathbf{X}, u_{ij}, S) = \rho_1 U(\mathbf{X}, 0, S) + C^S{}_{ij} u_{ij}$$
$$+ (C^S{}_{jl}\delta_{ik} + C^S{}_{ijkl}) u_{ij} u_{kl} + \cdots. \quad (3.18)$$

Since all nine u_{ij} are independent, the coefficients in (3.17) and (3.18) must be equal:

$$A^S{}_{ij} = C^S{}_{ij} = T_{ij}; \quad (3.19)$$

$$A^S{}_{ijkl} = T_{jl}\delta_{ik} + C^S{}_{ijkl}. \quad (3.20)$$

Again we have used (1.13) to introduce $T_{ij}(\mathbf{X})$. The relation (3.20) is the same as (3.6), and it is obvious from this equation that the $A^S{}_{ijkl}$ have Voigt symmetry if and only if $T_{ij} = 0$. Relations analogous to (3.19) and (3.20) hold for the isothermal coefficients $A^T{}_{ij}$ and $A^T{}_{ijkl}$.

Now in (3.17) the u_{ij} are arbitrary, and in particular not necessarily symmetric. Therefore (3.17) does not contain any information about the rotational invariance of U. But such invariance is completely contained in our original expansion (1.9), and hence also in (3.18). Thus the restrictions placed on the A coefficients through (3.19) and (3.20), along with the symmetry properties of T_{ij} and C_{ijkl}, should contain the rotational invariance conditions. For adiabatic or isothermal coefficients, these restrictions are found to be

$$A_{ij} = A_{ji},$$
$$A_{ijkl} - A_{klij} = 0, \quad (3.21)$$
$$A_{ijkl} - A_{jikl} + A_{il}\delta_{jk} - A_{jl}\delta_{ik} = 0.$$

These are just the relations which Huang[15] found by requiring the energy density to be invariant with respect to infinitesimal rotation.

4. Harmonic Generation

Brillouin[26] noted that, because of third-order terms in the energy density expansion, a traveling wave generates higher harmonics, and superposed waves generate sum- and difference-frequency waves. Thuras et al.[27] studied harmonic generation in air, both theoretically and experimentally. Recently the theory of harmonic generation has been developed for crystals at zero

---

[26] Brillouin,[2] Chapter 11, Section 8.
[27] A. L. Thuras, R. T. Jenkins, and H. T. O'Neil, *J. Acoust. Soc. Am.* **6**, 173 (1935).

stress[28,29]; here we give a general theory of resonance harmonic generation for solids in the presence of arbitrary initial stress.

It is now necessary to keep the first nonlinear term in the equation of motion; the derivation will be illustrated for the case of adiabatic motion. We can begin with (3.2) for the constant entropy stress derivative, and write this as

$$\partial T_{ji}/\partial x_j = (1/J)(\partial \alpha_{il}/\partial X_k)[\rho_1(\partial U/\partial \eta_{kl})]$$
$$+ (\alpha_{il}/J)(\partial/\partial X_k)[\rho_1(\partial U/\partial \eta_{kl})]. \quad (4.1)$$

The derivative of $\boldsymbol{\alpha}$ is a second-order derivative of $\mathbf{u}$:

$$(\partial \alpha_{il}/\partial X_k) = (\partial u_{il}/\partial X_k) = (\partial^2 u_i/\partial X_l \, \partial X_k). \quad (4.2)$$

From the internal energy expansion (1.9) we can write

$$\rho_1(\partial U/\partial \eta_{kl}) = C^S_{kl} + C^S_{klmn}\eta_{mn} + \tfrac{1}{2}C^S_{klmnpq}\eta_{mn}\eta_{pq} + \cdots,$$
$$= C^S_{kl} + C^S_{klmn}(u_{mn} + \tfrac{1}{2}u_{rm}u_{rn})$$
$$+ \tfrac{1}{2}C^S_{klmnpq}(u_{mn} + \tfrac{1}{2}u_{rm}u_{rn})(u_{pq} + \tfrac{1}{2}u_{sp}u_{sq}) + \cdots, \quad (4.3)$$

where we have used the symmetry of the elastic constants. The procedure is to carry out the differentiation indicated in (4.1), keeping terms to *third* order in derivatives of $\mathbf{u}$, evaluate the coefficients at $\mathbf{X}$ where $u_{ij} = 0$, and use the equation of motion $\rho_1 \ddot{u}_i = (\partial T_{ji}/\partial x_j)$ at $\mathbf{X}$. The result is

$$\rho_1 \ddot{u}_i = (\partial^2 u_k/\partial X_j \, \partial X_l)[A^S_{ijkl} + (\partial u_p/\partial X_q)A^S_{ijklpq} + \cdots], \quad (4.4)$$

where

$$A^S_{ijklpq} = C^S_{jlpq}\delta_{ik} + C^S_{ijql}\delta_{kp} + C^S_{jkql}\delta_{ip} + C^S_{ijklpq}. \quad (4.5)$$

An important general point is that while the initial stress components appear explicitly in the linear propagation coefficients A_{ijkl}, they do *not* appear explicitly in the nonlinear coefficients A_{ijklpq}.

This first-order nonlinear equation of motion is the basis of all the related theory, including harmonic generation, generation of sum- and difference-frequency waves by the interaction of two elastic waves (sometimes called three-phonon interactions), etc. For adiabatic (or isothermal) propagation, the coefficients in (4.4) are all-adiabatic (or all-isothermal); no mixed coefficients appear. The next order nonlinear terms, which were dropped in (4.4), are of fourth order in derivatives of $\mathbf{u}$ and contain second-, third-, and fourth-order elastic constants. The idea, of course, is that the

---

[28] O. Buck and D. O. Thompson, *Mater. Sci. Eng.* **1**, 117 (1966).
[29] A. C. Holt and J. Ford, *J. Appl. Phys.* **38**, 42 (1967).

displacements and their derivatives are small, and in (4.4) we have kept the first two terms of a rapidly converging perturbation expansion of the equation of motion. In this connection we note the coefficients $A^S{}_{ijkl}$, $A^S{}_{ijklpq}$, ... are all of the same order of magnitude for a given material.

Let us write a perturbation expansion for the displacement **u** which is a solution to the nonlinear equation of motion:

$$\mathbf{u} = \mathbf{u}^{(1)} + \mathbf{u}^{(2)} + \cdots. \tag{4.6}$$

As it will be clear below, the superscript refers to the frequency of the wave component, and not to the order of perturbation. The zeroth-order equation of motion is then

$$\rho_1 \ddot{u}_i{}^{(1)} = (\partial^2 u_k{}^{(1)}/\partial X_j \, \partial X_l) A^S{}_{ijkl}. \tag{4.7}$$

Solutions are the normal elastic waves discussed in Section 3; a plane wave solution with amplitude a_1 and polarization vector **w** may be written

$$\mathbf{u}^{(1)} = a_1 \mathbf{w} \cos(\mathbf{k} \cdot \mathbf{X} - \omega t), \tag{4.8}$$

and the eigenvalue–eigenvector equation is

$$\rho_1 v^2 w_i = L_{ik} w_k. \tag{4.9}$$

A consistent first-order perturbation equation of motion for $\mathbf{u}^{(2)}$ is now obtained by using the zeroth-order solution in the nonlinear term:

$$\rho_1 \ddot{u}_i{}^{(2)} = (\partial^2 u_k{}^{(2)}/\partial X_j \, \partial X_l) A^S{}_{ijkl}$$
$$+ (\partial^2 u_k{}^{(1)}/\partial X_j \, \partial X_l)(\partial u_p{}^{(1)}/\partial X_q) A^S{}_{ijklpq}. \tag{4.10}$$

Working out the details gives

$$\rho_1 \ddot{u}_i{}^{(2)} = (\partial^2 u_k{}^{(2)}/\partial X_j \, \partial X_l) A^S{}_{ijkl} + \tfrac{1}{2} a_1^2 k^3 D_i \sin[2(\mathbf{k} \cdot \mathbf{X} - \omega t)], \tag{4.11}$$

where

$$D_i = A^S{}_{ijklpq} \kappa_j \kappa_l \kappa_q w_k w_p. \tag{4.12}$$

Henceforth we will refer to $\mathbf{u}^{(1)}$, given by (4.8), as the input wave. Because of the nonlinear term in the equation of motion, the input wave generates a new displacement $\mathbf{u}^{(2)}$. The first-order perturbation equation (4.11) for $\mathbf{u}^{(2)}$ is just the zeroth-order equation with an added driving term; we will look only for resonance solutions. The driving term drives at wave vector $2\mathbf{k}$, frequency 2ω, and in the direction of the vector **D**. Hence the driving term will be in resonance with the first harmonic of the input wave, and also with the first harmonic of a mode which is degenerate with the input wave, i.e. a mode which has the same **k** and ω as $\mathbf{u}^{(1)}$ but a different displacement direction. These are called self-resonant and mutual-resonant cases, respectively.[29] Let us denote the generated-harmonic displacement direction by $\hat{\mathbf{w}}$, which may or may not be the same as $\mathbf{w}$ for the input wave. If and only if $\mathbf{D}$ has a component along $\hat{\mathbf{w}}$, then the harmonic will be continually enforced along the path of the wave, and its amplitude will increase with the path length. The component of $\mathbf{D}$ along $\hat{\mathbf{w}}$ is $\mathbf{D} \cdot \hat{\mathbf{w}}$, and the resonance equation of motion is therefore

$$\rho_1 \ddot{u}_i^{(2)} = (\partial^2 u_k^{(2)}/\partial X_j\, \partial X_l) A^S{}_{ijkl} + \tfrac{1}{2} a_1^2 k^3 (\mathbf{D} \cdot \hat{\mathbf{w}}) \hat{\mathbf{w}}_i \sin[2(\mathbf{k} \cdot \mathbf{X} - \omega t)]. \tag{4.13}$$

This equation holds for self- and mutual-resonant cases, where a_1, $\mathbf{k}$, ω, and $\mathbf{D}$ correspond to the input wave. The essential properties of the solution are determined by the vector $\mathbf{D}$; in particular no harmonic generation can occur if $\mathbf{D} \cdot \hat{\mathbf{w}} = 0$.

Let the input wave be generated at the plane $\mathbf{k} \cdot \mathbf{X} = 0$, and let l be the distance from this plane measured along $\mathbf{k}$, so that

$$\mathbf{k} \cdot \mathbf{X} = kl, \qquad l > 0. \tag{4.14}$$

The solution for the generated harmonic, which satisfies the boundary condition $\mathbf{u}^{(2)} = 0$ at $l = 0$, is

$$\mathbf{u}^{(2)} = b\hat{\mathbf{w}} (\mathbf{k} \cdot \mathbf{X}) \cos[2(\mathbf{k} \cdot \mathbf{X} - \omega t)], \tag{4.15}$$

where b is a constant to be determined. Since $\hat{\mathbf{w}}$ is a unit vector, the amplitude of $\mathbf{u}^{(2)}$ is

$$a_2 = b(\mathbf{k} \cdot \mathbf{X}) = bkl. \tag{4.16}$$

Solving for b, from the equation of motion (4.13) and with the aid of the eigenvalue–eigenvector equation (4.9) for the input wave, we obtain

$$b = a_1^2 k (\mathbf{D} \cdot \hat{\mathbf{w}})/8\rho_1 v^2. \tag{4.17}$$

The relation between the amplitudes a_1 and a_2 is conveniently expressed from (4.16) and (4.17) as

$$|\,a_2/a_1^2\,| = k^2 l\,|\,\mathbf{D} \cdot \hat{\mathbf{w}}\,|/8\rho_1 v^2. \tag{4.18}$$

The primary experimental measurement of harmonic generation is $|\,a_2/a_1^2\,|$ for a given input wave and generated harmonic; thus one determines $|\,\mathbf{D} \cdot \hat{\mathbf{w}}\,|$. Measurement of the amplitudes alone does not determine the sign of $\mathbf{D} \cdot \hat{\mathbf{w}}$. Also, the steady-state analysis we have presented does not determine the sign of $\mathbf{D} \cdot \hat{\mathbf{w}}$ since there is some arbitrariness in the phase between the input wave and the harmonic. For example, we could have chosen $\sin(\mathbf{k} \cdot \mathbf{X} - \omega t)$ for the phase of $\mathbf{u}^{(1)}$, with the result that the sign of b in (4.17) is changed. It should be possible to determine the sign of $\mathbf{D} \cdot \hat{\mathbf{w}}$ by a theoretical and experimental analysis of the total wave shape and the relative phase of the components.

In addition, the above analysis is for the *interior* of the elastic material; at a stress-free surface of the material all normal mode amplitudes are doubled upon reflection. Thus for displacement amplitudes *on* the surface, our result becomes

$$| a_2/a_1{}^2 | = k^2 l \, | \, \mathbf{D} \cdot \hat{\mathbf{w}} \, |/16\rho_1 v^2, \qquad \text{stress-free surface.} \qquad (4.19)$$

In the case of an initially stressed material, (4.18) holds for the interior, and (4.19) holds if the stresses are not applied to the surface on which the amplitudes are observed, or if that surface is only weakly coupled acoustically to the stress system, as in the case of a gas pressure.

It is of interest to compare the directions of energy flux for the input wave and the generated harmonic. The input wave displacement is given by (4.8), and according to (3.16) the adiabatic energy flux is

$$E^S{}_i = (a_1{}^2 \omega^2/2v) C^S{}_{ijkl} w_j w_k \kappa_l, \qquad \text{input wave.} \qquad (4.20)$$

The harmonic displacement is given by (4.15), and a calculation from the general formula (2.32) gives for the adiabatic energy flux, averaged over one cycle,

$$E^S{}_i = (4a_2{}^2 \omega^2/2v) C^S{}_{ijkl} \hat{w}_j \hat{w}_k \kappa_l, \qquad \text{harmonic wave.} \qquad (4.21)$$

We can therefore conclude that the energy flux direction is the same for the harmonic as for the input wave for all self-resonant cases, but not necessarily for mutual-resonant cases where $\hat{\mathbf{w}} \neq \mathbf{w}$.

There are other, nonresonant, displacements generated by the input wave, and these are also found as perturbation solutions of the nonlinear equation of motion (4.4). Buck and Thompson[28] have made a detailed study of nonresonant-displacement generation, and of reflections at stress-free surfaces, for a cubic crystal with $\mathbf{k}$ and $\mathbf{w}$ lying in the same $(1\bar{1}0)$ plane. Many of their conclusions should be valid for any crystal symmetry. They find that nonresonant displacements are generated by the input wave as it travels, by the interaction of the input wave with its reflection in the region where these two waves overlap, and by the stress-free boundary. All of these displacements are normally small and can be neglected compared to the resonant harmonic displacements; we make additional comments on this point in Section 18. They also find for self-resonant cases that after reflection from a stress-free surface, the harmonic amplitude *decreases* on the return path until $a_2 = 0$ again at $l = 0$. Then, if the surface $l = 0$ is also considered stress-free, the cycle repeats itself on every round trip of the input wave.

Holt[24] has pointed out that it is not possible to determine *all* the third-order elastic constants of a crystal from harmonic generation experiments. In particular, the motion during the propagation of the input wave and the generated harmonic is such to satisfy $\det[\eta_{ij}] = 0$, and this independent condition reduces by one the effective number of *independent* $C_{\alpha\beta\gamma}$ which appear in the thermodynamic functions. To show this we write the total displacement components $u_i = u_i^{(1)} + u_i^{(2)}$ from (4.8) and (4.15), calculate $u_{ij} = (\partial u_i/\partial X_j)$, and write the result as

$$u_{ij} = y_i k_j, \qquad (4.22)$$

where

$$y_i = -a_1 w_i \sin(\mathbf{k}\cdot\mathbf{X} - \omega t)$$
$$+ b\hat{w}_i\{\cos[2(\mathbf{k}\cdot\mathbf{X} - \omega t)] - 2(\mathbf{k}\cdot\mathbf{X})\sin[2(\mathbf{k}\cdot\mathbf{X} - \omega t)]\}. \quad (4.23)$$

The form of y_i is not important to our argument; we only need (4.22) for u_{ij}, where y_i is a function of $\mathbf{X}$ and t and $\mathbf{k}$ is constant everywhere. Then the Lagrangian strain parameters are

$$\eta_{ij} = \tfrac{1}{2}(y_i k_j + y_j k_i + y_k y_k k_i k_j). \qquad (4.24)$$

Now $\det[\eta_{ij}]$ is independent of rotation of the coordinate system, so we take the 1 axis to be along $\mathbf{k}$:

$$|\mathbf{k}| = k_1; \qquad k_2 = k_3 = 0.$$

Then since η_{ij} is now zero unless i or $j = 1$,

$$\det[\eta_{ij}] = 0, \qquad \text{at all } \mathbf{X}, t, \qquad (4.25)$$

or in Voigt notation

$$\eta_1\eta_2\eta_3 + 2\eta_4\eta_5\eta_6 - \eta_1\eta_4^2 - \eta_2\eta_5^2 - \eta_3\eta_6^2 = 0. \qquad (4.26)$$

Now let $\rho_1 U_{3d}$ denote those terms in the expansion (1.9) of $\rho_1 U$ which contain the strain terms appearing in $\det[\eta_{ij}]$:

$$\rho_1 U_{3d} = C^S{}_{123}\eta_1\eta_2\eta_3 + 8C^S{}_{456}\eta_4\eta_5\eta_6$$
$$+ 2C^S{}_{144}\eta_1\eta_4^2 + 2C^S{}_{255}\eta_2\eta_5^2 + 2C^S{}_{366}\eta_3\eta_6^2. \quad (4.27)$$

Solving (4.26) for one of its terms, say $\eta_1\eta_4^2$, and inserting in (4.27), we obtain

$$\rho_1 U_{3d} = (C^S{}_{123} + 2C^S{}_{144})\eta_1\eta_2\eta_3 + (8C^S{}_{456} + 4C^S{}_{144})\eta_4\eta_5\eta_6$$
$$+ 2(C^S{}_{255} - C^S{}_{144})\eta_2\eta_5^2 + 2(C^S{}_{366} - C^S{}_{144})\eta_3\eta_6^2. \quad (4.28)$$

Here five $C_{\alpha\beta\gamma}$ have been replaced, through the condition $\det[\eta_{ij}] = 0$, by the four combinations $C^S{}_{123} + 2C^S{}_{144}$, $2C^S{}_{456} + C^S{}_{144}$, $C^S{}_{255} - C^S{}_{144}$, and $C^S{}_{366} - C^S{}_{144}$; only these four combinations can be measured by harmonic generation experiments.[24]

A well-known property of the first-order-nonlinear equation of motion (4.4) is that its solutions exhibit discontinuous behavior at some point in space and time.[30] In particular, suppose the input wave of amplitude a_1 is produced at the plane $l = 0$, starting at time $t = 0$. The amplitude of the generated harmonic increases with the distance l, and the total displacement approaches a sawtooth profile, until at the discontinuity distance l_D the particle velocity becomes discontinuous as a function of l. The approximate expression for l_D has been derived for the one-dimensional problem of longitudinal motion only.[30-32] In the usual notation,[31] the one-dimensional equation of motion corresponding to (4.4) is

$$\rho_1(\partial^2 u/\partial t^2) = (\partial^2 u/\partial l^2)[K_2 + (3K_2 + K_3)(\partial u/\partial l)], \quad (4.29)$$

where $K_2 = \rho_1 v^2$ and K_3 is a combination of third-order elastic constants. The discontinuity distance is found to be

$$l_D \approx 2K_2 v^2/[a_1 \omega^2 \mid 3K_2 + K_3 \mid]. \quad (4.30)$$

At the discontinuity distance, $\mid \partial^2 u/\partial t \, \partial l \mid \to \infty$; it is presumed that a shock wave originates at this point and travels in both forward and backward directions.[30] Beyond l_D there is no simple-wave solution; furthermore, since the shock arrives back at $l = 0$ at a time $t_D \approx 2l_D/v$, then after t_D the simple-wave solution breaks down even at $l = 0$. It is not clear how well the expression (4.30) applies to real solids, where the effects of attenuation and of higher order terms in the equation of motion will be important; however, l_D probably gives an estimate of a limit of applicability of our simple-wave solutions. According to Breazeale,[33] for $a_1 = 10$ Å and $\omega = 30$ MHz, l_D is about 30–50 cm for Ge and about 100–200 cm for Si.

5. Attenuation of Elastic Waves

There are two standard ways to treat attenuation of elastic waves in solids; both are phenomenological and approximate. One procedure is to include a dissipation term in the equation of motion; we will illustrate this for the one-dimensional problem. Following Landau and Lifshitz,[34] the one-dimensional dissipative equation of motion is

$$\rho_1(\partial^2 u/\partial t^2) = \rho_1 v^2(\partial^2 u/\partial l^2) + \lambda(\partial^3 u/\partial t \, \partial l^2), \quad (5.1)$$

[30] R. N. Thurston and M. J. Shapiro, *J. Acoust. Soc. Am.* **41**, 1112 (1967).
[31] M. A. Breazeale and J. Ford, *J. Appl. Phys.* **36**, 3486 (1965).
[32] L. A. Pospelov, *Soviet Phys.—Acoust. (English Transl.)* **11**, 302 (1966).
[33] M. A. Breazeale, *J. Appl. Phys.* **37**, 3332 (1966).
[34] L. D. Landau and E. M. Lifshitz, "Fluid Mechanics," Section 15. Addison-Wesley, Reading, Massachusetts, 1959.

where λ is a small damping constant. Consider a trial solution

$$u = e^{i(kl-\omega t)}; \qquad (5.2)$$

the equation of motion (5.1) is satisfied by this trial solution if

$$k^2 = (\omega/v)^2[1 - i(\omega\lambda/\rho_1 v^2)]^{-1}. \qquad (5.3)$$

Now if $\omega\lambda \ll \rho_1 v^2$, the dispersion relation (5.3) may be expanded to give

$$k = (\omega/v) + i(\omega^2\lambda/2\rho_1 v^3) + \cdots. \qquad (5.4)$$

Thus to first order in λ, the wave amplitude is damped according to $e^{-\alpha l}$, where α is the attenuation coefficient[35]

$$\alpha = (\omega^2\lambda/2\rho_1 v^3). \qquad (5.5)$$

The characteristic result of this treatment is that α is proportional to ω^2.

A more general way of looking at attenuation is also discussed by Landau and Lifshitz.[36,37] Let $\mathcal{E}$ be the energy (adiabatic or isothermal) of the elastic wave, per unit initial volume of the material. The average rate at which energy is lost from the wave, due to any form of dissipation, is $d\mathcal{E}/dt$; since the energy loss in one cycle is small compared to the energy, $d\mathcal{E}/dt$ is averaged over one cycle. The variation of energy with the path length l is

$$d\mathcal{E}/dl = (d\mathcal{E}/dt)/v. \qquad (5.6)$$

We then assume $d\mathcal{E}/dl$ is proportional to $\mathcal{E}$ and write

$$d\mathcal{E}/dl = -2\alpha\mathcal{E}, \qquad (5.7)$$

or since $\mathcal{E}$ is proportional to a^2, where a is the wave amplitude,

$$da/dl = -\alpha a. \qquad (5.8)$$

Again the amplitude decreases as $e^{-\alpha l}$.

Mason[38] has discussed extensively the physical processes which contribute to attenuation in solids. In addition, Mason and Bateman[39] made detailed calculations of the temperature- and frequency-dependence of the attenuation of ultrasonic waves in several materials. They considered two sources of conversion of elastic wave energy to thermal energy, namely,

---

[35] There should be no confusion between attenuation coefficient α and the strain parameters α_{ij}.

[36] Landau and Lifshitz,[34] Section 77.

[37] L. D. Landau and E. M. Lifshitz, "Theory of Elasticity," Section 30. Addison-Wesley, Reading, Massachusetts, 1959.

[38] W. P. Mason, "Physical Acoustics and the Properties of Solids." Van Nostrand, Princeton, New Jersey, 1958.

[39] W. P. Mason and T. B. Bateman, *J. Acoust. Soc. Am.* **40**, 852 (1966).

the thermoelastic effect and the phonon viscosity effect. In the thermoelastic effect, thermal energy flows from the compressed (hotter) to the expanded (cooler) regions associated with an adiabatic compressional wave. To describe the phonon viscosity effect, one first considers the solid as containing a large number of thermally excited phonons. Then in the presence of the strain due to the elastic wave, the local phonon frequencies change but their occupation numbers do not change because the wave is adiabatic; hence the phonons are not in local equilibrium and they proceed toward equilibrium, dissipating energy. By approximate calculation of these effects, Mason and Bateman[39] obtained remarkably good agreement with measured values of attenuation coefficients.

We now wish to correct approximately the harmonic generation results of Section 4 for attenuation. We will follow Thuras et al.[27] and assume the input wave and generated harmonic are attenuated independently of one another. The first step is to calculate the rate at which energy is pumped into the harmonic by the driving term in the equation of motion (4.13). From (4.13), the driving force per unit initial volume is

$$\mathbf{f}(\mathbf{X}, t) = \tfrac{1}{2} a_1^2 k^3 (\mathbf{D} \cdot \hat{\mathbf{w}}) \hat{\mathbf{w}} \sin[2(\mathbf{k} \cdot \mathbf{X} - \omega t)], \quad (5.9)$$

while from (4.15) the harmonic displacement is

$$\mathbf{u}^{(2)}(\mathbf{X}, t) = a_2 \hat{\mathbf{w}} \cos[2(\mathbf{k} \cdot \mathbf{X} - \omega t)]. \quad (5.10)$$

Now with $\dot{\mathbf{u}}^{(2)} = \partial \mathbf{u}^{(2)}/\partial t$, and with brackets $\langle \ \rangle$ denoting an average over one period, the energy of the elastic wave per unit initial volume is

$$\mathcal{E} = \langle \rho_1 \dot{\mathbf{u}}^{(2)} \cdot \dot{\mathbf{u}}^{(2)} \rangle = 2\rho_1 \omega^2 a_2^2. \quad (5.11)$$

The average rate at which energy is pumped into the harmonic wave is

$$d\mathcal{E}/dt = \langle \mathbf{f} \cdot \dot{\mathbf{u}}^{(2)} \rangle = \tfrac{1}{2} a_2 a_1^2 \omega k^3 (\mathbf{D} \cdot \hat{\mathbf{w}}). \quad (5.12)$$

Finally, since $\mathcal{E}$ is proportional to a_2^2, we have

$$(da_2/dl)/a_2 = \tfrac{1}{2}(d\mathcal{E}/dl)/\mathcal{E}, \quad (5.13)$$

or

$$(da_2/dl) = \tfrac{1}{2} a_2 (d\mathcal{E}/dt)/v\mathcal{E}$$

$$= a_1^2 k^2 (\mathbf{D} \cdot \hat{\mathbf{w}})/8\rho_1 v^2. \quad (5.14)$$

It is of interest to note that integration of (5.14) gives again our result (4.18) for a_2 as a function of l, provided a_1 is constant.

Attenuation of the input wave is now represented according to (5.8), with attenuation coefficient α_1:

$$da_1/dl = -\alpha_1 a_1, \quad a_1 = a_0 \exp(-\alpha_1 l), \quad (5.15)$$

where a_0 is the input wave amplitude at the plane $l = 0$. At any point l the amplitude of the generated harmonic increases due to the pumping, according to (5.14) with a_1 evaluated at l, and decreases due to its independent attenuation α_2, with the net result

$$da_2/dl = [a_0^2 \exp(-2\alpha_1 l) k^2 (\mathbf{D} \cdot \hat{\mathbf{w}})/8\rho_1 v^2] - \alpha_2 a_2, \qquad (5.16)$$

where we have used (5.15) for a_1. This is a linear differential equation for $a_2(l)$; with the boundary condition $a_2 = 0$ at $l = 0$, the solution is

$$a_2 = \frac{a_0^2 k^2 (\mathbf{D} \cdot \hat{\mathbf{w}})}{8\rho_1 v^2} \left[\frac{\exp(-2\alpha_1 l) - \exp(-\alpha_2 l)}{\alpha_2 - 2\alpha_1} \right]. \qquad (5.17)$$

If (5.17) is expanded for small attenuation, i.e. $\alpha_1 l$, $\alpha_2 l \ll 1$, the leading term is just the nondissipative result (4.18), in which a_2 increases linearly with l. On the other hand, for larger l, a_2 does not continue to increase, but goes through a maximum and then decreases monotonically as l increases. From (5.17) the distance l_M at which a_2 is maximum is

$$l_M = (\alpha_2 - 2\alpha_1)^{-1} \ln(\alpha_2/2\alpha_1). \qquad (5.18)$$

It is interesting to note the results (5.17) and (5.18) were obtained for the one-dimensional problem by Pospelov,[32] by starting with a one-dimensional equation of motion which included a nonlinear term as in (4.29) and a dissipative term as in (5.1), and taking the solution for attenuation large compared to nonlinear effects. In Pospelov's treatment the attenuation coefficient was found to be proportional to ω^2, as in (5.5), and hence his results have $\alpha_2 = 4\alpha_1$.

It is not convenient to use (5.17) in analysis of experiments, since that equation contains a_0, the input wave amplitude at $l = 0$. We use (5.15) to eliminate a_0 and obtain the ratio $|a_2/a_1^2|$ at any l:

$$\left| \frac{a_2}{a_1^2} \right| = \frac{k^2 |\mathbf{D} \cdot \hat{\mathbf{w}}|}{8\rho_1 v^2} \frac{\{1 - \exp[-(\alpha_2 - 2\alpha_1)l]\}}{(\alpha_2 - 2\alpha_1)}. \qquad (5.19)$$

Again if the attenuation is small at l, (5.19) may be expanded to give the leading term

$$|a_2/a_1^2| = k^2 l |\mathbf{D} \cdot \hat{\mathbf{w}}|/8\rho_1 v^2, \quad \text{for} \quad (\alpha_2 - 2\alpha_1)l \ll 1. \qquad (5.20)$$

This is the same form as the nondissipative result (4.18), but now applies when both the input wave and harmonic are slightly attenuated. If the attenuation is large, (5.19) is applicable. Note there is no maximum in $|a_2/a_1^2|$ as a function of l. Also for amplitudes on a stress-free surface, the right-hand sides of (5.19) and (5.20) should be divided by 2, as in (4.19).

If repeated echoes are observed at the end of a sample, and if the reflecting surfaces are stress-free and there is no attenuation, the calculations of Buck and Thompson[28] for a cubic crystal show that $|a_2/a_1^2|$ is given by (4.19) for each echo, where l is just the length of the sample. If the attenuation is small during each pass through the sample, then we expect (5.20) to hold for each echo, again with l the sample length (and divided by 2 for surface amplitudes). Thus for repeated echoes, a_1 is attenuated according to the total distance traveled, and a_2 is attenuated simply as a_1^2. This behavior has been observed experimentally by Thompson et al.[40] Gedroits and Krasil'nikov[41] measured a_2 for repeated echoes, and assumed the decrease of a_2 was governed by Eq. (5.17) for large l, with l equal to the total distance traveled. Although this interpretation is incorrect for echoes after the first, it appears from their Fig. 1 that Gedroits and Krasil'nikov did see the maximum in a_2 vs l in their measurements of the first transmitted pulse for samples of various lengths.

6. Stress–Strain Relations and Compressibility

We will derive the relation between the variation of stress and the variation of strain from the initial configuration. For this work it is sufficient to consider infinitesimal strains from the initial configuration. The symmetric and antisymmetric infinitesimal displacement gradients, ϵ_{ij} measuring pure strain and ω_{ij} measuring pure rotation, are defined as follows:

$$\epsilon_{ij} = \tfrac{1}{2}(u_{ij} + u_{ji}) = \tfrac{1}{2}(\alpha_{ij} + \alpha_{ji} - 2\delta_{ij}), \qquad (6.1)$$

$$\omega_{ij} = \tfrac{1}{2}(u_{ij} - u_{ji}) = \tfrac{1}{2}(\alpha_{ij} - \alpha_{ji}), \qquad (6.2)$$

where u_{ij} and α_{ij} are defined in Section 1. Solving these for α_{ij} gives the following useful relations:

$$\alpha_{ij} = \tfrac{1}{2}(\epsilon_{ij} + \epsilon_{ji} + \omega_{ij} - \omega_{ji} + 2\delta_{ij}), \qquad (6.3)$$

$$(\partial\alpha_{ij}/\partial\epsilon_{mn}) = \tfrac{1}{2}(\delta_{im}\delta_{jn} + \delta_{jm}\delta_{in}), \qquad (6.4)$$

$$(\partial\alpha_{ij}/\partial\omega_{mn}) = \tfrac{1}{2}(\delta_{im}\delta_{jn} - \delta_{jm}\delta_{in}). \qquad (6.5)$$

The stress–strain relations will be calculated by taking the example of adiabatic variations. It is convenient to begin by combining equations

---

[40] D. O. Thompson, M. A. Tennison, and O. Buck, *J. Acoust. Soc. Am.* **44**, 435 (1968).
[41] A. A. Gedroits and V. A. Krasil'nikov, *Soviet Phys.—JETP (English Transl.)* **16**, 1122 (1963).

(3.1) and (3.3) to write the stress components as

$$T_{ij} = J^{-1}\alpha_{ik}\alpha_{jl}[C^S{}_{kl} + \tfrac{1}{2}C^S{}_{klmn}(\alpha_{pm}\alpha_{pn} - \delta_{mn}) + \cdots]. \quad (6.6)$$

Now (6.6) may be differentiated with respect to α_{rs} with the aid of Jacobi's identity (2.22); evaluating the result at **X** where $\alpha_{ij} = \delta_{ij}$, $\gamma_{ij} = \delta_{ij}$, $J = 1$, we obtain

$$(\partial T_{ij}/\partial \alpha_{rs}) = (T_{is}\delta_{jr} + T_{js}\delta_{ir} - T_{ij}\delta_{rs}) + C^S{}_{ijrs}. \quad (6.7)$$

As usual the $C^S{}_{ij}$ have been replaced by $T_{ij}(\mathbf{X})$, according to (1.13). Each side of (6.7) is symmetric in i, j, but not in r, s. Derivatives of T_{ij} with respect to ϵ_{kl} and ω_{kl} may be obtained by chain rule differentiation; e.g.

$$(\partial T_{ij}/\partial \epsilon_{kl}) = (\partial T_{ij}/\partial \alpha_{rs})(\partial \alpha_{rs}/\partial \epsilon_{kl}). \quad (6.8)$$

With the aid of (6.4) and (6.5), we find for adiabatic processes

$$(\partial T_{ij}/\partial \epsilon_{kl}) = B^S{}_{ijkl}; \quad (6.9)$$

$$B^S{}_{ijkl} = \tfrac{1}{2}(T_{ik}\delta_{jl} + T_{il}\delta_{jk} + T_{jk}\delta_{il} + T_{jl}\delta_{ik} - 2T_{ij}\delta_{kl}) + C^S{}_{ijkl}; \quad (6.10)$$

$$(\partial T_{ij}/\partial \omega_{kl}) = \tfrac{1}{2}(T_{il}\delta_{jk} - T_{ik}\delta_{jl} + T_{jl}\delta_{ik} - T_{jk}\delta_{il}). \quad (6.11)$$

The results (6.9)–(6.11) were previously derived by Wallace.[7] It follows from the derivation that for isothermal variations, $B^S{}_{ijkl}$ is replaced by $B^T{}_{ijkl}$ in (6.9), and $B^T{}_{ijkl}$ is defined by (6.10) with $C^S{}_{ijkl}$ replaced by $C^T{}_{ijkl}$. The symbol B_{ijkl} has been introduced because these coefficients are complete generalizations of the coefficients defined by Birch[42] for the special case of isothermal variations for a cubic material under initial isotropic pressure. The relation (6.10) displays the expected symmetry of B_{ijkl} in i, j and in k, l, although the B coefficients do not have complete Voigt symmetry since $B_{ijkl} \neq B_{klij}$ in general. According to (6.11), $(\partial T_{ij}/\partial \omega_{kl})$ has the expected symmetry in i, j and antisymmetry in k, l. In addition, (6.11) shows that even for infinitesimal elastic deformations from **X**, the stress depends on the rotation as well as on the pure strain. This is to be expected, since if the material undergoes a pure rotation, with no pure strain, then the initial stress must also be rotated to maintain zero pure strain. This circumstance has been pointed out by Truesdell and Noll.[43] There are of course special cases (e.g. initial isotropic pressure) when the stress does not depend on the rotation.

The stress–strain derivatives (6.9) and (6.11) are just the differential

---

[42] F. Birch, *Phys. Rev.* **71**, 809 (1947).
[43] Truesdell and Noll,[10] p. 250.

form of Hooke's law. In other words, a Taylor series expansion of the stress about the initial value may be written, for adiabatic or isothermal variations,

$$T_{ij}(\mathbf{x}, S) = T_{ij}(\mathbf{X}, S) + B^{S}{}_{ijkl}\epsilon_{kl} + (\partial T_{ij}/\partial \omega_{kl})\omega_{kl} + \cdots, \quad (6.12)$$

$$T_{ij}(\mathbf{x}, T) = T_{ij}(\mathbf{X}, T) + B^{T}{}_{ijkl}\epsilon_{kl} + (\partial T_{ij}/\partial \omega_{kl})\omega_{kl} + \cdots, \quad (6.13)$$

where $+ \cdots$ indicates terms of second and higher orders in the strain parameters.

It is useful to define the elastic compliances S_{ijkl} as coefficients of the tensor which is inverse to the tensor of B coefficients. Thus for adiabatic or isothermal coefficients,

$$B_{ijkl}S_{klmn} = S_{ijkl}B_{klmn} = \tfrac{1}{2}(\delta_{im}\delta_{jn} + \delta_{in}\delta_{jm}). \quad (6.14)$$

If the pure rotation is zero, the stress–strain derivatives (6.9) may be inverted to give

$$\begin{aligned}(\partial \epsilon_{ij}/\partial T_{kl}) &= S^{S}{}_{ijkl}, \quad \text{at constant } S; \\ (\partial \epsilon_{ij}/\partial T_{kl}) &= S^{T}{}_{ijkl}, \quad \text{at constant } T.\end{aligned} \quad (6.15)$$

Thus the compliances have their usual meaning as derivatives of the pure strain with respect to stress, evaluated at the arbitrary initial configuration. In general then the compliance tensor is inverse to the tensor of Birch coefficients B_{ijkl}, and not the tensor of elastic constants C_{ijkl}; only if the initial stress is zero, when $B_{ijkl} = C_{ijkl}$, is the compliance tensor inverse to the elastic constant tensor. Also the S_{ijkl}, like the B_{ijkl}, are symmetric in i, j and in k, l, but do not have complete Voigt symmetry since $S_{ijkl} \neq S_{klij}$ in general.

It is now quite simple to calculate the compressibility evaluated at the initial configuration. The adiabatic compressibility, e.g., is given by

$$K^{S} = -(1/V)(\partial V/\partial P)_{S}, \quad (6.16)$$

where V is the volume of the material and P is the pressure. We can use the equation of continuity (2.3) to write

$$J = \rho(\mathbf{X})/\rho(\mathbf{x}) = V(\mathbf{x})/V(\mathbf{X}), \quad (6.17)$$

so that K^{S} evaluated at $\mathbf{X}$ is just

$$K^{S} = -(\partial J/\partial P)_{S}. \quad (6.18)$$

Now Jacobi's identity (2.22), evaluated at $\mathbf{X}$, gives

$$(\partial J/\partial \alpha_{rs}) = \delta_{rs} \quad \text{at} \quad \mathbf{X}. \quad (6.19)$$

We transform (6.19) to derivatives with respect to ϵ_{ij} and ω_{ij}, with the help of (6.4) and (6.5), and find

$$(\partial J/\partial \epsilon_{ij}) = \delta_{ij}, \quad (\partial J/\partial \omega_{ij}) = 0, \quad \text{at } \mathbf{X}. \quad (6.20)$$

Therefore the compressibility (6.18) may be written

$$K^S = -(\partial J/\partial \epsilon_{ij})(\partial \epsilon_{ij}/\partial T_{kl})_S(\partial T_{kl}/\partial P). \quad (6.21)$$

If the pressure is varied while the nonisotropic part of the stress is maintained constant, the total stress varies according to

$$dT_{kl} = -dP\,\delta_{kl}, \quad \text{or} \quad (\partial T_{kl}/\partial P) = -\delta_{kl}. \quad (6.22)$$

We now have all the derivatives appearing in the compressibility (6.21); with the aid of (6.15), (6.20), and (6.22), the adiabatic compressibility is

$$K^S = -\delta_{ij}\,S^S{}_{ijkl}(-\delta_{kl})$$
$$= S^S{}_{iijj}. \quad (6.23)$$

Similarly the isothermal compressibility is

$$K^T = S^T{}_{iijj}. \quad (6.24)$$

Note there are nine terms in the sums in (6.23) and (6.24).

Although the B coefficients and S coefficients do not have complete Voigt symmetry in general, they do have enough symmetry to be written in Voigt notation. Thus all the independent B_{ijkl} are contained in the set $B_{\alpha\beta}$, and all the independent S_{ijkl} are contained in the set $S_{\alpha\beta}$, for $\alpha, \beta = 1, \ldots, 6$. However these 6×6 matrices are not symmetric in general:

$$B_{\alpha\beta} \neq B_{\beta\alpha}, \quad S_{\alpha\beta} \neq S_{\beta\alpha}, \quad \text{in general.} \quad (6.25)$$

Now we come to a subtle complication in our definitions. First, it is convenient to define each $B_{\alpha\beta}$ as equal to the corresponding B_{ijkl}; it is also convenient to define the compliances in Voigt notation, $S_{\alpha\beta}$, as elements of the matrix inverse to the matrix of $B_{\alpha\beta}$:

$$B_{\alpha\beta}S_{\beta\gamma} = S_{\alpha\beta}B_{\beta\gamma} = \delta_{\alpha\gamma}. \quad (6.26)$$

Once the set of B_{ijkl} or $B_{\alpha\beta}$ is given, then the S_{ijkl} are defined uniquely by the inverse relation (6.14), while the $S_{\alpha\beta}$ are defined uniquely by the inverse relation (6.26), and each resulting S_{ijkl} and corresponding $S_{\alpha\beta}$ are *not* equal.

We can relate $S_{\alpha\beta}$ and S_{ijkl} as follows. Write out the 6×6 matrix $[B_{\alpha\beta}]$ as

$$[B_{\alpha\beta}] = \begin{pmatrix} b_{11} & b_{12} \\ b_{21} & b_{22} \end{pmatrix}, \quad (6.27)$$

where in Voigt notation

$$b_{11} = \begin{pmatrix} B_{11} & B_{12} & B_{13} \\ B_{21} & B_{22} & B_{23} \\ B_{31} & B_{32} & B_{33} \end{pmatrix}, \quad b_{12} = \begin{pmatrix} B_{14} & B_{15} & B_{16} \\ B_{24} & B_{25} & B_{26} \\ B_{34} & B_{35} & B_{36} \end{pmatrix},$$

$$b_{21} = \begin{pmatrix} B_{41} & B_{42} & B_{43} \\ B_{51} & B_{52} & B_{53} \\ B_{61} & B_{62} & B_{63} \end{pmatrix}, \quad b_{22} = \begin{pmatrix} B_{44} & B_{45} & B_{46} \\ B_{54} & B_{55} & B_{56} \\ B_{64} & B_{65} & B_{66} \end{pmatrix}$$

(6.28)

Similarly, write out the 6×6 matrix $S_{\alpha\beta}$ as

$$[S_{\alpha\beta}] = \begin{pmatrix} s_{11} & s_{12} \\ s_{21} & s_{22} \end{pmatrix}, \tag{6.29}$$

where s_{11}, etc. are 3×3 matrices of $S_{\alpha\beta}$ in Voigt notation, just like the 3×3 matrices of $B_{\alpha\beta}$ in (6.28). Now, writing out the condition $B_{\alpha\beta}S_{\beta\gamma} = \delta_{\alpha\gamma}$ we get the four matrix equations

$$\begin{array}{ll} b_{11}s_{11} + b_{12}s_{21} = I, & b_{21}s_{11} + b_{22}s_{21} = 0, \\ b_{11}s_{12} + b_{12}s_{22} = 0, & b_{21}s_{12} + b_{22}s_{22} = I, \end{array} \tag{6.30}$$

where I is a 3×3 unit matrix, and the matrix products are inner products. These four equations determine uniquely the four matrices s_{11}, s_{12}, s_{21}, s_{22}, and hence determine uniquely the 6×6 matrix $[S_{\alpha\beta}]$ as inverse to $[B_{\alpha\beta}]$.

In the full notation, B_{ijkl} and S_{ijkl} form 9×9 matrices; by definition the B_{ijkl} are equal to the corresponding $B_{\alpha\beta}$, but the S_{ijkl} must be written as $g_{\alpha\beta}S_{\alpha\beta}$, where $g_{\alpha\beta}$ are numerical counting factors to be determined. Then with consistent interchanges of rows and columns, the inverse condition (6.14) may be written

$$\begin{pmatrix} b_{11} & b_{12} & b_{12} \\ b_{21} & b_{22} & b_{22} \\ b_{21} & b_{22} & b_{22} \end{pmatrix} \begin{pmatrix} g_{11}s_{11} & g_{12}s_{12} & g_{12}s_{12} \\ g_{21}s_{21} & g_{22}s_{22} & g_{22}s_{22} \\ g_{21}s_{21} & g_{22}s_{22} & g_{22}s_{22} \end{pmatrix} = \begin{pmatrix} I & 0 & 0 \\ 0 & \frac{1}{2}I & \frac{1}{2}I \\ 0 & \frac{1}{2}I & \frac{1}{2}I \end{pmatrix}. \tag{6.31}$$

Of the nine matrix equations obtained by writing out (6.31), only the following four are independent:

$$\begin{array}{ll} b_{11}g_{11}s_{11} + 2b_{12}g_{21}s_{21} = I, & b_{21}g_{11}s_{11} + 2b_{22}g_{21}s_{21} = 0, \\ b_{11}g_{12}s_{12} + 2b_{12}g_{22}s_{22} = 0, & b_{21}g_{12}s_{12} + 2b_{22}g_{22}s_{22} = \frac{1}{2}I. \end{array} \tag{6.32}$$

These four matrix equations are consistent with (6.30) if and only if

$$g_{11} = 1, \qquad g_{12} = g_{21} = \tfrac{1}{2}, \qquad g_{22} = \tfrac{1}{4}. \tag{6.33}$$

In other words, with $ij = \alpha$, $kl = \beta$, the relations between S_{ijkl} and $S_{\alpha\beta}$ are

$$S_{ijkl} = S_{\alpha\beta}; \qquad \alpha, \beta = 1, 2, 3. \tag{6.34a}$$

$$S_{ijkl} = \tfrac{1}{2} S_{\alpha\beta}; \qquad \alpha = 1, 2, 3, \quad \beta = 4, 5, 6 \quad \text{and} \quad \alpha = 4, 5, 6, \quad \beta = 1, 2, 3. \tag{6.34b}$$

$$S_{ijkl} = \tfrac{1}{4} S_{\alpha\beta}; \qquad \alpha, \beta = 4, 5, 6. \tag{6.34c}$$

These relations apply to either adiabatic or isothermal compliances.

The compressibility was expressed in terms of the compliances in (6.23) and (6.24). In Voigt notation for the compliances, K^S for example is

$$K^S = \sum_{\alpha=1}^{3} \sum_{\beta=1}^{3} S^S{}_{\alpha\beta}. \tag{6.35}$$

Therefore, in order to calculate the compressibility, it is only necessary to invert the 6×6 matrix $B^S{}_{\alpha\beta}$, or $B^T{}_{\alpha\beta}$.

Finally we wish to calculate the difference between adiabatic and isothermal Birch coefficients in the arbitrary initial configuration. We consider only symmetric strains and define the thermal strain tensor

$$\beta_{ij} = (\partial \epsilon_{ij}/\partial T)_{T_{ij}}, \tag{6.36}$$

the thermal stress tensor

$$\tau_{ij} = (\partial T_{ij}/\partial T)_{\epsilon_{ij}}, \tag{6.37}$$

and the heat capacity per unit mass at constant configuration

$$C = T(\partial S/\partial T)_{\epsilon_{ij}}. \tag{6.38}$$

These quantities are all evaluated at the initial configuration **X**. If the stress is taken as a function of T and ϵ_{ij}, the stress variation is

$$dT_{ij} = \tau_{ij}\, dT + B^T{}_{ijkl}\, d\epsilon_{kl}. \tag{6.39}$$

Dividing (6.39) by dT and taking processes at constant stress gives

$$\tau_{ij} = -B^T{}_{ijkl} \beta_{kl}. \tag{6.40}$$

Inversion of this relation gives

$$\beta_{ij} = -S^T{}_{ijkl} \tau_{kl}. \tag{6.41}$$

Now the zero-stress derivation of Callen[44] is easily extended to the

[44] H. B. Callen, "Thermodynamics." Wiley, New York, 1960.

present case of arbitrary initial stress. Divide (6.39) by $d\epsilon_{kl}$ and evaluate at constant S to obtain

$$B^S_{ijkl} = \tau_{ij}(\partial T/\partial \epsilon_{kl})_S + B^T_{ijkl}, \qquad (6.42)$$

where the definition (6.9) of B^S_{ijkl} has been used. The basic equations (2.1) for the dependent variables may be written

$$T = (\partial U/\partial S)_{\epsilon_{ij}}, \qquad T_{ij} = \rho_1(\partial U/\partial \epsilon_{ij})_S, \qquad \text{at } \mathbf{X};$$

these equations give the Maxwell relation

$$(\partial T_{ij}/\partial S)_{\epsilon_{ij}} = \rho_1(\partial T/\partial \epsilon_{ij})_S, \qquad \text{at } \mathbf{X}. \qquad (6.43)$$

By transforming variables, (6.43) may be written

$$(\partial T/\partial \epsilon_{ij})_S = (1/\rho_1)(\partial T_{ij}/\partial T)_{\epsilon_{ij}}(\partial T/\partial S)_{\epsilon_{ij}} = T\tau_{ij}/\rho_1 C. \qquad (6.44)$$

When (6.44) is used in (6.42), the difference between adiabatic and isothermal B coefficients is

$$B^S_{ijkl} - B^T_{ijkl} = (T/\rho_1 C)\tau_{ij}\tau_{kl}. \qquad (6.45)$$

It will be recalled from (6.10) that B^S_{ijkl} and C^S_{ijkl} differ only by a function of the stresses in the initial configuration; the same is true for B^T_{ijkl} and C^T_{ijkl}, and hence the adiabatic–isothermal differences are the same for the elastic constants as for the stress–strain coefficients. A similar result holds for the wave propagation coefficients, as noted in (3.8). Therefore, for any initial configuration,

$$C^S_{ijkl} - C^T_{ijkl} = A^S_{ijkl} - A^T_{ijkl} = B^S_{ijkl} - B^T_{ijkl}. \qquad (6.46)$$

7. Case of Initial Isotropic Pressure

When the initial stress is an isotropic pressure P, so that

$$T_{ij}(\mathbf{X}) = -P\,\delta_{ij}, \qquad (7.1)$$

there are certain simplifications of our general results which are worth noting. The propagation coefficients (3.6) and (3.7) reduce to

$$A_{ijkl} = -P\,\delta_{jl}\,\delta_{ik} + C_{ijkl}, \qquad (7.2)$$

for either adiabatic or isothermal coefficients. These A_{ijkl} still do not have complete Voigt symmetry unless P vanishes. The stress–strain relations (6.9)–(6.11), for either adiabatic or isothermal processes, reduce to

$$B_{ijkl} = -P(\delta_{jl}\,\delta_{ik} + \delta_{il}\,\delta_{jk} - \delta_{ij}\,\delta_{kl}) + C_{ijkl}, \qquad (7.3)$$

$$(\partial T_{ij}/\partial \omega_{kl}) = 0. \qquad (7.4)$$

Now the B_{ijkl} have complete Voigt symmetry, for arbitrary P. This means the compliances will also have complete Voigt symmetry, and we have

$$B_{\alpha\beta} = B_{\beta\alpha}, \qquad S_{\alpha\beta} = S_{\beta\alpha}, \qquad \text{arbitrary } P. \tag{7.5}$$

The result (7.4) is to be expected: the stress is invariant under pure rotation of the material since the stress itself is rotation-invariant. Alternately, a pure rotation of the material, in the presence of isotropic pressure, induces no pure strain.

The 6×6 matrix $[B_{\alpha\beta}]$ is not only symmetric, but has the same symmetry as the matrix $[C_{\alpha\beta}]$ for any of the crystal classes, including isotropic materials. To show this we write

$$[B_{\alpha\beta}] = [C_{\alpha\beta}] + \Delta, \tag{7.6}$$

and from (7.3) the matrix Δ is

$$\Delta = \left(\begin{array}{ccc|ccc} -P & P & P & & & \\ P & -P & P & & 0 & \\ P & P & -P & & & \\ \hline & & & -P & 0 & 0 \\ & 0 & & 0 & -P & 0 \\ & & & 0 & 0 & -P \end{array} \right). \tag{7.7}$$

Since Δ is of no lower symmetry than $[C_{\alpha\beta}]$ for any solid (see e.g. Section 11 for elastic constant symmetries), then the symmetry of $[B_{\alpha\beta}]$ is determined by that of $[C_{\alpha\beta}]$. This means the inversion of $[B_{\alpha\beta}]$ may be accomplished by the usual equations for the inverse of $[C_{\alpha\beta}]$ for arbitrary initial pressure. Inverses of $[C_{\alpha\beta}]$ for several crystal classes are given by Hearmon[45] and by Boas and Mackenzie.[46]

In small-amplitude wave propagation experiments, one measures wave velocities and hence the eigenvalues of the propagation matrices **L**, whose components are given by (3.13):

$$L_{ik} = A_{ijkl}K_jK_l. \tag{7.8}$$

Since the right-hand-side contains a sum over j and l, only the symmetric combination $(A_{ijkl} + A_{ilkj})$ contributes to the propagation matrix, and thus only this symmetric part is observed in wave propagation experiments. But for initial isotropic pressure, from (7.2) and (7.3), it is seen that the combination of A_{ijkl} symmetric in j, l is the same as the combination of

[45] R. F. S. Hearmon, *Rev. Mod. Phys.* **18**, 409 (1946).
[46] W. Boas and J. K. Mackenzie, *in* "Progress in Metal Physics" (B. Chalmers, ed.), Vol. 2, p. 90. Wiley (Interscience), New York, 1950.

B_{ijkl} symmetric in j, l:

$$A_{ijkl} + A_{ilkj} = B_{ijkl} + B_{ilkj}. \tag{7.9}$$

This means the coefficients measured in experiments on wave propagation in materials under isotropic pressure may be interpreted to be either A_{ijkl} or B_{ijkl}, since it is only the combination (7.9) which is observed. This special result for isotropic pressure was pointed out by Thurston[47] and by Barron and Klein.[48]

It is of interest to construct the propagation matrices from the elastic constants for the case of initial isotropic pressure. With A_{ijkl} expressed by (7.2), the L_{ik} of (7.8) become

$$L_{ik} = C_{ijkl}\kappa_j\kappa_l - P\,\delta_{ik}, \tag{7.10}$$

since $\kappa_j\kappa_l\,\delta_{jl} = \kappa_j\kappa_j = 1$. From the results of Section 3, e.g. from (3.15), the eigenvalue for a wave with polarization vector **w** can be written

$$\rho_1 v^2 = w_i(C_{ijkl}\kappa_j\kappa_l)w_k - P. \tag{7.11}$$

III. Expansions for Small Initial Stress

8. Expansion of Elastic Constants

In Part II the thermoelastic motion of a material about the initial (stressed) configuration was studied. In particular, the propagation of small amplitude waves and the variation of stress with strain was formulated in terms of the initial stress components and the second-order elastic constants evaluated at the initial configuration. Harmonic generation was found to depend also on the third-order elastic constants evaluated at the initial configuration. In most laboratory experiments the initial stress is small compared to the elastic constants, or equivalently, the Lagrangian strain parameters which measure the deformation from a configuration of zero stress to the initial stressed configuration are small compared to 1. For this reason it is convenient to express the various elastic coefficients of a material in the arbitrary initial configuration by means of Maclaurin expansions of these coefficients about a configuration of zero stress. In this way the results of Part II are easily and directly transformed to power series in the stress components, or in the strains from a zero-stress configuration, and we may study the stress- or strain-dependence of thermoelastic motion to any desired order. This procedure provides, e.g., a simple way of calculating the stress derivatives of the small-amplitude wave

---

[47] R. N. Thurston, *J. Acoust. Soc. Am.* **37**, 348 (1965).
[48] T. H. K. Barron and M. L. Klein, *Proc. Phys. Soc.* (*London*) **85**, 523 (1965).

velocities; these stress derivatives are in fact just combinations of third-order elastic constants.

To begin, the definitions of Section 1 are specialized to the case in which the initial configuration is the zero-stress configuration $\bar{\mathbf{X}}$. The position of a material particle in the zero-stress configuration is $\bar{\mathbf{X}}$, while the position of the same material particle in the final configuration is still $\mathbf{x}$. A superscript bar is used to denote strains measured from $\bar{\mathbf{X}}$:

$$\bar{\alpha}_{ij} = \partial x_i / \partial \bar{X}_j, \tag{8.1}$$

$$\bar{u}_{ij} = \partial \bar{u}_i / \partial \bar{X}_j, \quad \text{where} \quad \bar{u}_i = x_i - \bar{X}_i, \tag{8.2}$$

$$\bar{\eta}_{ij} = \tfrac{1}{2}(\bar{\alpha}_{ki}\bar{\alpha}_{kj} - \delta_{ij}) = \tfrac{1}{2}(\bar{u}_{ij} + \bar{u}_{ji} + \bar{u}_{ki}\bar{u}_{kj}). \tag{8.3}$$

The functional dependence of the state functions is

$$U(\mathbf{x}, S) = U(\bar{\eta}_{ij}, S), \quad F(\mathbf{x}, T) = F(\bar{\eta}_{ij}, T). \tag{8.4}$$

The density at $\bar{\mathbf{X}}$ is labeled ρ_0 for abbreviation; ρ_0 is a function only of S or T. In the expansions of U and F, the terms linear in $\bar{\eta}_{ij}$ vanish because the stress vanishes at $\bar{\eta}_{ij} = 0$. Up to fourth order in $\bar{\eta}_{ij}$, the internal energy, e.g., is

$$\rho_0 U(\bar{\eta}_{ij}, S) = \rho_0 U(0, S) + \tfrac{1}{2} \bar{C}^S{}_{ijkl} \bar{\eta}_{ij} \bar{\eta}_{kl}$$

$$+ (1/3!) \bar{C}^S{}_{ijklmn} \bar{\eta}_{ij} \bar{\eta}_{kl} \bar{\eta}_{mn}$$

$$+ (1/4!) \bar{C}^S{}_{ijklmnpq} \bar{\eta}_{ij} \bar{\eta}_{kl} \bar{\eta}_{mn} \bar{\eta}_{pq} + \cdots, \tag{8.5}$$

where the bar on the elastic constants means they are evaluated at $\bar{\mathbf{X}}$. These adiabatic elastic constants are derivatives of $\rho_0 U$ evaluated at zero stress, and are functions only of the entropy:

$$\bar{C}^S{}_{ijkl} = \rho_0 (\partial^2 U / \partial \bar{\eta}_{ij}\, \partial \bar{\eta}_{kl}),$$

$$\bar{C}^S{}_{ijklmn} = \rho_0 (\partial^3 U / \partial \bar{\eta}_{ij}\, \partial \bar{\eta}_{kl}\, \partial \bar{\eta}_{mn}), \tag{8.6}$$

$$\bar{C}^S{}_{ijklmnpq} = \rho_0 (\partial^4 U / \partial \bar{\eta}_{ij}\, \partial \bar{\eta}_{kl}\, \partial \bar{\eta}_{mn}\, \partial \bar{\eta}_{pq}).$$

The isothermal elastic constants are defined similarly in terms of derivatives of F. Again the elastic constants have complete Voigt symmetry.

Now we can express the elastic constants at $\mathbf{X}$ as a power series in the strains from $\bar{\mathbf{X}}$ to $\mathbf{X}$. At this point our notation becomes rather cumbersome. We are using α_{ij}, η_{ij} to denote strains about the stressed configuration $\mathbf{X}$, and $\bar{\alpha}_{ij}$, $\bar{\eta}_{ij}$ to denote strains about the zero-stress configuration $\bar{\mathbf{X}}$. In order to relate the properties at $\mathbf{X}$ to those at $\bar{\mathbf{X}}$, we need to specify the strain from $\bar{\mathbf{X}}$ to $\mathbf{X}$; these strains are just $\bar{\alpha}_{ij}$, $\bar{\eta}_{ij}$ evaluated at $\mathbf{X}$, but to avoid confusion on this we will introduce the new notation

$$a_{ij} = (\partial X_i/\partial \bar{X}_j) = \bar{\alpha}_{ij} \qquad \text{at } \mathbf{X}; \qquad (8.7)$$

$$n_{ij} = \tfrac{1}{2}(a_{ki}a_{kj} - \delta_{ij}) = \bar{\eta}_{ij} \qquad \text{at } \mathbf{X}. \qquad (8.8)$$

The relation between η_{ij}, defined by (1.5), and $\bar{\eta}_{ij}$, defined by (8.3), is given by

$$\bar{\eta}_{ij} + \tfrac{1}{2}\delta_{ij} = \tfrac{1}{2}(\partial x_k/\partial \bar{X}_i)(\partial x_k/\partial \bar{X}_j) = \tfrac{1}{2}(\partial x_k/\partial X_r)a_{ri}(\partial x_k/\partial X_s)a_{sj}$$

$$= (\eta_{rs} + \tfrac{1}{2}\delta_{rs})a_{ri}a_{sj}. \qquad (8.9)$$

From this relation follows

$$(\partial \bar{\eta}_{ij}/\partial \eta_{rs}) = a_{ri}a_{sj}. \qquad (8.10)$$

The adiabatic second-order elastic constants at $\mathbf{X}$ may now be transformed as follows:

$$C^S{}_{ijkl} = \rho_1(\partial^2 U/\partial \eta_{ij}\,\partial \eta_{kl}), \qquad \text{at } \mathbf{X}$$

$$= (\rho_1/\rho_0)a_{im}a_{jn}a_{kp}a_{lq}\rho_0(\partial^2 U/\partial \bar{\eta}_{mn}\,\partial \bar{\eta}_{pq}), \qquad \text{at } \mathbf{X}$$

$$= (\rho_1/\rho_0)a_{im}a_{jn}a_{kp}a_{lq}[\bar{C}^S{}_{mnpq} + \bar{C}^S{}_{mnpqrs}n_{rs}$$

$$+ \tfrac{1}{2}\bar{C}^S{}_{mnpqrstu}n_{rs}n_{tu} + \cdots]. \qquad (8.11)$$

The expansion (8.5) of $\rho_0 U$ has been used to obtain the last form in (8.11), and the expression has been evaluated at $\mathbf{X}$ by setting $\bar{\eta}_{ij} = n_{ij}$. Exactly the same procedure may be followed to express the third-order elastic constants at $\mathbf{X}$, with the result

$$C^S{}_{ijklmn} = (\rho_1/\rho_0)a_{ip}a_{jq}a_{kr}a_{ls}a_{mt}a_{nu}$$

$$\times [\bar{C}^S{}_{pqrstu} + \bar{C}^S{}_{pqrstuvw}n_{vw} + \cdots]. \qquad (8.12)$$

Equations (8.11) and (8.12) provide general expressions for $C^S{}_{ijkl}$ and $C^S{}_{ijklmn}$ evaluated at $\mathbf{X}$, in terms of adiabatic second-, third-, and fourth-order elastic constants evaluated at $\bar{\mathbf{X}}$, and the parameters which measure the deformation from $\bar{\mathbf{X}}$ to $\mathbf{X}$, namely (ρ_1/ρ_0), a_{ij}, and n_{ij}. Similar expansions hold for the isothermal constants $C^T{}_{ijkl}$ and $C^T{}_{ijklmn}$ at $\mathbf{X}$. An important point needs to be emphasized here. In deriving (8.11) and (8.12), all differentiations were carried out at constant S. This implies that (ρ_1/ρ_0), a_{ij}, and n_{ij} in these equations must be evaluated at constant S. Likewise, if (8.11) and (8.12) are used to relate isothermal constants, then the parameters measuring strain from $\bar{\mathbf{X}}$ to $\mathbf{X}$ must be evaluated for a constant T strain.

We have derived (8.11) previously,[7] and this proves to be a convenient expansion for our subsequent work. It is also of value to eliminate the parameters (ρ_1/ρ_0) and a_{ij} in (8.11) in favor of the Lagrangian strains n_{ij}. We will do this for the case where the strain from $\bar{\mathbf{X}}$ to $\mathbf{X}$ is symmetric, i.e. no rotations. It will be useful for later work to go back to the general strains (8.1)–(8.3) and derive some results for the symmetric strain case:

$$\bar{u}_{ij} = \bar{u}_{ji}, \quad \text{symmetric strains;} \tag{8.13}$$

$$\bar{\eta}_{ij} = \bar{u}_{ij} + \tfrac{1}{2}\bar{u}_{ki}\bar{u}_{kj}; \tag{8.14}$$

$$\bar{\alpha}_{ij} = \delta_{ij} + \bar{u}_{ij} = \delta_{ij} + \bar{\eta}_{ij} - \tfrac{1}{2}\bar{\eta}_{ki}\bar{\eta}_{kj} + \cdots. \tag{8.15}$$

The Jacobian of the strain will be needed only to first order in the strains, and is

$$J = (\rho_0/\rho) = \det[\bar{\alpha}_{ij}] = 1 + \bar{\eta}_{ii} + \cdots, \tag{8.16}$$

so that

$$(\rho/\rho_0) = 1 - \bar{\eta}_{ii} + \cdots. \tag{8.17}$$

The strains from $\bar{\mathbf{X}}$ to $\mathbf{X}$ can all be expressed in terms of n_{ij} by evaluating (8.15) and (8.17) at $\mathbf{X}$ to give

$$a_{ij} = \delta_{ij} + n_{ij} - \tfrac{1}{2}n_{ki}n_{kj} + \cdots, \tag{8.18}$$

$$(\rho_1/\rho_0) = 1 - n_{ii} + \cdots. \tag{8.19}$$

With these results, the expansion (8.11) for C^S_{ijkl} becomes

$$C^S_{ijkl} = \bar{C}^S_{ijkl} + [-\bar{C}^S_{ijkl}n_{pp} + \bar{C}^S_{ijkp}n_{lp} + \bar{C}^S_{ijpl}n_{kp}$$
$$+ \bar{C}^S_{ipkl}n_{jp} + \bar{C}^S_{pjkl}n_{ip} + \bar{C}^S_{ijklpq}n_{pq}] + \cdots. \tag{8.20}$$

9. Expansion of Strains and Propagation Matrices

Consider now the strains $\bar{\eta}_{ij}$ measured from the zero-stress configuration $\bar{\mathbf{X}}$ to the arbitrary final configuration $\mathbf{x}$. The relation of the stresses $T_{ij}(\mathbf{x})$ to these strains is given by the basic equations (2.1) and (2.2) as

$$T_{ij} = \rho\bar{\alpha}_{ik}\bar{\alpha}_{jl}(\partial U/\partial\bar{\eta}_{kl}) = \rho\bar{\alpha}_{ik}\bar{\alpha}_{jl}(\partial F/\partial\bar{\eta}_{kl}). \tag{9.1}$$

With the results of Section 8 this may be expanded into a power series in $\bar{\eta}_{ij}$, and the series may be inverted to give the strains as power series in the stresses. The calculation will be illustrated for adiabatic processes; (9.1) may be rewritten as

$$T_{ij} = (\rho/\rho_0)\bar{\alpha}_{ik}\bar{\alpha}_{jl}\rho_0(\partial U/\partial\bar{\eta}_{kl}). \tag{9.2}$$

The expansions of (ρ/ρ_0) and $\bar{\alpha}_{ij}$ are given by (8.17) and (8.15), respectively, and from the energy expansion (8.5) we have

$$\rho_0(\partial U/\partial\bar{\eta}_{kl}) = \bar{C}^S_{klmn}\bar{\eta}_{mn} + \tfrac{1}{2}\bar{C}^S_{klmnpq}\bar{\eta}_{mn}\bar{\eta}_{pq} + \cdots. \tag{9.3}$$

Then, to second order in $\bar{\eta}_{ij}$, the stress expansion is calculated from (9.2) as

$$T_{ij} = \bar{C}^S{}_{ijmn}\bar{\eta}_{mn} + [\bar{C}^S{}_{ipmn}\bar{\eta}_{mn}\bar{\eta}_{jp} + \bar{C}^S{}_{pjmn}\bar{\eta}_{mn}\bar{\eta}_{ip}$$
$$- \bar{C}^S{}_{ijmn}\bar{\eta}_{mn}\bar{\eta}_{pp} + \tfrac{1}{2}\bar{C}^S{}_{ijmnpq}\bar{\eta}_{mn}\bar{\eta}_{pq}]. \qquad (9.4)$$

This is the general expansion of the stress at **x**, to second order in the adiabatic strains measured from the zero-stress configuration to **x**.

At zero stress, the stress–strain coefficients of Section 6 are equal to the second-order elastic constants:

$$\bar{B}^S{}_{ijkl} = \bar{C}^S{}_{ijkl}; \qquad \bar{B}^T{}_{ijkl} = \bar{C}^T{}_{ijkl}. \qquad (9.5)$$

Therefore the compliance tensor at zero stress is inverse to the elastic constant tensor:

$$\bar{C}^S{}_{ijkl}\bar{S}^S{}_{klmn} = \bar{S}^S{}_{ijkl}\bar{C}^S{}_{klmn} = \tfrac{1}{2}(\delta_{im}\delta_{jn} + \delta_{in}\delta_{jm}). \qquad (9.6)$$

Multiplying the stress expansion (9.4) on the left by $\bar{S}^S{}_{klij}$ and summing over i,j gives

$$\bar{S}^S{}_{klij}T_{ij} = \bar{\eta}_{kl} + \bar{S}^S{}_{klij}[\bar{C}^S{}_{ipmn}\bar{\eta}_{mn}\bar{\eta}_{jp} + \bar{C}^S{}_{pjmn}\bar{\eta}_{mn}\bar{\eta}_{ip}$$
$$- \bar{C}^S{}_{ijmn}\bar{\eta}_{mn}\bar{\eta}_{pp} + \tfrac{1}{2}\bar{C}^S{}_{ijmnpq}\bar{\eta}_{mn}\bar{\eta}_{pq}], \qquad (9.7)$$

or with the aid of the Voigt symmetries,

$$\bar{\eta}_{kl} = \bar{S}^S{}_{klij}T_{ij} + \bar{S}^S{}_{klij}[\bar{C}^S{}_{ijmn}\bar{\eta}_{mn}\bar{\eta}_{pp} - 2\bar{C}^S{}_{ipmn}\bar{\eta}_{mn}\bar{\eta}_{jp}$$
$$- \tfrac{1}{2}\bar{C}^S{}_{ijmnpq}\bar{\eta}_{mn}\bar{\eta}_{pq}]. \qquad (9.8)$$

To convert the right-hand side to a power series in the stresses, correct to second order, it is only necessary to replace the $\bar{\eta}_{ij}$ with the leading term linear in the stresses

$$\bar{\eta}_{kl} = \bar{S}^S{}_{klij}T_{ij} + \cdots;$$

the result is

$$\bar{\eta}_{ij} = \bar{S}^S{}_{ijkl}T_{kl} + [\bar{S}^S{}_{ijkl}\bar{S}^S{}_{ppmn} - 2\bar{S}^S{}_{ijkp}\bar{S}^S{}_{plmn}$$
$$- \tfrac{1}{2}\bar{S}^S{}_{ijpq}\bar{C}^S{}_{pqrstu}\bar{S}^S{}_{rskl}\bar{S}^S{}_{tumn}]T_{kl}T_{mn}. \qquad (9.9)$$

In (9.9) both sides are evaluated at **x** and the process is presumed to be adiabatic. The same expansion holds for isothermal processes, with all elastic constants and compliances replaced by isothermal coefficients. This result is very useful since it allows the calculation of strains for an arbitrary solid due to the application of an arbitrary stress, to second order in the stress. The terms of third order in stresses, omitted above, will contain fourth-order elastic constants. The special case of (9.9) evaluated at **X** is

$$n_{ij} = \bar{S}^S{}_{ijkl}T_{kl}(\mathbf{X}) + [ijklmn]T_{kl}(\mathbf{X})T_{mn}(\mathbf{X}), \qquad (9.10)$$

where $[ijklmn]$ is exactly the same as the bracket in (9.9).

To prepare for the interpretation of wave propagation experiments in stressed crystals, we should examine more closely what the experimentalist measures. The directly measured quantity is either the transit time t required for the wave to transit the sample once, or the frequency f of repetition of echoes. Let l_1 be the length of the sample in the stressed configuration, measured along the propagation direction κ, so that these experimental quantities are given by

$$t = l_1/v; \tag{9.11}$$

$$f = (1/2t) = (v/2l_1). \tag{9.12}$$

In these quantities both l_1 and the wave velocity v vary with the applied stress. It is convenient to eliminate the dependence on l_1 by studying not the velocity but the quantity W introduced by Thurston and Brugger,[49] and defined by

$$W = (l_0/l_1)v, \tag{9.13}$$

where l_0 is the length of the sample along the propagation direction at zero stress. In addition, instead of studying the eigenvalues $\rho_1 v^2$ of the propagation matrices, where ρ_1 also varies with the applied stress, it is convenient to calculate $\rho_0 W^2$. From (9.13) we can write

$$\rho_0 W^2 = (\rho_0/\rho_1)(l_0/l_1)^2 (\rho_1 v^2). \tag{9.14}$$

We therefore define the reduced propagation matrices **M** as the matrices whose eigenvalues are $\rho_0 W^2$:

$$M_{ik} = (\rho_0/\rho_1)(l_0/l_1)^2 L_{ik}, \tag{9.15}$$

where the L_{ik} are given by (3.13). For each propagation direction and initial stress, **M** is just a number times **L**; therefore **M** has the same eigenvectors as **L**. The eigenvalue problem in diagonal form is then

$$\rho_0 W^2 = w_i M_{ik} w_k, \tag{9.16}$$

and the experimental quantities at any initial stress are simply related to W by

$$W = 2l_0 f = l_0/t. \tag{9.17}$$

When stress is applied to the solid, to go from the configuration $\bar{\mathbf{X}}$ to $\mathbf{X}$, the wave propagation direction κ will also change. We presume the surfaces of constant phase remain parallel to the sample surface at which the input wave is generated, and calculate the change of κ with the strain, for the coordinate axes fixed in space. Let $\bar{\mathbf{X}}$ be a plane of constant phase in the zero-stress configuration, and let $\bar{\kappa} \cdot \bar{\mathbf{X}}$ be the perpendicular distance from

---

[49] R. N. Thurston and K. Brugger, *Phys. Rev.* **133**, A1604 (1964).

the input face to this plane, where $\bar{\kappa}$ is the propagation direction at zero stress. In the stressed configuration the distance to this same material plane is $\kappa \cdot \mathbf{X}$, and is changed in the ratio (l_1/l_0); that is,

$$\kappa \cdot \mathbf{X} = \kappa_i X_i = (l_1/l_0) \bar{\kappa} \cdot \bar{\mathbf{X}} = (l_1/l_0) \bar{\kappa}_j \bar{X}_j. \tag{9.18}$$

We can relate $\bar{\kappa}$ and κ through the strain from $\bar{\mathbf{X}}$ to $\mathbf{X}$ as follows:

$$\kappa_i X_i = \kappa_i (\partial X_i/\partial \bar{X}_j) \bar{X}_j = \kappa_i a_{ij} \bar{X}_j$$

from (8.7) for a_{ij}, or by comparing with (9.18),

$$\kappa_i a_{ij} = (l_1/l_0) \bar{\kappa}_j. \tag{9.19}$$

Now define $[b_{ij}]$ as the matrix inverse to $[a_{ij}]$,

$$a_{ij} b_{jk} = \delta_{ik}, \tag{9.20}$$

and solve (9.19) for κ_i to get

$$\kappa_i = (l_1/l_0) \bar{\kappa}_j b_{ji}. \tag{9.21}$$

This is the expression for the unit propagation vector κ in the configuration $\mathbf{X}$, in terms of the unit propagation vector $\bar{\kappa}$ at $\bar{\mathbf{X}}$ and the strain from $\bar{\mathbf{X}}$ to $\mathbf{X}$.

We are now ready to expand the reduced propagation matrices $\mathbf{M}$ about the configuration of zero stress. From (9.15) for $\mathbf{M}$ and (3.13) for $\mathbf{L}$, the elements of $\mathbf{M}$ evaluated at the stressed configuration are

$$M_{ik} = (\rho_0/\rho_1)(l_0/l_1)^2 A_{ijkl} \kappa_j \kappa_l, \tag{9.22}$$

for either adiabatic or isothermal cases, with

$$A_{ijkl} = C_{ijkl} + T_{jl}(\mathbf{X}) \delta_{ik}. \tag{9.23}$$

It is convenient to separate $\mathbf{M}$ into two contributions

$$M_{ik} = M_{ik}^{(1)} + M_{ik}^{(2)}, \tag{9.24}$$

where $M_{ik}^{(1)}$ is the term involving C_{ijkl}, and $M_{ik}^{(2)}$ the term involving T_{jl}, from (9.23). The first term is written out with the aid of the expansion (8.11) for the elastic constants at stress, and (9.21) for κ_i at stress, to give

$$M_{ik}^{(1)} = a_{im} a_{jn} a_{kp} a_{lq} [\bar{C}_{mnpq} + \cdots] \bar{\kappa}_r \bar{\kappa}_s b_{rj} b_{sl},$$
$$= a_{im} a_{kp} [\bar{C}_{mnpq} + \bar{C}_{mnpqrs} \eta_{rs} + \cdots] \bar{\kappa}_n \bar{\kappa}_q, \tag{9.25}$$

since $b_{rj} a_{jn} = \delta_{rn}$, etc. The second-order elastic constants appear in (9.25) in the form $\bar{C}_{mnpq} \bar{\kappa}_n \bar{\kappa}_q$, and this quantity can be replaced by $\bar{M}_{mp}$, since evaluation of (9.22) at zero stress gives

$$\bar{M}_{ik} = \bar{C}_{ijkl} \bar{\kappa}_j \bar{\kappa}_l. \tag{9.26}$$

The strain parameters a_{ij} are expressed in terms of n_{ij} to second order in (8.18); after a little algebra (9.25) is, to second order in n_{ij},

$$M_{ik}{}^{(1)} = \bar{M}_{ik} + [\bar{M}_{ip}n_{kp} + \bar{M}_{kp}n_{ip} + \bar{C}_{inkqrs}\bar{\kappa}_n\bar{\kappa}_q n_{rs}]$$
$$+ [\bar{M}_{pq}n_{ip}n_{kq} - \tfrac{1}{2}\bar{M}_{ip}n_{sk}n_{sp} - \tfrac{1}{2}\bar{M}_{kp}n_{si}n_{sp} + (\bar{C}_{inpqrs}n_{rs}n_{kp}$$
$$+ \bar{C}_{pnkqrs}n_{rs}n_{ip} + \tfrac{1}{2}\bar{C}_{inkqrs\,tu}n_{rs}n_{tu})\bar{\kappa}_n\bar{\kappa}_q]. \quad (9.27)$$

The second contribution is written with the aid of (9.21) for κ_i at stress:

$$M_{ik}{}^{(2)} = (\rho_0/\rho_1)T_{jl}(\mathbf{X})\,\delta_{ik}\,\bar{\kappa}_r\bar{\kappa}_s b_{rj}b_{sl}. \quad (9.28)$$

Since this term is of lowest order one in the stress, we only need the strains (ρ_0/ρ_1) and b_{ij} to first order in the n_{ij}. Inverting (8.19) gives

$$(\rho_0/\rho_1) = 1 + n_{ii} + \cdots, \quad (9.29)$$

and it is easily verified that

$$b_{ij} = \delta_{ij} - n_{ij} + \cdots \quad (9.30)$$

satisfies $b_{ij}a_{jk} = \delta_{ik}$ to first order in n_{ij}. With these results, (9.28) is

$$M_{ik}{}^{(2)} = T_{jl}(\mathbf{X})\,\delta_{ik}\,[\bar{\kappa}_j\bar{\kappa}_l(1 + n_{pp}) - \bar{\kappa}_j\bar{\kappa}_p n_{pl} - \bar{\kappa}_p\bar{\kappa}_l n_{pj} + \cdots]. \quad (9.31)$$

Eliminating the stresses gives

$$M_{ik}{}^{(2)} = \delta_{ik}\bar{C}_{jlmn}n_{mn}[\bar{\kappa}_j\bar{\kappa}_l(1 + n_{pp}) - \bar{\kappa}_j\bar{\kappa}_p n_{pl} - \bar{\kappa}_p\bar{\kappa}_l n_{pj}]. \quad (9.32)$$

The above equations give the reduced propagation matrices **M** at any configuration **X** as a power series in the strains n_{ij} from the zero-stress configuration $\bar{\mathbf{X}}$ to **X**. These equations are valid for adiabatic propagation when the adiabatic elastic constants are used and the strains are evaluated along an adiabatic line; similarly for isothermal propagation. The strains can be eliminated in favor of the stresses at **X** with (9.10). The matrices **M** are convenient for analysis of experimental data since their eigenvalues are $\rho_0 W^2$. In a general problem the eigenvalues of **M** may be calculated to second order in the strains n_{ij}, or the corresponding stresses $T_{ij}(\mathbf{X})$, by using second-order perturbation theory. The eigenvalues so calculated are then directly comparable to measurements of $\rho_0 W^2$ as a function of strain, or stress, and are functions of second-, third-, and fourth-order elastic constants. We will illustrate these calculations in Part VI for a few simple cases where the eigenvectors of **M** do not change with the applied stress, and hence the second-order perturbation calculation is very much simplified.

10. Stress Derivatives of the Transit Time

In recent years the experiment of measuring the stress derivatives of ultrasonic wave velocities, evaluated at zero stress, has been well developed.

The equations required for analyzing this experiment are obtained from a first-order perturbation treatment of the results of Section 9. To first order in the stresses and strains, and for adiabatic or isothermal propagation, the reduced propagation matrices **M** are taken from (9.27) and (9.31) as

$$M_{ik} = \bar{M}_{ik} + \delta M_{ik}, \tag{10.1}$$

$$\delta M_{ik} = \bar{M}_{ip}n_{kp} + \bar{M}_{kp}n_{ip} + \bar{C}_{ijklpq}\bar{\kappa}_j\bar{\kappa}_l n_{pq} + \delta_{ik} T_{jl}(\mathbf{X}) \bar{\kappa}_j\bar{\kappa}_l. \tag{10.2}$$

Let $\bar{\mathbf{w}}$ be an eigenvector of $\bar{\mathbf{M}}$, and $\rho_0 \bar{W}^2$ the corresponding eigenvalue, so that

$$\rho_0 \bar{W}^2 = \bar{w}_i \bar{M}_{ik} \bar{w}_k. \tag{10.3}$$

Then to first order in the stresses and strains, the correction to the eigenvalue is

$$\delta(\rho_0 W^2) = \bar{w}_i \, \delta M_{ik} \, \bar{w}_k. \tag{10.4}$$

To evaluate (10.4) we note the zero-stress eigenvalue–eigenvector equation (10.3) can also be written in the form

$$\rho_0 \bar{W}^2 \bar{w}_i = \bar{M}_{ik} \bar{w}_k = \bar{w}_k \bar{M}_{ki}, \tag{10.5}$$

and

$$\bar{w}_i \bar{w}_i = 1. \tag{10.6}$$

The terms in (10.4) are then evaluated as follows:

$$\bar{w}_i \bar{M}_{ip} n_{kp} \bar{w}_k = \rho_0 \bar{W}^2 \bar{w}_p n_{kp} \bar{w}_k; \tag{10.7}$$

$$\bar{w}_i \, \delta_{ik} \, T_{jl}(\mathbf{X}) \, \bar{\kappa}_j \bar{\kappa}_l \bar{w}_k = T_{jl}(\mathbf{X}) \, \bar{\kappa}_j \bar{\kappa}_l. \tag{10.8}$$

With these results, (10.4) becomes

$$\delta(\rho_0 W^2) = 2\rho_0 \bar{W}^2 n_{ij} \bar{w}_i \bar{w}_j + \bar{C}_{ijklpq} \bar{\kappa}_j \bar{\kappa}_l \bar{w}_i \bar{w}_k n_{pq} + T_{jl}(\mathbf{X}) \bar{\kappa}_j \bar{\kappa}_l; \tag{10.9}$$

or, eliminating the strains with the aid of (9.10),

$$\delta(\rho_0 W^2) = [\bar{\kappa}_p \bar{\kappa}_q + 2\rho_0 \bar{W}^2 \bar{S}_{ijpq} \bar{w}_i \bar{w}_j + \bar{C}_{ijklmn} \bar{S}_{mnpq} \bar{\kappa}_j \bar{\kappa}_l \bar{w}_i \bar{w}_k] T_{pq}(\mathbf{X}). \tag{10.10}$$

This is a completely general first-order perturbation result, and is valid for any solid, any zero-stress eigenvector, and any applied stress no matter how the symmetry of the material changes with the stress. In addition, this result gives exactly the stress derivative of the eigenvalue, evaluated at zero stress:

$$\partial(\rho_0 W^2)/\partial T_{pq} = \bar{\kappa}_p \bar{\kappa}_q + 2\rho_0 \bar{W}^2 \bar{S}_{ijpq} \bar{w}_i \bar{w}_j + \bar{C}_{ijklmn} \bar{S}_{mnpq} \bar{\kappa}_j \bar{\kappa}_l \bar{w}_i \bar{w}_k. \tag{10.11}$$

If the wave propagation is adiabatic *and* the stress is applied to the sample adiabatically, then (10.11) is obviously

$$[\partial(\rho_0 W^2)_S/\partial T_{pq}]_S = \bar{\kappa}_p \bar{\kappa}_q + 2(\rho_0 \bar{W}^2)_S \bar{S}^S{}_{ijpq} \bar{w}_i \bar{w}_j$$
$$+ \bar{C}^S{}_{ijklmn} \bar{S}^S{}_{mnpq} \bar{\kappa}_j \bar{\kappa}_l \bar{w}_i \bar{w}_k. \qquad (10.12)$$

A similar equation holds for all-isothermal processes. However, in the usual experimental arrangement, the wave propagation is adiabatic but the stress is applied isothermally; keeping this process in mind through the above derivation leads to

$$[\partial(\rho_0 W^2)_S/\partial T_{pq}]_T = \bar{\kappa}_p \bar{\kappa}_q + 2(\rho_0 \bar{W}^2)_S \bar{S}^T{}_{ijpq} \bar{w}_i \bar{w}_j$$
$$+ \bar{C}^*{}_{ijklmn} \bar{S}^T{}_{mnpq} \bar{\kappa}_j \bar{\kappa}_l \bar{w}_i \bar{w}_k, \qquad (10.13)$$

where the "mixed" third-order elastic constant $\bar{C}^*{}_{ijklmn}$ is defined by[49a]

$$\bar{C}^*{}_{ijklmn} = \rho_0 [\partial(\partial^2 U/\partial \bar{\eta}_{ij}\,\partial \bar{\eta}_{kl})_S/\partial \bar{\eta}_{mn}]_T, \quad \text{at} \quad \bar{\mathbf{X}}. \qquad (10.14)$$

For most solids the differences between $\bar{C}^T{}_{ijklmn}$, $\bar{C}^*{}_{ijklmn}$, and $\bar{C}^S{}_{ijklmn}$ are small except at very high temperatures. The procedure for calculating isothermal and adiabatic constants from the mixed constants is illustrated for an example in Section 16. The appearance of mixed third-order elastic constants in (10.13) was recognized by Thurston and Brugger,[49] who also derived that equation.

For problems dealing with thermoelastic behavior of materials in the presence of arbitrary initial stress, or for evaluation of the fourth-order elastic constants at zero stress, it is useful to extend the above results to the case of arbitrary initial stress. Suppose an elastic material is under an initial stress $T_{ij}(\mathbf{X})$, and a given small amplitude wave has propagation direction κ, eigenvector $\mathbf{w}$, and eigenvalue $\rho_1 W^2$, in the initial configuration. Then if a stress δT_{ij} is added to the stress system, the eigenvalue changes by an amount $\delta(\rho_1 W^2)$ in first order, where

$$\delta(\rho_1 W^2) = [\kappa_p \kappa_q + 2\rho_1 W^2 S_{ijpq} w_i w_j + C_{ijklmn} S_{mnpq} \kappa_j \kappa_l w_i w_k] \delta T_{pq}$$
$$+ T_{jl}(\mathbf{X}) [(S_{mmpq} - 2 S_{mnpq} w_m w_n) \kappa_j \kappa_l$$
$$- S_{mlpq} \kappa_j \kappa_m - S_{mjpq} \kappa_l \kappa_m] \delta T_{pq}. \qquad (10.15)$$

Note if $T_{jl}(\mathbf{X})$ is taken to be zero, the zero-stress result (10.10) is recovered. We shall have occasion to use this result in the analysis of experiments which measure fourth-order elastic constants in Part VI.

[49a] If one is concerned with the several different types of mixed higher-order elastic constants, a more descriptive notation is in order, such as $\bar{C}^{ST}{}_{ijklmn}$ for the quantities defined in (10.14).

At zero temperature the elastic wave velocities are equal to the long-wavelength acoustic phonon velocities, for arbitrary configuration of the crystal, or equivalently for arbitrary initial stress.[49b] This means the long-wavelength acoustic phonons are simply quantized elastic waves, i.e. elastic waves whose wave vectors $\mathbf{k}$ are quantized according to periodic boundary conditions. For a given $\mathbf{k}$, the long-wavelength acoustic frequency is

$$\omega_{\mathbf{k}} = v \, | \, \mathbf{k} \, |, \qquad (10.16)$$

where v is the elastic wave velocity; note for each $\mathbf{k}$ there are three polarizations for the elastic waves and the corresponding phonons (e.g. one longitudinal and two transverse in the pure mode directions). Since $\mathbf{k}$ is quantized, for an arbitrary initial configuration $|\mathbf{k}|$ is proportional to l_1^{-1}, where l_1 is the length of the material along the direction of $\mathbf{k}$; therefore from (9.13) we can write

$$\omega_{\mathbf{k}} \propto W/l_0. \qquad (10.17)$$

The generalized Gruneisen parameters are strain derivatives of the phonon frequencies, defined by

$$\gamma_{\mathbf{k},ij} = -d \ln \omega_{\mathbf{k}}/d\eta_{ij}. \qquad (10.18)$$

Therefore the Grüneisen parameters for the long-wavelength acoustic phonons at $T = 0$ can be calculated from the strain derivatives of the reduced propagation velocities at $T = 0$. At finite temperatures this is still a good approximation, the error being of the order of the differences in the strain derivatives of the adiabatic and isothermal velocities (see Section 22).

In view of (10.17), the Grüneisen parameters may be written

$$\gamma_{\mathbf{k},ij} = -\tfrac{1}{2} d \ln \rho_0 W^2 / d\eta_{ij}. \qquad (10.19)$$

The first-order variation of $\rho_0 W^2$ with the strains n_{ij} measured from the zero-stress configuration is obtained by eliminating the stresses from (10.9):

$$\delta(\rho_0 W^2) = [2\rho_0 \bar{W}^2 \bar{w}_p \bar{w}_q + (\bar{C}_{jlpq} + \bar{C}_{ijklpq}\bar{w}_i\bar{w}_k)\bar{\kappa}_j\bar{\kappa}_l] n_{pq}. \qquad (10.20)$$

Hence the Grüneisen parameters (10.19) evaluated at zero stress are

$$\bar{\gamma}_{\mathbf{k},pq} = -(1/2\rho_0 \bar{W}^2)[2\rho_0 \bar{W}^2 \bar{w}_p \bar{w}_q + (\bar{C}_{jlpq} + \bar{C}_{ijklpq}\bar{w}_i\bar{w}_k)\bar{\kappa}_j\bar{\kappa}_l]. \qquad (10.21)$$

When only isotropic pressure is considered, the appropriate Grüneisen parameters are the volume derivatives of the phonon frequencies, and may

---

[49b] This is true to a high order in lattice dynamics perturbation theory, and might be true to all orders; see Section 22.

be calculated as follows:

$$\gamma_k = -d \ln \omega_k / d \ln V,$$
$$= -\tfrac{1}{2} d \ln \rho_0 W^2 / d \ln V,$$
$$= \tfrac{1}{2} B \, d \ln \rho_0 W^2 / dP, \qquad (10.22)$$

where B is the bulk modulus. With the stress given by $T_{pq} = -P\,\delta_{pq}$, (10.22) is evaluated at zero stress with the aid of (10.11) to give

$$\bar{\gamma}_k = -(B/2\rho_0 \bar{W}^2)[1 + 2\rho_0 \bar{W}^2 \bar{S}_{ijpp}\bar{w}_i\bar{w}_j + \bar{C}_{ijklmn}\bar{S}_{mnpp}\bar{k}_j\bar{k}_l\bar{w}_i\bar{w}_k]. \qquad (10.23)$$

Again (10.23) is valid for long-wavelength acoustic phonons, and is correct at $T = 0$ and approximate at finite T. Brugger[49c] has assumed the phonon velocities and elastic wave velocities are equal at arbitrary temperature, and has defined temperature-dependent Grüneisen parameters and derived equations similar to (10.21) and (10.23). We prefer not to make this approximation since in the theory of lattice dynamics the errors involved are of the same order as the temperature dependence itself.

IV. Crystal Symmetry Properties

11. Schemes of Independent Elastic Constants

The total number of second-order elastic constants in Voigt notation is 36, and they may be arranged in a 6 × 6 matrix. This matrix is symmetric, however, so the maximum number of *independent* second-order elastic constants is 21. Because of the crystal symmetry for a given material, certain of these 21 second-order elastic constants may be equal, or otherwise simply related, since they represent derivatives of the thermodynamic state functions with respect to strains which are crystallographically equivalent. Thus the crystal symmetry may further reduce the number of independent second-order elastic constants. These observations apply also to the set of third-order elastic constants, the set of fourth-order, and so on.

The crystal symmetry properties apply to adiabatic and isothermal constants alike, so the notation of constant S or constant T will generally be dropped in the present Part IV. It is most convenient to analyze the effect of crystal symmetry on the elastic constants, since these constants are most simply related to the crystal symmetry. In fact the elastic constants are not explicitly dependent on the initial stress, so our discussion will be valid for stressed crystals so long as the crystal symmetry is taken

[49c] K. Brugger, *Phys. Rev.* **137**, A1826 (1965).

TABLE I. LAUE GROUPS AND THE NUMBER OF INDEPENDENT ELASTIC CONSTANTS[a]

Laue group	Point groups	Crystal system	$C_{\alpha\beta}$	$C_{\alpha\beta\gamma}$	$C_{\alpha\beta\gamma\delta}$
N	$1, \bar{1}$	Triclinic	21	56	126
M	$m, 2, \dfrac{2}{m}$	Monoclinic	13	32	70
O	$2mm, 222, \dfrac{2\ 2\ 2}{m\,m\,m}$	Orthorhombic	9	20	42
T II	$4, \bar{4}, \dfrac{4}{m}$	Tetragonal	7	16	36
T I	$4mm, \bar{4}2m, 422, \dfrac{4\ 2\ 2}{m\,m\,m}$	Tetragonal	6	12	25
R II	$3, \bar{3}$	Rhombohedral	7	20	42
R I	$3m, 32, \bar{3}\dfrac{2}{m}$	Rhombohedral	6	14	28
H II	$\bar{6}, 6, \dfrac{6}{m}$	Hexagonal	5	12	24
H I	$\bar{6}2m, 6mm, 622, \dfrac{6\ 2\ 2}{m\,m\,m}$	Hexagonal	5	10	19
C II	$23, \dfrac{2}{m}\bar{3}$	Cubic	3	8	14
C I	$\bar{4}3m, 432, \dfrac{4}{m}\bar{3}\dfrac{2}{m}$	Cubic	3	6	11
I		Isotropic	2	3	4

[a] The second column gives the Hermann–Mauguin symbol for each class in the group, and the last three columns give the number of independent second-, third-, and fourth-order elastic constants, respectively.

to be that of the given crystal in the given initial stress system. Once the set of independent elastic constants has been determined, the stress-strain coefficients, wave propagation coefficients, and compliances can then be obtained from the elastic constants and the stress components.

Thermoelasticity is a centrosymmetric property, which is to say all equations remain invariant with respect to inversion of the coordinate system. Therefore we are not concerned directly with the 32 crystal classes,

TABLE II. SCHEME OF INDEPENDENT SECOND-ORDER ELASTIC CONSTANTS FOR EACH GROUP[a]

N	M	O	T II	T I	R II	R I	H	C	I
11	11	11	11	11	11	11	11	11	11
12	12	12	12	12	12	12	12	12	12
13	13	13	13	13	13	13	13	12	12
14	0	0	0	0	14	14	0	0	0
15	15	0	0	0	15	0	0	0	0
16	0	0	16	0	0	0	0	0	0
22	22	22	11	11	11	11	11	11	11
23	23	23	13	13	13	13	13	12	12
24	0	0	0	0	−14	−14	0	0	0
25	25	0	0	0	−15	0	0	0	0
26	0	0	−16	0	0	0	0	0	0
33	33	33	33	33	33	33	33	11	11
34	0	0	0	0	0	0	0	0	0
35	35	0	0	0	0	0	0	0	0
36	0	0	0	0	0	0	0	0	0
44	44	44	44	44	44	44	44	44	44[c]
45	0	0	0	0	0	0	0	0	0
46	46	0	0	0	−15	0	0	0	0
55	55	55	44	44	44	44	44	44	44[c]
56	0	0	0	0	14	14	0	0	0
66	66	66	66	66	66[b]	66[b]	66[b]	44	44[c]

[a] Only the subscripts $\alpha\beta$ for each $C_{\alpha\beta}$ are listed. The scheme is the same for the two hexagonal groups (denoted as H) and the two cubic groups (denoted as C).
[b] $C_{66} = \frac{1}{2}(C_{11} - C_{12})$.
[c] $C_{44} = \frac{1}{2}(C_{11} - C_{12})$.

but only with the 11 Laue groups, since all classes in a given group have a common array of elastic constants of each order.[50] The number of independent $C_{\alpha\beta}$ and $C_{\alpha\beta\gamma}$ for each group was calculated group-theoretically and tabulated by Bhagavantam and Suryanarayana[50] and by Jahn.[51] The actual scheme of independent $C_{\alpha\beta\gamma}$ was worked out for each group by Fumi[52] and by Hearmon.[53] These numbers and schemes of independent $C_{\alpha\beta\gamma}$ are all

[50] S. Bhagavantam and D. Suryanarayana, *Acta Cryst.* **2**, 21 (1949).
[51] H. A. Jahn, *Acta Cryst.* **2**, 30 (1949).
[52] F. G. Fumi, *Phys. Rev.* **83**, 1274 (1951); **86**, 561 (1952). Note the first paper was in error for the two trigonal (rhombohedral) groups; this was corrected in the second paper.
[53] R. F. S. Hearmon, *Acta Cryst.* **6**, 331 (1953).

in agreement. The number of independent $C_{\alpha\beta\gamma\delta}$ for each group, as well as the scheme of these constants for cubic I and isotropic materials, was given by Krishnamurty.[54] A listing of the Laue groups and the number of independent second-, third-, and fourth-order elastic constants for each, as well as for isotropic materials, is given in Table I. We also list the scheme of independent second-order elastic constants for each group in Table II; note these tables are valid for arbitrary initial stress.

Several definitions of third-order elastic constants, differing only by numerical factors, have been used previously. The definitions are based on the total third-order contributions to the energy, as follows.

(a) The present definition is that of Brugger[9]:

$$(\rho_1 U)_3 = \tfrac{1}{6} C_{ijklmn} \eta_{ij} \eta_{kl} \eta_{mn}. \qquad (11.1)$$

(b) Hearmon[53] used the Voigt notation sum

$$(\rho_1 U)_3 = C^{\mathrm{H}}{}_{\alpha\beta\gamma} \eta_\alpha \eta_\beta \eta_\gamma. \qquad (11.2)$$

(c) For a cubic crystal, Birch[42] used the definition

$$(\rho_1 U)_3 = C^{\mathrm{B}}{}_{\alpha\beta\gamma} \eta_\alpha \eta_\beta \eta_\gamma, \qquad \alpha \leq \beta \leq \gamma, \qquad (11.3)$$

except for the C_{456} term, which Birch counted twice in his equation (12). For a cubic I crystal, e.g., the present constants $C_{\alpha\beta\gamma}$ are related to the Birch definition according to

$$\begin{array}{lcccccc}
\text{Present:} & C_{111} & C_{112} & C_{123} & C_{456} & C_{144} & C_{166} \\
\text{Birch:} & 6C^{\mathrm{B}}_{111} & 2C^{\mathrm{B}}_{112} & C^{\mathrm{B}}_{123} & \tfrac{1}{4}C^{\mathrm{B}}_{456} & \tfrac{1}{2}C^{\mathrm{B}}_{144} & \tfrac{1}{2}C^{\mathrm{B}}_{166}
\end{array} \qquad (11.4)$$

For the rhombohedral, hexagonal, and isotropic systems, there are relations among the third-order elastic constants, which reduce the number of independent constants. This is analogous to the relations among second-order elastic constants, noted in Table II. These relations among $C_{\alpha\beta\gamma}$ for a given group depend on the exact definition used for the $C_{\alpha\beta\gamma}$. Brugger[21] has also tabulated the scheme of independent $C_{\alpha\beta\gamma}$ for each group in the present notation; this tabulation agrees with that of Hearmon,[53] although the relations among certain elastic constants are different because of the differences in the definitions (11.1) and (11.2).

In the following sections we consider certain crystal symmetries in more detail, and illustrate how the scheme of independent elastic constants is used to construct the proper description of the thermoelastic properties.

[54] T. S. G. Krishnamurty, *Acta Cryst.* **16**, 839 (1963).

12. Cubic I Crystal under Isotropic Pressure

The initial stress is an isotropic pressure P, so that

$$T_{ij}(\mathbf{X}) = -P\,\delta_{ij},$$

and P can be set equal to zero to obtain zero-stress results at any point in our discussion. We presume the crystal symmetry is cubic I, and remains unchanged as a function of the pressure. There are three independent second-order elastic constants C_{11}, C_{12}, and C_{44}; six independent third-order elastic constants

$$C_{111} = C_{222} = C_{333}, \qquad C_{144} = C_{255} = C_{366} = \cdots,$$

$$C_{112} = C_{113} = C_{223} = \cdots, \qquad C_{166} = C_{155} = C_{244} = \cdots, \qquad (12.1)$$

$$C_{123} = C_{132} = \cdots, \qquad C_{456} = C_{465} = \cdots;$$

and eleven independent fourth-order elastic constants C_{1111}, C_{1112}, C_{1122}, C_{1123}, C_{1144}, C_{1166}, C_{1244}, C_{1266}, C_{1456}, C_{4444}, and C_{4466}. For the special case of initial isotropic pressure, the results of Section 7 apply, and the matrix of $B_{\alpha\beta}$ is just the matrix of $C_{\alpha\beta}$ plus the matrix Δ given by (7.7). Hence, there are only three independent $B_{\alpha\beta}$, given by

$$B_{11} = C_{11} - P, \qquad (12.2\text{a})$$

$$B_{12} = C_{12} + P, \qquad (12.2\text{b})$$

$$B_{44} = C_{44} - P. \qquad (12.2\text{c})$$

As pointed out in Section 7 for initial isotropic pressure, the inverse to the matrix $[B_{\alpha\beta}]$ is given by the usual formulas for the inverse to $[C_{\alpha\beta}]$, with the elements $C_{\alpha\beta}$ replaced by the corresponding $B_{\alpha\beta}$. The compliances are then

$$S_{11} = (B_{11} + B_{12})/[(B_{11} - B_{12})(B_{11} + 2B_{12})], \qquad (12.3\text{a})$$

$$S_{12} = -B_{12}/[(B_{11} - B_{12})(B_{11} + 2B_{12})], \qquad (12.3\text{b})$$

$$S_{44} = 1/B_{44}. \qquad (12.3\text{c})$$

Note the compliance matrix has exactly the same symmetry and zero elements as $[B_{\alpha\beta}]$. The relation between $S_{\alpha\beta}$ and S_{ijkl}, according to (6.34), is

$$S_{1111} = S_{11}, \qquad S_{1122} = S_{12}, \qquad S_{2323} = \tfrac{1}{4}S_{44}. \qquad (12.4)$$

The compressibility K (adiabatic or isothermal) at arbitrary pressure is, from (6.35),

$$K = 3(S_{11} + 2S_{12}) = 3/(B_{11} + 2B_{12}) = 3/(C_{11} + 2C_{12} + P). \quad (12.5)$$

There are four independent propagation coefficients, since there are three independent second-order elastic constants and one independent stress component. The A_{ijkl} cannot be written in Voigt notation; they are defined by (7.2) in the case of initial isotropic pressure, and for a cubic crystal reduce to

$$\begin{aligned}
A_{1111} &= A_{2222} = A_{3333} = C_{11} - P, \\
A_{1122} &= A_{1133} = A_{2233} = \cdots = C_{12}, \\
A_{1212} &= A_{1313} = A_{2323} = \cdots = C_{44} - P, \\
A_{1221} &= A_{1331} = A_{2332} = \cdots = C_{44}.
\end{aligned} \quad (12.6)$$

These coefficients can now be used to construct the propagation matrices. We will write out the matrices **L** given by (3.13),

$$L_{ik} = A_{ijkl}\kappa_j\kappa_l,$$

for propagation directions along those crystal axes which are pure mode directions. We use the notation $\mathbf{w}(l)$ for a longitudinal eigenvector, and $\mathbf{w}(t_1)$, $\mathbf{w}(t_2)$ for the two transverse eigenvectors. It will be recalled from Section 7 that the matrix elements L_{ik} can always be written as combinations of $B_{\alpha\beta}$ since the stress is isotropic pressure; it will be convenient to do this.

$\boldsymbol{\kappa} = [100]$:

$$\mathbf{L} = \begin{pmatrix} A_{1111} & 0 & 0 \\ 0 & A_{2121} & 0 \\ 0 & 0 & A_{3131} \end{pmatrix} = \begin{pmatrix} B_{11} & 0 & 0 \\ 0 & B_{44} & 0 \\ 0 & 0 & B_{44} \end{pmatrix} \quad (12.7)$$

$$\mathbf{w}(l) = [100], \quad \rho_1 v^2 = B_{11} = C_{11} - P; \quad (12.8\text{a})$$
$$\mathbf{w}(t_1) = [010], \quad \rho_1 v^2 = B_{44} = C_{44} - P; \quad (12.8\text{b})$$
$$\mathbf{w}(t_2) = [001], \quad \rho_1 v^2 = B_{44} = C_{44} - P. \quad (12.8\text{c})$$

For this propagation direction the two transverse modes are degenerate. In fact, any two independent linear combinations of $\mathbf{w}(t_1)$ and $\mathbf{w}(t_2)$ can serve as the two transverse eigenvectors, with eigenvalues the same in any case. For other equivalent propagation directions, such as $\kappa = [010]$, the matrix elements and eigenvector components are interchanged accordingly.

$\kappa = (1/\sqrt{2})[110]$:

$$\mathbf{L} = \frac{1}{2}\begin{pmatrix} A_{1111} + A_{1212} & A_{1122} + A_{1221} & 0 \\ A_{2211} + A_{2112} & A_{2222} + A_{2121} & 0 \\ 0 & 0 & A_{3131} + A_{3232} \end{pmatrix}$$

$$= \frac{1}{2}\begin{pmatrix} B_{11} + B_{44} & B_{12} + B_{44} & 0 \\ B_{12} + B_{44} & B_{11} + B_{44} & 0 \\ 0 & 0 & 2B_{44} \end{pmatrix} \quad (12.9)$$

$\mathbf{w}(l) = (1/\sqrt{2})[110], \quad \rho_1 v^2 = \frac{1}{2}(B_{11} + B_{12} + 2B_{44})$
$\qquad\qquad\qquad\qquad\qquad = \frac{1}{2}(C_{11} + C_{12} + 2C_{44} - 2P);$ (12.10a)

$\mathbf{w}(t_1) = (1/\sqrt{2})[1\bar{1}0], \quad \rho_1 v^2 = \frac{1}{2}(B_{11} - B_{12}) = \frac{1}{2}(C_{11} - C_{12} - 2P);$
$\qquad\qquad\qquad\qquad\qquad\qquad\qquad\qquad\qquad$ (12.10b)

$\mathbf{w}(t_2) = [001], \quad \rho_1 v^2 = B_{44} = C_{44} - P.$ (12.10c)

$\kappa = (1/\sqrt{3})[111]$:

$$\mathbf{L} = \frac{1}{3}\begin{pmatrix} B_{11} + 2B_{44} & B_{12} + B_{44} & B_{12} + B_{44} \\ B_{12} + B_{44} & B_{11} + 2B_{44} & B_{12} + B_{44} \\ B_{12} + B_{44} & B_{12} + B_{44} & B_{11} + 2B_{44} \end{pmatrix} \quad (12.11)$$

$\mathbf{w}(l) = (1/\sqrt{3})[111], \quad \rho_1 v^2 = \frac{1}{3}(B_{11} + 2B_{12} + 4B_{44})$
$\qquad\qquad\qquad\qquad\qquad = \frac{1}{3}(C_{11} + 2C_{12} + 4C_{44} - 3P);$ (12.12a)

$\mathbf{w}(t_1) = (1/\sqrt{2})[1\bar{1}0], \quad \rho_1 v^2 = \frac{1}{3}(B_{11} - B_{12} + B_{44})$
$\qquad\qquad\qquad\qquad\qquad = \frac{1}{3}(C_{11} - C_{12} + C_{44} - 3P);$ (12.12b)

$\mathbf{w}(t_2) = (1/\sqrt{6})[11\bar{2}], \quad \rho_1 v^2 = \frac{1}{3}(B_{11} - B_{12} + B_{44})$
$\qquad\qquad\qquad\qquad\qquad = \frac{1}{3}(C_{11} - C_{12} + C_{44} - 3P).$ (12.12c)

Again for this propagation direction the two transverse modes are degenerate.

The reduced propagation matrices **M**, and their eigenvalues $\rho_0 W^2$, are obtained from the above results simply by multiplying them by $(\rho_0/\rho_1)(l_0/l_1)^2$, according to (9.14) and (9.15). Note that wave propagation experiments on a cubic crystal under isotropic pressure can determine at most three constants, e.g., B_{11}, B_{12}, and B_{44}, and hence cannot determine all six independent third-order elastic constants. Also the symmetries of the $C_{\alpha\beta}$, $B_{\alpha\beta}$, $S_{\alpha\beta}$, and the propagation matrices are the same for cubic I and cubic II at arbitrary pressure, but the higher-order elastic constants have different symmetries for these two groups.

In order to make an expansion of our results for small pressures, we need first to expand the crystal strains as a function of the pressure. The general strain expansion was derived in Section 9; the term linear in the stress in (9.10) is evaluated as

$$\bar{S}_{ijkl} T_{kl}(\mathbf{X}) = -P \bar{S}_{ijkl} \delta_{kl} = -P \bar{S}_{ijkk} = -P \delta_{ij}(\bar{S}_{11} + 2\bar{S}_{12}). \quad (12.13)$$

The term of second order in the stress is similarly evaluated, with the result

$$n_{ij} = n \delta_{ij}, \quad (12.14)$$

$$n = -gP + \tfrac{1}{2}g^2(2 - gh)P^2, \quad (12.15)$$

where we have introduced the abbreviations for a cubic I crystal

$$g = \bar{S}_{11} + 2\bar{S}_{12} = \tfrac{1}{3}\bar{K} \quad (\bar{K} = \text{compressibility at } P = 0); \quad (12.16)$$

$$h = \bar{C}_{111} + 6\bar{C}_{112} + 2\bar{C}_{123}. \quad (12.17)$$

The other strain parameters which measure strain from $\bar{\mathbf{X}}$ (zero pressure) to $\mathbf{X}$ (pressure P) are given by (8.18) and (8.19):

$$a_{ij} = (1 + 2n)^{1/2} \delta_{ij} = (1 + n - \tfrac{1}{2}n^2 + \cdots) \delta_{ij}; \quad (12.18)$$

$$(\rho_0/\rho_1) = (1 + 2n)^{3/2} = 1 + 3n + \tfrac{3}{2}n^2 + \cdots. \quad (12.19)$$

The general expansion of the second-order elastic constants is (8.11); with the above strains this becomes

$$C_{ijkl} = \bar{C}_{ijkl} + n(\bar{C}_{ijkl} + \bar{C}_{ijklmm})$$
$$+ \tfrac{1}{2}n^2(-\bar{C}_{ijkl} + 2\bar{C}_{ijklmm} + \bar{C}_{ijklmmpp}) + \cdots. \quad (12.20)$$

We have previously worked out this equation for the three independent elastic constants for a cubic I crystal[7]; the results to order n^2 are

$$C_{11} = \bar{C}_{11} + n[\bar{C}_{11} + (\bar{C}_{111} + 2\bar{C}_{112})]$$
$$+ \tfrac{1}{2}n^2[-\bar{C}_{11} + 2(\bar{C}_{111} + 2\bar{C}_{112}) + (\bar{C}_{1111} + 4\bar{C}_{1112} + 2\bar{C}_{1122} + 2\bar{C}_{1123})],$$
(12.21a)

$$C_{12} = \bar{C}_{12} + n[\bar{C}_{12} + (2\bar{C}_{112} + \bar{C}_{123})]$$
$$+ \tfrac{1}{2}n^2[-\bar{C}_{12} + 2(2\bar{C}_{112} + \bar{C}_{123}) + (2\bar{C}_{1112} + 2\bar{C}_{1122} + 5\bar{C}_{1123})],$$
(12.21b)

$$C_{44} = \bar{C}_{44} + n[\bar{C}_{44} + (\bar{C}_{144} + 2\bar{C}_{166})]$$
$$+ \tfrac{1}{2}n^2[-\bar{C}_{44} + 2(\bar{C}_{144} + 2\bar{C}_{166}) + (\bar{C}_{1144} + 2\bar{C}_{1166} + 4\bar{C}_{1244} + 2\bar{C}_{1266})].$$
(12.21c)

The expansion (8.12) for third-order elastic constants is worked out in a similar way for a cubic I crystal under isotropic pressure to give, correct to order n,

$$\begin{aligned}
C_{111} &= \bar{C}_{111} + n(3\bar{C}_{111} + \bar{C}_{1111} + 2\bar{C}_{1112}), \\
C_{112} &= \bar{C}_{112} + n(3\bar{C}_{112} + \bar{C}_{1112} + \bar{C}_{1122} + \bar{C}_{1123}), \\
C_{123} &= \bar{C}_{123} + n(3\bar{C}_{123} + 3\bar{C}_{1123}), \\
C_{144} &= \bar{C}_{144} + n(3\bar{C}_{144} + \bar{C}_{1144} + 2\bar{C}_{1244}), \\
C_{166} &= \bar{C}_{166} + n(3\bar{C}_{166} + \bar{C}_{1166} + \bar{C}_{1244} + \bar{C}_{1266}), \\
C_{456} &= \bar{C}_{456} + n(3\bar{C}_{456} + 3\bar{C}_{1456}).
\end{aligned}$$
(12.22)

All of these results may be further transformed to power series in P by eliminating n for P with the aid of (12.15).

It is also useful to expand the compliances for small pressure. This may be done from the definitions (12.3) of $S_{\alpha\beta}$, by replacing $B_{\alpha\beta}$ by $C_{\alpha\beta} \pm P$ according to (12.2), then inserting the expansions (12.21) for $C_{\alpha\beta}$, and carrying out the power series for small P. The results to order P are

$$S_{11} = \bar{S}_{11} - d_1 P, \qquad (12.23a)$$

$$S_{12} = \bar{S}_{12} - d_2 P, \qquad (12.23b)$$

$$S_{44} = \bar{S}_{44} + [1 + g(\bar{C}_{44} + \bar{C}_{144} + 2\bar{C}_{166})]\bar{S}_{44}^2 P, \qquad (12.23c)$$

where

$$d_1 = 2\tilde{S}_{11}(\tilde{S}_{12} - \tilde{S}_{11})$$
$$- g[\tilde{C}_{111}(\tilde{S}_{11}^2 + 2\tilde{S}_{12}^2) + 2\tilde{C}_{112}g^2 + 2\tilde{C}_{123}\tilde{S}_{12}(2\tilde{S}_{11} + \tilde{S}_{12})], \quad (12.24)$$
$$d_2 = \tilde{S}_{11}(\tilde{S}_{11} - \tilde{S}_{12})$$
$$- g[\tilde{C}_{111}\tilde{S}_{12}(2\tilde{S}_{11} + \tilde{S}_{12}) + 2\tilde{C}_{112}g^2 + \tilde{C}_{123}(\tilde{S}_{11}^2 + 2\tilde{S}_{11}\tilde{S}_{12} + 3\tilde{S}_{12}^2)]. \quad (12.25)$$

Finally, we wish to calculate the difference between adiabatic and isothermal second-order elastic constants for a cubic crystal under isotropic pressure. The general theory is provided in Section 6. Let V be the volume per unit mass, so that

$$V = 1/\rho; \quad (12.26)$$

the symmetric infinitesimal strain parameters ϵ_{ij} may be written

$$\epsilon_{ij} = \epsilon\, \delta_{ij}; \qquad \epsilon = (V - V_1)/3V_1. \quad (12.27)$$

The thermal strains are

$$\beta_{ij} = (\partial \epsilon/\partial T)_P\, \delta_{ij} = \tfrac{1}{3}\beta\, \delta_{ij}, \quad (12.28)$$

where β is the volume thermal expansion coefficient evaluated at the initial configuration

$$\beta = [V^{-1}(\partial V/\partial T)_P]_{V=V_1}. \quad (12.29)$$

The thermal stresses are found with the aid of (6.40);

$$\tau_{ij} = -B^T{}_{ijkl}\beta_{kl} = -\beta B^T\, \delta_{ij}, \quad (12.30)$$

where B^T is the isothermal bulk modulus evaluated at the initial configuration

$$B^T = \tfrac{1}{3}(B^T{}_{11} + 2B^T{}_{12}) = 1/K^T. \quad (12.31)$$

Now the adiabatic–isothermal difference (6.45) becomes

$$B^S{}_{ijkl} - B^T{}_{ijkl} = (V_1 T/C_V)(\beta B^T)^2\, \delta_{ij}\, \delta_{kl}, \quad (12.32)$$

where C_V is the heat capacity per unit mass at constant volume, the same as C defined by (6.38). This result is easily worked out to give

$$B^S{}_{11} - B^T{}_{11} = B^S{}_{12} - B^T{}_{12} = TV_1(\beta B^T)^2/C_V, \quad (12.33)$$
$$B^S{}_{44} - B^T{}_{44} = 0. \quad (12.34)$$

The adiabatic bulk modulus is

$$B^S = \tfrac{1}{3}(B^S{}_{11} + 2B^S{}_{12}) = 1/K^S, \qquad (12.35)$$

and from (12.33) we then have the usual thermodynamic result

$$B^S - B^T = TV_1(\beta B^T)^2/C_V, \qquad (12.36)$$

or

$$K^T - K^S = TV_1\beta^2/C_P, \qquad (12.37)$$

since $C_P/C_V = K^T/K^S$.[55]

It should be emphasized that these adiabatic–isothermal differences were derived under the assumption that pressure is the *only* stress variable. The results are valid for cubic I and cubic II at arbitrary pressure; they also give the adiabatic–isothermal differences for the second-order elastic constants and the wave propagation coefficients, in view of the general equalities (6.46).

13. Tetragonal I Crystal under Uniaxial Stress

We consider tetragonal I crystals with uniaxial stress applied along the tetragonal axis (3 axis) so that

$$T_{ij}(\mathbf{X}) = -P\,\delta_{i3}\,\delta_{j3}, \qquad (13.1)$$

and our discussion applies only so long as the tetragonal I symmetry remains in the presence of the stress. In addition, our discussion applies to a cubic I crystal in the presence of [001] stress, unless for some reason the symmetry of the stressed crystal is reduced from tetragonal I. The stress P is applied uniformly to (001) faces of the crystal, and is the force per unit area of these faces in the stressed configuration. In the usual experimental arrangement, however, the directly measured quantity is the total force applied to the crystal face, and not the force per unit area. To prepare for analysis of experiments it is therefore convenient to introduce the reduced stress Q, defined by

$$Q = P(s_1/s_0), \qquad (13.2)$$

where s_0 and s_1 are the areas of the (001) faces in the zero-stress and stressed configurations, respectively. Since the total force on the (001) faces is $Ps_1 = Qs_0$, then Q is easily calculated from the total force divided by the zero-stress area s_0. Interpretation of experimental results is then simplified by eliminating P in favor of Q, with the aid of (13.2), in our final results. We note this step is *not* necessary in the analysis of first stress derivatives,

[55] M. W. Zemansky, "Heat and Thermodynamics," 3rd ed. McGraw-Hill, New York, 1951.

to obtain third-order elastic constants according to the general theory of Section 10, since $(s_1/s_0) = 1 + O(P)$. However, this point is important in analysis of experiments designed to measure fourth-order elastic constants from second stress derivatives.

For a tetragonal I crystal there are six independent second-order elastic constants C_{11}, C_{12}, C_{13}, C_{33}, C_{44}, and C_{66}, and twelve independent third-order elastic constants C_{111}, C_{112}, C_{113}, C_{123}, C_{133}, C_{333}, C_{144}, C_{155}, C_{166}, C_{344}, C_{366}, and C_{456}. There are seven independent stress–strain coefficients $B_{\alpha\beta}$, which are related to the elastic constants according to (6.10):

$$B_{11} = B_{22} = C_{11}, \quad B_{31} = B_{32} = C_{13} + P, \quad B_{44} = B_{55} = C_{44} - \tfrac{1}{2}P,$$
$$B_{12} = B_{21} = C_{12}, \quad B_{33} = C_{33} - P, \quad B_{66} = C_{66}. \quad (13.3)$$
$$B_{13} = B_{23} = C_{13},$$

These coefficients illustrate the general theory of Section 6, in that the matrix $[B_{\alpha\beta}]$ is nonsymmetric since $B_{13} \neq B_{31}$, $B_{23} \neq B_{32}$.

The matrix of compliance coefficients will also be nonsymmetric in general. In the analysis of experiments measuring third-order elastic constants evaluated at zero stress, for which examples are given in Part V, the compliances are required at zero stress only; we therefore list here the inverse of $[\bar{B}_{\alpha\beta}] = [\bar{C}_{\alpha\beta}]$, where P vanishes and the 6×6 matrix is symmetric:

$$\bar{S}_{11} = (\bar{C}_{11}\bar{C}_{33} - \bar{C}_{13}^2)/D, \qquad \bar{S}_{33} = (\bar{C}_{11} + \bar{C}_{12})(\bar{C}_{11} - \bar{C}_{12})/D,$$
$$\bar{S}_{12} = -(\bar{C}_{12}\bar{C}_{33} - \bar{C}_{13}^2)/D, \qquad \bar{S}_{44} = 1/\bar{C}_{44}, \quad (13.4)$$
$$\bar{S}_{13} = -\bar{C}_{13}(\bar{C}_{11} - \bar{C}_{12})/D, \qquad \bar{S}_{66} = 1/\bar{C}_{66},$$

where

$$D = (\bar{C}_{11} - \bar{C}_{12})[\bar{C}_{33}(\bar{C}_{11} + \bar{C}_{12}) - 2\bar{C}_{13}^2]. \quad (13.5)$$

There are seven independent wave propagation coefficients, which are related to the elastic constants by (3.6) or (3.7):

$$A_{1111} = A_{2222} = C_{11},$$
$$A_{1122} = A_{2211} = C_{12},$$
$$A_{1133} = A_{2233} = \cdots = C_{13},$$
$$A_{1212} = A_{1221} = \cdots = C_{66}, \quad (13.6)$$
$$A_{1313} = A_{2323} = C_{44} - P,$$
$$A_{1331} = A_{3131} = A_{3113} = A_{2332} = \cdots = C_{44},$$
$$A_{3333} = C_{33} - P.$$

From these coefficients we will construct the propagation matrices **L** for propagation directions along those crystal axes which are pure mode directions. Again $\mathbf{w}(l)$ stands for a longitudinal eigenvector, and $\mathbf{w}(t_1)$ and $\mathbf{w}(t_2)$ stand for the two transverse eigenvectors.

$\kappa = [100]$:

$$\mathbf{L} = \begin{pmatrix} A_{1111} & 0 & 0 \\ 0 & A_{2121} & 0 \\ 0 & 0 & A_{3131} \end{pmatrix} = \begin{pmatrix} C_{11} & 0 & 0 \\ 0 & C_{66} & 0 \\ 0 & 0 & C_{44} \end{pmatrix} \quad (13.7)$$

$$\mathbf{w}(l) = [100], \quad \rho_1 v^2 = C_{11}; \quad (13.8a)$$
$$\mathbf{w}(t_1) = [010], \quad \rho_1 v^2 = C_{66}; \quad (13.8b)$$
$$\mathbf{w}(t_2) = [001], \quad \rho_1 v^2 = C_{44}. \quad (13.8c)$$

$\kappa = (1/\sqrt{2})[110]$:

In terms of A_{ijkl}, **L** is given by (12.9) for cubic I; in terms of $C_{\alpha\beta}$,

$$\mathbf{L} = \frac{1}{2}\begin{pmatrix} C_{11} + C_{66} & C_{12} + C_{66} & 0 \\ C_{12} + C_{66} & C_{11} + C_{66} & 0 \\ 0 & 0 & 2C_{44} \end{pmatrix} \quad (13.9)$$

$$\mathbf{w}(l) = (1/\sqrt{2})[110], \quad \rho_1 v^2 = \tfrac{1}{2}(C_{11} + C_{12} + 2C_{66}); \quad (13.10a)$$
$$\mathbf{w}(t_1) = (1/\sqrt{2})[1\bar{1}0], \quad \rho_1 v^2 = \tfrac{1}{2}(C_{11} - C_{12}); \quad (13.10b)$$
$$\mathbf{w}(t_2) = [001], \quad \rho_1 v^2 = C_{44}. \quad (13.10c)$$

$\kappa = [001]$:

$$\mathbf{L} = \begin{pmatrix} A_{1313} & 0 & 0 \\ 0 & A_{2323} & 0 \\ 0 & 0 & A_{3333} \end{pmatrix} = \begin{pmatrix} C_{44} - P & 0 & 0 \\ 0 & C_{44} - P & 0 \\ 0 & 0 & C_{33} - P \end{pmatrix}$$

$$(13.11)$$

$$\mathbf{w}(l) = [001], \quad \rho_1 v^2 = C_{33} - P; \quad (13.12a)$$
$$\mathbf{w}(t_1) = [010], \quad \rho_1 v^2 = C_{44} - P; \quad (13.12b)$$
$$\mathbf{w}(t_2) = [100], \quad \rho_1 v^2 = C_{44} - P. \quad (13.12c)$$

This is another example of degenerate transverse modes. In addition, this example of $\kappa = [001]$ shows how the stress components appear explicitly in the propagation matrices, and hence in the eigenvalues, only when **k** has a component parallel to the uniaxial stress.

We note for hexagonal crystals that the second-order elastic constants are of the same symmetry as for tetragonal I, with the additional condition

$$C_{66} = \tfrac{1}{2}(C_{11} - C_{12}), \qquad \text{hexagonal.} \tag{13.13}$$

Therefore the $B_{\alpha\beta}$ relations of (13.3), and the $\bar{S}_{\alpha\beta}$ relations of (13.4), apply also for hexagonal crystals, with the additional condition

$$\bar{S}_{66} = 2(\bar{S}_{11} - \bar{S}_{12}), \qquad \text{hexagonal.} \tag{13.14}$$

Also, the propagation matrices, eigenvectors, and eigenvalues listed above for tetragonal I apply to hexagonal in the presence of [001] stress.

14. Isotropic Material

Consider an isotropic material in the presence of arbitrary isotropic pressure P. There are two independent second-order elastic constants, three independent third-order constants, and so on. It is customary to dispense with the notation $C_{\alpha\beta}$ for the second-order elastic constants in favor of the simpler Lamé constants λ and μ. We will define the third-order elastic constants ζ, ξ, and ν from the Murnaghan expansion[6] of the thermodynamic state functions below; in fact these constants are equal to Murnaghan's constants:

$$\zeta, \xi, \nu = l, m, n \qquad \text{of Murnaghan;} \tag{14.1}$$

we have changed the notation only to avoid confusion with our other quantities.

The three strain functions I_1, I_2, and I_3 are invariant with respect to rotation of the material, or the coordinate system, where

$$I_1 = \text{tr}[\eta_{ij}], \tag{14.2}$$

$$I_2 = \sum_j \eta^{jj}, \tag{14.3}$$

$$I_3 = \det[\eta_{ij}]. \tag{14.4}$$

Here $\text{tr}[\eta_{ij}]$ is the trace of the 3×3 matrix $[\eta_{ij}]$ and η^{jj} is the cofactor of η_{jj}. In Voigt notation the three strain invariants may be written

$$I_1 = \eta_1 + \eta_2 + \eta_3, \tag{14.5}$$

$$I_2 = (\eta_1\eta_2 - \eta_6^2) + (\eta_1\eta_3 - \eta_5^2) + (\eta_2\eta_3 - \eta_4^2), \tag{14.6}$$

$$I_3 = \eta_1\eta_2\eta_3 + 2\eta_4\eta_5\eta_6 - \eta_1\eta_4^2 - \eta_2\eta_5^2 - \eta_3\eta_6^2. \tag{14.7}$$

Since an isotropic material is invariant with respect to rotation, the state functions may be expanded in terms of the three strain invariants. This invariance condition is just a more restrictive analog, appropriate for isotropic materials, of our general invariance conditions (1.7). The internal energy, e.g., is

$$\rho_1 U(\mathbf{X}, \eta_{ij}, S) = \rho_1 U(\mathbf{X}, 0, S) + \alpha^S I_1 + [\tfrac{1}{2}(\lambda^S + 2\mu^S)I_1^2 - 2\mu^S I_2]$$
$$+ [\tfrac{1}{3}(\zeta^S + 2\xi^S)I_1^3 - 2\xi^S I_1 I_2 + \nu^S I_3] + \cdots. \quad (14.8)$$

The Helmholtz free energy at constant T is similarly expanded in terms of isothermal constants.

At the initial configuration $\mathbf{X}$, the stress is given by the equilibrium condition (1.12)

$$T_{ij}(\mathbf{X}) = \rho_1(\partial U/\partial \eta_{ij}) = \alpha^S \delta_{ij}, \quad \text{at} \quad \mathbf{X}. \quad (14.9)$$

Therefore the stress must be isotropic, and the pressure is

$$P = -\alpha^S. \quad (14.10)$$

The second-order terms in the expansion of $\rho_1 U$ are denoted as $(\rho_1 U)_2$, and in Voigt notation are obtained from (14.8) as

$$(\rho_1 U)_2 = \tfrac{1}{2}(\lambda^S + 2\mu^S)(\eta_1^2 + \eta_2^2 + \eta_3^2)$$
$$+ \lambda^S(\eta_1\eta_2 + \eta_1\eta_3 + \eta_2\eta_3) + 2\mu^S(\eta_4^2 + \eta_5^2 + \eta_6^2). \quad (14.11)$$

Likewise the third-order terms are

$$(\rho_1 U)_3 = \tfrac{1}{3}(\zeta^S + 2\xi^S)(\eta_1^3 + \eta_2^3 + \eta_3^3) + (2\zeta^S - 2\xi^S + \nu^S)\eta_1\eta_2\eta_3$$
$$+ \zeta^S(\eta_1^2\eta_2 + \eta_1^2\eta_3 + \eta_2^2\eta_1 + \eta_2^2\eta_3 + \eta_3^2\eta_1 + \eta_3^2\eta_2)$$
$$+ 2\xi^S(\eta_1\eta_6^2 + \eta_1\eta_5^2 + \eta_2\eta_6^2 + \eta_2\eta_4^2 + \eta_3\eta_5^2 + \eta_3\eta_4^2)$$
$$+ (2\xi^S - \nu^S)(\eta_1\eta_4^2 + \eta_2\eta_5^2 + \eta_3\eta_6^2) + 2\nu^S\eta_4\eta_5\eta_6. \quad (14.12)$$

Now it is useful to consider an isotropic material as a special case of a cubic I crystal, with additional relations among the elastic constants. This description allows all the theoretical results for a cubic I crystal to be applied to the isotropic case. To relate the isotropic elastic constants to those of cubic I, we simply write out the energy contributions $(\rho_1 U)_2$ and $(\rho_1 U)_3$ for a cubic I crystal and compare them with (14.11) and (14.12), respectively. The results are valid for adiabatic or isothermal elastic constants, and can be considered as giving the isotropic elastic constants $C_{\alpha\beta}$ in terms of λ and μ, and $C_{\alpha\beta\gamma}$ in terms of ζ, ξ, and ν. We find

$$C_{11} = \lambda + 2\mu, \quad (14.13\text{a})$$

$$C_{12} = \lambda, \quad (14.13\text{b})$$

$$C_{44} = \mu = \tfrac{1}{2}(C_{11} - C_{12}). \quad (14.13\text{c})$$

Note (14.13c) is a relation among the elastic constants, reducing the number of independent second-order elastic constants from three to two. For the third-order elastic constants, of which three are independent, we find

$$C_{111} = 2\zeta + 4\xi, \tag{14.14a}$$

$$C_{112} = 2\zeta, \tag{14.14b}$$

$$C_{123} = 2\zeta - 2\xi + \nu, \tag{14.14c}$$

$$C_{144} = \xi - \tfrac{1}{2}\nu = \tfrac{1}{2}(C_{112} - C_{123}), \tag{14.14d}$$

$$C_{166} = \xi = \tfrac{1}{4}(C_{111} - C_{112}), \tag{14.14e}$$

$$C_{456} = \tfrac{1}{4}\nu = \tfrac{1}{8}(C_{111} - 3C_{112} + 2C_{123}). \tag{14.14f}$$

As an application of this description of an isotropic material, the propagation matrices of Section 12 for cubic I under isotropic pressure may be evaluated for isotropic materials under pressure. Although the matrix elements L_{ik} are different for the different propagation directions κ, the eigenvalues are the same, as they must be for any κ. The eigenvectors are always one pure longitudinal, two pure transverse, and the transverse modes are degenerate:

$$\rho_1 v^2 = \lambda + 2\mu - P, \quad \text{longitudinal}; \tag{14.15}$$

$$\rho_1 v^2 = \mu - P, \quad \text{transverse}. \tag{14.16}$$

The stress-strain coefficients may also be evaluated from Eqs. (12.2) for cubic I:

$$B_{11} = \lambda + 2\mu - P, \tag{14.17a}$$

$$B_{12} = \lambda + P, \tag{14.17b}$$

$$B_{44} = \mu - P = \tfrac{1}{2}(B_{11} - B_{12}). \tag{14.17c}$$

From (12.3) the compliances at arbitrary pressure are

$$S_{11} = (\lambda + \mu)/[(\mu - P)(3\lambda + 2\mu + P)], \tag{14.18a}$$

$$S_{12} = -(\lambda + P)/[2(\mu - P)(3\lambda + 2\mu + P)], \tag{14.18b}$$

$$S_{44} = 1/(\mu - P). \tag{14.18c}$$

Finally the adiabatic or isothermal compressibility is obtained from (12.5).

$$K = 3/(3\lambda + 2\mu + P). \tag{14.19}$$

It is interesting to note that if one measures small-amplitude wave velocities, or stress–strain relations, for an isotropic material under pressure P, the quantities which are directly measured according to (14.15)–(14.17) are just combinations of $\lambda + P$ and $\mu - P$, and not λ and μ alone. Brillouin[56]

---

[56] Brillouin,[2] Chapter 10, Section 12.

arrived at this conclusion for the case of stress–strain measurements, and pointed out the same was recognized by Poincaré in 1892.

V. Measurement of Third-Order Elastic Constants

15. Wave Propagation in Stressed Crystals—Tables of Results

Here it is only necessary to work out the general equations (10.10) or (10.11) for any desired crystal and any stress system. The stresses will be described by the following notation:

$$T_{ij}(\mathbf{X}) = -P\,\delta_{ij}, \quad \text{isotropic pressure;} \tag{15.1}$$

$$T_{ij}(\mathbf{X}) = -Pt_i t_j, \quad \text{uniaxial stress along } \mathbf{t}; \tag{15.2}$$

where $\mathbf{t}$ is a unit vector.

For illustration, consider a tetragonal I crystal with uniaxial stress applied along $[1\bar{1}0]$. It is irrelevant that the stressed crystal may have rhombohedral symmetry, since the first-order perturbation effects are expressed in terms of properties of the zero-stress state. Thus the first-order variation of $\rho_0 W^2$ with stress is given by (10.10) with the stress (15.2), where $\mathbf{t} = (1/\sqrt{2})[1\bar{1}0]$, and with the tetragonal I elastic constants:

$$\delta(\rho_0 W^2) = -\tfrac{1}{2}[\,(\bar{\kappa}_1{}^2 + \bar{\kappa}_2{}^2 - 2\bar{\kappa}_1\bar{\kappa}_2) + 2\rho_0 \bar{W}^2(\bar{S}_{ij11} + \bar{S}_{ij22} - 2\bar{S}_{ij12})\,\bar{w}_i \bar{w}_j$$
$$+ \bar{C}_{ijklmn}(\bar{S}_{mn11} + \bar{S}_{mn22} - 2\bar{S}_{mn12})\bar{\kappa}_j\bar{\kappa}_l\bar{w}_i\bar{w}_k\,]P. \tag{15.3}$$

Choosing the zero-stress pure mode propagation direction $\bar{\kappa} = (1/\sqrt{2})[110]$, this simplifies further with the aid of the elastic constant symmetries:

$$\delta(\rho_0 W^2) = -\tfrac{1}{2}[\,2\rho_0 \bar{W}^2(\bar{S}_{ij11} + \bar{S}_{ij22} - 2\bar{S}_{ij12})\,\bar{w}_i \bar{w}_j$$
$$+ (\bar{C}_{i1k1mn} + \bar{C}_{i1k2mn})(\bar{S}_{mn11} + \bar{S}_{mn22} - 2\bar{S}_{mn12})\,\bar{w}_i\bar{w}_k\,]P. \tag{15.4}$$

The three zero-stress eigenvectors are $\bar{\mathbf{w}}(l) = (1/\sqrt{2})[110]$, $\bar{\mathbf{w}}(t_1) = (1/\sqrt{2})[1\bar{1}0]$, and $\bar{\mathbf{w}}(t_2) = [001]$. For the transverse $\bar{\mathbf{w}}(t_1) = (1/\sqrt{2})[1\bar{1}0]$, the zero stress eigenvalue is $\rho_0 \bar{W}^2 = \tfrac{1}{2}(\bar{C}_{11} - \bar{C}_{12})$, and (15.4) becomes

$$\delta(\rho_0 W^2) = -\tfrac{1}{4}[\,(\bar{C}_{11} - \bar{C}_{12})(2\bar{S}_{1111} + 2\bar{S}_{1122} + 4\bar{S}_{1212})$$
$$+ (\bar{C}_{1111mm} + \bar{C}_{2121mm} - \bar{C}_{1122mm} - \bar{C}_{2112mm})(\bar{S}_{mm11} + \bar{S}_{mm22})$$
$$+ (-2\bar{C}_{1121mn} + 2\bar{C}_{1121mn})(-2\bar{S}_{mn12})\,]P$$
$$= -\tfrac{1}{4}[\,(\bar{C}_{11} - \bar{C}_{12})(2\bar{S}_{11} + 2\bar{S}_{12} + \bar{S}_{66})$$
$$+ (\bar{C}_{1111mm} - \bar{C}_{1122mm})(\bar{S}_{mm11} + \bar{S}_{mm22})\,]P$$
$$= -\tfrac{1}{4}[\,(\bar{C}_{11} - \bar{C}_{12})(2\bar{S}_{11} + 2\bar{S}_{12} + \bar{S}_{66})$$
$$+ (\bar{C}_{111} - \bar{C}_{112})(\bar{S}_{11} + \bar{S}_{12}) + 2(\bar{C}_{113} - \bar{C}_{123})\bar{S}_{13}\,]P. \tag{15.5}$$

Any other examples may be worked out in the same way.

TABLE III. STRESS DERIVATIVES OF $\rho_0 W^2$, EVALUATED AT ZERO STRESS, FOR CUBIC I CRYSTALS[a]

Case	Stress	$\bar{\kappa}$	$\bar{\mathbf{w}}$	$\rho_0 \bar{W}^2 = \rho_0 \bar{v}^2$	$-d(\rho_0 W^2)/dP$
1	Isotropic	[100]	[100]	$\bar{C}_{11}$	$1 + g(2\rho_0 \bar{W}^2 + \bar{C}_{111} + 2\bar{C}_{112})$
2	Isotropic	[100]	[010] or [001]	$\bar{C}_{44}$	$1 + g(2\rho_0 \bar{W}^2 + \bar{C}_{144} + 2\bar{C}_{166})$
3	Isotropic	$(1/\sqrt{2})[110]$	$(1/\sqrt{2})[110]$	$\frac{1}{2}(\bar{C}_{11} + \bar{C}_{12} + 2\bar{C}_{44})$	$1 + \frac{1}{2}g(4\rho_0\bar{W}^2 + \bar{C}_{111} + 4\bar{C}_{112} + \bar{C}_{123} + 2\bar{C}_{144} + 4\bar{C}_{166})$
4	Isotropic	$(1/\sqrt{2})[110]$	$(1/\sqrt{2})[1\bar{1}0]$	$\frac{1}{2}(\bar{C}_{11} - \bar{C}_{12})$	$1 + \frac{1}{2}g(4\rho_0\bar{W}^2 + \bar{C}_{111} - \bar{C}_{123})$
5	Isotropic	$(1/\sqrt{2})[110]$	[001]	$\bar{C}_{44}$	$1 + \frac{1}{2}g(2\rho_0\bar{W}^2 + \bar{C}_{144} + 2\bar{C}_{166})$
6	Isotropic	$(1/\sqrt{3})[111]$	$(1/\sqrt{3})[111]$	$\frac{1}{3}(\bar{C}_{11} + 2\bar{C}_{12} + 4\bar{C}_{44})$	$1 + \frac{1}{3}g(6\rho_0\bar{W}^2 + \bar{C}_{111} + 6\bar{C}_{112} + 2\bar{C}_{123} + 4\bar{C}_{144} + 8\bar{C}_{166})$
7	Isotropic	$(1/\sqrt{3})[111]$	$(1/\sqrt{2})[1\bar{1}0]$	$\frac{1}{3}(\bar{C}_{11} - \bar{C}_{12} + \bar{C}_{44})$	$1 + \frac{1}{3}g(6\rho_0\bar{W}^2 + \bar{C}_{111} - \bar{C}_{123} + \bar{C}_{144} + 2\bar{C}_{166})$
8	[001]	[100]	[100]	$\bar{C}_{11}$	$\bar{S}_{12}(2\rho_0\bar{W}^2 + \bar{C}_{111} + 2\bar{C}_{112}) + \bar{S}_{11}\bar{C}_{112}$
9	[001]	[100]	[010]	$\bar{C}_{44}$	$\bar{S}_{12}(2\rho_0\bar{W}^2 + 2\bar{C}_{166}) + \bar{S}_{11}\bar{C}_{144}$
10	[001]	[100]	[001]	$\bar{C}_{44}$	$\bar{S}_{11}(2\rho_0\bar{W}^2 + \bar{C}_{111}) + \bar{S}_{12}(\bar{C}_{144} + \bar{C}_{166})$
11	[001]	$(1/\sqrt{2})[110]$	$(1/\sqrt{2})[110]$	$\frac{1}{2}(\bar{C}_{11} + \bar{C}_{12} + 2\bar{C}_{44})$	$\frac{1}{2}[\bar{S}_{12}(4\rho_0\bar{W}^2 + \bar{C}_{111} + 3\bar{C}_{112} + 4\bar{C}_{166}) + \bar{S}_{11}(\bar{C}_{112} + \bar{C}_{123} + 2\bar{C}_{144})]$
12	[001]	$(1/\sqrt{2})[110]$	$(1/\sqrt{2})[1\bar{1}0]$	$\frac{1}{2}(\bar{C}_{11} - \bar{C}_{12})$	$\frac{1}{2}\bar{S}_{12}(4\rho_0\bar{W}^2 + \bar{C}_{111} - \bar{C}_{112}) + \bar{S}_{11}(\bar{C}_{112} - \bar{C}_{123})$
13	[001]	$(1/\sqrt{2})[110]$	[001]	$\bar{C}_{44}$	$\bar{S}_{11}(2\rho_0\bar{W}^2 + 2\bar{C}_{166}) + \bar{S}_{12}(\bar{C}_{144} + \bar{C}_{166})$
14	[001]	[001]	[001]	$\bar{C}_{11}$	$1 + \bar{S}_{11}(2\rho_0\bar{W}^2 + \bar{C}_{111}) + 2\bar{S}_{12}\bar{C}_{112}$
15	[001]	[001]	[100] or [010]	$\bar{C}_{44}$	$1 + \bar{S}_{12}(2\rho_0\bar{W}^2 + \bar{C}_{144} + \bar{C}_{166}) + \bar{S}_{11}\bar{C}_{166}$
16	$(1/\sqrt{2})[110]$	$(1/\sqrt{2})[110]$	$(1/\sqrt{2})[110]$	$\frac{1}{2}(\bar{C}_{11} + \bar{C}_{12} + 2\bar{C}_{44})$	$\frac{1}{4}[(2g_1 - \bar{S}_{44})2\rho_0\bar{W}^2 + g_1\bar{C}_{111} + (g_3 + 2g_1)\bar{C}_{112} + 2\bar{S}_{12}(\bar{C}_{123} + 2\bar{C}_{144}) + 4(g_1 - \bar{S}_{44})\bar{C}_{166}]$
17	$(1/\sqrt{2})[110]$	$(1/\sqrt{2})[110]$	$(1/\sqrt{2})[1\bar{1}0]$	$\frac{1}{2}(\bar{C}_{11} - \bar{C}_{12})$	$\frac{1}{4}[(2g_1 + \bar{S}_{44})2\rho_0\bar{W}^2 + g_1\bar{C}_{111} + (g_3 - 2g_1)\bar{C}_{112} - 2\bar{S}_{12}\bar{C}_{123}]$
18	$(1/\sqrt{2})[1\bar{1}0]$	$(1/\sqrt{2})[110]$	[001]	$\bar{C}_{44}$	$\frac{1}{2}[4\bar{S}_{12}\rho_0\bar{W}^2 + g_1\bar{C}_{144} + g_3\bar{C}_{166} - \bar{S}_{44}\bar{C}_{456}]$
19	$(1/\sqrt{2})[1\bar{1}0]$	[001]	[001]	$\bar{C}_{11}$	$\bar{S}_{12}(2\rho_0\bar{W}^2 + \bar{C}_{111} + \bar{C}_{112}) + \bar{S}_{11}\bar{C}_{112}$
20	$(1/\sqrt{2})[1\bar{1}0]$	[001]	[100] or [010]	$\bar{C}_{44}$	$\frac{1}{2}[g_1(2\rho_0\bar{W}^2 + \bar{C}_{111} + \bar{C}_{144}) + g_3\bar{C}_{166}]$
21	$(1/\sqrt{2})[1\bar{1}0]$	[001]	$(1/\sqrt{2})[110]$	$\bar{C}_{44}$	$\frac{1}{2}[(2g_1 - \bar{S}_{44})\rho_0\bar{W}^2 + g_1\bar{C}_{144} + g_3\bar{C}_{166} - \bar{S}_{44}\bar{C}_{456}]$
22	$(1/\sqrt{2})[1\bar{1}0]$	[001]	$(1/\sqrt{2})[1\bar{1}0]$	$\bar{C}_{44}$	$\frac{1}{2}[(2g_1 + \bar{S}_{44})\rho_0\bar{W}^2 + g_1\bar{C}_{144} + g_3\bar{C}_{166} + \bar{S}_{44}\bar{C}_{456}]$

[a] The abbreviations are $g_1 = (\bar{S}_{11} + \bar{S}_{12})$, $g = (\bar{S}_{11} + 2\bar{S}_{12})$, $g_3 = (\bar{S}_{11} + 3\bar{S}_{12})$.

Table IV. Stress Derivatives of $\rho_0 W^2$, Evaluated at Zero Stress, for Isotropic Materials[a]

Case	Stress	$\bar{\kappa}$	$\bar{\mathbf{w}}$	$\rho_0 \bar{W}^2$	$-d(\rho_0 W^2)/dP$
1	Isotropic	Any	$\bar{\mathbf{w}} \parallel \bar{\kappa}$	$\lambda + 2\mu$	$1 + g(2\lambda + 4\mu + 6\zeta + 4\xi)$
2	Isotropic	Any	$\bar{\mathbf{w}} \perp \bar{\kappa}$	μ	$1 + g(2\mu + 3\xi - \tfrac{1}{2}\nu)$
3	Uniaxial	$\bar{\kappa} \perp \mathbf{t}$	$\bar{\mathbf{w}} \parallel \bar{\kappa}$	$\lambda + 2\mu$	$g[2\zeta - (\lambda/\mu)(\lambda + 2\mu + 2\xi)]$
4	Uniaxial	$\bar{\kappa} \perp \mathbf{t}$	$\mathbf{t} \parallel \bar{\mathbf{w}} \perp \bar{\kappa}$	μ	$g[2\mu + \xi + (\lambda/\mu)(\mu + \tfrac{1}{4}\nu)]$
5	Uniaxial	$\bar{\kappa} \perp \mathbf{t}$	$\mathbf{t} \perp \bar{\mathbf{w}} \perp \bar{\kappa}$	μ	$g[\xi - \tfrac{1}{2}\nu - (\lambda/\mu)(\mu + \tfrac{1}{2}\nu)]$

[a] The abbreviations are $g = 1/(3\lambda + 2\mu)$, and for uniaxial stress t is the stress direction. All elastic constants here are evaluated at zero stress.

Several results for cubic I crystals under isotropic pressure and uniaxial stress are listed in Table III. Under isotropic pressure the cubic symmetry remains, and it is possible to measure at most three independent quantities. Therefore in cases 1–7 of Table III, only three independent combinations of third-order elastic constants appear; these can be taken as $\bar{C}_{111} + 2\bar{C}_{112}$, $2\bar{C}_{112} + \bar{C}_{123}$, and $\bar{C}_{144} + 2\bar{C}_{166}$. For uniaxial stress it is not convenient to make measurements with propagation direction along the stress direction; however, some examples with $\kappa \parallel \mathbf{t}$ are included in Table III, since such measurements must be considered as potentially useful. There are sufficient experimental cases listed to determine uniquely all six third-order elastic constants for a cubic I crystal, without including $\kappa \parallel \mathbf{t}$ cases.

The results for cubic I crystal are specialized to an isotropic material in Table IV. Here there are only a few independent cases since there are no crystallographically significant directions. Note with uniaxial stress perpendicular to the propagation direction, the three pure modes allow determination of all three independent third-order elastic constants.

Results for tetragonal I crystals are listed in Table V. Again it is not possible to determine all the third-order elastic constants by measurements under isotropic pressure, but with inclusion of uniaxial stresses all twelve can be determined from the cases tabulated.

Thurston and Brugger[49] presented tables of $d(\rho_0 W^2)/dP$ evaluated at $P = 0$ for isotropic materials and for cubic I crystals. Table IV agrees with their results for isotropic materials, and Table III agrees (with two slight exceptions) with their results for cubic I. Brugger[57] has presented extensive tables for orthorhombic, tetragonal, rhombohedral, hexagonal, and cubic crystal systems; Table V for tetragonal I agrees with Brugger's results, but Table III for cubic I contains a number of differences from Brugger's results.

[57] K. Brugger, *J. Appl. Phys.* **36**, 768 (1965).

TABLE V. Stress Derivatives of $\rho_0 W^2$, Evaluated at Zero Stress, for Tetragonal I Crystals[a]

Case	Stress	$\boldsymbol{\kappa}$	$\bar{\mathbf{w}}$	$\rho_0 \bar{W}^2 = \rho_0 \bar{v}^2$	$-d(\rho_0 W^2)/dP$
1	Isotropic	[100]	[100]	$\bar{C}_{11}$	$1 + g_2(2\rho_0 \bar{W}^2 + \bar{C}_{111} + \bar{C}_{112}) + g_3 \bar{C}_{113}$
2	Isotropic	[100]	[010]	$\bar{C}_{66}$	$1 + g_2(2\rho_0 \bar{W}^2 + 2\bar{C}_{166}) + g_3 \bar{C}_{366}$
3	Isotropic	[100]	[001]	$\bar{C}_{44}$	$1 + g_3(2\rho_0 \bar{W}^2 + \bar{C}_{344}) + g_2(\bar{C}_{144} + \bar{C}_{155})$
4	Isotropic	[001]	[100] or [010]	$\bar{C}_{44}$	$1 + g_2(2\rho_0 \bar{W}^2 + \bar{C}_{144} + \bar{C}_{155}) + g_3 \bar{C}_{344}$
5	Isotropic	[001]	[001]	$\bar{C}_{33}$	$1 + g_3(2\rho_0 \bar{W}^2 + \bar{C}_{333}) + 2g_2 \bar{C}_{133}$
6	Isotropic	$(1/\sqrt{2})[110]$	$(1/\sqrt{2})[1\bar{1}0]$	$\tfrac{1}{2}(\bar{C}_{11} + \bar{C}_{12} + 2\bar{C}_{66})$	$1 + \tfrac{1}{2}g_2(4\rho_0 \bar{W}^2 + \bar{C}_{111} + 3\bar{C}_{112} + 4\bar{C}_{166}) + \tfrac{1}{2}g_3(\bar{C}_{113} + \bar{C}_{123} + 2\bar{C}_{366})$
7	Isotropic	$(1/\sqrt{2})[110]$	$(1/\sqrt{2})[1\bar{1}0]$	$\tfrac{1}{2}(\bar{C}_{11} - \bar{C}_{12})$	$1 + \tfrac{1}{2}g_2(2\rho_0 \bar{W}^2 + \bar{C}_{111} - \bar{C}_{112}) + \tfrac{1}{2}g_3(\bar{C}_{113} - \bar{C}_{123})$
8	Isotropic	$(1/\sqrt{2})[110]$	[001]	$\bar{C}_{44}$	$1 + g_3(2\rho_0 \bar{W}^2 + \bar{C}_{344}) + g_2(\bar{C}_{144} + \bar{C}_{155})$
9	[001]	[100]	[100]	$\bar{C}_{11}$	$\bar{S}_{13}(2\rho_0 \bar{W}^2 + \bar{C}_{111} + \bar{C}_{112}) + \bar{S}_{33} \bar{C}_{113}$
10	[001]	[100]	[010]	$\bar{C}_{66}$	$\bar{S}_{13}(2\rho_0 \bar{W}^2 + 2\bar{C}_{166}) + \bar{S}_{33} \bar{C}_{366}$
11	[001]	[100]	[001]	$\bar{C}_{44}$	$\bar{S}_{33}(2\rho_0 \bar{W}^2 + \bar{C}_{344}) + \bar{S}_{13}(\bar{C}_{144} + \bar{C}_{155})$
12	[001]	$(1/\sqrt{2})[110]$	$(1/\sqrt{2})[1\bar{1}0]$	$\tfrac{1}{2}(\bar{C}_{11} + \bar{C}_{12} + 2\bar{C}_{66})$	$\tfrac{1}{2}\bar{S}_{13}(4\rho_0 \bar{W}^2 + \bar{C}_{111} + 3\bar{C}_{112} + 4\bar{C}_{166}) + \tfrac{1}{2}\bar{S}_{33}(\bar{C}_{113} + \bar{C}_{123} + 2\bar{C}_{366})$
13	[001]	$(1/\sqrt{2})[110]$	$(1/\sqrt{2})[1\bar{1}0]$	$\tfrac{1}{2}(\bar{C}_{11} - \bar{C}_{12})$	$\tfrac{1}{2}\bar{S}_{13}(4\rho_0 \bar{W}^2 + \bar{C}_{111} - \bar{C}_{112}) + \tfrac{1}{2}\bar{S}_{33}(\bar{C}_{113} - \bar{C}_{123})$
14	[001]	$(1/\sqrt{2})[110]$	[001]	$\bar{C}_{44}$	$\bar{S}_{33}(2\rho_0 \bar{W}^2 + \bar{C}_{344}) + \bar{S}_{13}(\bar{C}_{144} + \bar{C}_{155})$
15	[010]	[100]	[100]	$\bar{C}_{11}$	$\bar{S}_{12}(2\rho_0 \bar{W}^2 + \bar{C}_{111}) + \bar{S}_{11} \bar{C}_{112} + \bar{S}_{13} \bar{C}_{113}$
16	[010]	[100]	[010]	$\bar{C}_{66}$	$\bar{S}_{11}(2\rho_0 \bar{W}^2 + \bar{C}_{166}) + \bar{S}_{12} \bar{C}_{166} + \bar{S}_{13} \bar{C}_{366}$

TABLE V.—Continued

Case	Stress	κ	$\bar{\mathbf{W}}$	$\rho_0 \bar{W}^2 = \rho_0 \bar{v}^2$	$-d(\rho_0 W^2)/dP$
17	[010]	[100]	[001]	$\bar{C}_{44}$	$\bar{S}_{13}(2\rho_0\bar{W}^2 + \bar{C}_{344}) + \bar{S}_{11}\bar{C}_{144} + \bar{S}_{12}\bar{C}_{155}$
18	[010]	[001]	[100]	$\bar{C}_{44}$	$\bar{S}_{12}(2\rho_0\bar{W}^2 + \bar{C}_{155}) + \bar{S}_{11}\bar{C}_{144} + \bar{S}_{13}\bar{C}_{344}$
19	[010]	[001]	[010]	$\bar{C}_{44}$	$\bar{S}_{11}(2\rho_0\bar{W}^2 + \bar{C}_{155}) + \bar{S}_{12}\bar{C}_{144} + \bar{S}_{13}\bar{C}_{344}$
20	[010]	[001]	[001]	$\bar{C}_{33}$	$\bar{S}_{13}(2\rho_0\bar{W}^2 + \bar{C}_{333}) + g_1\bar{C}_{133}$
21	[1$\bar{1}$0]	[0C1]	$(1/\sqrt{2})[110]$	$\bar{C}_{44}$	$\frac{1}{2}[(2g_1 - \bar{S}_{66})\rho_0\bar{W}^2 + g_1(\bar{C}_{144} + \bar{C}_{155}) + 2\bar{S}_{13}\bar{C}_{344} - \bar{S}_{66}\bar{C}_{456}]$
22	[1$\bar{1}$0]	[0C1]	$(1/\sqrt{2})[1\bar{1}0]$	$\bar{C}_{44}$	$\frac{1}{2}[(2g_1 + \bar{S}_{66})\rho_0\bar{W}^2 + g_1(\bar{C}_{144} + \bar{C}_{155}) + 2\bar{S}_{13}\bar{C}_{344} + \bar{S}_{66}\bar{C}_{456}]$
23	[1$\bar{1}$0]	[001]	[001]	$\bar{C}_{33}$	$\bar{S}_{13}(2\rho_0\bar{W}^2 + \bar{C}_{333}) + g_1\bar{C}_{133}$
24	[1$\bar{1}$0]	$(1/\sqrt{2})[110]$	$(1/\sqrt{2})[110]$	$\frac{1}{2}(\bar{C}_{11} + \bar{C}_{12} + 2\bar{C}_{66})$	$\frac{1}{4}[(2g_1 - \bar{S}_{66})2\rho_0\bar{W}^2 + g_1(\bar{C}_{111} + 3\bar{C}_{112} + 4\bar{C}_{166}) + 2\bar{S}_{13}(\bar{C}_{113} + \bar{C}_{123} + 2\bar{C}_{366}) - 4\bar{S}_{66}\bar{C}_{166}]$
25	[1$\bar{1}$0]	$(1/\sqrt{2})[110]$	$(1/\sqrt{2})[1\bar{1}0]$	$\frac{1}{2}(\bar{C}_{11} - \bar{C}_{12})$	$\frac{1}{4}[(2g_1 + \bar{S}_{66})2\rho_0\bar{W}^2 + g_1(\bar{C}_{111} - \bar{C}_{112}) + 2\bar{S}_{13}(\bar{C}_{113} - \bar{C}_{123})]$
26	[1$\bar{1}$0]	$(1/\sqrt{2})[110]$	[001]	$\bar{C}_{44}$	$\frac{1}{2}[2\bar{S}_{13}(2\rho_0\bar{W}^2 + \bar{C}_{344}) + g_1(\bar{C}_{144} + \bar{C}_{155}) - \bar{S}_{66}\bar{C}_{456}]$

[a] The abbreviations are $g_1 = \bar{S}_{11} + \bar{S}_{12}$, $g_2 = \bar{S}_{11} + \bar{S}_{12} + \bar{S}_{13}$, $g_3 = \bar{S}_{33} + 2\bar{S}_{13}$.

16. Wave Propagation in Stressed Crystals—Interpretation of Experiments

Experimental procedures for measuring the pressure derivatives of the three independent wave velocities for cubic materials have been described in the literature. Lazarus[58] made measurements on several cubic materials; Daniels and Smith[59] made measurements on Cu, Ag, Au; Daniels[60] on Na; and Bartels and Schuele[61] made measurements on NaCl and KCl at 195° and 295°K. These authors all listed their results as $dC_{\alpha\beta}/dP$; in the present notation this corresponds to $dB_{\alpha\beta}/dP$. Hughes and Kelly[62] measured the three third-order elastic constants for several isotropic materials by application of pressure and uniaxial stress; Bateman et al.[63] measured the six $C_{\alpha\beta\gamma}$ for Ge; Hiki and Granato[64] measured $C_{\alpha\beta\gamma}$ for Cu, Ag, Au; Thomas[65] measured $C_{\alpha\beta\gamma}$ for Al; Drabble and Strathen[66] measured $C_{\alpha\beta\gamma}$ for KCl, NaCl, and LiF; and Salama and Alers[67] measured $C_{\alpha\beta\gamma}$ for Cu at 4.2°, 77°, and 295°K. These and many other papers describe the experimental procedures. For interpretation of these experiments, we also wish to draw attention to the work of Waterman,[23] who derived corrections for the ultrasonic measurements of elastic constants due to slight misorientation of the crystal from pure mode directions.

Two methods for measuring very small changes in sound velocities have been discussed recently, and although they have not yet been applied to measurement of stress-induced variations, there is every reason to believe they soon will be. We will therefore relate these experiments to the present theory. First is the phase shift method, in which sending and receiving transducers are placed on opposite ends of a highly attenuating sample, and the sending transducer is operated continuously at a fixed frequency ω. If the sample is of sufficiently high attenuation, the echoes are negligible, and the phase of the signal which passes once through the sample may be monitored by the receiving transducer. Variations in the phase of 1 part in 10^7, due to the application of a magnetic field to the sample, were observed by Beattie and Uehling.[68,69] For a wave vector **k**, and sample length along

---

[58] D. Lazarus, *Phys. Rev.* **76**, 545 (1949).
[59] W. B. Daniels and C. S. Smith, *Phys. Rev.* **111**, 713 (1958).
[60] W. B. Daniels, *Phys. Rev.* **119**, 1246 (1960).
[61] R. A. Bartels and D. E. Schuele, *J. Phys. Chem. Solids* **26**, 537 (1965).
[62] D. S. Hughes and J. L. Kelly, *Phys. Rev.* **92**, 1145 (1953).
[63] T. Bateman, W. P. Mason, and H. J. McSkimin, *J. Appl. Phys.* **32**, 928 (1961).
[64] Y. Hiki and A. V. Granato, *Phys. Rev.* **144**, 411 (1966).
[65] J. F. Thomas, Jr., *Phys. Rev.* **175**, 955 (1968).
[66] J. R. Drabble and R. E. B. Strathen, *Proc. Phys. Soc. (London)* **92**, 1090 (1967).
[67] K. Salama and G. A. Alers, *Phys. Rev.* **161**, 673 (1967).
[68] A. G. Beattie and E. H. Uehling, *Phys. Rev.* **148**, 657 (1966).
[69] A. G. Beattie, *Phys. Rev.* **174**, 721 (1968).

k given by l_1 in the stressed configuration, and neglecting the stress-variation of the phase in the transducer and bonds, the phase may be taken to be

$$\varphi = |\mathbf{k}| l_1 = \omega(l_1/v), \quad \text{in radians.} \quad (16.1)$$

Just as in Section 9, we eliminate the l_1 dependence by introducing the reduced velocity W of (9.13), and write

$$\varphi = \omega(l_0/W), \quad (16.2)$$

where l_0 is the sample length along **k** at zero stress. Since ω is constant, differentiation of (16.2) with respect to the magnitude P of an arbitrary stress gives

$$d\varphi/dP = -(\omega l_0/W^2)(dW/dP) = -\tfrac{1}{2}(\omega l_0/\rho_0 W^3)[d(\rho_0 W^2)/dP]$$

$$= -\tfrac{1}{2}(\varphi/\rho_0 W^2)[d(\rho_0 W^2)/dP], \quad (16.3)$$

or

$$(1/\varphi)(d\varphi/dP) = -(1/2\rho_0 W^2)[d(\rho_0 W^2)/dP]. \quad (16.4)$$

When (16.4) is evaluated at $P = 0$, where $\varphi_0 = \omega(l_0/\bar{v})$, then the left-hand side is just the experimental quantity directly measured by the phase shift method, while the right-hand side is given in general by the equations of Section 10, and for particular examples by Tables III–V.

Another experimental technique, suitable for low-attenuating samples, is the mechanical resonance method. Here one sets up a mechanical resonance in the sample-transducer system, and then varies the frequency ω to maintain resonance as the external stress is varied. Variations in the resonance frequency of 1 part in 10^7, due to the application of a magnetic field rather than a stress, were observed by Melcher et al.[70] There is a slight difficulty in interpreting this experiment, since one measures a system phase velocity, which is not equal to the sound velocity of the sample, and which also has dispersion, i.e. is not constant as ω varies.[71,72] These effects are small at high frequencies, where the wavelength is small compared to all sample dimensions; neglecting these effects the resonance condition is

$$kl_1 = \text{constant}, \quad (16.5)$$

where k is the magnitude $|\mathbf{k}|$. The resonance frequency is then written

$$\omega = vk = kl_1(v/l_1) = kl_1(W/l_0). \quad (16.6)$$

[70] R. L. Melcher, D. I. Bolef, and J. B. Merry, *Rev. Sci. Instr.* **39**, 1618 (1968).
[71] H. J. McSkimin, *J. Acoust. Soc. Am.* **28**, 484 (1956).
[72] M. Redwood and J. Lamb, *Proc. Phys. Soc. (London)* **B70**, 136 (1957).

Since kl_1 is constant, the variation of the resonance frequency with applied stress is

$$d\omega/dP = (kl_1/l_0)(dW/dP) = \tfrac{1}{2}(kl_1/l_0)(1/\rho_0 W)[d(\rho_0 W^2)/dP]$$
$$= \tfrac{1}{2}(\omega/\rho_0 W^2)[d(\rho_0 W^2)/dP], \qquad (16.7)$$

or

$$(1/\omega)(d\omega/dP) = (1/2\rho_0 W^2)[d(\rho_0 W^2)/dP]. \qquad (16.8)$$

Now when (16.8) is evaluated at $P = 0$, where $\omega_0 = \bar{v}k$, then the left-hand side is the directly measured quantity of the mechanical resonance method, and the right-hand side is given by our present theory. Note in the analysis of the phase shift and mechanical resonance experiments, the stress may be isotropic pressure *or* uniaxial, with the notation of Section 15.

As it was pointed out in Section 10, if the wave propagation is adiabatic and the stress is applied isothermally, the third-order elastic constants appearing in the stress derivatives of $\rho_0 W^2$ are the mixed adiabatic–isothermal constants. Whenever the difference between these constants and the adiabatic or isothermal constants is of importance, one may convert to the adiabatic or isothermal constants by thermodynamics; we illustrate this procedure by treating a cubic I crystal under isotropic pressure. First, the stress derivatives given in Table III need to be interpreted more carefully if adiabatic–isothermal differences are being considered. With the aid of (10.13), some examples from Table III are:

Case	$(\rho_0 W^2)_S$	$-[\partial(\rho_0 W^2)_S/\partial P]_T$	
1	$\bar{C}^S_{11}$	$1 + \tfrac{1}{3}K^T(2\bar{C}^S_{11} + \bar{C}^*_{111} + 2\bar{C}^*_{112})$	(16.9a)
2	$\bar{C}^S_{44}$	$1 + \tfrac{1}{3}K^T(2\bar{C}^S_{44} + \bar{C}^*_{144} + 2\bar{C}^*_{166})$	(16.9b)
4	$\tfrac{1}{2}(\bar{C}^S_{11} - \bar{C}^S_{12})$	$1 + \tfrac{1}{3}K^T[\bar{C}^S_{11} - \bar{C}^S_{12} + \tfrac{1}{2}(\bar{C}^*_{111} - \bar{C}^*_{123})]$	(16.9c)

Here $\tfrac{1}{3}K^T = (\bar{S}^T_{11} + 2\bar{S}^T_{12})$, and the mixed constants $\bar{C}^*_{\alpha\beta\gamma}$ are defined by (10.14). The three independent combinations of $\bar{C}^*_{\alpha\beta\gamma}$ which can be determined by measurements under isotropic pressure, e.g., $\bar{C}^*_{111} + 2\bar{C}^*_{112}$, $2\bar{C}^*_{112} + \bar{C}^*_{123}$, and $\bar{C}^*_{144} + 2\bar{C}^*_{166}$, may be determined from these three cases. We presume these three combinations of $\bar{C}^*_{\alpha\beta\gamma}$ have been determined, and proceed now to relate them to similar combinations of adiabatic and isothermal constants.

When the thermodynamic variables are P, V, and T, so that no anisotropic stresses are considered, two thermodynamic results are[55]

$$K^T - K^S = (TV\beta^2/C_P);\tag{16.10}$$

$$(\partial/\partial P)_S = (\partial/\partial P)_T + (TV\beta/C_P)(\partial/\partial T)_P.\tag{16.11}$$

Here, as in Section 12, V and C_P may be taken per unit mass, or for any other common unit since they appear as V/C_P. Now write out the small pressure expansion (12.21) for say $C^S{}_{11}$ for an adiabatic strain, and also for an isothermal strain:

$$C^S{}_{11} = \bar{C}^S{}_{11} - \tfrac{1}{3}K^S(\bar{C}^S{}_{11} + \bar{C}^S{}_{111} + 2\bar{C}^S{}_{112})P + \cdots, \quad \text{constant } S;\tag{16.12}$$

$$C^S{}_{11} = \bar{C}^S{}_{11} - \tfrac{1}{3}K^T(\bar{C}^S{}_{11} + \bar{C}^*{}_{111} + 2\bar{C}^*{}_{112})P + \cdots, \quad \text{constant } T.\tag{16.13}$$

Differentiating (16.12) with respect to P at constant S, and (16.13) at constant T, using (16.11) and evaluating everything at $P = 0$ gives

$$(\partial C^S{}_{11}/\partial P)_S - (\partial C^S{}_{11}/\partial P)_T = (TV\beta/C_P)(\partial C^S{}_{11}/\partial T)_{P=0}$$
$$= \tfrac{1}{3}K^T(\bar{C}^S{}_{11} + \bar{C}^*{}_{111} + 2\bar{C}^*{}_{112})$$
$$- \tfrac{1}{3}K^S(\bar{C}^S{}_{11} + \bar{C}^S{}_{111} + 2\bar{C}^S{}_{112}).\tag{16.14}$$

The experimental quantity appearing in (16.9a) may be extracted from (16.14):

$$\tfrac{1}{3}K^T(\bar{C}^S{}_{11} + \bar{C}^*{}_{111} + 2\bar{C}^*{}_{112}) = \tfrac{1}{3}K^S(\bar{C}^S{}_{11} + \bar{C}^S{}_{111} + 2\bar{C}^S{}_{112})$$
$$+ (TV\beta/C_P)(\partial \bar{C}^S{}_{11}/\partial T)_P.\tag{16.15}$$

The other two experimental quantities in (16.9b) and (16.9c) are similarly determined by starting at Eqs. (16.12) and (16.13) with $C^S{}_{12}$ and $C^S{}_{44}$ in place of $C^S{}_{11}$. On the other hand, (16.14) may be solved for the combination $\bar{C}^S{}_{111} + 2\bar{C}^S{}_{112}$; this and the other two independent combinations of $\bar{C}^S{}_{\alpha\beta\gamma}$ are found to be

$$\bar{C}^S{}_{111} + 2\bar{C}^S{}_{112} = (K^T/K^S)(\bar{C}^*{}_{111} + 2\bar{C}^*{}_{112})$$
$$+ (TV\beta/K^S C_P)[\beta\bar{C}^S{}_{11} - 3(\partial\bar{C}^S{}_{11}/\partial T)_P],\tag{16.16a}$$

$$2\bar{C}^S{}_{112} + \bar{C}^S{}_{123} = (K^T/K^S)(2\bar{C}^*{}_{112} + \bar{C}^*{}_{123})$$
$$+ (TV\beta/K^S C_P)[\beta\bar{C}^S{}_{12} - 3(\partial\bar{C}^S{}_{12}/\partial T)_P],\tag{16.16b}$$

$$\bar{C}^S{}_{144} + 2\bar{C}^S{}_{166} = (K^T/K^S)(\bar{C}^*{}_{144} + 2\bar{C}^*{}_{166})$$
$$+ (TV\beta/K^S C_P)[\beta\bar{C}^S{}_{44} - 3(\partial\bar{C}^S{}_{44}/\partial T)_P],\tag{16.16c}$$

where we have used (16.10) also.

A similar procedure can be used to calculate the three combinations of $C^T{}_{\alpha\beta\gamma}$. Define B' by

$$B' = [\partial(B^S - B^T)/\partial P]_{T;P=0}, \qquad (16.17)$$

where $B^S = 1/K^S$, $B^T = 1/K^T$. From Section 12 we have

$$C^S{}_{11} - C^T{}_{11} = C^S{}_{12} - C^T{}_{12} = B^S - B^T, \qquad C^S{}_{44} - C^T{}_{44} = 0. \quad (16.18)$$

Now write out the small pressure expansion (12.21) for say $C^T{}_{11}$ for an isothermal strain:

$$C^T{}_{11} = \tilde{C}^T{}_{11} - \tfrac{1}{3}K^T(\tilde{C}^T{}_{11} + \tilde{C}^T{}_{111} + 2\tilde{C}^T{}_{112})P + \cdots, \qquad \text{constant } T. \quad (16.19)$$

Differentiating (16.19) and (16.13) with respect to P at constant T, using (16.17) and (16.18), and evaluating at $P = 0$ gives

$$(\partial C^S{}_{11}/\partial P)_T - (\partial C^T{}_{11}/\partial P)_T = B'$$
$$= \tfrac{1}{3}K^T(\tilde{C}^T{}_{11} + \tilde{C}^T{}_{111} + 2\tilde{C}^T{}_{112})$$
$$- \tfrac{1}{3}K^T(\tilde{C}^S{}_{11} + \tilde{C}^*{}_{111} + 2\tilde{C}^*{}_{112}). \quad (16.20)$$

This equation, and the two others obtained by starting at (16.19) with $C^T{}_{12}$ and $C^T{}_{44}$ in place of $C^T{}_{11}$, may be solved to give

$$\tilde{C}^T{}_{111} + 2\tilde{C}^T{}_{112} = \tilde{C}^*{}_{111} + 2\tilde{C}^*{}_{112} + (B^S - B^T + 3B^T B'), \quad (16.21\text{a})$$

$$2\tilde{C}^T{}_{112} + \tilde{C}^T{}_{123} = 2\tilde{C}^*{}_{112} + \tilde{C}^*{}_{123} + (B^S - B^T + 3B^T B'), \quad (16.21\text{b})$$

$$\tilde{C}^T{}_{144} + 2\tilde{C}^T{}_{166} = \tilde{C}^*{}_{144} + 2\tilde{C}^*{}_{166}. \quad (16.21\text{c})$$

Barsch[73] has derived the results (16.16) and (16.21) for a cubic I crystal under isotropic pressure. In addition, numerical values for the three combinations of $\tilde{C}^*{}_{\alpha\beta\gamma}$, of $\tilde{C}^S{}_{\alpha\beta\gamma}$, and of $\tilde{C}^T{}_{\alpha\beta\gamma}$ were tabulated by Barsch and Chang[74] for a number of cubic materials at room temperature.

17. Harmonic Generation—Tables of Results

The general theory of resonance harmonic generation for a crystal in the presence of initial stress is provided in Section 4. We will apply this theory to a few simple examples at zero stress to show how the experiments may be interpreted in terms of combinations of third-order elastic constants. The input wave has propagation direction $\bar{\kappa}$, eigenvector $\bar{w}$, and eigenvalue $\rho_0 \bar{v}^2 = \rho_0 \bar{W}^2$ at zero stress; the generated harmonic has dis-

---

[73] G. R. Barsch, *Phys. Status Solidi* **19**, 129 (1967).
[74] G. R. Barsch and Z. P. Chang, *Phys. Status Solidi* **19**, 139 (1967).

placement direction $\hat{\mathbf{w}}$, where $\hat{\mathbf{w}} = \bar{\mathbf{w}}$ for self-resonance and $\hat{\mathbf{w}} \neq \bar{\mathbf{w}}$ for mutual resonance. At any distance from the input face the amplitude of the input wave is a_1, that of the harmonic is a_2, and a measurement of $|a_2/a_1^2|$ yields the quantity $|\bar{\mathbf{D}} \cdot \hat{\mathbf{w}}|$, which depends on third-order elastic constants. In the next section we will discuss in more detail how to extract $|\bar{\mathbf{D}} \cdot \hat{\mathbf{w}}|$ from the experimental results; here we concentrate on calculating $|\bar{\mathbf{D}} \cdot \hat{\mathbf{w}}|$ for any given experimental situation.

The driving vector $\bar{\mathbf{D}}$ is given by (4.12) and (4.5):

$$\bar{D}_i = \bar{A}_{ijklpq} \bar{\kappa}_j \bar{\kappa}_l \bar{\kappa}_q \bar{w}_k \bar{w}_p; \tag{17.1}$$

$$\bar{A}_{ijklpq} = \bar{C}_{jlpq} \delta_{ik} + \bar{C}_{ijql} \delta_{kp} + \bar{C}_{jkql} \delta_{ip} + \bar{C}_{ijklpq}; \tag{17.2}$$

where for adiabatic propagation the elastic constants are all-adiabatic.

Take the example of a tetragonal I crystal with input wave propagation direction $\bar{\boldsymbol{\kappa}} = [001]$. Then from Section 13, especially from (13.11) and (13.12), the eigenvectors are $\bar{\mathbf{w}}(l) = [001]$, $\bar{\mathbf{w}}(t_1) = [010]$, and $\bar{\mathbf{w}}(t_2) = [100]$. Since the two transverse modes are degenerate, an arbitrary normalized vector $\bar{\mathbf{w}}(t) = [ab0]$ is still a pure transverse mode eigenvector with the same eigenvalue. For the longitudinal and arbitrary transverse input waves, (17.1) and (17.2) are easily worked out as follows.

$\bar{\boldsymbol{\kappa}} = [001]$, $\bar{\mathbf{w}} = [001]$:

$$\bar{D}_i = \bar{A}_{i33333} = \delta_{i3}(3\bar{C}_{33} + \bar{C}_{333}). \tag{17.3}$$

$\bar{\boldsymbol{\kappa}} = [001]$, $\bar{\mathbf{w}} = [ab0]$:

$$\bar{D}_i = a^2 \bar{A}_{i31313} + b^2 \bar{A}_{i32323} + ab(\bar{A}_{i31323} + \bar{A}_{i32313})$$
$$= \delta_{i3}[a^2 \bar{A}_{331313} + b^2 \bar{A}_{332323}]$$
$$= \delta_{i3}[a^2(\bar{C}_{33} + \bar{C}_{355}) + b^2(\bar{C}_{33} + \bar{C}_{344})]$$
$$= \delta_{i3}(\bar{C}_{33} + \bar{C}_{344}), \quad \text{since} \quad a^2 + b^2 = 1, \text{ and} \quad \bar{C}_{355} = \bar{C}_{344}. \tag{17.4}$$

Thus for the longitudinal input wave, and also for the arbitrary transverse input wave, $\bar{\mathbf{D}}$ is directed along the propagation direction $[001]$, and hence will pump only a longitudinal mode, i.e. $\hat{\mathbf{w}} = [001]$. The longitudinal input wave therefore generates a self-resonant harmonic, with

$$\bar{\mathbf{D}} \cdot \hat{\mathbf{w}} = \bar{\mathbf{D}} \cdot \bar{\mathbf{w}} = 3\bar{C}_{33} + \bar{C}_{333}, \quad \text{for} \quad \bar{\mathbf{w}} = [001]. \tag{17.5}$$

On the other hand, the transverse input wave drives a longitudinal displacement, but the driving frequency and wave vector are not in resonance with the frequency and wave vector of any longitudinal mode, so no resonant longitudinal harmonic is generated. Because of the degeneracy of

TABLE VI. RESONANCE HARMONIC GENERATION RESULTS FOR PURE MODE DIRECTIONS IN CUBIC I CRYSTALS AT ZERO STRESS

Case	$\bar{\kappa}$	$\hat{\bar{w}}$	$\rho_0 \bar{v}^2$	$\bar{D}$ direction	$\hat{\bar{w}}$	$\bar{D} \cdot \hat{\bar{w}}$
1	[100]	[100]	$\bar{C}_{11}$	$\bar{D} \parallel \bar{\kappa}$	[100]	$3\bar{C}_{11} + \bar{C}_{111}$
2	$(1/\sqrt{2})[110]$	$(1/\sqrt{2})[110]$	$\frac{1}{2}(\bar{C}_{11} + \bar{C}_{12} + 2\bar{C}_{44})$	$\bar{D} \parallel \bar{\kappa}$	$(1/\sqrt{2})[110]$	$3\rho_0\bar{v}^2 + \frac{1}{4}(\bar{C}_{111} + 3\bar{C}_{112} + 12\bar{C}_{166})$
3	$(1/\sqrt{3})[111]$	$(1/\sqrt{3})[111]$	$\frac{1}{3}(\bar{C}_{11} + 2\bar{C}_{12} + 4\bar{C}_{44})$	$\bar{D} \parallel \bar{\kappa}$	$(1/\sqrt{3})[111]$	$3\rho_0\bar{v}^2 + \frac{1}{9}(\bar{C}_{111} + 6\bar{C}_{112} + 2\bar{C}_{123} + 12\bar{C}_{144} + 24\bar{C}_{166} + 16\bar{C}_{456})$
4	$(1/\sqrt{3})[111]$	$(1/\sqrt{6})[11\bar{2}]$	$\frac{1}{3}(\bar{C}_{11} - \bar{C}_{12} + \bar{C}_{44})$	$\bar{D} \parallel \hat{\bar{w}}^a$	$(1/\sqrt{6})[11\bar{2}]$	$-(1/9\sqrt{2})(\bar{C}_{111} - 3\bar{C}_{112} + 2\bar{C}_{123} + 3\bar{C}_{144} - 3\bar{C}_{166} - 2\bar{C}_{456})$
5	$(1/\sqrt{3})[111]$	$(1/\sqrt{2})[1\bar{1}0]$	$\frac{1}{3}(\bar{C}_{11} - \bar{C}_{12} + \bar{C}_{44})$	$\bar{D} \perp \hat{\bar{w}}$	$(1/\sqrt{6})[11\bar{2}]$	$(1/9\sqrt{2})(\bar{C}_{111} - 3\bar{C}_{112} + 2\bar{C}_{123} + 3\bar{C}_{144} - 3\bar{C}_{166} - 2\bar{C}_{456})$

[a] $\bar{D}$ has a component $\parallel \hat{\bar{w}}$, and lies in the $\bar{\kappa}, \bar{w}$ plane.

the transverse modes, any transverse mode with displacement vector [$cd0$] is mutually resonant with the input transverse wave, but such a mode cannot be driven since $\bar{\mathbf{D}} \cdot \hat{\mathbf{w}} = 0$ for this case, and hence no resonant transverse harmonic is generated.

We have worked out $\bar{\mathbf{D}} \cdot \hat{\mathbf{w}}$ for all the pure mode directions considered in Sections 12 and 13 for cubic I and tetragonal I crystals. The results for the resonant cases only are given in Table VI for cubic I, and in Table VII for tetragonal I. For cubic I crystals, the longitudinal self-resonant cases 1, 2, and 3 were previously worked out by Gauster and Breazeale.[75] Buck and Thompson[28] worked out the three longitudinal self-resonant cases and the transverse self-resonant case 4 for cubic I crystals. For the transverse mutual-resonance case 5, $|\bar{\mathbf{D}} \cdot \hat{\mathbf{w}}|$ is the same as for the self-resonance case 4. In addition, for a cubic I crystal, Buck and Thompson[28] tabulated the

TABLE VII. RESONANCE HARMONIC GENERATION RESULTS FOR PURE MODE DIRECTIONS IN TETRAGONAL I CRYSTALS AT ZERO STRESS[a]

Case	$\bar{\boldsymbol{\kappa}}$	$\bar{\mathbf{w}}$	$\rho_0 \bar{v}^2$	$\bar{\mathbf{D}} \cdot \bar{\mathbf{w}}$
1	[100]	[100]	$\bar{C}_{11}$	$3\bar{C}_{11} + \bar{C}_{111}$
2	[001]	[001]	$\bar{C}_{33}$	$3\bar{C}_{33} + \bar{C}_{333}$
3	$(1/\sqrt{2})$[110]	$(1/\sqrt{2})$[110]	$\frac{1}{2}(\bar{C}_{11} + \bar{C}_{12} + 2\bar{C}_{66})$	$3\rho_0 \bar{v}^2 + \frac{1}{4}(\bar{C}_{111} + 3\bar{C}_{112} + 12\bar{C}_{166})$

[a] For each case $\bar{\mathbf{D}}$ is parallel to $\bar{\boldsymbol{\kappa}}$.

self-resonant harmonic amplitudes for an input wave which is arbitrary within the restriction that $\bar{\boldsymbol{\kappa}}$ and $\bar{\mathbf{w}}$ both lie in the same $(1\bar{1}0)$ plane; this includes nonpure mode directions. Note Buck and Thompson wrote their results in terms of the Birch third-order elastic constants, and for the amplitudes measured on a stress-free surface. Cases 1–4 of Table VI agree with the results of Buck and Thompson, when the relation (11.4) between Birch and Brugger third-order elastic constants is taken into account. Holt and Ford[29] considered arbitrary propagation directions for the examples of Cu and NaCl, found the normal mode (not necessarily pure mode) eigenvectors, and tabulated expressions for the self-resonant harmonic amplitudes; all this was done numerically in terms of measured values of the second-order elastic constants. Their results for pure mode directions agree with Table VI.

[75] W. B. Gauster and M. A. Breazeale, *Phys. Rev.* **168**, 655 (1968).

18. Harmonic Generation—Interpretation of Experiments

There are several clearly defined experimental situations, for which slightly different equations should be used for interpreting harmonic generation experiments. Let l_0 be the distance traveled by the waves, i.e. l_0 is the distance from the input face, measured along $\bar{\kappa}$ at zero stress. Then if the attenuation in the distance l_0 is small, Eq. (5.20) applies:

$$| a_2/a_1^2 | = k^2 l_0 | \bar{\mathbf{D}} \cdot \hat{\mathbf{w}} |/8\rho_0 \bar{v}^2, \qquad \text{interior of material;} \qquad (18.1)$$

$$| a_2/a_1^2 | = k^2 l_0 | \bar{\mathbf{D}} \cdot \hat{\mathbf{w}} |/16\rho_0 \bar{v}^2, \qquad \text{stress free surface;} \qquad (18.2)$$

where k is the wave vector magnitude and $\bar{v}$ the velocity of the input wave, and a_1 and a_2 are the amplitudes of the input wave and the generated harmonic, respectively, at l_0. If the attenuation is not small, we simply replace l_0 in the above two equations by

$$l_0 \to \{1 - \exp[-(\alpha_2 - 2\alpha_1)l_0]\}/(\alpha_2 - 2\alpha_1), \qquad \text{large attenuation,} \qquad (18.3)$$

according to (5.19).

The first question that arises in the interpretation of harmonic generation experiments is how well does the first-order perturbation treatment apply, or how does one know the experiment is in the range of applicability of the first-order perturbation results. The formal perturbation requirement is $a_2 \ll a_1$; this has been well satisfied in the experiments. Gedroits and Krasil'nikov[41] found for a Mg–Al alloy at the distance of maximum a_2 that $a_2 \approx 0.02 a_1$, while in the experiments of Gauster and Breazeale[75] on Cu, the condition was $a_2 \lesssim 0.04 a_1$. However, the question is not answered quite so simply. The theory has neglected other effects such as higher-order perturbation contributions and nonresonant displacements; in order to use the present results these displacements should have amplitudes small compared to a_2. These effects are, of course, derivable from the general equation of motion (4.4) with higher-order terms included.

With regard to higher-order perturbation effects, if the perturbation series is well behaved, then $a_2 \ll a_1$ implies higher-harmonic amplitudes $\ll a_2$. A more direct test of the first-order results is to make experimental studies of the relations implied by (18.1) and (18.2):

$$a_2 \propto a_1^2, \qquad \text{at fixed } l_0; \qquad (18.4)$$

$$a_2 \propto l_0, \qquad \text{at fixed } a_1. \qquad (18.5)$$

The first of these has been observed for a large variation in the amplitudes by Breazeale and Thompson[76]; the second was observed by Breazeale and

---

[76] M. A. Breazeale and D. O. Thompson, *Appl. Phys. Letters* **3**, 77 (1963).

Ford,[31] where the attenuation was small and a_1 was taken as the input amplitude at $l_0 = 0$; and both relations were again observed for three propagation directions in single-crystal Cu by Gauster and Breazeale.[75] We further note that if the attenuation is large, (18.5) will break down but (18.4) will still hold. Actually (18.5) will still hold with l_0 replaced according to (18.3), and this relation could certainly be tested experimentally.

On the other hand, if higher-order perturbation effects are important, the relation (18.4) will break down; this may be concluded as follows. Second-order perturbation contributions arise from second-order contributions of the nonlinear term we have studied, and from the second-order nonlinear term which we dropped in the equation of motion (4.4). Melngailis et al.[77] have iterated once again the first-order nonlinear one-dimensional equation of motion (4.29); they find this second-order perturbation contribution contains some additional fundamental displacement a_1, plus some higher harmonics. This has the effect of destroying (18.4), and we expect similar effects from the second-order nonlinear terms. Thus it appears most likely that as long as (18.4) is satisfied experimentally, then higher-order perturbation effects can be safely neglected.

The generation of nonresonant displacements was discussed extensively by Buck and Thompson.[28] If the distance l_0 through which a resonant mode is pumped contains many wavelengths, then the nonresonant displacements generated by the input wave at any point, as well as those generated at a stress-free surface, will be small compared to the resonant harmonic displacement. Further, if the distance l_0 is large compared to the length of the pulse train, then the displacements generated by the interaction of the input wave and its reflected component, in the region where these two waves overlap, will also be small compared to the resonant harmonic displacement. Finally, in order to apply the present steady-state theory, the pulse train should be long enough to contain many wavelengths; otherwise a wave packet analysis is appropriate.

Hikata and Elbaum[78,79] pointed out that the input wave will cause motion of dislocations, and this motion will also generate harmonics; in other words, the input wave couples to the harmonic through the dislocation field, as well as through the elastic nonlinearity. For small attenuation and uniform distribution of dislocations, Hikata and Elbaum found the dislocation coupling also leads to Eqs. (18.4) and (18.5). Therefore to measure third-order elastic constants by harmonic generation, it is desirable to eliminate dislocation effects experimentally, e.g. by neutron irradiation,[75] or by prestressing.[64]

[77] J. Melngailis, A. A. Maradudin, and A. Seeger, *Phys. Rev.* **131**, 1972 (1963).
[78] A. Hikata and C. Elbaum, *Phys. Rev.* **144**, 469 (1966).
[79] A. Hikata, F. A. Sewell, Jr., and C. Elbaum, *Phys. Rev.* **151**, 442 (1966).

Finally it should be recalled from Section 5 that Eqs. (18.1)–(18.3) apply for the first pass of the input wave through the sample, and the first reflection from the end of the sample; for multiple passes or echoes the interpretation is more complicated. In addition, a further analysis is required to determine the sign of $\bar{\mathbf{D}} \cdot \hat{\mathbf{w}}$.

The measurement of the amplitudes at the output end of the sample can be accomplished by transducers,[31,76] for either longitudinal or transverse waves. However, the sample surface is not stress-free, since the transducer is mechanically coupled to the sample, and it is difficult to know what effect this has on the surface vibration amplitudes. The capacitive microphone developed by Gauster and Breazeale[80] appears to be the most sensitive device at present for measuring surface amplitudes of longitudinal displacements; in this experimental arrangement the surface is essentially stress-free. This device has been used for several accurate measurements.[40,75] Mikhailov and Shutilov[81] used an electric current generator to measure longitudinal displacements of the input wave at the output end of the sample. They deposited a thin metal strip across the output face, attached leads, applied a magnetic field along the propagation direction, and measured the current at the appropriate frequency. It appears this procedure should be accurate enough to measure the harmonic amplitudes also; it has the advantage that the surface is essentially stress-free, and transverse displacements can be measured if the magnetic field is turned around. Unfortunately the presence of the magnetic field may complicate the interpretation of the measurements. Melngailis et al.[77] suggested measuring amplitudes of longitudinal waves in the interior of transparent materials by light diffraction experiments (Raman–Nath diffraction). This method was studied by Parker et al.[82] who experimentally verified the relations (18.4) and (18.5) and also determined C_{111} for NaCl.

An interesting extension of harmonic generation experiments is the observation of the interaction of two noncollinear input elastic waves to generate sum- and difference-frequency waves. This experiment is also governed by the nonlinear equation of motion (4.4), and a perturbation solution is generally valid, where one starts with two input waves instead of one. If a_1 and a_2 are the amplitudes of the two input waves, and a_3 is the amplitude of the wave which is generated in the region where the two input waves interact, then a_3 is proportional to $a_1 a_2 G$, where G is a function of second- and third-order elastic constants. Again there are resonance conditions which give ω and $\mathbf{k}$ for the generated wave; these conditions arise

---

[80] W. B. Gauster and M. A. Breazeale, *Rev. Sci. Instr.* **37,** 1544 (1966).

[81] I. G. Mikhailov and V. A. Shutilov, *Soviet Phys.—Acoust.* (*English Transl.*) **10,** 77 (1964).

[82] J. H. Parker, Jr., E. F. Kelly, and D. I. Bolef, *Appl. Phys. Letters* **5,** 7 (1964).

from conservation of energy and momentum requirements. This experiment is potentially a versatile method for measuring third-order elastic constants, but has not yet been developed enough to give these results.

The theory for this experiment was studied by Jones and Kobett,[83] who found the resonance conditions for an isotropic material when the input waves are both longitudinal, both transverse, or one longitudinal and one transverse. For several isotropic materials, the scattered wave was observed by Rollins,[84] and the resonance conditions were experimentally verified. The theory for isotropic materials was reformulated in terms of wave packets, instead of plane waves, by Childress and Hambrick.[85] Again for isotropic materials, Taylor and Rollins[86] represented the elastic waves as incoherent beams of phonons, calculated the rate of generation of sum- and difference-frequency waves with the aid of quantum mechanical transition probabilities, and then recovered the amplitudes of the elastic waves by classically relating the phonon energy density to the wave amplitude. The resonance conditions found by Taylor and Rollins are in agreement with those of Jones and Kobett. These conditions were verified experimentally by Rollins et al.[87] for the frequencies, propagation vectors, polarization vectors, and amplitudes of the three elastic waves involved in each resonance. Barrett and Matsinger[88] did the calculations by considering the elastic waves as coherent beams of phonons, and presented experimental data which again verified the resonance conditions. Finally, Holt and Ford[89] carried out a calculation similar to that of Taylor and Rollins[86] for the case of anisotropic materials, and pointed out the third-order elastic constants may be determined from measurements of the amplitudes of the three elastic waves for a number of resonant situations in a given crystal. For the example of Cu, Holt and Ford worked out numerically the combinations of third-order elastic constants which may be determined from ten resonance conditions; these are sufficient to determine all $C_{\alpha\beta\gamma}$ except C_{111}.

We would like to comment that the theoretical procedure of representing the elastic waves as beams of phonons, and then using quantum mechanical perturbation theory, is quite cumbersome and is not required for the analysis of the experiments. In the experiment, one generates two noncollinear ultrasonic waves and observes a third sum- or difference-

---

[83] G. L. Jones and D. R. Kobett, *J. Acoust. Soc. Am.* **35**, 5 (1963).
[84] F. R. Rollins, Jr., *Appl. Phys. Letters* **2**, 147 (1963).
[85] J. D. Childress and C. G. Hambrick, *Phys. Rev.* **136**, A411 (1964).
[86] L. H. Taylor and F. R. Rollins, Jr., *Phys. Rev.* **136**, A591 (1964).
[87] F. R. Rollins, Jr., L. H. Taylor, and P. H. Todd, Jr., *Phys. Rev.* **136**, A597 (1964).
[88] H. H. Barrett and J. H. Matsinger, *Phys. Rev.* **154**, 877 (1967).
[89] A. C. Holt and J. Ford, *J. Appl. Phys.* **40**, 142 (1969).

frequency wave with a definite propagation direction. Note the propagation directions of all three waves lie in a plane. The properties of ultrasonic waves are governed by the general theory of Part II, and classical perturbation theory for the interactions among ultrasonic waves, as in the harmonic generation theory of Section 4, may be employed. Wave packets may be used in place of plane waves.

VI. Fourth-Order Elastic Constants of Cubic Crystals

19. Application of Isotropic Pressure or Uniaxial Stress

The fourth-order elastic constants of a crystal can be determined from measurements of the second stress derivatives of small-amplitude elastic wave velocities, provided of course the second- and third-order elastic constants are known. Although such experiments are very difficult, and have not yet been performed, it is only a matter of time until they are carried out. The analysis of the experiments is based on the second-order expansion of the reduced propagation matrices **M**, given in Section 9. We will illustrate the analysis for a cubic I crystal under isotropic pressure, and also for a cubic I crystal under uniaxial stress along [001]. The differences between adiabatic, isothermal, and mixed higher-order elastic constants will be neglected.

For isotropic pressure, the stress is

$$T_{ij}(\mathbf{X}) = -P\,\delta_{ij}, \tag{19.1}$$

and from (12.14) and (12.15) the Lagrangian strain parameters n_{ij} measuring strain from zero pressure to the pressure P are

$$n_{ij} = n\,\delta_{ij}; \qquad n = -gP + \tfrac{1}{2}g^2(2 - gh)P^2. \tag{19.2}$$

Here n is correct to order P^2, and g and h are given by (12.16) and (12.17). Equations (9.27) and (9.31) give the contributions to the reduced propagation matrices to second order in the stresses and strains; for the present example these contributions simplify to

$$M_{ik} = \bar{M}_{ik}(1 + 2n) - P\,\delta_{ik}(1 + n) \\ + [\bar{C}_{ijklrr}(n + 2n^2) + \tfrac{1}{2}\bar{C}_{ijklrrss}n^2]\bar{\kappa}_j\bar{\kappa}_l. \tag{19.3}$$

Take the example of $\bar{\kappa} = [100]$. Then from the cubic crystal results of Section 12 evaluated at zero stress,

$$\bar{M}_{11} = \bar{C}_{11}, \qquad \bar{M}_{22} = \bar{M}_{33} = \bar{C}_{44}, \qquad \bar{M}_{12} = \bar{M}_{23} = \bar{M}_{31} = 0. \tag{19.4}$$

Let N_{ik} denote the term in (19.3) containing third- and fourth-order elastic constants. For $\bar{\kappa} = [100]$ this is

$$N_{ik} = \tilde{C}_{i1k1rr}(n + 2n^2) + \tfrac{1}{2}\tilde{C}_{i1k1rrss}n^2. \tag{19.5}$$

Each matrix element may be worked out and transformed to Voigt notation as follows:

$$N_{11} = \tilde{C}_{1111rr}(n + 2n^2) + \tfrac{1}{2}\tilde{C}_{1111rrss}n^2,$$
$$= (\tilde{C}_{111} + 2\tilde{C}_{112})(n + 2n^2) + \tfrac{1}{2}(\tilde{C}_{1111} + 2\tilde{C}_{1122} + 4\tilde{C}_{1112} + 2\tilde{C}_{1123})n^2; \tag{19.6}$$

$$N_{22} = N_{33} = \tilde{C}_{2121rr}(n + 2n^2) + \tfrac{1}{2}\tilde{C}_{2121rrss}n^2,$$
$$= (2\tilde{C}_{166} + \tilde{C}_{144})(n + 2n^2) + \tfrac{1}{2}(2\tilde{C}_{1166} + \tilde{C}_{1144} + 2\tilde{C}_{1266} + 4\tilde{C}_{1244})n^2; \tag{19.7}$$

$$N_{12} = N_{23} = N_{31} = 0. \tag{19.8}$$

The strain n can be eliminated for the pressure with (19.2), and the total reduced propagation matrix for $\bar{\kappa} = [100]$ is written

$$\mathbf{M} = \begin{pmatrix} E_{11} & 0 & 0 \\ 0 & E_{44} & 0 \\ 0 & 0 & E_{44} \end{pmatrix}, \quad \text{for } \bar{\kappa} = [100], \tag{19.9}$$

where

$$E_{11} = \tilde{C}_{11} - gP(g^{-1} + 2\tilde{C}_{11} + \tilde{C}_{111} + 2\tilde{C}_{112})$$
$$+ \tfrac{1}{2}g^2P^2[2g^{-1} - 8\tilde{C}_{11} + (2\tilde{C}_{11} + \tilde{C}_{111} + 2\tilde{C}_{112})(6 - gh)$$
$$+ (\tilde{C}_{1111} + 4\tilde{C}_{1112} + 2\tilde{C}_{1122} + 2\tilde{C}_{1123})]; \tag{19.10}$$

$$E_{44} = \tilde{C}_{44} - gP(g^{-1} + 2\tilde{C}_{44} + \tilde{C}_{144} + 2\tilde{C}_{166})$$
$$+ \tfrac{1}{2}g^2P^2[2g^{-1} - 8\tilde{C}_{44} + (2\tilde{C}_{44} + \tilde{C}_{144} + 2\tilde{C}_{166})(6 - gh)$$
$$+ (\tilde{C}_{1144} + 2\tilde{C}_{1166} + 4\tilde{C}_{1244} + 2\tilde{C}_{1266})]; \tag{19.11}$$

$$g = \tilde{S}_{11} + 2\tilde{S}_{12}, \quad g^{-1} = 1/g = \tilde{C}_{11} + 2\tilde{C}_{12}, \quad h = \tilde{C}_{111} + 6\tilde{C}_{112} + 2\tilde{C}_{123}.$$

Here the reduced matrix $\mathbf{M}$ is of exactly the same form as $\bar{\mathbf{M}}$, with $\tilde{C}_{11}$ and $\tilde{C}_{44}$ replaced by E_{11} and E_{44}. Therefore for this propagation direction, the eigenvectors are the same as at zero pressure, and the eigenvalues are simply E_{11} for the longitudinal mode and E_{44} for the two transverse modes, at any pressure. This is to be expected since by assumption the cubic I symmetry remains as a function of P. The same result holds for the other propagation directions, and the reduced eigenvalues may be summarized as follows.

$\bar{\kappa} = [100]$:

$$\mathbf{w}(l) = [100], \quad \rho_0 W^2 = E_{11}; \tag{19.12a}$$
$$\mathbf{w}(t_1) = [010], \quad \mathbf{w}(t_2) = [001], \quad \rho_0 W^2 = E_{44}. \tag{19.12b}$$

$\bar{\kappa} = (1/\sqrt{2})[110]$:

$$\mathbf{w}(l) = (1/\sqrt{2})[110], \quad \rho_0 W^2 = \tfrac{1}{2}(E_{11} + E_{12} + 2E_{44}); \tag{19.13a}$$
$$\mathbf{w}(t_1) = (1/\sqrt{2})[1\bar{1}0], \quad \rho_0 W^2 = \tfrac{1}{2}(E_{11} - E_{12}); \tag{19.13b}$$
$$\mathbf{w}(t_2) = [001], \quad \rho_0 W^2 = E_{44}. \tag{19.13c}$$

$\bar{\kappa} = (1/\sqrt{3})[111]$:

$$\mathbf{w}(l) = (1/\sqrt{3})[111], \quad \rho_0 W^2 = \tfrac{1}{3}(E_{11} + 2E_{12} + 4E_{44}); \tag{19.14a}$$
$$\mathbf{w}(t_1) = (1/\sqrt{2})[1\bar{1}0],$$
$$\mathbf{w}(t_2) = (1/\sqrt{6})[11\bar{2}], \quad \rho_0 W^2 = \tfrac{1}{3}(E_{11} - E_{12} + E_{44}). \tag{19.14b}$$

Here also we have used the additional matrix element

$$E_{12} = \tilde{C}_{12} - gP(-g^{-1} + 2\tilde{C}_{12} + 2\tilde{C}_{112} + \tilde{C}_{123})$$
$$+ \tfrac{1}{2}g^2P^2[-2g^{-1} - 8\tilde{C}_{12} + (2\tilde{C}_{12} + 2\tilde{C}_{112} + \tilde{C}_{123})(6-gh)$$
$$+ (2\tilde{C}_{1112} + 2\tilde{C}_{1122} + 5\tilde{C}_{1123})]. \tag{19.15}$$

Note these reduced eigenvalues $\rho_0 W^2$ in terms of $E_{\alpha\beta}$ are of the same form as the eigenvalues $\rho_1 v^2$ in terms of the stress–strain coefficients $B_{\alpha\beta}$, listed in Section 12. This is to be expected since from (9.14) the two sets of eigenvalues are related by

$$\rho_0 W^2 = (\rho_0/\rho_1)(l_0/l_1)^2(\rho_1 v^2). \tag{19.16}$$

This means

$$E_{\alpha\beta} = (\rho_0/\rho_1)(l_0/l_1)^2 B_{\alpha\beta}, \tag{19.17}$$

and indeed we can recover the above results for $E_{\alpha\beta}$ by expanding this equation in powers of the pressure.

The example of uniaxial stress is more interesting. Let the stress be along [001], so that

$$T_{ij}(\mathbf{X}) = -P\,\delta_{i3}\,\delta_{j3}, \tag{19.18}$$

and for convenience in interpreting experimental results introduce the reduced stress Q defined by

$$Q = (s_1/s_0)P. \tag{19.19}$$

Here s_1 and s_0 are the areas of the (001) faces at stress P and at zero stress, respectively, so that Q is just the total uniaxial force divided by the zero-stress area s_0.

The first step is to calculate the strains corresponding to the stress P, including the ratio s_1/s_0. The general expansion of the strains n_{ij} from $\bar{\mathbf{X}}$ to $\mathbf{X}$ is given by (9.10); with the stress (19.18) and the scheme of elastic constants for a cubic I crystal this is worked out to give

$$n_{11} = n_{22} = -\bar{S}_{12}P + \cdots, \tag{19.20}$$

$$n_{33} = -\bar{S}_{11}P + \cdots. \tag{19.21}$$

We defer writing the higher-order terms until s_1/s_0 is calculated. To do this let $|\Delta\bar{\mathbf{X}}|$ be the length of a material line at zero stress, $|\Delta\mathbf{X}|$ be the length of the same material line at stress P, and apply the general relation (1.6) for the change in line lengths under the strain from $\bar{\mathbf{X}}$ to $\mathbf{X}$:

$$|\Delta\mathbf{X}|^2 = |\Delta\bar{\mathbf{X}}|^2 + 2n_{ij}\,\Delta\bar{X}_i\,\Delta\bar{X}_j. \tag{19.22}$$

If $\Delta\bar{\mathbf{X}} = [100]$ or $[010]$, or any vector in the (001) plane, since $n_{11} = n_{22}$ from (19.20), we have

$$|\Delta\mathbf{X}|^2 = |\Delta\bar{\mathbf{X}}|^2(1 + 2n_{11}), \qquad \Delta\bar{\mathbf{X}} \text{ in (001)}. \tag{19.23}$$

Therefore the ratio of areas is just

$$s_1/s_0 = 1 + 2n_{11} = 1 - 2\bar{S}_{12}P + \cdots,$$

$$= 1 - 2\bar{S}_{12}Q + \cdots, \tag{19.24}$$

correct to order P or Q. Then from (19.19) we can write

$$P = Q + 2\bar{S}_{12}Q^2 + \cdots. \tag{19.25}$$

Now the strains n_{ij}, which are expanded in powers of P in (19.20) and (19.21), can be rewritten in terms of Q; the result to second order in Q is

$$n_{11} = n_{22} = -\bar{S}_{12}Q - \tfrac{1}{2}e_1 Q^2, \tag{19.26}$$

$$n_{33} = -\bar{S}_{11}Q - \tfrac{1}{2}e_2 Q^2, \tag{19.27}$$

$$n_{12} = n_{23} = n_{31} = 0, \tag{19.28}$$

where

$$e_1 = 2\tilde{S}_{11}\tilde{S}_{12} + (\tilde{C}_{111} + 2\tilde{C}_{123})\tilde{S}_{12}(\tilde{S}_{11}^2 + \tilde{S}_{11}\tilde{S}_{12} + \tilde{S}_{12}^2)$$
$$+ \tilde{C}_{112}(\tilde{S}_{11}^3 + 3\tilde{S}_{11}^2\tilde{S}_{12} + 9\tilde{S}_{11}\tilde{S}_{12}^2 + 5\tilde{S}_{12}^3) \qquad (19.29)$$

$$e_2 = 2\tilde{S}_{11}^2 + \tilde{C}_{111}(\tilde{S}_{11}^3 + 2\tilde{S}_{12}^3) + 6\tilde{C}_{112}\tilde{S}_{12}(\tilde{S}_{11}^2 + \tilde{S}_{11}\tilde{S}_{12} + \tilde{S}_{12}^2)$$
$$+ 6\tilde{C}_{123}\tilde{S}_{11}\tilde{S}_{12}^2. \qquad (19.30)$$

It is a straightforward calculation to work out the matrix elements M_{ik} from the general equations (9.27) and (9.31), with the above strains as functions of Q. The general symmetry of the matrices $\mathbf{M}$ is the same as that of tetragonal I under uniaxial stress along [001]; for propagation directions [100] and $(1/\sqrt{2})$[110] we have the following results.

$\bar{\kappa} = [100]$:

$$\mathbf{M} = \begin{pmatrix} F_{11} & 0 & 0 \\ 0 & F_{66} & 0 \\ 0 & 0 & F_{44} \end{pmatrix} \qquad (19.31)$$

$$\mathbf{w}(l) = [100], \qquad \rho_0 W^2 = F_{11}; \qquad (19.32a)$$
$$\mathbf{w}(t_1) = [010], \qquad \rho_0 W^2 = F_{66}; \qquad (19.32b)$$
$$\mathbf{w}(t_2) = [001], \qquad \rho_0 W^2 = F_{44}. \qquad (19.32c)$$

Note in this example the stress along [001] breaks the degeneracy of the two transverse modes.

$\bar{\kappa} = (1/\sqrt{2})[110]$:

$$\mathbf{M} = \frac{1}{2}\begin{pmatrix} F_{11} + F_{66} & F_{12} + F_{66} & 0 \\ F_{12} + F_{66} & F_{11} + F_{66} & 0 \\ 0 & 0 & 2F_{44} \end{pmatrix} \qquad (19.33)$$

$$\mathbf{w}(l) = (1/\sqrt{2})[110], \qquad \rho_0 W^2 = \tfrac{1}{2}(F_{11} + F_{12} + 2F_{66}); \qquad (19.34a)$$
$$\mathbf{w}(t_1) = (1/\sqrt{2})[1\bar{1}0], \qquad \rho_0 W^2 = \tfrac{1}{2}(F_{11} - F_{12}); \qquad (19.34b)$$
$$\mathbf{w}(t_2) = [001], \qquad \rho_0 W^2 = F_{44}. \qquad (19.34c)$$

The matrix elements are defined as follows, to second order in Q:

$$F_{11} = \bar{C}_{11} - Q[(2\bar{C}_{11} + \bar{C}_{111} + \bar{C}_{112})\bar{S}_{12} + \bar{C}_{112}\bar{S}_{11}]$$
$$+ \tfrac{1}{2}Q^2[(2\bar{C}_{11} + \bar{C}_{111} + \bar{C}_{112})(4\bar{S}_{12}^2 - e_1) + \bar{C}_{112}(4\bar{S}_{11}\bar{S}_{12} - e_2) - 8\bar{C}_{11}\bar{S}_{12}^2$$
$$+ (\bar{C}_{1111} + 2\bar{C}_{1112} + \bar{C}_{1122})\bar{S}_{12}^2 + 2(\bar{C}_{1112} + \bar{C}_{1123})\bar{S}_{11}\bar{S}_{12} + \bar{C}_{1122}\bar{S}_{11}^2],$$
(19.35)

$$F_{66} = \bar{C}_{44} - Q[2(\bar{C}_{44} + \bar{C}_{166})\bar{S}_{12} + \bar{C}_{144}\bar{S}_{11}]$$
$$+ \tfrac{1}{2}Q^2[2(\bar{C}_{44} + \bar{C}_{166})(4\bar{S}_{12}^2 - e_1) + \bar{C}_{144}(4\bar{S}_{11}\bar{S}_{12} - e_2) - 8\bar{C}_{44}\bar{S}_{12}^2$$
$$+ 2(\bar{C}_{1166} + \bar{C}_{1266})\bar{S}_{12}^2 + 4\bar{C}_{1244}\bar{S}_{11}\bar{S}_{12} + \bar{C}_{1144}\bar{S}_{11}^2], \quad (19.36)$$

$$F_{44} = \bar{C}_{44} - Q[(\bar{C}_{144} + \bar{C}_{166})\bar{S}_{12} + (2\bar{C}_{44} + \bar{C}_{166})\bar{S}_{11}]$$
$$+ \tfrac{1}{2}Q^2[(\bar{C}_{144} + \bar{C}_{166})(4\bar{S}_{11}\bar{S}_{12} - e_1)$$
$$+ (2\bar{C}_{44} + \bar{C}_{166})(4\bar{S}_{11}^2 - e_2) - 8\bar{C}_{44}\bar{S}_{11}^2 + (\bar{C}_{1144} + \bar{C}_{1166} + 2\bar{C}_{1244})\bar{S}_{12}^2$$
$$+ 2(\bar{C}_{1244} + \bar{C}_{1266})\bar{S}_{11}\bar{S}_{12} + \bar{C}_{1166}\bar{S}_{11}^2], \quad (19.37)$$

$$F_{12} = \bar{C}_{12} - Q[2(\bar{C}_{12} + \bar{C}_{112})\bar{S}_{12} + \bar{C}_{123}\bar{S}_{11}]$$
$$+ \tfrac{1}{2}Q^2[2(\bar{C}_{12} + \bar{C}_{112})(4\bar{S}_{12}^2 - e_1) + \bar{C}_{123}(4\bar{S}_{11}\bar{S}_{12} - e_2) - 8\bar{C}_{12}\bar{S}_{12}^2$$
$$+ 2(\bar{C}_{1112} + \bar{C}_{1122})\bar{S}_{12}^2 + 4\bar{C}_{1123}\bar{S}_{11}\bar{S}_{12} + \bar{C}_{1123}\bar{S}_{11}^2]. \quad (19.38)$$

An interesting point is that the stressed crystal here may be regarded as tetragonal I with uniaxial stress along [001]. The propagation matrices and eigenvalues can be written out as in Section 13 in terms of the stress and the tetragonal I second-order elastic constants. These elastic constants can then be expanded in terms of the strains and the zero-stress elastic constants of a cubic I crystal, and the above results verified.

20. Application of Isotropic Pressure plus Uniaxial Stress

It is presumably possible to measure the second pressure derivatives of the ultrasonic reduced-wave velocities $\rho_0 W^2$, by making extremely accurate measurements up to high pressures. Such measurements give only three independent combinations of the eleven fourth-order elastic constants for cubic I. Unfortunately the measurement of second stress derivatives for a uniaxial stress is much more difficult since crystals flow at a fairly low uniaxial stress. One way around this difficulty would be to measure the pressure variation of the first-uniaxial-stress derivative of $\rho_0 W^2$, i.e. to measure the pressure derivatives of the third-order elastic constants. Since there are six independent third-order elastic constants for a cubic I crystal, such measurements would give six independent combinations of the fourth-order elastic constants; note these six combinations contain the three combinations observable as second pressure derivatives. Additional stress systems, such as independently varying uniaxial stresses in two directions, would be required to measure all the independent fourth-order elastic constants.

In order to analyze this experiment, we take isotropic pressure plus uniaxial stress along [001]:

$$T_{ij}(\mathbf{X}) = -P\,\delta_{ij} - Q(s_0/s_1)\,\delta_{i3}\,\delta_{j3}. \tag{20.1}$$

Here Q is the reduced stress, i.e. the total uniaxial force on the (001) faces divided by the area s_0 of the (001) faces at zero stress. Note the *total* stress on (001) faces is the pressure *plus* the uniaxial stress; this corresponds to an experimental arrangement, e.g., with a uniaxial press inside a pressure bomb. Alternate experimental arrangements may be worked out along the same lines as the following.

We will keep all terms of order P, Q, and PQ, dropping the P^2 and Q^2 terms since these contain the same combinations of fourth-order elastic constants as the corresponding terms in Section 19. The first step is to use the stress (20.1) to work out the strain expansions (9.10) in powers of P and $(s_0/s_1)Q$. Then s_0/s_1 may be eliminated since $s_1/s_0 = 1 + 2n_{11}$, just as in (19.24), except now n_{11} has a term in P and one in Q. The final results for the strains are

$$n_{11} = n_{22} = -gP - \tilde{S}_{12}Q + d_2PQ, \tag{20.2}$$

$$n_{33} = -gP - \tilde{S}_{11}Q + d_1PQ, \tag{20.3}$$

$$n_{12} = n_{23} = n_{31} = 0, \tag{20.4}$$

where d_1 and d_2 are the quantities defined by (12.24) and (12.25). The second step is to use these strains to evaluate the reduced dynamical matrices (9.27) and (9.31), for given zero-stress propagation directions $\bar{\varkappa}$. The matrices have exactly the same symmetry as the **M** matrices for tetragonal I under uniaxial [001] stress. The results for two propagation directions are given here.

$\bar{\varkappa} = [100]$:

$$\mathbf{M} = \begin{pmatrix} G_{11} & 0 & 0 \\ 0 & G_{66} & 0 \\ 0 & 0 & G_{44} \end{pmatrix} \tag{20.5}$$

$$\mathbf{w}(l) = [100], \quad \rho_0 W^2 = G_{11}; \tag{20.6a}$$

$$\mathbf{w}(t_1) = [010], \quad \rho_0 W^2 = G_{66}; \tag{20.6b}$$

$$\mathbf{w}(t_2) = [001], \quad \rho_0 W^2 = G_{44}. \tag{20.6c}$$

Here again the uniaxial stress removes the degeneracy of the two transverse modes, and the transverse eigenvectors at stress cannot be rotated.

$\bar{\kappa} = (1/\sqrt{2})[110]$:

$$\mathbf{M} = \frac{1}{2} \begin{pmatrix} G_{11} + G_{66} & G_{12} + G_{66} & 0 \\ G_{12} + G_{66} & G_{11} + G_{66} & 0 \\ 0 & 0 & 2G_{44} \end{pmatrix} \quad (20.7)$$

$\mathbf{w}(l) = (1/\sqrt{2})[110], \quad \rho_0 W^2 = \tfrac{1}{2}(G_{11} + G_{12} + 2G_{66}); \quad (20.8\text{a})$

$\mathbf{w}(t_1) = (1/\sqrt{2})[1\bar{1}0], \quad \rho_0 W^2 = \tfrac{1}{2}(G_{11} - G_{12}); \quad (20.8\text{b})$

$\mathbf{w}(t_2) = [001], \quad \rho_0 W^2 = G_{44}. \quad (20.8\text{c})$

The matrix elements appearing here are rather lengthy, and are given as follows:

$$G_{11} = \bar{C}_{11} - p_{11}P - q_{11}Q + r_{11}PQ, \quad (20.9\text{a})$$

$$G_{12} = \bar{C}_{12} - p_{12}P - q_{12}Q + r_{12}PQ, \quad (20.9\text{b})$$

$$G_{44} = \bar{C}_{44} - p_{44}P - q_{44}Q + r_{44}PQ, \quad (20.9\text{c})$$

$$G_{66} = \bar{C}_{44} - p_{66}P - q_{66}Q + r_{66}PQ, \quad (20.9\text{d})$$

$$p_{11} = 1 + g(2\bar{C}_{11} + \bar{C}_{111} + 2\bar{C}_{112}), \quad (20.10\text{a})$$

$$p_{12} = -1 + g(2\bar{C}_{12} + 2\bar{C}_{112} + \bar{C}_{123}), \quad (20.10\text{b})$$

$$p_{44} = p_{66} = 1 + g(2\bar{C}_{44} + \bar{C}_{144} + 2\bar{C}_{166}), \quad (20.10\text{c})$$

$$q_{11} = (2\bar{C}_{11} + \bar{C}_{111} + \bar{C}_{112})\bar{S}_{12} + \bar{C}_{112}\bar{S}_{11}, \quad (20.11\text{a})$$

$$q_{12} = 2(\bar{C}_{12} + \bar{C}_{112})\bar{S}_{12} + \bar{C}_{123}\bar{S}_{11}, \quad (20.11\text{b})$$

$$q_{44} = (\bar{C}_{144} + \bar{C}_{166})\bar{S}_{12} + (2\bar{C}_{44} + \bar{C}_{166})\bar{S}_{11} \quad (20.11\text{c})$$

$$q_{66} = 2(\bar{C}_{44} + \bar{C}_{166})\bar{S}_{12} + \bar{C}_{144}\bar{S}_{11}, \quad (20.11\text{d})$$

$r_{11} = \bar{S}_{11} + (2\bar{C}_{11} + \bar{C}_{111} + \bar{C}_{112})(4g\bar{S}_{12} + d_2) + \bar{C}_{112}[2gg_1 + d_1] - 8g\bar{C}_{11}\bar{S}_{12}$
$\quad + g[(\bar{C}_{1111} + 3\bar{C}_{1112} + \bar{C}_{1122} + \bar{C}_{1123})\bar{S}_{12} + (\bar{C}_{1112} + \bar{C}_{1122} + \bar{C}_{1123})\bar{S}_{11}],$
$\quad\quad\quad\quad\quad\quad\quad\quad\quad\quad\quad\quad\quad\quad\quad\quad\quad\quad\quad\quad (20.12\text{a})$

$r_{12} = -\bar{S}_{11} + 2(\bar{C}_{12} + \bar{C}_{112})(4g\bar{S}_{12} + d_2) + \bar{C}_{123}[2gg_1 + d_1] - 8g\bar{C}_{12}\bar{S}_{12}$
$\quad + g[2(\bar{C}_{1112} + \bar{C}_{1122} + \bar{C}_{1123})\bar{S}_{12} + 3\bar{C}_{1123}\bar{S}_{11}], \quad (20.12\text{b})$

$$r_{44} = \bar{S}_{11} + (\bar{C}_{144} + \bar{C}_{166})[2gg_1 + d_2] + (2\bar{C}_{44} + \bar{C}_{166})(4g\bar{S}_{11} + d_1) - 8g\bar{C}_{44}\bar{S}_{11}$$
$$+ g[(\bar{C}_{1144} + \bar{C}_{1166} + 3\bar{C}_{1244} + \bar{C}_{1266})\bar{S}_{12} + (\bar{C}_{1166} + \bar{C}_{1244} + \bar{C}_{1266})\bar{S}_{11}],$$
(20.12c)

$$r_{66} = \bar{S}_{11} + 2(\bar{C}_{44} + \bar{C}_{166})(4g\bar{S}_{12} + d_2) + \bar{C}_{144}[2gg_1 + d_1] - 8g\bar{C}_{44}\bar{S}_{12}$$
$$+ g[2(\bar{C}_{1166} + \bar{C}_{1244} + \bar{C}_{1266})\bar{S}_{12} + (\bar{C}_{1144} + 2\bar{C}_{1244})\bar{S}_{11}]. \quad (20.12d)$$

Here we have also used the abbreviations

$$g = \bar{S}_{11} + 2\bar{S}_{12}, \qquad g_1 = \bar{S}_{11} + \bar{S}_{12}.$$

The first stress derivatives of the reduced eigenvalues, namely, $d(\rho_0 W^2)/dP$ and $d(\rho_0 W^2)/dQ$ evaluated at zero stress, are the same as the corresponding cases given in Table III for cubic I. To measure the

TABLE VIII. Second Stress Derivatives of $\rho_0 W^2$, Evaluated at Zero Stress, for Cubic I Crystals

κ	W	$d^2(\rho_0 W^2)/dP\,dQ$
[100]	[100]	r_{11}
[100]	[010]	r_{66}
[100]	[001]	r_{44}
$(1/\sqrt{2})[110]$	$(1/\sqrt{2})[110]$	$\frac{1}{2}(r_{11} + r_{12} + 2r_{66})$
$(1/\sqrt{2})[110]$	$(1/\sqrt{2})[1\bar{1}0]$	$\frac{1}{2}(r_{11} - r_{12})$
$(1/\sqrt{2})[110]$	[001]	r_{44}

second stress derivative, one could measure the pressure derivative $[d(\rho_0 W^2)/dP]_Q = \Delta_Q$ at various values of Q; then Δ_Q should be linear with Q and the slope $d\Delta_Q/dQ$ is the second stress derivative. Alternatively $[d(\rho_0 W^2)/dQ]_P = \Delta_P$ should be a linear function of P and the slope $d\Delta_P/dP$ is the same second stress derivative. For pure modes propagating in [100] and $(1/\sqrt{2})[110]$ directions, the second stress derivatives of $\rho_0 W^2$, evaluated at zero stress, are listed in Table VIII.

We have verified the above calculations by the following alternate procedure. Starting with a cubic I crystal under isotropic pressure P, the first stress derivatives of $\rho_1 W^2$ for additional uniaxial stress along [001], evaluated at P, were calculated by the theory of Section 10, using Eq. (10.15) valid at arbitrary initial stress. The combinations of second- and third-order elastic constants so obtained were then expanded for small P, with the cubic I expansions (12.21) and (12.22), to get the desired results.

21. Harmonic Generation in Stressed Crystals

An interesting possibility for measuring fourth-order elastic constants is to carry out harmonic generation experiments on stressed crystals. Since harmonic generation experiments measure directly the third-order elastic constants, one can then measure the stress derivatives of the third-order constants. The harmonic generation theory of Section 4 is valid for a crystal in the presence of arbitrary initial stress, and it is only necessary to apply the theory to a given crystal under a given stress, and then expand the results to first order in the stress. We illustrate the procedure for the example of a cubic I crystal under isotropic pressure.

Suppose the frequency of the input wave is ω, and is maintained constant independent of the pressure P. Suppose also the amplitudes a_1 and a_2 of the input wave and generated harmonic are measured at the surface of the sample after one pass of the waves through the sample, and this surface can be considered stress-free in the acoustic sense, i.e. it is not coupled acoustically to the pressure transmitting medium. Finally, take the case of small attenuation; then the amplitudes are related by [see (4.19) and (5.20)]

$$|a_2/a_1{}^2| = k^2 l_1 |\mathbf{D}\cdot\hat{\mathbf{w}}|/16\rho_1 v^2. \tag{21.1}$$

Here k is the magnitude of the wave vector of the input wave, v is its velocity, and l_1 is the length of the sample measured along $\mathbf{k}$, all evaluated in the configuration at arbitrary pressure P. The quantity $\mathbf{D}\cdot\hat{\mathbf{w}}$ is a combination of second- and third-order elastic constants evaluated at P; for cubic I crystals the combination is just the same as the zero-stress results listed in Table VI. In fact we note a general result that the stress does not appear explicitly in $\mathbf{D}\cdot\hat{\mathbf{w}}$. Since $\omega = vk$, a more convenient form for (21.1) is

$$|a_2/a_1{}^2| = \rho_1 l_1 \omega^2 |\mathbf{D}\cdot\hat{\mathbf{w}}|/16(\rho_1 v^2)^2. \tag{21.2}$$

Now it is only necessary to expand the quantities $\rho_1 l_1$, $|\mathbf{D}\cdot\hat{\mathbf{w}}|$, and $\rho_1 v^2$ to first order in the pressure; we will assume the third-order elastic constants are known, and hence the sign of $\mathbf{D}\cdot\hat{\mathbf{w}}$ is known. From (12.15) and (12.19) for a cubic crystal under isotropic pressure we have

$$\rho_1 l_1 = \rho_0 (1+2n)^{-3/2} l_0 (1+2n)^{1/2} = \rho_0 l_0 (1+2n)^{-1}$$
$$= \rho_0 l_0 (1 + 2gP + \cdots); \tag{21.3}$$
$$g = \bar{S}_{11} + 2\bar{S}_{12}. \tag{21.4}$$

Define the coefficients r and d in terms of the expansions

$$\rho_1 v^2 = \rho_0 \bar{v}^2 - grP, \tag{21.5}$$

$$\mathbf{D}\cdot\hat{\mathbf{w}} = \bar{\mathbf{D}}\cdot\hat{\mathbf{w}} - g\,d\,P, \tag{21.6}$$

TABLE IX. PARAMETERS WHICH APPEAR IN THE FIRST PRESSURE DERIVATIVE OF THE AMPLITUDE RATIO $|a_2/a_1^2|$ IN HARMONIC GENERATION EXPERIMENTS FOR CUBIC I CRYSTALS[a]

Case	κ	w	$\rho_0 \bar{v}^2$	r and d
1	[100]	[100]	$\bar{C}_{11}$	$r = 2(\bar{C}_{11} + \bar{C}_{12}) + \bar{C}_{111} + 2\bar{C}_{112}$ $d = 3\bar{C}_{11} + 6\bar{C}_{111} + 6\bar{C}_{112} + \bar{C}_{1111} + 2\bar{C}_{1112}$
2	$(1/\sqrt{2})[110]$	$(1/\sqrt{2})[110]$	$\frac{1}{2}(\bar{C}_{11} + \bar{C}_{12} + 2\bar{C}_{44})$	$r = \frac{1}{2}[3\bar{C}_{11} + 5\bar{C}_{12} + 2\bar{C}_{44} + \bar{C}_{111} + 4\bar{C}_{112} + \bar{C}_{123} + 2\bar{C}_{144} + 4\bar{C}_{166}]$ $d = 3\rho_0\bar{v}^2 + \frac{1}{4}[3(3\bar{C}_{111} + 11\bar{C}_{112} + 2\bar{C}_{123} + 4\bar{C}_{144} + 20\bar{C}_{166}) + (\bar{C}_{1111} + 5\bar{C}_{1112} + 3\bar{C}_{1122} + 3\bar{C}_{1123} + 12\bar{C}_{1166} + 12\bar{C}_{1244} + 12\bar{C}_{1266})]$
3	$(1/\sqrt{3})[111]$	$(1/\sqrt{3})[111]$	$\frac{1}{3}(\bar{C}_{11} + 2\bar{C}_{12} + 4\bar{C}_{44})$	$r = \frac{1}{3}[4(\bar{C}_{11} + 2\bar{C}_{12} + \bar{C}_{44}) + \bar{C}_{111} + 6\bar{C}_{112} + 2\bar{C}_{123} + 4\bar{C}_{144} + 8\bar{C}_{166}]$ $d = 3\rho_0\bar{v}^2 + \frac{1}{9}[12(\bar{C}_{111} + 6\bar{C}_{112} + 2\bar{C}_{123} + 6\bar{C}_{144} + 12\bar{C}_{166} + 4\bar{C}_{456}) + (\bar{C}_{1111} + 8\bar{C}_{1112} + 6\bar{C}_{1122} + 12\bar{C}_{1123} + 12\bar{C}_{1144} + 24\bar{C}_{1166} + 48\bar{C}_{1244} + 24\bar{C}_{1266} + 48\bar{C}_{1456})]$
4	$(1/\sqrt{3})[111]$	$(1/\sqrt{6})[11\bar{2}]$	$\frac{1}{3}(\bar{C}_{11} - \bar{C}_{12} + \bar{C}_{44})$	$r = \frac{1}{3}[4\bar{C}_{11} + 5\bar{C}_{12} + \bar{C}_{44} + \bar{C}_{111} - \bar{C}_{123} + \bar{C}_{144} + 2\bar{C}_{166}]$ $d = -(1/9\sqrt{2})[3(\bar{C}_{111} - 3\bar{C}_{112} + 2\bar{C}_{123} + 3\bar{C}_{144} - 3\bar{C}_{166} - 2\bar{C}_{456}) + (\bar{C}_{1111} - \bar{C}_{1112} - 3\bar{C}_{1122} + 3\bar{C}_{1123} + 3\bar{C}_{1144} - 3\bar{C}_{1166} + 3\bar{C}_{1244} + 3\bar{C}_{1123} + 3\bar{C}_{1144} - 3\bar{C}_{1166} + 3\bar{C}_{1244} - 3\bar{C}_{1266} - 6\bar{C}_{1456})]$

[a] The pressure derivative is given by Eq. (21.8).

correct to order P, where as usual $\rho_0 \bar{v}^2$ and $\bar{\mathbf{D}} \cdot \hat{\mathbf{w}}$ are evaluated at $P = 0$. Then a straightforward expansion of (21.2) gives to order P

$$| a_2/a_1{}^2 | = | \bar{a}_2/\bar{a}_1{}^2 | [1 + 2gP - (gd/\bar{\mathbf{D}} \cdot \hat{\mathbf{w}})P + (2gr/\rho_0 \bar{v}^2)P], \tag{21.7}$$

or

$$(d \ln | a_2/a_1{}^2 |/dP)_{P=0} = g[2 - (d/\bar{\mathbf{D}} \cdot \hat{\mathbf{w}}) + (2r/\rho_0 \bar{v}^2)]. \tag{21.8}$$

Take, e.g., the self-resonant case 1 of Table VI, where $\kappa = [100]$ and $\hat{\mathbf{w}} = \mathbf{w} = [100]$. Then from Table VI applied at pressure P, $\rho_1 v^2 = B_{11} = C_{11} - P$, and $\mathbf{D} \cdot \hat{\mathbf{w}} = 3C_{11} + C_{111}$. These may be expanded for small pressure with the aid of Eqs. (12.21) and (12.22):

$$\rho_1 v^2 = \tilde{C}_{11} - g(g^{-1} + \tilde{C}_{11} + \tilde{C}_{111} + 2\tilde{C}_{112})P; \tag{21.9}$$

$$\mathbf{D} \cdot \hat{\mathbf{w}} = 3\tilde{C}_{11} + \tilde{C}_{111} - g(3\tilde{C}_{11} + 6\tilde{C}_{111} + 6\tilde{C}_{112} + \tilde{C}_{1111} + 2\tilde{C}_{1112})P. \tag{21.10}$$

Comparing (21.9) with (21.5) and (21.10) with (21.6) gives r and d for this example. For a cubic crystal under isotropic pressure and with propagation along the three pure-mode directions, there are four self-resonant cases and one mutually resonant case of harmonic generation. The parameters r and d are listed in Table IX for the four self-resonant cases. The mutually resonant case gives no new information since for that case both d and $\bar{\mathbf{D}} \cdot \hat{\mathbf{w}}$ are of opposite sign to those of case 4, and hence $(d/\bar{\mathbf{D}} \cdot \hat{\mathbf{w}})$ is the same as for case 4.

Thus Eq. (21.8) may be used to interpret the pressure derivatives of the amplitude ratio $| a_2/a_1{}^2 |$, and to obtain the function d of fourth-order elastic constants. For the example of a cubic I crystal under uniaxial stress along [001], the harmonic generation theory of a tetragonal I crystal under [001] stress applies, and the stress derivatives of $| a_2/a_1{}^2 |$ can be worked out similarly to (21.8).

Peters and Breazeale[89a] have recently observed the generated third harmonic in single crystal copper; such observations may prove to be useful for measuring the stress derivatives of third-order elastic constants, or for measuring the fourth-order elastic constants directly.

VII. Interpretation of Measurements in Terms of Atomic Theories

22. Application of Lattice Dynamics

In this final section we present a brief outline of the theory of elastic properties of crystals, as derived from the description of a crystal as a

[89a] R. D. Peters and M. A. Breazeale, *Appl. Phys. Letters* **12**, 106 (1968).

collection of interacting ions and electrons. This theory is developed with the aid of lattice dynamics, in which the anharmonicity is treated as a perturbation.[90] The thermodynamic state functions are then expressed as rapidly converging perturbation series, and we shall examine the properties of the elastic constants in each successive order of approximation. It should be emphasized that the various elastic coefficients are *defined* by the thermoelastic theory of Part II; lattice-dynamical expressions for elastic coefficients are obtained by *comparison* of lattice-dynamical results with those definitions.

Consider a large but finite crystal containing N unit cells; the equilibrium lattice position of ion J is $\mathbf{r}(J)$, and the displacement of this ion from equilibrium at any time is $\mathbf{u}(J)$. The total potential energy of the crystal is presumed to be an analytic function of the positions of all the ions, and is expanded as

$$\Phi(\mathbf{r}(J) + \mathbf{u}(J)) = \Phi(\mathbf{r}(J)) + \sum_{J,j} \Phi_j(J) u_j(J)$$
$$+ \tfrac{1}{2} \sum_{JK,jk} \Phi_{jk}(JK) u_j(J) u_k(K) + \cdots, \quad (22.1)$$

where the subscripts, $j, k, \ldots$ indicate Cartesian components. The expansion (22.1) is justified for all crystals except metals by the adiabatic approximation,[91] which says essentially that the vibrational energy of the ions is too small to excite the electronic system from its ground state, so that the electronic contribution to the lattice potential is just the total energy (including kinetic) of the electronic ground state. For metals this is still true except for a few electrons near the Fermi energy, so that one may make the adiabatic approximation and write the lattice potential in the form (22.1), and then treat nonadiabatic effects as a perturbation.[92] We will neglect nonadiabatic effects in the following discussion.

The theory may be developed for a stressed crystal, in analogy with the thermodynamic theory of Part II, by allowing for externally applied forces $\mathbf{f}(J)$ on the atoms J at the surface.[18] If the ions undergo virtual displacements $\mathbf{u}(J)$, in which the $\mathbf{f}(J)$ remain constant, the total potential

---

[90] We omit consideration of the "quantum crystals" He, H_2, and displacive ferroelectrics near the Curie temperature, where the anharmonicity is large and the perturbation approach is inconvenient. The self-consistent lattice dynamics approach has been developed for treatment of such examples; see e.g. N. R. Werthamer, *Am. J. Phys.* **37**, 763 (1969).

[91] M. Born and K. Huang, "Dynamical Theory of Crystal Lattices." Oxford Univ. Press (Clarendon), London and New York, 1954.

[92] See e.g. G. V. Chester, *Advan. Phys.* **10**, 357 (1961).

of the system of crystal plus external forces is

$$\chi(\mathbf{r}(J) + \mathbf{u}(J)) = \chi(\mathbf{r}(J)) + \sum_{J,j} [\Phi_j(J) - f_j(J)] u_j(J)$$
$$+ \tfrac{1}{2} \sum_{JK,jk} \Phi_{jk}(JK) u_j(J) u_k(K) + \cdots. \quad (22.2)$$

The equilibrium condition for the static lattice is that the net force on each ion must vanish when all ions are located at their equilibrium positions; that is, $-\partial\chi/\partial u_j(J)$ vanishes when evaluated at all $u_j(J) = 0$. This gives

$$\Phi_j(J) - f_j(J) = 0, \quad \text{all } J, j. \quad (22.3)$$

Additional conditions on the potential energy coefficients $\Phi_j(J)$, $\Phi_{jk}(JK)$, ... arise from the requirements of translational and rotational invariance of the potential energy.[11,18,91,93] The invariance conditions may be derived equivalently from the lattice potential Φ or from the system potential χ.[19]

Our finite model crystal is presumed to be large compared to the effective range of interactions among the ions.[94] Near the surface, within a distance of the order of the effective range of interactions, the crystal is distorted from perfect periodicity; in the interior, however, the lattice is perfectly periodic. Surface effects are presumed to be negligible in the bulk properties of the crystal, and they may be eliminated according to a prescription given earlier.[18] With the aid of the surface forces, which are of arbitrary sign, the finite crystal may be brought to any desired macroscopic configuration, and provided the equilibrium and invariance conditions are satisfied, we may proceed to develop the theory of lattice dynamics at arbitrary lattice configuration. An alternate procedure is to consider a subcrystal in the perfectly periodic interior of the large finite crystal, or in the interior of an infinite lattice; then there is no distortion near the surface of the subcrystal. It might appear that such a model automatically eliminates surface effects, but this is not the case. The reason is that the lattice outside the subcrystal exerts "externally applied" forces on the surface region of the subcrystal, so that the picture is not really changed at all.

To proceed with the small oscillation problem of the ion motion, the crystal Hamiltonian is the kinetic energy of the ions plus the crystal potential. Imposing periodic boundary conditions and neglecting surface

---

[93] G. Leibfried and W. Ludwig, *Z. Physik* **160**, 80 (1960).

[94] By effective range of interactions we mean the range over which all the interactions need to be counted in the perfectly periodic lattice to get an accurate representation of the energy per unit cell; this is for example the range of an Ewald sum in ionic crystals, or a screened ion–ion interaction in metals.

effects leads to the phonon Hamiltonian for N unit cells of the crystal[95]:

$$\mathcal{H} = \Phi(\mathbf{r}(J)) + \sum_{\kappa} \hbar\omega_{\kappa}(A_{\kappa}^{\dagger}A_{\kappa} + \tfrac{1}{2}) + \mathcal{H}_A. \tag{22.4}$$

Here κ stands for both the wave vector $\mathbf{k}$ and the polarization index s, where there are N values of $\mathbf{k}$ distributed uniformly over the first Brillouin zone and the number of different polarizations (branches) is three times the number of ions in each unit cell. The quasi-harmonic part of $\mathcal{H}$, i.e. the middle term of (22.4), is obtained by diagonalizing the homogeneous quadratic form composed of the ion kinetic energy plus the potential energy quadratic in ion displacements. $A_{\kappa}^{\dagger}A_{\kappa}$ is the number operator for phonon κ, and the phonon frequencies ω_{κ} are obtained in the diagonalization as the eigenvalues of the dynamical matrices.[91] The anharmonic term $\mathcal{H}_A$ in (22.4) represents the phonon–phonon interactions, which arise from terms cubic, quartic, etc., in the expansion (22.1) of Φ.

The phonons are bosons, and the crystal free energy may be calculated by treating the interaction term as a perturbation. The total Helmholtz free energy $\mathcal{F}$ for N unit cells of the crystal in arbitrary macroscopic configuration is then

$$\mathcal{F} = \Phi(\mathbf{r}(J)) + \sum_{\kappa}\{\tfrac{1}{2}\hbar\omega_{\kappa} + KT \ln[1 - \exp(-\hbar\omega_{\kappa}/KT)]\} + \mathcal{F}_A. \tag{22.5}$$

Here K is the Boltzmann constant, and we should have no occasion to confuse this with the compressibility defined earlier. The leading contribution to $\mathcal{F}_A$ is obtained by treating the three-phonon interactions in second-order perturbation and the four-phonon interactions in first-order; the quantum mechanical expressions were first worked out by Ludwig.[96] Now (22.5) is a rapidly converging series of contributions to the Helmholtz free energy, and leads to a similar converging series of contributions to all thermodynamic functions. We can take $\sqrt{\hbar}$ as the expansion parameter[97]; in this case ω_{κ} is a zero-order quantity, and K is second order since $\hbar\omega_{\kappa}/KT \sim 1$. The first term $\Phi(\mathbf{r}(J))$ is the largest contribution to $\mathcal{F}$; this is a zero-order quantity and is a function only of the macroscopic configuration of the crystal. The quasi-harmonic contribution in (22.5) is of second order, and depends on the macroscopic configuration through the configuration dependence of the ω_{κ}, and also depends explicitly on the temperature T.

---

[95] The Hamiltonian for the system of crystal plus externally applied forces contains in addition the work done by the external forces, measured from an arbitrary zero.

[96] W. Ludwig, *J. Phys. Chem. Solids* **4**, 283 (1958).

[97] An alternate, but internally consistent, procedure is to let the potential energy expansion (22.1) determine the order of the potential energy coefficients, and hence of all other quantities; in this case ω_{κ} is of second order, while the three contributions to $\mathcal{F}$ in (22.5) are still of zero-, second-, and fourth-order, respectively.

The anharmonic contribution $\mathfrak{F}_A$ is of fourth order (and higher orders), and is also a function of the macroscopic configuration and the temperature. We note that for metals the Hamiltonian (22.4) should contain an additional term representing the excitation of the conduction electrons above the ground state, and the Helmholtz free energy (22.5) then contains also the electronic excitation contribution $-\frac{1}{2}\Gamma T^2$, where Γ is a function of the macroscopic configuration.

The thermodynamic functions may be derived from $\mathfrak{F}$ by means of well-known formulas; e.g., the internal energy $\mathfrak{U}$ is

$$\mathfrak{U} = \Phi(\mathbf{r}(J)) + \sum_\kappa \hbar\omega_\kappa(n_\kappa + \tfrac{1}{2}) + \mathfrak{U}_A, \qquad (22.6)$$

and the volume thermal expansion coefficient β is given by

$$\beta B^T V = K \sum_\kappa \gamma_\kappa(\hbar\omega_\kappa/KT)^2 n_\kappa(n_\kappa + 1) + (\beta B^T V)_A. \qquad (22.7)$$

Here the explicit anharmonic contributions are denoted by a subscript A, B^T is the isothermal bulk modulus, V is the crystal volume, and

$$n_\kappa = [\exp(\hbar\omega_\kappa/KT) - 1]^{-1}, \qquad (22.8)$$

$$\gamma_\kappa = -d \ln \omega_\kappa / d \ln V. \qquad (22.9)$$

In order to make comparisons with the thermodynamic theory of Part II, the crystal volume in the arbitrary initial configuration is denoted by V_1, and the relation to the state functions per unit mass is

$$\rho_1 F = V_1^{-1}\mathfrak{F}, \qquad \rho_1 U = V_1^{-1}\mathfrak{U}. \qquad (22.10)$$

We are now ready to discuss the properties of the elastic constants, and their relations to other thermodynamic functions, in each order of approximation to the complete lattice-dynamical expression for $\mathfrak{F}$. The zeroth approximation is

$$\mathfrak{F} \approx \Phi(\mathbf{r}(J)), \qquad \text{potential approximation.} \qquad (22.11)$$

In this approximation $\mathfrak{F} = \mathfrak{U}$, and the second-, third-, and higher-order elastic constants have the properties that (a) they depend only on the macroscopic configuration and not explicitly on the temperature, and (b) there is no difference between adiabatic and isothermal constants. For most materials the potential approximation is accurate to a few percent at low temperatures, so comparison of theoretical and experimental elastic constants can provide significant tests of a model for the total lattice potential. If one has a model for the lattice potential, it is straightforward to eliminate surface effects from the expression for $\Phi(\mathbf{r}(J))$ and calculate the elastic constants from the strain derivatives.

For illustration consider a primitive lattice with a central potential interaction $\varphi(IJ)$ between all pairs of ions I, J. Then

$$\Phi = \tfrac{1}{2} \sum_{IJ}{}' \varphi(IJ). \tag{22.12}$$

To eliminate surface effects we let the ion I be at the origin of coordinates in the interior of the crystal, sum the central potential $\varphi(J)$ over ions located at positions $\mathbf{r}(J)$ measured from the origin of coordinates, and multiply by N for the sum on I:

$$\Phi = \tfrac{1}{2} N \sum_{J}{}' \varphi(J). \tag{22.13}$$

Now under a homogeneous deformation corresponding to the Lagrangian strains η_{ij}, each lattice site $\mathbf{r}(J)$ goes to $\hat{\mathbf{r}}(J)$, where from (1.6)

$$|\hat{\mathbf{r}}(J)|^2 = |\mathbf{r}(J)|^2 + 2 \sum_{ij} \eta_{ij} r_i(J) r_j(J). \tag{22.14}$$

A Taylor series expansion of $\varphi(J)$ about the initial equilibrium configuration then gives

$$\Phi(\hat{\mathbf{r}}(J)) = \tfrac{1}{2} N \sum_{J}{}' [\varphi(J) + 2\varphi'(J) \sum_{ij} \eta_{ij} r_i(J) r_j(J)$$
$$+ 2\varphi''(J) \sum_{ijkl} \eta_{ij} \eta_{kl} r_i(J) r_j(J) r_k(J) r_l(J) + \cdots]. \tag{22.15}$$

Here the coefficients are

$$\varphi'(J) = \partial \varphi(r^2)/\partial r^2, \qquad \text{at} \quad r = |\mathbf{r}(J)|, \tag{22.16a}$$

$$\varphi''(J) = \partial^2 \varphi(r^2)/\partial (r^2)^2, \qquad \text{at} \quad r = |\mathbf{r}(J)|, \tag{22.16b}$$

and so on, and they are evaluated in the initial configuration. The initial volume per unit cell is $V_{1C} = V_1/N$, and by comparing the expansion (22.15) with the free energy expansion (1.10) we can immediately write the lattice dynamical expressions for the stresses and the elastic constants, evaluated in the arbitrary initial configuration:

$$T_{ij} = V_{1C}^{-1} \sum_{J}{}' \varphi'(J) r_i(J) r_j(J),$$

$$C_{ijkl} = 2 V_{1C}^{-1} \sum_{J}{}' \varphi''(J) r_i(J) r_j(J) r_k(J) r_l(J), \tag{22.17}$$

$$C_{ijklmn} = 4 V_{1C}^{-1} \sum_{J}{}' \varphi'''(J) r_i(J) r_j(J) r_k(J) r_l(J) r_m(J) r_n(J),$$

and so forth.

For a primitive cubic lattice, the lattice sums in (22.17) vanish unless the Cartesian indices are equal in pairs. This means the stress must be isotropic pressure:

$$T_{ij} = -P\,\delta_{ij}; \qquad P = -V_{1C}^{-1} \sum_{J}{}' \varphi'(J)[r_x(J)]^2. \tag{22.18}$$

There are only two independent second-order elastic constants; in Voigt notation these are

$$C_{11} = 2V_{1C}^{-1} \sum_{J}{}' \varphi''(J)[r_x(J)]^4, \tag{22.19a}$$

$$C_{12} = C_{44} = 2V_{1C}^{-1} \sum_{J}{}' \varphi''(J)[r_x(J)]^2[r_y(J)]^2. \tag{22.19b}$$

There are three independent third-order elastic constants:

$$C_{111} = 4V_{1C}^{-1} \sum_{J}{}' \varphi'''(J)[r_x(J)]^6, \tag{22.20a}$$

$$C_{112} = C_{166} = 4V_{1C}^{-1} \sum_{J}{}' \varphi'''(J)[r_x(J)]^4[r_y(J)]^2, \tag{22.20b}$$

$$C_{123} = C_{144} = C_{456} = 4V_{1C}^{-1} \sum_{J}{}' \varphi'''(J)[r_x(J)]^2[r_y(J)]^2[r_z(J)]^2; \tag{22.20c}$$

and there are four independent fourth-order elastic constants:

$$C_{1111} = 8V_{1C}^{-1} \sum_{J}{}' \varphi''''(J)[r_x(J)]^8, \tag{22.21a}$$

$$C_{1112} = C_{1166} = 8V_{1C}^{-1} \sum_{J}{}' \varphi''''(J)[r_x(J)]^6[r_y(J)]^2, \tag{22.21b}$$

$$C_{1122} = C_{1266} = C_{4444} = 8V_{1C}^{-1} \sum_{J}{}' \varphi''''(J)[r_x(J)]^4[r_y(J)]^4, \tag{22.21c}$$

$$C_{1123} = C_{1144} = C_{1244} = C_{1456} = C_{4466}$$
$$= 8V_{1C}^{-1} \sum_{J}{}' \varphi''''(J)[r_x(J)]^4[r_y(J)]^2[r_z(J)]^2. \tag{22.21d}$$

The relations among the elastic constants of each order are the Cauchy relations.

This central potential example illustrates the straightforward way to calculate stresses and elastic constants in the potential approximation. On the other hand, it is possible to proceed with the theory for a general potential, given only by the expansion (22.1) and subject to equilibrium and invariance conditions; although the calculations are rather involved, they lead to some very important consequences which are valid for any potential. For a primitive lattice, e.g., if the crystal undergoes a homogeneous deformation with the displacement gradients u_{ij} defined by (1.3), the ion displacements are given by

$$u_i(J) = \sum_j u_{ij} r_j(J). \tag{22.22}$$

In the method of homogeneous deformation, the displacements (22.22) are substituted into the potential energy expansion (22.1) to obtain the Huang expansion of the potential in powers of the displacement gradients u_{ij} [see e.g. (3.17)]. The coefficients of the Huang expansion contain surface effects, and to derive general lattice-dynamical expressions for the elastic constants of each order it is necessary to transform out the surface effects and also to transform to an expansion in the Lagrangian strains η_{ij}. In the case of nonprimitive lattices there is the additional complication of sublattice displacements which must be eliminated; these details have been discussed in the literature.[18,93,98] The stresses, e.g., are the coefficients of the terms linear in u_{ij} and are obviously

$$T_{ij} = V_1^{-1} \sum_J \Phi_i(J) r_j(J); \tag{22.23}$$

in view of the equilibrium condition (22.3) this is just the mechanical stress applied to the crystal by the external forces:

$$T_{ij} = V_1^{-1} \sum_J f_i(J) r_j(J). \tag{22.24}$$

An important point is that we have evaluated the general lattice-dynamical expressions for the stresses and the second-order elastic constants for a primitive lattice with central potential interactions, and have obtained precisely the results given in (22.17).[19]

An alternate procedure developed by Begbie and Born[91,99] is the method of long waves; here one begins with the dynamical matrix, eliminates surface effects, expands the eigenvalue–eigenvector equation for long wavelength (small $|\mathbf{k}|$), and finds the phonon frequencies $\omega_\kappa = \omega_{ks}$ are proportional to $|\mathbf{k}|$ for a primitive lattice:

$$\omega_{ks} = c_s(\theta, \varphi) |\mathbf{k}|, \quad \text{small } |\mathbf{k}|. \tag{22.25}$$

The phonon velocity $c_s(\theta, \varphi)$ depends on the direction (θ, φ) of $\mathbf{k}$, but not on the magnitude. For nonprimitive lattices the result (22.25) still holds for the three acoustic branches, and the process of separating optical and acoustic modes in the method of long waves is essentially the same as the process of eliminating sublattice displacements in the method of homoge-

[98] G. Leibfried, in "Handbuch der Physik" (S. Flügge, ed.), Vol. VII/1, p. 104. Springer-Verlag, Berlin, 1955.
[99] G. H. Begbie and M. Born, Proc. Roy. Soc. (London) **A188**, 179 (1947).

neous deformation. The result that ω_{ks} is proportional to $|\mathbf{k}|$ has important consequences in the low-temperature thermodynamics of crystals, as mentioned in more detail below.

The method of long waves may be interpreted by comparing the equation of motion for the long-wavelength acoustic phonons with the equation of motion (3.10) for thermoelastic waves; this comparison leads to exactly the same lattice-dynamical expressions for the stresses and second-order elastic constants as does the method of homogeneous deformation.[18] In other words, in the potential approximation, the long-wavelength acoustic phonons are identical to the elastic waves, for any lattice and any initial configuration. This derivation has also been carried through by Srinivasan[100] for piezoelectric crystals in the presence of a macroscopic electric field.

Lattice dynamical expressions for second- and third-order elastic constants, in terms of the potential energy coefficients coupling nearest and next-nearest neighbors, have been written for some cubic lattices by Pfleiderer,[101] Coldwell-Horsfall,[102] and Srinivasan,[103] to mention a few. Of greater interest are the works which attempt to calculate the lattice potential, and then the elastic constants in the potential approximation, from a knowledge of the interactions among the ions and electrons in the crystal. The classic calculation of the second-order elastic constants of the alkali metals and copper by Fuchs,[104] in the adiabatic approximation, provided support for the picture of a metal as a lattice of rigid ion cores with a band of conduction electrons. More recently the pseudopotential theory, which is based on the same picture of a metal, has been used to calculate the second- and third-order elastic constants of the alkali metals by Suzuki et al.,[105] and the second-order elastic constants and their pressure derivatives for Na and K by Wallace.[106] Comparison of these calculations with experimental results provides support for the pseudopotential models, and indicates directions for improvement of the theory. We note in the pseudopotential theory there is a difference between the longitudinal elastic constants as calculated from homogeneous deformation and from long waves; this difference is a failure of the pseudopotential perturbation treatment, and results essentially from treating the screening to a different order in each of the two calculations.[106]

[100] R. Srinivasan, Phys. Rev. **165**, 1041 (1968).
[101] H. Pfleiderer, Phys. Status Solidi **2**, 1539 (1962).
[102] R. A. Coldwell-Horsfall, Phys. Rev. **129**, 22 (1963).
[103] R. Srinivasan, J. Phys. Chem. Solids **28**, 2385 (1967).
[104] K. Fuchs, Proc. Roy. Soc. (London) **A153**, 622 (1936); **A157**, 444 (1936).
[105] T. Suzuki, A. V. Granato, and J. F. Thomas, Jr., Phys. Rev. **175**, 766 (1968).
[106] D. C. Wallace, Phys. Rev. **182**, 778 (1969).

Before discussing the higher-order lattice dynamical corrections to the elastic constants, let us investigate the relations between the leading expressions for the elastic constants (i.e. the potential approximation) and the other thermodynamic functions. The leading explicit temperature dependence of the free energy appears in the quasi-harmonic approximation. At very low temperatures, only the low energy phonons contribute to the temperature-dependent part of $\mathfrak{F}$; these low energy phonons are just the long-wavelength acoustic modes, and because the ω_{ks} are proportional to $|\mathbf{k}|$ for these modes, the free energy goes as T^4 at low T:

$$\mathfrak{F} = \mathfrak{F}_0 - NP(\pi^4/5)(KT)(T/\Theta_{H_0})^3, \quad \text{low } T, \quad (22.26)$$

where $\mathfrak{F}_0$ is the value at $T = 0$ and we have dropped the anharmonic contribution. Here P is the number of atoms in each unit cell and Θ_{H_0} is the quasi-harmonic Debye temperature at $T = 0$, given by

$$(\hbar/K\Theta_{H_0})^3 = (V/18NP\pi^2)\sum_s{}' \langle c_s^{-3} \rangle, \quad (22.27)$$

where $\sum_s'$ is over the three acoustic modes only and the brackets $\langle\ \rangle$ indicate an average over all angles θ, φ of the phonon velocities $c_s(\theta, \varphi)$. Now since the thermoelastic sound wave velocities are equal to the long-wavelength acoustic phonon velocities c_s, in the potential approximation, the Debye temperature Θ_{H_0} is given by (22.27) with the measured sound velocities in place of the c_s, to leading approximation. Note the sound velocities may be calculated from the propagation coefficients, i.e. the elastic constants and the stresses, by means of the equation of motion (3.12) or (3.15).

Furthermore, the above results hold for arbitrary initial configuration of the lattice; this means the strain derivatives of the temperature-dependent part of the free energy can be calculated from the strain derivatives of the thermoelastic sound velocities, in leading approximation and in the low temperature limit. For example, considering only volume changes, the macroscopic Grüneisen parameter γ is defined by

$$\gamma = \beta B^T V/C_V, \quad (22.28)$$

where C_V is the heat capacity at constant volume. In the quasi-harmonic approximation and the low temperature limit, γ is obtained from (22.26) as

$$\gamma = -d \ln \Theta_{H_0}/d \ln V, \quad \text{low } T. \quad (22.29)$$

In leading approximation, then, this volume derivative may be evaluated from the measured volume derivatives of the sound velocities, and with the aid of the definition (22.27) of Θ_{H_0}. A convenient way of writing the volume derivative is

$$\gamma = \langle \gamma_{ks} c_s^{-3} \rangle / \langle c_s^{-3} \rangle, \quad \text{low } T, \quad (22.30)$$

where γ_{ks} are the long-wavelength acoustic phonon Grüneisen parameters, defined by (22.9), and the averages are again over angles. These γ_{ks} may be evaluated, with the aid of (10.23), from the measured third-order elastic constants or the pressure derivatives of the second-order elastic constants.

For metals, of course, the electronic excitation contribution to the thermodynamic functions is important at low temperatures; this contribution can be separated from the lattice-dynamical contribution on account of its different temperature dependence.

At intermediate temperatures $(T \lesssim \Theta_{H_0})$ the thermal expansion coefficient, given by (22.7), depends on the Grüneisen parameters of all the phonons; in fact, at high temperatures the quasi-harmonic approximation gives

$$\gamma = \langle \gamma_\kappa \rangle, \quad (22.31)$$

where the average is over all modes. In the Einstein approximation all ω_κ are the same; correspondingly, in the Grüneisen approximation[107] all γ_κ are the same, and γ becomes independent of the temperature in the quasi-harmonic approximation. In the Debye approximation the phonons are dispersionless, and the ω_κ and γ_κ can all be calculated from the measured sound velocities and their pressure derivatives. Low- and high-temperature limits of γ, and in some cases the temperature dependence at intermediate temperatures, were calculated in Debye approximations for several cubic materials by Sheard,[108] Seeger and Buck,[109] Collins,[110] Sharma and Joshi,[111] and Hiki et al.[112] These calculations are in reasonably good agreement with the measured γ, perhaps better than would be expected in view of the large dispersion of the phonon Grüneisen parameters found in recent pseudopotential calculations for simple metals.[113]

Let us now examine the next higher correction to the elastic constants, which arises from the strain derivatives of the quasi-harmonic contributions to the Helmholtz free energy and the internal energy. From (22.5), neglecting the explicit anharmonic term $\mathfrak{F}_A$, we have for arbitrary initial con-

---

[107] E. Grüneisen, in "Handbuch der Physik" (H. Geiger and K. Scheel, eds.), Vol. 10, p. 1. Springer-Verlag, Berlin, 1926.
[108] F. W. Sheard, Phil. Mag. **3**, 1381 (1958).
[109] A. Seeger and O. Buck, Z. Naturforsch. **15A**, 1056 (1960).
[110] J. G. Collins, Phil. Mag. **8**, 323 (1963).
[111] K. C. Sharma and S. K. Joshi, Phil. Mag. **9**, 507 (1964).
[112] Y. Hiki, J. F. Thomas, Jr., and A. V. Granato, Phys. Rev. **153**, 764 (1967).
[113] D. C. Wallace, Phys. Rev. **176**, 832 (1968); **178**, 900 (1969).

figuration of the lattice

$$C^T_{ijkl} = V_1^{-1} d^2\Phi/d\eta_{ij}\,d\eta_{kl}$$
$$+ V_1^{-1}\sum_{\kappa}[(n_\kappa + \tfrac{1}{2})(\hbar\,d^2\omega_\kappa/d\eta_{ij}\,d\eta_{kl})$$
$$- n_\kappa(n_\kappa + 1)(KT)^{-1}(\hbar\,d\omega_\kappa/d\eta_{ij})(\hbar\,d\omega_\kappa/d\eta_{kl})]; \quad (22.32)$$
$$C^S_{ijkl} = C^T_{ijkl} + (TV_1C)^{-1}\sum_{\kappa}(\hbar\omega_\kappa/KT)n_\kappa(n_\kappa + 1)(\hbar d\omega_\kappa/d\eta_{ij})$$
$$\times \sum_{\kappa'}(\hbar\omega_{\kappa'}/KT)n_{\kappa'}(n_{\kappa'} + 1)(\hbar d\omega_{\kappa'}/d\eta_{kl}), \quad (22.33)$$

where C is the total constant configuration heat capacity. This last equation was derived from the adiabatic-isothermal difference (6.45). Now, on account of the quasi-harmonic contributions, the elastic constants are explicitly temperature dependent, and at $T \neq 0$ the adiabatic and isothermal elastic constants are different. In addition, any Cauchy relations which may be found to hold for the leading contributions to the elastic constants, as e.g. those in (22.19)–(22.21), found for primitive cubic lattices with central potential interactions, will break down because of the quasi-harmonic contributions. These conclusions apply also to the higher-order elastic constants.

Approximate calculation of the sums involving strain derivatives of the phonon frequencies has been discussed, for primitive cubic lattices with central potential interactions, by Barron and Klein[114]; their results were applied to the rare gas crystals. Leibfried and Hahn[115] introduced an Einstein approximation to treat the alkali halides: if each phonon frequency ω_κ is replaced by the average over all modes $\omega_E = \langle\omega_\kappa^2\rangle^{1/2}$, the quasi-harmonic free energy may be written

$$\mathcal{F} = \Phi + 6NKT\ln[2\sinh(\hbar\omega_E/2KT)] \quad (22.34)$$

for N unit cells with two atoms in each unit cell. The reason for choosing this Einstein approximation is that $\langle\omega_\kappa^2\rangle$ is easily calculated as a total trace of the dynamical matrices, and hence the strain derivatives of $\langle\omega_\kappa^2\rangle$ are easily calculated. Leibfried and Hahn took Coulomb interactions and short-range Born–Mayer repulsion among the ions, and calculated the second-order elastic constants as a function of temperature from the approximate free energy (22.34) for a number of alkali halides. They also accounted for thermal expansion of the crystals by calculating at the volume

---

[114] T. H. K. Barron and M. L. Klein, *Proc. Phys. Soc. (London)* **82**, 161 (1963); **85**, 533 (1965).

[115] G. Leibfried and H. Hahn, *Z. Physik* **150**, 497 (1958).

which minimized the approximate $\mathfrak{F}$ at each temperature. A similar model was used by Nran'yan[116] and by Ghate and co-workers[117,118] to calculate the third-order elastic constants of the alkali halides as a function of temperature, and the latter calculation is in good agreement with experiment for NaCl, KCl, and LiF at room temperature.[66]

A comment is in order on the atomic theory for the alkali halides. For a static lattice of the NaCl or CsCl structure, the ions are at centers of inversion symmetry, and have no dipole moments even though they are polarizable ions; this remains true under homogeneous deformation of the lattice (i.e. there is no sublattice displacement). Therefore the central potentials consisting of Coulomb interactions and Born–Mayer repulsion should be an accurate model from which to construct the static lattice potential $\Phi(\mathbf{r}(J))$. When the ions vibrate, however, they polarize, and the strong multipole interactions become important; hence the central potentials give a very poor representation of the phonon frequencies.[119,120] Therefore, although the Einstein model may give an adequate approximation to the free energy, the central potential model is not satisfactory in principle for calculating $\langle \omega_\kappa^2 \rangle$ and its strain derivatives.

We noted that in the potential approximation the sound velocities and the long-wavelength acoustic phonon velocities are the same, for any lattice and any macroscopic configuration. In the next higher lattice dynamical approximation, the sound velocities are temperature dependent and are different for adiabatic and isothermal waves, according to (22.32) and (22.33). The same order of approximation for the phonon frequencies takes the phonon–phonon interactions into account, with the result that the phonon frequencies become explicitly temperature dependent.[121] From the calculations of Cowley,[122] and some additional calculations carried out by the author, we are able to conclude the following for any lattice and any macroscopic configuration, and to second order in lattice-dynamical perturbation. At $T = 0$, the adiabatic and isothermal sound velocities and the long-wavelength acoustic phonon velocities are the same. In fact this result has been found to all orders of perturbation for primitive lattices by

[116] A. A. Nran'yan, *Soviet Phys.—Solid State* (*English Transl.*) **5**, 129 (1963); **5**, 1361 (1963).
[117] P. B. Ghate, *Phys. Rev.* **139**, A1666 (1965).
[118] R. C. Lincoln, K. M. Koliwad, and P. B. Ghate, *Phys. Status Solidi* **18**, 265 (1966).
[119] A. D. B. Woods, B. N. Brockhouse, R. A. Cowley, and W. Cochran, *Phys. Rev.* **131**, 1025 (1963).
[120] A. M. Karo and J. R. Hardy, *Phys. Rev.* **129**, 2024 (1963).
[121] See, e.g., D. C. Wallace, *Phys. Rev.* **152**, 247 (1966).
[122] R. A. Cowley, *Proc. Phys. Soc.* (*London*) **90**, 1127 (1967).

Götze.[123] For a given propagation mode at finite temperatures, the adiabatic sound velocity is greater than the isothermal sound velocity, and the long-wavelength acoustic phonon velocity is equal to the isothermal sound velocity plus a function which is most likely positive in all cases.

Now there is one more step required to relate the renormalized long-wavelength acoustic phonons to the temperature-dependent part of the free energy at low temperatures; that is, to determine the $|\mathbf{k}|$ dependence of these phonon frequencies. We have shown[124] that the renormalized frequencies are proportional to $|\mathbf{k}|$ at small $\mathbf{k}$; this means the free energy, including the leading anharmonic correction $\mathfrak{F}_A$, still goes as T^4 at low T. All of our previous conclusions relating elastic wave velocities to the low temperature thermodynamics are valid to the next order of lattice-dynamical perturbation; in particular the measured sound velocities and their strain derivatives at $T = 0$ can be used to calculate the renormalized zero-temperature Debye temperature and its strain derivatives.

ACKNOWLEDGMENTS

The author appreciates encouragement from Professor Andy Granato during the preparation of this article; in addition many helpful discussions with Professor Granato and with Dr. Will Gauster are gratefully acknowledged.

[123] W. Götze, *Phys. Rev.* **156**, 951 (1967).
[124] D. C. Wallace, *Phys. Rev.* **133**, A153 (1964).

Author Index

Numbers in parentheses are reference numbers and indicate that an author's work is referred to although his name is not cited in the text.

A

Adams, A., Jr., 108(220)
Adlington, R. E., 293
Aerts, E., 14(49), 50(111), 53(49, 111, 116), 54
Albon, N., 278, 279
Alder, B. J., 167
Alers, C. A., 368
Alfintzev, G. A., 247(197), 249
Allen, G. F., 93(142, 177–179), 103(142, 177–179), 104(177–179), 106(197, 202, 203), 109(202, 203), 112(202), 113 (177, 178, 202), 114(177–179, 202), 115(177, 178, 202), 116, 117, 120(202), 122(257), 125(261), 126(177–179, 264), 128, 129(142, 261), 130, 132, 133(261), 134(197, 277), 135(177–179, 271), 139(178), 143(177–179, 277), 147(203)
Alstrup, I., 30(83a)
Amos, A. T., 8(46), 21(72), 19(65), 119
Anderson, S., 111(237), 112(237)
Antončik, E., 4
Arakawa, K., 271
Arata, H., 94(181), 108(181, 219), 113 (219), 129(181, 219), 131(181, 219), 132(219)
Arfken, G., 18(64), 25(64), 46(64), 48(64)
Arthur, J. R., 106
Artmann, K., 4, 73(22)
Aspnes, D.E., 94(180), 103(180), 108(180), 113(180), 114(180), 115(180), 118, 119(180), 131
Aven, M., 134(274)

B

Baldock, G. R., 19(67), 22(75), 30, 76(67)

Balkanski, M., 134(275)
Ball, J. G., 254
Bamforth, A. W., 285
Banbury, P. C., 127
Bankoff, S. G., 172, 195
Bardeen, J., 4
Bardsley, W., 294
Barker, J. A., 167
Barnes, G. A., 127
Barrett, H. H., 379
Barron, T. H. K., 336, 402
Barsh, G. R., 372
Bartels, R. A., 368
Bartini, G. R., 253
Bartoš, I., 122, 124
Batchelor, G. K., 296(312)
Bateman, T. B., 325, 326, 368
Beattie, A. G., 368
Becker, G. E., 89(168)
Becker, R., 200
Begbie, G. H., 398
Bennema, P., 241, 242, 286, 288, 289
Benson, G. C., 112(239), 222
Bergstresser, T. K., 58(124), 79(124)
Berlad, A. L., 193, 248
Bernstein, B., 315
Bethce, H., 238
Bhagavantam, S., 349(50)
Birch, F., 329, 350
Birman, J. L., 88(162)
Blakely, J. M., 192
Bliznakov, G., 278
Bloch, F., 5(36), 38(36), 41(36)
Bloot, E. I., 133
Boas, W., 335
Bockris, J. O'M., 244, 245
Bolef, D. I., 369(70), 378(82)
Bolling, G. F., 190
Borgnis, F. E., 316

Borisov, V. T., 176(58)
Born, M., 2(10), 392(91), 393(91), 394(91), 398
Boudreaux, D. S., 77(132a, 135)
Brand, L., 7(44)
Brattain, W. H., 130
Breazeale, M. A., 324(31, 33), 357, 376, 377, 378(31, 76), 391
Brian, P. L. T., 194(108)
Brice, J. C., 286, 288, 289, 294
Brillouin, L., 31(86, 89), 303(2, 4), 318(2, 26), 362(56)
Brockhouse, B. N., 403(119)
Brout, R., 166
Brower, W. S., 295(306), 296(306)
Brown, A. G., 36(103a)
Brown, D. M., 107(213), 147(213)
Brugger, K., 308, 317, 350, 341, 345, 347, 365
Brust, D., 106(201), 107(201)
Buck, O., 318(28), 322, 328(28, 40) 375, 377, 401
Buckle, E. R., 206
Buckley, H. E., 273(255)
Budurov, S. I., 253
Bunn, C. W., 179(65)
Burke, J., 206
Burton, W. K., 167, 215, 223, 224(40, 41), 225, 226(41), 228(41), 239, 276, 277(41), 280(41), 289(41), 290(254, 289, 290), 291, 292, 294
Burger, G., 109(225)

C

Cabrera, N., 159, 167(40, 41), 209(13), 213, 215, 216(146), 220, 223, 224(40, 41), 225(41), 226(41), 228(41), 232 234, 235, 237(175), 239, 250, 276(41), 277(41), 278, 280(41, 204), 281(13, 204, 274), 289(41)
Cahn, J. W., 159, 213, 160(23), 181, 182, 183, 189, 191(73), 198, 206, 207, 247(198, 199), 248, 249, 263, 264(198), 265, 266
Callan, J. M., 294(299)
Callaway, J., 32(100)
Callen, H. B., 333
Campbell, B. D., 110(227), 111(227, 227a), 134(227, 227a), 143, 146

Cardona, M., 77(137), 78(137), 122(137), 123(137)
Carlson, A., 286, 288(286), 289
Carruthers, J. R., 293
Carslaw, H. S., 172(51), 173(51), 174(51), 178, 187, 195
Chakerian, G. D., 211
Chalmers, B., 183(79), 187, 198, 205, 206, 246, 261(80, 221), 270(221), 272, 273, 283(277)
Chambré, P. L., 298
Chandrasekhar, S., 254, 296(213), 297(213)
Chang, Z. P., 372
Chaves, C. M., 77(137), 78(137), 122, 123
Chedzey, H. A., 294(299)
Cheng, Y. C., 5(35)
Chernov, A. A., 159, 188, 218, 220, 237, 240, 241, 253(17, 209, 210), 264, 265, 273(259), 275, 276, 277, 284, 289
Chester, G. V., 392(92)
Chiarotti, G., 107(207, 209), 128(209)
Childress, J. D. 379
Choi, J. Y., 233(165)
Chow, P. C., 32(97), 41(97)
Christian, J. W., 160(19), 161, 169(44), 198, 199(19), 201, 202(19)
Chung, M. F., 108(216–218), 111(216, 217), 115(216, 217), 119, 120(216, 217), 145(216)
Cochran, W.
Cohen, E. G. D., 162
Cohen, M. L., 58(124), 79(124)
Cole, G. S., 295(308)
Coleman, R. V., 159, 209(13), 213, 220, 232, 235, 281(13), 283
Coldwell-Horsfall, R. A., 399
Collins, J. G., 401
Conklin, J. B., Jr., 32(93, 95)
Coriell, S. R., 192(101), 193(101), 196 (109), 236(101), 259(109), 259, 262 (109, 226), 263, 265, 268(109, 239, 240), 269, 270(240)
Cosgrove, G. J., 188
Courtney, W. G., 206
Cowley, R. A., 403(119, 122)
Crank, J., 178

D

Daniels, W. B., 368
Davies, E. A., 127(265)

Davis, J. L., 147(297)
Davis, L., Jr., 108(220)
Davison, S. G., 2(4, 6, 11), 5(31, 35), 8(146), 10(31, 33), 11(31, 33), 14(31), 15, 16(56, 60a), 17(31, 63), 19(65), 21(11, 32), 22(4, 32, 78, 80), 30(78, 83b), 31(78, 80), 32(101), 35(103), 36(103, 103a, b, c), 42(107), 45, 75(33), 76(33), 87(161), 119, 143(33)
Dauenbaugh, C. E., 108(221), 120(251), 125(221, 263), 127(221, 263), 132
de Boer, J., 167(36)
Defay, R., 270(241)
deGroot, S. R., 160(18)
Delves, R. T., 268
Del Signore, G., 107(209), 128(209)
de Mars, G. A., 108(220)
Dempsey, E., 212(239)
Derco, G. L., 22(78), 30(78), 31(78)
Devonshire, A. F., 167
Dillon, J. A., Jr., 111(230), 112(230), 113(230), 114(230), 117(247), 120, 121, 125(220), 133(230)
DiMarzio, E. A., 273(260), 274(260, 261), 275
Dirac, P. A. M., 5(40)
Dittmar, W., 192, 236
Domb, C., 164(31), 165(31), 207(31)
Donnelly, R. J., 254, 298(215)
Doremus, R. H., 179, 243(187, 189)
Döring, W., 200
Drabble, J. R., 368
Drake, R. M., Jr., 286, 287(284), 288(284) 296(284)
Dropkin, D., 298(315, 316)
Ducros, P., 119
Dufour, L., 270(241)
Dukova, E. D., 253(209)
Dunning, W. J., 278, 279
Duty, R. L., 172(48), 184(48)

E

Early, J. G., 295(306), 296(306)
Eckert, E. R. G., 286, 287(284), 288(284), 296(284)
Eisenhour, S., 94(183), 127
Elbaum, C., 377
Emanuel, A. S., 292

Emde, F., 55(122), 56(122), 60(122), 67(122), 73(122), 182(76)

F

Fairbairn, W. M., 22(79), 25, 30, 31(84)
Farnsworth, H. E., 15(55), 110(227), 109(225), 111(55, 227, 227a, 230), 112(230), 113(230, 243), 114 (230), 120, 121, 125(230), 133(230), 134 (227, 227a), 143, 146(230, 291)
Feder, J., 198
Fein, A. E., 317
Feinstein, L. P., 111(238)
Feshbach, H., 46(110), 48(110), 180(72)
Feuer, P., 24(82)
Finn, M. C., 82(146)
Fischer, T. E., 106(198), 107(205), 114, 115, 119, 134(198, 276, 277), 135, 138(198, 276), 142, 143(198, 205, 276, 277), 145(245)
Fisher, J. C., 183
Flemings, M. C., 295(305, 307), 297(305, 307)
Flinn, I., 147(301)
Foldy, L. L., 32(99)
Ford, G. W., 156(7)
Ford, J., 318(29), 320(29), 324(31), 378(31), 375, 379
Forman, R., 129, 130, 132
Forty, A. J., 243
Fowler, A. B., 81(142), 93(142), 103(142), 129(142), 130
Frank, F. C., 154, 159, 167(41), 173, 176, 178, 185, 211, 224(41), 225(41, 160), 226(41), 228(41), 230, 232, 237(176), 239, 250, 251, 252, 276(41), 277(41), 280(41), 281(203), 283, 289(41)
Frankl, D. R., 2, 15(8), 94(184), 105, 109(8), 115(8), 117, 130, 131, 132
Freeman, P. I., 112(239)
Freeman, S., 79(140), 136(140), 140(140, 281)
Frenkel, J., 207, 224
Freund, T., 146(292)
Friedman, A., 172
Fuchs, K., 399
Fujita, S., 162(25)
Fukushima, M., 30(83)
Fullmer, L. D., 293
Fumi, F. G., 349

G

Galloway, T. R., 295
Garrett, G. G. B., 130
Gatos, H. C., 5(28, 29), 81(29), 82(144, 146), 146(29), 147(29, 300), 209
Gauster, W. B., 375, 376, 377, 378
Gedroits, A. A., 328
Geus, J. W., 89(167)
Ghate, P. B., 403
Ghosh, S. K., 54
Ghosh, U. S., 17(63)
Gibbs, J. W., 198
Gillman, J., 295(311), 298(311)
Gjostein, N. A., 222(156), 233(164, 166)
Glasser, M. L., 32(101), 35(103)
Glicksman, M. E., 188
Gobeli, G. W., 106(197, 202, 203), 107(202), 109(202, 203), 111(54, 152), 112(54, 152, 202), 113(54, 117, 178, 202), 114(54, 177–179, 202), 115(177, 178, 202), 116, 117, 120(202), 125(541, 261), 126(177–179, 264), 128, 129(261), 132, 133(54, 261), 134(152, 197, 277), 135(277), 139(178), 143(277), 147(203)
Götze, W., 404(123)
Goldstein, H., 312(13)
Goldstein, S., 286, 288(282), 290(282)
Goldstein, Y., 2(7), 98(7), 104(7), 105(189), 109(7), 115(7), 127(7), 132
Gomer, R., 192
Goodman, A. M., 90(171), 147(171)
Goodwin, E. T., 3, 10(16), 71, 72, 73(16), 78(16)
Goss, A. J., 293
Granato, A. V., 368, 399(105), 401(112)
Grant, J. T. P., 111(235), 119(235)
Gray, P. V., 107(213), 147(213)
Green, A. E., 315
Green, G. W., 127(265)
Green, M., 90(173)
Gregg, J. L., 292
Gretz, R. D., 204(122)
Grimley, T. B., 2(1), 5(1), 10(1), 15(1, 51), 28(1), 31(84), 76(1)
Grindlay, J., 22(80), 31
Grover, N. B., 2(7), 98(7), 104(7), 105(189), 109(7), 115(7), 127(7), 132
Gruber, E. E., 220, 221, 222(152)
Grüneisen, E., 401

Gubanov, A. I., 107(206)
Gurney, R. A., 84
Gyftopoulous, E. P., 89(170)

H

Hagstrum, H. D., 89(168)
Hahn, H., 402
Ham, F. S., 179, 180(70), 181, 254
Hambrick, G. G., 379
Handler, P., 81, 94(180, 183), 103(180, 188), 108(180, 188), 109(223), 113(180), 114(180), 115(180), 118, 119(180), 120, 127, 129, 130, 131, 132, 133(223), 146(223), 147(223)
Haneman, D., 82(147), 83, 91, 108(215–217), 111(150, 215–217, 235, 236), 114(215), 115(216, 127), 119(50, 233, 235), 120(216, 217), 145(216)
Hanneman, R. E., 82(145)
Hansen, N. R., 111(233), 119(233)
Hardy, J. R., 403(120)
Hardy, S. C., 204(121), 268, 269, 270(240)
Harrick, N. J., 107(208)
Hartke, J. H., 102(186)
Hartman, P., 225, 273
Harp, J., 295(311), 298(311)
Hauge, C. Z., 111(227a), 134(227a)
Hearmon, R. F. S., 335, 349
Heck, C. M., 243
Heiland, G., 109(222), 113(241a), 114, 118, 120, 121, 134(222), 148(222)
Heine, V., 76(130), 77, 147
Henderson, D., 167
Henish, H. K., 109(224), 146(224)
Henzler, M., 17, 104(190), 114, 118, 125(61, 262), 126, 127, 128(61), 129, 132, 133(262)
Herman, F., 32(94), 139(94)
Herman, R., 254(215), 298(215)
Herring, C., 155, 159, 208, 209(11), 210, 211, 213
Heyer, H., 237(177), 238(178)
Hikata, A., 377
Hiki, Y., 368, 401
Hill, T. L., 207
Hilliard, J. E., 159, 198, 206, 213
Hillig, W. B., 193(105), 194, 247(105, 198), 248(105), 263(198), 264(198)
Himmelblau, D. M., 254(217)

Hirth, J. P., 192, 198, 204, 213, 231(112, 144), 233(144), 234, 237, 252, 253
Hoang, A. T., 111(238a)
Hoffman, T. A., 4, 10(23, 48), 15(23), 17(62)
Holt, A. C., 317, 318(29), 320(29), 322, 323(24), 375, 379
Holt, D. B., 83
Horvay, G., 172, 181, 182, 183(79), 184, 185, 187, 189, 191(73), 298(84, 317)
Howard, R. E., 192(100)
Huang, K., 161, 164, 170(24), 315, 317(15, 25), 318, 392(91), 393(91), 394(91)
Huber, A., 176(59), 178(59)
Huff, H. R., 5(29), 81(29), 146, 147(29)
Hughel, T. J., 166(32)
Hughes, D. S., 368
Hulett, L. D., Jr., 281, 282(275)
Humphreys-Owen, S. P. F., 179(66)
Hunt, M. D., 271(242)
Hunt, J. D., 205, 206, 245(192), 246(192), 248(192), 249(192)
Hurle, D. T. J., 273, 294, 295, 298
Hutner, R. A., 31, 32(85), 35(85), 37(85), 38(85), 46(85)

I

Igras, E., 113
Ivantsov, G. P., 176, 177(63), 178, 182–184, 186, 187, 189, 191
Ives, J., 252

J

Jackson, K. A., 168, 169, 192, 205, 206, 216, 218, 245, 246, 247(193), 248(192, 193), 249(192), 261(221), 264, 270 (221), 283(276, 277)
Jackson, P. J., 192(93)
Jackson, R. W., 279(270)
Jacobson, J. D., 167(39)
Jaeger, J. C., 172(51), 173(51), 175(51), 178, 187, 195
Jaffee, R. I., 167(38)
Jahn, H. A., 349
Jahnke, E., 55(122), 56(122), 60(122), 67(122), 73(122), 182
James, D. W., 249
James, T. H., 90(172)
Jaynes, E. T., 162(25)
Jenkins, R. T., 318(27), 326(27)

Johansen, K., 30(83a)
Johnson, C. A., 211
Johnson, L. E., 32(93, 95)
Jones, H., 39(105), 52(105), 76(105, 131)
Jones, G. L., 379
Jones, R. O., 77(139), 78(139), 122(139), 124, 136(139)
Joshi, S. K., 401
Jost, W., 271(243)
Juretschke, H. J., 89(167), 209

K

Kahn, B., 162
Kaischew, R., 225
Kane, E. O., 106(199), 116, 126(264)
Karo, A. M., 403(120)
Katzir, A., 145(290), 146(290)
Kawaji, S., 5(29), 81(29), 108(219), 113(219), 118, 129(219), 131(219), 132(219), 146(29), 147(29, 300)
Kelly, E. F., 378(82)
Kelly, J. L., 368
Kendall, E. J. M., 148(302)
Kerr, H. W., 218
Kikuchi, R., 207(135–138), 208
Kindig, N. B., 106(200)
Kirkaldy, J. S., 160(22)
Kirtisinghe, D., 193(106), 248
Klein, M. L., 336, 402
Kleinman, I., 72
Kobayashi, A., 94(181), 108(181), 129, 131, 208(219), 113(219), 132(219)
Kobett, D. R., 379
Kolb, E. D., 291(290), 292
Koliwad, K. M., 403(118)
Kortum, R. L., 32(94), 139(94)
Korshunov, I. P., 253(209)
Kotler, G. R., 265
Kossel, W., 223
Koster, G. F., 19(68), 22(68), 30
Koutecký, J., 2(2), 4, 5(31, 32), 10(31), 11(31), 14, 15(2, 50), 16(25), 17(31), 21(2, 32), 22(2, 32, 76, 77), 24(25, 76), 28(2), 30, 31(76), 54, 73(2), 77(2), 87(161), 122, 136(26, 279, 258)
Kozlovskii, M. I., 243
Knacke, O., 227, 233(167)
Knor, Z., 89(169), 110(169)
Kramer, J. J., 248(201)
Krasil'nikov, V. A., 328

Krishnamurty, T. S. G., 350
Kronig, D. de. L., 3, 16(13), 31, 32(13), 37(13)
Kuglin, G. D., 32(94), 139(94)
Kuntzmann, P., 109(222), 113(241a), 134(222), 148(22)
Kurik, M. V., 117(247)
Kurtin, S., 142(286), 147(286)

L

Lamatsch, H., 120, 121,
Lamb, J., 369(72)
Landau, L. D., 324, 325(36, 37), 254, 286, 298
Lander, J. J., 15(54), 111(54, 230), 112(54, 230), 113(54, 230), 114(54, 230), 125(54), 133(54, 230)
Langford, D., 175(55), 178(55)
Lasker, A. K., 54
Lauritzen, J. I., Jr., 273(260) 274, 275
Lavine, M. C., 5(28), 82
Law, J. T., 120, 121(253)
Lazarus, D., 368
Lee, R. N., 146(291)
Lee, T. D., 163, 164, 167
Leibfried, G., 310, 393(11, 93), 398(98, 93), 402
Lennard-Jones, J. E., 167
Levich, V. G., 294
Levine, J. D., 2(5), 5(30, 33, 34), 7(33), 9(33), 10(33), 11(33), 55, 60(34), 66(34), 67(34), 69(125), 73(5, 127), 74(34), 75(33), 76(33), 79(33, 140), 84, 87(30), 88(127), 89(170), 90(174), 95(30), 111(231), 134(30, 231), 135(278), 136(140), 140(140, 278), 143(30, 33, 34)
Levine, M. M., 234
Levine, R. D., 21(72)
Lester, J. E., 107(210), 144
Lewis, J., 264
Lighthill, M. J., 250
Lifshitz, I. M., 19(66)
Lifshitz, E. M., 254, 286, 298, 324, 325, 326(36, 37)
Lin, C. C., 254
Lincoln, R. C., 403(118)
Ling, T. Y., 17(63)
Lippmann, B. A., 4, 72

Liu, L., 32(97), 41(97)
Love, A. E. H., 313
Long, D., 138(280)
Losch, F., 182(76)
Lothe, J., 198(114)
Löwdin, P.-O., 2(12), 19(69), 21(71)
Loucks, T. L., 32(98), 41(98), 76(98)
Ludwig, W., 310, 393(11, 93), 394, 398(93)
Lyubov, Ya, B., 176(56–58), 177(57), 179(57), 187, 188, 193, 194

M

McCabe, W. L., 289
Mackenzie, J. K., 221, 222, 335
McLachlan, N. W., 31(92), 55(92), 56(92), 60(92), 66(92), 67(92), 73(92)
McSkimin, H. J., 368(63), 369(71)
MacRae, A. U., 83(152, 153), 111(152, 153), 112(152, 153), 134(152)
McRae, E. G., 110(229), 134(229)
Majlis, N., 77(137), 78(137), 122 (137), 123(137)
Mann, E., 304
Many, A., 2., 98, 104(104), 105(189), 109(7), 115(7), 127, 132, 145(290), 147
Maradudin, A. N., 377(77), 378(77)
Margoninski, Y., 105(196), 129, 131, 132
Mariano, A. N., 82(145)
Mark, P., 2(3, 5), 5(30), 10(3), 73(5), 74, 84, 87(30), 88(164), 95(30), 107(214), 110(226a), 111(231), 134(30, 231), 142(30), 143(30), 147(214, 298), 148
Marsh, J. B., 15(55), 111(55), 113(243)
Mason, B. J., 272
Mason, W. P., 325, 326, 368(63)
Mathieu, E., 31
Matsinger, J. H., 379
Maue, A. W., 3, 71
Mazur, P., 160(18)
Mead, C. A., 142(284, 286), 147(284, 286)
Mead, D. G., 279(270)
Mees, C. E. K., 90(172)
Melcher, R. L., 369
Melngailias, J., 377, 378
Melnikov, A. M., 284
Memming, R., 147(295)
Merry, J. B., 369(70)

Merzbacher, E., 38(104)
Messiah, A., 20(70)
Meyerhof, W. E., 4(19)
Michaels, A. S., 194
Mikhailov, I. G., 378
Milne-Thompson, L. M., 7(43)
Mil'vidskii, M. G., 293
Moazed, K. L., 204(123)
Moesta, H., 107(211)
Moore, A. J. W., 212, 221(153), 222
Morrison, J., 15(54), 111(54, 230), 112 (54, 230), 113(54, 230), 114(54, 230), 120(251), 125(54), 133(54, 230)
Morrison, S. R., 108(221), 120(251) 125(221, 263), 127(221, 263), 132, 146(292)
Morse, P. M., 31(87), 46(110), 48(110), 162, 180(72)
Mott, N. F., 76(131), 84(157, 158), 154
Mourad, P., 73(128)
Müller, A., 298(314)
Müller, E. W., 89(169), 110(169)
Müller, K., 134(272)
Mullin, J. W., 285, 289, 290
Mullins, W. W., 159, 160(23), 209(12), 210(12), 213, 216, 219, 220, 222(152), 253, 254, 257(211), 258(211), 260 (211), 261(218), 262(218), 266(211), 267, 268(218)
Murkland, I., 111(237), 112(237)
Murnaghan, F. D., 303, 306, 307, 309(6), 311, 360(6)
Mutaftschiev, B., 219

N

Nabarro, F. R. N., 192(93)
Nakaya, U., 155(5), 272(5)
Nancollas, G. H., 245
Nannarone, S., 107(209), 128(209)
Nesterenko, B. A., 105(193–195), 120(194, 195), 121, 129, 131, 132
Neumann, K., 192, 236
Newkirk, J. B., 243
Newman, J., 294, 295
Nicholas, J. F., 221(153), 222
Nichols, F. A., 262
Nielsen, A. E., 198, 245, 295
Noble, W. P., Jr., 109(224), 146(224)
Noll, W., 309(10), 329(43)

Nosker, R., 87(161), 111(231), 134(231)
Nran'yan, A. A., 403
Nucciotti, A., 107(207)

O

Oda, Z., 94(181), 108(181, 219), 113(219), 129(181, 219), 131(181, 219), 132(219)
Olander, D. R., 292
Oman, R. M., 121, 145
O'Neil, H. T., 318(27), 326(27)
Onsager, L., 160, 163, 164, 215
Oppenheimer, J. R., 2(10)
Oser, H., 192(101), 193(101), 236(101)
Ovsienko, D. E., 247(197), 249

P

Pagnia, H., 147(296)
Palmberg, P. W., 111(234)
Palmer, D. R., 108(221), 120, 125, 127(221, 263), 132
Papapetrou, A., 181, 272
Parker, J. H., Jr., 378
Parker, R. L., 192(100), 196(109), 258 (109), 259, 262(109), 263, 265, 268(109)
Pashley, D. W., 198
Passaglia, E., 273(260), 274(260, 261), 275
Pauling, L., 86(160)
Pearson, G. L., 4, 112
Penney, W. G., 3, 16(13), 31, 32(13), 37(13)
Perdok, W. G., 225
Peria, W. T., 110(226), 111(234)
Perry, J. J., 77(135)
Peters, R. D., 391
Pfann, W. G., 273, 285
Pfleiderer, H., 399
Phariseau, P., 16(58), 51(113), 52(114), 53, 54(58, 118–120), 77(136), 78(136)
Phillips, J. C., 72
Pines, D., 90(175)
Pollard, E. R., 102(186)
Portnoy, W. M., 129, 130, 131, 132
Pospelov, L. A., 324(32), 327
Pound, G. M., 198(114), 204(122, 123), 231(112), 234, 237
Pratt, G. W., Jr., 32(93, 95)

Prener, J. S., 134(274)
Pretzer, D. D., 89(168)
Price, P. B., 280
Prigogine, I., 254(215), 298(215)
Prim, R. C., 273(254), 290(254), 294(254)
Priolo, M. J., 145
Proix, F., 109(223), 133(223), 146(223), 147(223)
Prutton, M., 112(239)
Pugh, D., 77(138), 122, 123
Purdie, N., 245

R

Raether, H., 89(165)
Razumney, G. A., 244, 245
Read, W. T., Jr., 112(240)
Redwood, M., 369(72)
Revesz, A. G., 147(292a), 148(292a)
Rhoderick, E. H., 143
Rhodin, T. N., 205(125)
Rhys-Roberts, C., 77(138)
Rice, S. A., 170
Rieck, 176, 178(59)
Ritchie, R. H., 89(165)
Roberts, B. W., 179(71), 243(187)
Robertson, D. S., 293
Robertson, W. M., 222
Roitburo, A. L., 176(57), 177(57), 179(57)
Rollings, F. R., Jr., 379
Roots, W. D., 111(235), 119(235)
Rose, A., 95(185)
Rosenzweig, N., 162(25)
Rossby, H. T., 296
Rozumnyuk, V. T., 105(195), 120(195)
Rudman, P. S., 167
Ruppel, W., 142(285), 147(285)
Russell, K. C., 198(114)
Ruth, V., 192
Rutter, J. W., 187, 261(221–223), 271
Rymer, T. B., 15

S

Sachs, M., 5(38), 6(38), 32(38), 35(38), 38(38), 41(38)
Sage, B. H., 295
Salem, L., 6(42)
Salama, K., 368
Samoggia, G., 107(207)
Sangster, R. G., 83(148)
Saratovkin, D. D., 272
Sawamoto, K., 147(299)
Saxon, D. S., 31, 32(85), 35(85), 37(85), 38(85), 46(85)
Schaefer, R. J., 188
Scheer, J. T., 103(187a–c), 106(204), 109(187a–c), 114, 117(187a–c), 143(187c), 144(204), 148(187c)
Schiff, L. T., 32(102)
Schlier, R. E., 109(225), 111(230), 112(230), 113(230), 114(230), 125(230), 133(230)
Schlichting, H., 286, 287, 288
Schrieffer, J. R., 94(182)
Schwandt, G., 147(295)
Schwoebel, R. L., 236, 238(179, 181)
Schuele, D. E., 368
Sears, G. W., 192, 235, 247(198), 248, 263(198), 264(198), 277
Sebenne, C., 134(275)
Seeger, A., 183, 265, 304, 377(77), 378(77), 401
Seidensticker, R. G., 264
Seitz, F., 84(155)
Seiwatz, R., 90(173), 111(232)
Sekerka, R. F., 254, 257(211), 258(211), 260(211), 261(218, 219), 262(218), 266(211), 267, 268(218)
Sewell, F. A., Jr., 377(79)
Shaked, U., 145(290), 146(290)
Shapiro, M. J., 324(30)
Shappir, J., 145(290), 146(290)
Sharma, K. C., 401
Sheard, F. W., 401
Shepherd, W. B., 106
Shewmon, P. G., 222, 233(165), 262, 265
Shimaoka, G., 117(247)
Shockley, W., 3, 4, 30, 112(241)
Shoemaker, D. P., 111(238)
Shooten, D., 134(273)
Shutilov, V. A., 378
Siegert, A. J. F., 162(25)
Sih, P. H., 294, 295
Simmons, J. A., 192, 193(101), 236(101)
Simnad, M. T., 204(120)
Slater, J. C., 19(68), 22(68), 30, 31(88)
Slichter, W. P., 273(254), 290(254), 292, 294(254)
Smith, C. S., 317, 368

Smith, R. A., 5(39), 6(99), 41(39)
Smith, R. W., 271(242)
Snitko, O. V., 105(193–195), 120(194, 195), 121, 129, 131, 132
Sholuchowski, R., 130
Somerscales, E., 298(315, 316)
Somorjai, G. A., 107(210), 144
Soven, P., 32(95)
Sparrow, E. M., 292
Sperry, P. R., 194(108)
Spicer, W. E., 106(200), 107(200)
Spittle, J. A., 271(242)
Srinivasan, R., 399
Statz, H., 4, 71, 108(220)
Stefan, J., 171, 172, 175(46)
Stern, R. M., 77
Stęślicka, M., 2, 15(59, 60, 60a), 16(57), 17(57, 59, 60), 36(103a, b, c), 42(107), 44(108), 45
Stevens, R. P., 289
Stillinger, F. H., 207
Stranski, I. N., 225, 233(167)
Strathen, R. E. B., 368
Streitwieser, A., Jr., 6(41)
Strickland-Constable, R. F., 193(106)
Stringer, J., 167(38)
Struthers, J. D., 291(290), 292
Stuart, J. T., 254(216)
Sugiyama, K., 94(181), 108(181, 219), 113(219), 129(181, 219), 131(181, 219), 132(219)
Sundqvist, B. E., 222
Suryanarayana, D., 349(50)
Suzuki, T., 399
Swank, R., 139(282), 142(159, 282), 143(282)
Swank, R. K., 84, 106(159), 107(159), 113(159), 116(159), 134(159), 135, 139(159), 143, 144, 146(159)

T

Takeishi, Y., 89(168), 118
Taloni, A., 82(147), 108(217), 111(217, 236), 115(217), 220(217)
Tamm, I., 3, 10(14), 28(14), 30, 32, 43, 53, 70(14), 81(14)
Tammann, G., 193(107)
Tarshis, L. A., 264
Taylor, L. H., 379(87)

Taylor, N. F., 30(83b)
Temkin, D. E., 176(57, 58), 177(57), 179(57, 67), 187, 189, 190, 191, 216(147), 217, 218, 262
Temperley, H. N. V., 157, 162
Tennison, M. A., 328(40)
Thomas, J. F., Jr., 368, 399(105), 401(112)
Thomas, M. T., 117(247)
Thompson, D. O., 318(28), 322, 328, 375, 376, 377, 378(31)
Thuras, A. L., 318, 326
Thurston, R. N., 324(30), 336, 341, 345
Tiller, W. A., 160(21), 190, 191, 248(201), 261(221–223), 264, 265, 270, 271, 272
Todd, P. H., Jr., 379(87)
Tokutaka, H., 212(239)
Tomášek, M., 4, 22(76), 25(76), 119, 122, 136(26, 279)
Toots, J., 113(243)
Toupin, R. A., 312, 315
Toyoda, H., 147(299)
Trecolo, R. H., 77(138)
Trilling, L., 234
Truesdell, C., 309(10), 312, 329(43)
Turnbull, D., 154, 157(9), 158(9), 159, 161(9), 179(71), 198(9), 202(90), 203, 243(187), 247, 248, 283(9)
Turnbull, H. W., 40(106)
Turner, M. J., 143
Turovskii, B. M., 293

U

Ubbelohde, A. R., 166, 206
Uehling, E. H., 368
Uhlenbeck, G. E., 156, 162
Uhlmann, D. R., 198, 206, 245(192), 246(192), 248(192), 249(192), 283
Utech, H. P., 295, 296(306), 297(304, 305, 307)

V

Van Damme, M. A., 281
Van Hook, A., 273
Van Laar, J., 103(187a–c), 106(204), 109(187a–c), 114, 117(187a–c), 143(187c), 144(204), 148(187c)
Verma, A. R., 243
Vermilyea, D. A., 250, 278, 280(271), 281(204)

Vitrikhovskii, N. I., 117(247)
Voigt, W., 302, 306
Vol'kenshtein, F. F., 2
Volmer, M., 199(118)
Voronkov, V. V., 218, 260, 261

W

Wainwright, T. E., 167
Walker, J. L., 298
Wallace, D. C., 306, 315, 316(19), 338(7), 355(7), 393(18, 19), 398(18, 19), 399(18, 106), 401(113), 403(121), 404(124)
Wallis, G., 131
Walton, A. G., 198, 245
Walton, D., 205(124, 125), 261, 270(222), 271
Wang, S., 131
Wannier, G. H., 22(73, 74)
Warfield, G., 90(171), 147(171)
Waterman, P. C., 317, 368
Watson, G. N., 31(91)
Webb, M. B., 280(271)
Weber, A., 199(118)
Weber, H., 172
Weber, R. E., 110(226)
Wells, A. F., 84(154)
Werthamer, N. R., 392(90)
Whitham, G. B., 250
Whittaker, E. T., 31(91)
Widom, B., 156, 162, 163, 164(8), 166, 213
Wilcox, W. R., 172(47), 184(48), 273, 285, 292, 293

Wilhelm, M., 298(314)
Williams, R., 107(212), 148(212)
Williamson, R. R., 272
Willis, A., 107(212), 148(212)
Wilson, H. A., 246
Winecard, W. C., 218, 261(222), 270(222), 271, 295
Winterbottom, W. L., 222(156), 233
Wojciechowski, K. E., 16(59), 17(59)
Wolff, G. A., 88(163), 133(163)
Wolsky, S. P., 134(273)
Wood, R. D., 32(100)
Wood, W. W., 167(39)
Woods, A. D. B., 403(119)
Wouthuysen, S. A., 32(99)

Y

Yang, C. N., 163, 164, 167
Yang, L., 204(120)
Young, F. W., Jr., 281, 282(275)
Yun, K. S., 222

Z

Zaininger, K., 147(292a), 148(292a)
Zdanuk, E. J., 134(273)
Zdhanov, G. S., 15
Zemansky, M. W., 357(55), 371(55)
Zener, C., 176, 178
Zief, M., 273, 285, 292, 293
Ziman, J. M., 5(37), 6(37), 41(37)
Zuehlke, R. W., 89(166)
Zwanzig, R. W., 234, 273(169)
Zyrianov, G. K., 111(238a)

Subject Index

A

Acoustic phonons, velocity, 346, 403
Adiabatic displacement, energy flux, 314
Adiabatic processes, stress-strain relations, 329
Alkali halides, elastic constants, 403
Amorphous semiconductors, band structure, 54
Anharmonicity, lattice dynamics effects, 392 ff.
Attenuation coefficient, 325
Auger spectroscopy, surface impurity studies, 110

B

Band bending
  compound semiconductors, 138 ff.
  semiconductor surfaces, 103
Band gap
  compound semiconductors, 135
  surface states, correlation, 135 ff.
Birch coefficients, 333
Bloch's theorem, 41
Boundary layer, fluid flow, 287

C

Cadmium compounds, surface states, 142
Capacitive microphone, 378
Cauchy relations, 397, 402
Cauchy stress tensor, 308
Chemisorption, surface states effects, 145
Chemisorption states, AB linear crystal, 15
Clausius-Clapeyron equation, 157
Cloud chamber, condensation, 201
Compound semiconductors, surface states, 133 ff.
Compressibility, elastic compliances, relation, 330 ff.
Condensation, van der Waals theory, 162

Constitutional supercooling, 261
  fluid flow theory, 294
  shape stability, 270
Cooperative phenomena, 165
Copper
  electrocrystallization, 282
  etch pits, 281
  harmonic generation, 376
Crystal, equilibrium form, 211
Crystal growth
  Chernov model, 240
  cylinder, 178
  Czochralski method, 290 ff.
  ellipsoidal, 179
  filamentary, 183 ff., 192, 235 ff.
  fluid flow effects, 284 ff.
  horizontal boats, 295
  impurity effects, 186 ff., 272 ff.
    one-dimensional theory, 273
    step pinning, 279 ff.
  kinematic theory, 250 ff.
  kink poisoning effects, 276
  Kossel model, 223, 227
  mechanisms, 151–299
  melt phase, mechanisms, 245 ff.
  particle effects, 283
  polyhedral, 183
  screw dislocation
    mechanism, 226 ff.
    melts, 247
  shape stability, 154, 179
    theory, 253 ff.
  solutions, 239 ff.
  spheres, fluid flow effects, 294
  spherical form, 255 ff.
  spirals, 234, 243
  Stefan problem, 170 ff.
  step motion, 229
  step pinning, 278 ff.
  step profiles, 251 ff.
  surface nucleation
    mechanism, 231
    melts, 247
  vapor phase, 237 ff.

Crystal orbital theory, surface states, 5 ff., 78 ff.
Crystal potential methods, surface states, 31 ff.
Crystal symmetry, elastic constants, 347 ff.
Crystals, dissolution, 281 ff.
Cubic crystal
  elastic constants, 351 ff.
    fourth order, 380 ff.
    third order, 364, 370, 374
  propagation coefficients, 352 ff.
Cylinder, shape stability, 258 ff.
Czochralski method, crystal growth, 184, 290 ff.

D

Dangling bond model, surfaces, 81 ff.
Debye temperature, quasi-harmonic, 400
Dendrites
  growth, 272
    density effects, 191
    impurity effects, 190
    paraboloidal, 181 ff.
    shape effects, 189 ff.
Density of states, surfaces, 79 ff.
Diamond structure, surface states, theory, 122
Dirac equation, 32
Dirac function, linear crystal, 51
Dirac surface states, 44
Dislocations
  crystal growth mechanism, 226
  harmonic generation effects, 377
Dynamic nucleation, 205

E

Einstein model, 403
Elastic compliances, 330
  compressibility, relation, 330 ff.
  elastic constants, relation, 340
Elastic constants
  crystal symmetry, 347 ff.
  cubic crystal, I, 351 ff.
    third order, 364, 374
  definition, 308
  elastic compliances, relation, 340

  fourth order, 345
    measurement, 389
  hexagonal crystals, 360
  higher order, 301–404
  isotropic materials, 360 ff.
    third order, 365
  microscopic theory, 391 ff.
  pseudopotential theory, 399
  second-order, 338
  tetragonal crystal, I, 357 ff.
    third order, 366 ff.
  third order
    definition, 350
    harmonic generation, 322
    measurements, 363 ff.
    mixed, 345
    pressure derivative, 385 ff.
    theory, 336 ff.
Elastic waves
  attenuation, 324 ff.
  energy, 326
  energy flux, 317
  propagation, 314 ff.
    isotropic pressure, 334 ff.
    stressed crystals, 363 ff., 368 ff.
  propagation matrices, 341
  velocity, 346
Elasticity, thermodynamics, 305 ff.
Electrocrystallization, 245
  copper, 282
Energy band theory, relativistic corrections, 32
Entropy, production rate, 160
Equation of motion, dissipative, 324
ESR, studies, surface states, 108

F

Fluid flow, magnetic field effects, 297
Fourth-order elastic constants, cubic crystals, 380 ff.
Free energy, anharmonic contribution, 394

G

Gallium arsenide
  LEED studies, 111
  surface properties, 82 ff.

Germanium
  crystal growth, Czochralski method, 292
  dislocations, 112
  LEED studies, 111, 125
  surface properties, 82
  surface states, 125 ff.
Gibbs free energy, temperature dependence, 158
Gibbs–Thomson equation, 213, 257
Grashof number, 298
Green function technique, 46 ff.
  surface states, 54
Grüneisen parameters, 346, 400 ff.

H

Hard spheres, equation of state, 166 ff.
Harmonic generation
  experiments, 376 ff.
  solids, theory, 318 ff.
  stressed crystals, 389 ff.
  theory, 372 ff.
Harmonic wave, energy input, 326
Heterophase fluctuations, 199
Hexagonal crystals, elastic constants, 360
Hooke's law, differential form, 330
Huang coefficients, 317
Hückel approximation, 6
Hydrodynamic stability, 254

I

Ice crystals, shape stability, 269 ff.
Interface states, metal-semiconductors, 147
Interfaces
  Burton–Cabrera theory, 215
  diffuseness, 206, 213 ff.
  faceting, 211
  liquid-solid
    diffuseness, 247
    shape stability, 260
  structure, 208 ff.
    hill and valley, 211
  thermodynamics, 158 ff., 209
Interfacial tension, definition, 158
Ion bombardment, surface preparation, 109

Ionic crystals, surface states, 73 ff.
Ionic surface states, Madelung method, 84
Irreversible processes, thermodynamics, 160
Ising model, 162
  surface structures, 167 ff.

J

Jackson theory
  crystal growth, melt, 245
  surface roughness, 216

K

Kinematic theory, crystal growth, 250 ff.
Kinks, crystal growth, 276
Kossel model, crystal growth, 223
Kronig–Penney Model, 3, 31 ff.
Kronig–Penney potential, 37

L

Lagrange's equations, 312
Lagrangian strain parameters, 307, 323, 380
Langmuir adsorption isotherm, 278
Lattice dynamics, anharmonic effects, 392 ff.
Lattice gas model, phase changes, 163
Laue groups, elastic constants, 348
LCAO method, 6
LEED
  surface structure studies, 110 ff.
  zinc blende, 83
Lennard–Jones–Devonshire theory, melting, 165 ff.
Low energy electron diffraction, see LEED

M

Madelung constant, surfaces, 86 ff.
Madelung method, surface states, 84
Magnetic field, fluid flow, effects, 297
Mathieu equation, 31
  surface states, calculation, 54 ff.
Mathieu functions, properties, 57
Mathieu model, ionic surface states, 73 ff.

Mechanical resonance, sound velocity, measurements, 369
Melting, Lennard–Jones–Devonshire theory, 165 ff.
Molecular dynamics, hard spheres, 166 ff.
Morphological stability, see Shape stability
Morphology, crystal, 155
Murnaghan's constants, 360

N

Navier–Stokes equation, 188, 286
Nucleation
  crystals from vapor, 204
  droplets in vapor, 198
  dynamic, 205
  heterogeneous, 198, 203
  phase changes, 197 ff.
    classical theory, 198 ff.
  solid from liquid, 202
  time-dependent, 202, 206
  two-dimensional, 231

O

Onsager relations, 160

P

Paraffin, crystallization, 275
Peclet number, 183
Phase changes
  continuity, 153 ff.
  lattice gas model, 163
  nucleation, 197 ff.
  thermodynamics, 155 ff.
Phase shift method, sound velocity, 368
Photoeffect studies, surface states, 106
Photoelectric threshold, anion electronegativity, correlation, 144
Prandtl number, 288, 298
Pseudopotential theory, elastic constants, 399

R

Raman–Nath diffraction, 378
Regular solution, theory, 169
Relativistic corrections, energy band theory, 32

Resolvent crystal orbital theory, surface states, 16 ff.
Reynolds number, 287
Rock salt, dissolution, impurity effects, 281

S

Salol, crystallization rate, 264
Schmidt number, 288
Screw dislocation
  crystal growth
    mechanism, 226 ff.
    melts, 247
Semiconductor(s)
  p-type, space charge layer, 92
  surface density of states, 79 ff.
  surface states, 90 ff.
  III–V, surface states, 54
Semiconductor compounds
  surface states, 133 ff.
    ionic, 144
Semiconductor surfaces, dangling bond model, 81 ff.
Shape stability
  anisotropy effect, 265 ff.
  constitutional supercooling, 270
  crystal growth, 154, 179
  cylinder, 258 ff.
  experimental studies, 268 ff.
  ice crystals, 269 ff.
  interface kinetics effect, 263
  planar interface, 260 ff.
  sphere, 255 ff.
  surface diffusion effect, 262
  time-dependent effects, 267
Shockley states, 3, 53
  Mathieu model, 72
  Tamm states, interrelation, 30
Silicon
  LEED studies, 111, 114
  surface states, 114 ff.
    conductivity, 118
    energy distribution, 121 ff.
    EPR, 119
Silver, whisker growth impurity effects, 281
Sodium chlorate, crystal growth, 242
Solid, equilibrium form, 222
Solid–liquid interface, diffuseness, 168 ff.

Solidification, density differences effects, 187
Solution theory, regular, 169
Sound velocity, measurement, crystals, 368 ff.
Sphere, shape stability, 255 ff.
Spinodal decomposition, 206
Stefan problem
  approximate solutions, 195 ff.
  crystal growth, 170 ff.
  ellipsoid, 179
  plane interface, 172 ff., 195
  polyhedral growth, 183
  special cases, 185 ff.
  spherical interface, 176, 196
  Tammann tube, 193
Step pinning, crystal growth, 278 ff.
Step potential, electron scattering, 35
Stokes problem, 294
Strain tensor, thermal, 333
Stress-stain relations, 328 ff.
  adiabatic processes, 339 ff.
Stress tensor, thermal, 333
Sucrose, crystal growth, impurity effects, 278 ff.
Surface conductivity, 53
Surface diffusion, crystal growth, 232
Surface preparation, techniques, 108 ff.
Surface roughness, 167
  Jackson theory, 216
  Temkin theory, 216
Surface states
  AB linear crystal, 10 ff.
  abbreviations, 149
  adsorption effects, 145
  applied field studies, 103 ff.
  band gap, correlation, 135 ff.
  cadmium compounds, 142
  charge mobility, 93
  compound semiconductors, 133 ff.
  crystal orbital theory, 5 ff., 78 ff.
  crystal potential methods, 31 ff.
  density, 79 ff.
  diamond structure, theory, 122
  distortion effects, 15 ff.
  doping effect, 102
  energy, 63 ff.
    distribution determination, 94 ff., 102 ff.
  energy classification, 44

ESR, studies, 108
existence condition, 52, 62
extrinsic, 89, 145 ff.
finite crystal, 67
germanium, 125 ff.
Green function technique, 54
ionic, compound semiconductors, 144
ionic crystals, 73 ff.
Madelung method, 84
Mathieu equation, calculation, 54 ff.
merged band model, 99 ff.
metals, 89
monatomic crystals, 16
  relativistic treatment, 32
photoeffect studies, 106
resolvent crystal orbital theory, 16 ff.
semiconductor, 90 ff.
silicon, 114 ff.
theory, 1–149
three-dimensional models, 75 ff.
two-band model, 98 ff.
zinc compounds, 142
Surface stress
  definition, 159
  solids, 209
Surface structure, Ising model, 167 ff.
Surface tension
  definition, 159
  solids, 209 ff.
    anisotropy, 219 ff.

T

Tamm equation, 43
Tamm states, 3
  existence condition, 28
  Mathieu model, 70 ff.
  relativistic, 44
  Shockley states, interrelation, 30
Temkin theory, surface roughness, 216
Tetragonal crystal, I
  elastic constants, 357 ff.
    third order, 366 ff., 375
  propagation coefficients, 358 ff.
Thermodynamics, irreversible processes, 160
Thermoelasticity, theory, 301–404
Tight-binding approximation, 6
Tin-lead alloy, shape stability, 271
Transport, Stefan problem, 170 ff.

U

Ultrasonic waves
  solids, attenuation, 325
  velocity-stress relations, 343 ff.

V

van der Waals theory, condensation, 162
Voigt symmetry, 309, 317, 335, 337

W

Whiskers
  growth, 192
  growth mechanism, 235 ff.
Wulff construction, 210, 220
Wulff's theorem, 210

Z

Zinc blende, surface structure, 83
Zinc compounds, surface states, 142

QC
173
S477
v.25
1970